# APPLICATIONS OF COMBINATORIAL MATRIX THEORY TO LAPLACIAN MATRICES OF GRAPHS

# DISCRETE MATHEMATICS AND ITS APPLICATIONS

Series Editor
## Kenneth H. Rosen, Ph.D.

*R. B. J. T. Allenby and Alan Slomson*, How to Count: An Introduction to Combinatorics, Third Edition

*Juergen Bierbrauer*, Introduction to Coding Theory

*Katalin Bimbó*, Combinatory Logic: Pure, Applied and Typed

*Donald Bindner and Martin Erickson*, A Student's Guide to the Study, Practice, and Tools of Modern Mathematics

*Francine Blanchet-Sadri*, Algorithmic Combinatorics on Partial Words

*Richard A. Brualdi and Dragoš Cvetković*, A Combinatorial Approach to Matrix Theory and Its Applications

*Kun-Mao Chao and Bang Ye Wu*, Spanning Trees and Optimization Problems

*Charalambos A. Charalambides,* Enumerative Combinatorics

*Gary Chartrand and Ping Zhang,* Chromatic Graph Theory

*Henri Cohen, Gerhard Frey, et al.,* Handbook of Elliptic and Hyperelliptic Curve Cryptography

*Charles J. Colbourn and Jeffrey H. Dinitz,* Handbook of Combinatorial Designs, Second Edition

*Martin Erickson,* Pearls of Discrete Mathematics

*Martin Erickson and Anthony Vazzana,* Introduction to Number Theory

*Steven Furino, Ying Miao, and Jianxing Yin,* Frames and Resolvable Designs: Uses, Constructions, and Existence

*Mark S. Gockenbach,* Finite-Dimensional Linear Algebra

*Randy Goldberg and Lance Riek,* A Practical Handbook of Speech Coders

*Jacob E. Goodman and Joseph O'Rourke,* Handbook of Discrete and Computational Geometry, Second Edition

*Jonathan L. Gross,* Combinatorial Methods with Computer Applications

*Jonathan L. Gross and Jay Yellen,* Graph Theory and Its Applications, Second Edition

*Titles (continued)*

*Jonathan L. Gross and Jay Yellen,* Handbook of Graph Theory

*David S. Gunderson,* Handbook of Mathematical Induction: Theory and Applications

*Richard Hammack, Wilfried Imrich, and Sandi Klavžar,* Handbook of Product Graphs, Second Edition

*Darrel R. Hankerson, Greg A. Harris, and Peter D. Johnson,* Introduction to Information Theory and Data Compression, Second Edition

*Darel W. Hardy, Fred Richman, and Carol L. Walker,* Applied Algebra: Codes, Ciphers, and Discrete Algorithms, Second Edition

*Daryl D. Harms, Miroslav Kraetzl, Charles J. Colbourn, and John S. Devitt,* Network Reliability: Experiments with a Symbolic Algebra Environment

*Silvia Heubach and Toufik Mansour,* Combinatorics of Compositions and Words

*Leslie Hogben,* Handbook of Linear Algebra

*Derek F. Holt with Bettina Eick and Eamonn A. O'Brien*, Handbook of Computational Group Theory

*David M. Jackson and Terry I. Visentin,* An Atlas of Smaller Maps in Orientable and Nonorientable Surfaces

*Richard E. Klima, Neil P. Sigmon, and Ernest L. Stitzinger,* Applications of Abstract Algebra with Maple™ and MATLAB®, Second Edition

*Patrick Knupp and Kambiz Salari,* Verification of Computer Codes in Computational Science and Engineering

*William Kocay and Donald L. Kreher*, Graphs, Algorithms, and Optimization

*Donald L. Kreher and Douglas R. Stinson,* Combinatorial Algorithms: Generation Enumeration and Search

*Hang T. Lau,* A Java Library of Graph Algorithms and Optimization

*C. C. Lindner and C. A. Rodger,* Design Theory, Second Edition

*Nicholas A. Loehr,* Bijective Combinatorics

*Alasdair McAndrew,* Introduction to Cryptography with Open-Source Software

*Elliott Mendelson,* Introduction to Mathematical Logic, Fifth Edition

*Alfred J. Menezes, Paul C. van Oorschot, and Scott A. Vanstone,* Handbook of Applied Cryptography

*Stig F. Mjølsnes,* A Multidisciplinary Introduction to Information Security

*Jason J. Molitierno,* Applications of Combinatorial Matrix Theory to Laplacian Matrices of Graphs

*Richard A. Mollin,* Advanced Number Theory with Applications

*Richard A. Mollin,* Algebraic Number Theory, Second Edition

*Richard A. Mollin*, Codes: The Guide to Secrecy from Ancient to Modern Times

*Richard A. Mollin,* Fundamental Number Theory with Applications, Second Edition

*Richard A. Mollin,* An Introduction to Cryptography, Second Edition

**Titles (continued)**

*Richard A. Mollin*, Quadratics

*Richard A. Mollin*, RSA and Public-Key Cryptography

*Carlos J. Moreno and Samuel S. Wagstaff, Jr.*, Sums of Squares of Integers

*Goutam Paul and Subhamoy Maitra*, RC4 Stream Cipher and Its Variants

*Dingyi Pei*, Authentication Codes and Combinatorial Designs

*Kenneth H. Rosen*, Handbook of Discrete and Combinatorial Mathematics

*Douglas R. Shier and K.T. Wallenius*, Applied Mathematical Modeling: A Multidisciplinary Approach

*Alexander Stanoyevitch*, Introduction to Cryptography with Mathematical Foundations and Computer Implementations

*Jörn Steuding*, Diophantine Analysis

*Douglas R. Stinson*, Cryptography: Theory and Practice, Third Edition

*Roberto Togneri and Christopher J. deSilva*, Fundamentals of Information Theory and Coding Design

*W. D. Wallis*, Introduction to Combinatorial Designs, Second Edition

*W. D. Wallis and J. C. George*, Introduction to Combinatorics

*Lawrence C. Washington*, Elliptic Curves: Number Theory and Cryptography, Second Edition

DISCRETE MATHEMATICS AND ITS APPLICATIONS

Series Editor KENNETH H. ROSEN

# APPLICATIONS OF COMBINATORIAL MATRIX THEORY TO LAPLACIAN MATRICES OF GRAPHS

## Jason J. Molitierno

Sacred Heart University
Fairfield, Connecticut, USA

CRC Press is an imprint of the
Taylor & Francis Group, an **informa** business

A CHAPMAN & HALL BOOK

The author would like to thank Kimberly Polauf for her assistance in designing the front cover.

CRC Press
Taylor & Francis Group
6000 Broken Sound Parkway NW, Suite 300
Boca Raton, FL 33487-2742

© 2012 by Taylor & Francis Group, LLC
CRC Press is an imprint of Taylor & Francis Group, an Informa business

No claim to original U.S. Government works

Version Date: 20111129

International Standard Book Number: 978-1-4398-6337-4 (Hardback)

This book contains information obtained from authentic and highly regarded sources. Reasonable efforts have been made to publish reliable data and information, but the author and publisher cannot assume responsibility for the validity of all materials or the consequences of their use. The authors and publishers have attempted to trace the copyright holders of all material reproduced in this publication and apologize to copyright holders if permission to publish in this form has not been obtained. If any copyright material has not been acknowledged please write and let us know so we may rectify in any future reprint.

Except as permitted under U.S. Copyright Law, no part of this book may be reprinted, reproduced, transmitted, or utilized in any form by any electronic, mechanical, or other means, now known or hereafter invented, including photocopying, microfilming, and recording, or in any information storage or retrieval system, without written permission from the publishers.

For permission to photocopy or use material electronically from this work, please access www.copyright.com (http://www.copyright.com/) or contact the Copyright Clearance Center, Inc. (CCC), 222 Rosewood Drive, Danvers, MA 01923, 978-750-8400. CCC is a not-for-profit organization that provides licenses and registration for a variety of users. For organizations that have been granted a photocopy license by the CCC, a separate system of payment has been arranged.

**Trademark Notice:** Product or corporate names may be trademarks or registered trademarks, and are used only for identification and explanation without intent to infringe.

**Visit the Taylor & Francis Web site at**
**http://www.taylorandfrancis.com**

**and the CRC Press Web site at**
**http://www.crcpress.com**

# Dedication

This book is dedicated to my Ph.D. advisor, Dr. Michael "Miki" Neumann, who passed away unexpectedly as this book was nearing completion. In addition to teaching me the fundamentals of combinatorial matrix theory that made writing this book possible, Miki always provided much encouragement and emotional support throughout my time in graduate school and throughout my career. Miki not only treated me as an equal colleague, but also as family. I thank Miki Neumann for the person that he was and for the profound effect he had on my career and my life. Miki was a great advisor, mentor, colleague, and friend.

# Contents

**Preface**

**Acknowledgments**

**Notation**

**1 Matrix Theory Preliminaries**   **1**
    1.1 Vector Norms, Matrix Norms, and the Spectral Radius of a Matrix . 1
    1.2 Location of Eigenvalues . . . . . . . . . . . . . . . . . . . . . . . 8
    1.3 Perron-Frobenius Theory . . . . . . . . . . . . . . . . . . . . . . . 15
    1.4 M-Matrices . . . . . . . . . . . . . . . . . . . . . . . . . . . . . . . 24
    1.5 Doubly Stochastic Matrices . . . . . . . . . . . . . . . . . . . . . . 28
    1.6 Generalized Inverses . . . . . . . . . . . . . . . . . . . . . . . . . . 34

**2 Graph Theory Preliminaries**   **39**
    2.1 Introduction to Graphs . . . . . . . . . . . . . . . . . . . . . . . . 39
    2.2 Operations of Graphs and Special Classes of Graphs . . . . . . . . 46
    2.3 Trees . . . . . . . . . . . . . . . . . . . . . . . . . . . . . . . . . . 55
    2.4 Connectivity of Graphs . . . . . . . . . . . . . . . . . . . . . . . . 61
    2.5 Degree Sequences and Maximal Graphs . . . . . . . . . . . . . . . 66
    2.6 Planar Graphs and Graphs of Higher Genus . . . . . . . . . . . . . 81

**3 Introduction to Laplacian Matrices**   **91**
    3.1 Matrix Representations of Graphs . . . . . . . . . . . . . . . . . . 91
    3.2 The Matrix Tree Theorem . . . . . . . . . . . . . . . . . . . . . . . 97
    3.3 The Continuous Version of the Laplacian . . . . . . . . . . . . . . 104
    3.4 Graph Representations and Energy . . . . . . . . . . . . . . . . . . 108
    3.5 Laplacian Matrices and Networks . . . . . . . . . . . . . . . . . . . 114

**4 The Spectra of Laplacian Matrices**   **119**
    4.1 The Spectra of Laplacian Matrices under Certain Graph Operations 119
    4.2 Upper Bounds on the Set of Laplacian Eigenvalues . . . . . . . . . 126
    4.3 The Distribution of Eigenvalues Less than One and Greater than One 136
    4.4 The Grone-Merris Conjecture . . . . . . . . . . . . . . . . . . . . . 145
    4.5 Maximal (Threshold) Graphs and Integer Spectra . . . . . . . . . . 151

|  |  |  |
|---|---|---|
|  | 4.6 Graphs with Distinct Integer Spectra | 163 |
| **5** | **The Algebraic Connectivity** | **173** |
|  | 5.1 Introduction to the Algebraic Connectivity of Graphs | 174 |
|  | 5.2 The Algebraic Connectivity as a Function of Edge Weight | 180 |
|  | 5.3 The Algebraic Connectivity with Regard to Distances and Diameters | 187 |
|  | 5.4 The Algebraic Connectivity in Terms of Edge Density and the Isoperimetric Number | 192 |
|  | 5.5 The Algebraic Connectivity of Planar Graphs | 197 |
|  | 5.6 The Algebraic Connectivity as a Function Genus $k$ Where $k \geq 1$ | 205 |
| **6** | **The Fiedler Vector and Bottleneck Matrices for Trees** | **211** |
|  | 6.1 The Characteristic Valuation of Vertices | 211 |
|  | 6.2 Bottleneck Matrices for Trees | 219 |
|  | 6.3 Excursion: Nonisomorphic Branches in Type I Trees | 235 |
|  | 6.4 Perturbation Results Applied to Extremizing the Algebraic Connectivity of Trees | 239 |
|  | 6.5 Application: Joining Two Trees by an Edge of Infinite Weight | 256 |
|  | 6.6 The Characteristic Elements of a Tree | 263 |
|  | 6.7 The Spectral Radius of Submatrices of Laplacian Matrices for Trees | 273 |
| **7** | **Bottleneck Matrices for Graphs** | **283** |
|  | 7.1 Constructing Bottleneck Matrices for Graphs | 283 |
|  | 7.2 Perron Components of Graphs | 290 |
|  | 7.3 Minimizing the Algebraic Connectivity of Graphs with Fixed Girth | 308 |
|  | 7.4 Maximizing the Algebraic Connectivity of Unicyclic Graphs with Fixed Girth | 322 |
|  | 7.5 Application: The Algebraic Connectivity and the Number of Cut Vertices | 328 |
|  | 7.6 The Spectral Radius of Submatrices of Laplacian Matrices for Graphs | 346 |
| **8** | **The Group Inverse of the Laplacian Matrix** | **361** |
|  | 8.1 Constructing the Group Inverse for a Laplacian Matrix of a Weighted Tree | 361 |
|  | 8.2 The Zenger Function as a Lower Bound on the Algebraic Connectivity | 370 |
|  | 8.3 The Case of the Zenger Equalling the Algebraic Connectivity in Trees | 378 |
|  | 8.4 Application: The Second Derivative of the Algebraic Connectivity as a Function of Edge Weight | 388 |
| **Bibliography** |  | **395** |
| **Index** |  | **401** |

# Preface

On the surface, matrix theory and graph theory are seemingly very different branches of mathematics. However, these two branches of mathematics interact since it is often convenient to represent a graph as a matrix. Adjacency, Laplacian, and incidence matrices are commonly used to represent graphs. In 1973, Fiedler [28] published his first paper on Laplacian matrices of graphs and showed how many properties of the Laplacian matrix, especially the eigenvalues, can give us useful information about the structure of the graph. Since then, many papers have been published on Laplacian matrices. This book is a compilation of many of the exciting results concerning Laplacian matrices that have been developed since the mid 1970s. Papers written by well-known mathematicians such as (alphabetically) Fallat, Fiedler, Grone, Kirkland, Merris, Mohar, Neumann, Shader, Sunder, and several others are consolidated here. Each theorem is referenced to its appropriate paper so that the reader can easily do more in-depth research on any topic of interest. However, the style of presentation in this book is not meant to be that of a journal but rather a reference textbook. Therefore, more examples and more detailed calculations are presented in this book than would be in a journal article. Additionally, most sections are followed by exercises to aid the reader in gaining a deeper understanding of the material. Some exercises are routine calculations that involve applying the theorems presented in the section. Other exercises require a more in-depth analysis of the theorems and require the reader to prove theorems that go beyond what was presented in the section. Many of these exercises are taken from relevant papers and they are referenced accordingly.

Only an undergraduate course in linear algebra and experience in proof writing are prerequisites for reading this book. To this end, Chapter 1 gives the necessities of matrix theory beyond that found in an undergraduate linear algebra course that are needed throughout this book. Topics such as matrix norms, mini-max principles, nonnegative matrices, M-matrices, doubly stochastic matrices, and generalized inverses are covered. While no prior knowledge of graph theory is required, it is helpful. Chapter 2 provides a basic overview of the necessary topics in graph theory that will be needed. Topics such as trees, special classes of graphs, connectivity, degree sequences, and the genus of graphs are covered in this chapter.

Once these basics are covered, we begin with a gentle approach to Laplacian matrices in which we motivate their study. This is done in Chapter 3. We begin with a brief study of other types of matrix representations of graphs, namely the adjacency and incidence matrices, and use these matrices to define the Laplacian matrix of a graph. Once the Laplacian matrix is defined, we present one of the most famous theorems in matrix-graph theory, the Matrix-Tree Theorem, which tells us the number of spanning trees in a given graph. Its proof is combinatoric in nature and the concepts in linear algebra that are employed are well within the grasp of a student who has a solid background in linear algebra. Chapter 3 continues to motivate the study of Laplacian matrices by deriving their construction from the continuous version of the Laplacian matrix which is used often in differential equations to study heat and energy flow through a region. We adopt these concepts to the study of energy flow on a graph. We further investigate these concepts at the end of Chapter 3 when we discuss networks which, historically, is the reason mathematicians began studying Laplacian matrices.

Once the motivation of studying Laplacian matrices is completed, we begin with a more rigorous study of their spectrum in Chapter 4. Since Laplacian matrices are symmetric, all eigenvalues are real numbers. Moreover, by the Gersgorin Disc Theorem, all of the eigenvalues are nonnegative. Since the row sums of a Laplacian matrix are all zero, it follows that zero is an eigenvalue since $e$, the vector of all ones, is an eigenvector corresponding to zero. We then explore the effects of the spectrum of the Laplacian matrix when taking the unions, joins, products, and complements of graphs. Once these results are established, we can then find upper bounds on the largest eigenvalue, and hence the entire spectrum, of the Laplacian matrix in terms of the structure of the graph. For example, an unweighted graph on $n$ vertices cannot have an eigenvalue greater than $n$, and will have an eigenvalue of $n$ if and only if the graph is the join of two graphs. Sharper upper bounds in terms of the number and the location of edges are also derived. Once we have upper bounds for the spectrum of the Laplacian matrix, we continue our study of its spectrum by illustrating the distribution of the eigenvalues less than, equal to, and greater than one. Additionally, the multiplicity of the eigenvalue $\lambda = 1$ gives us much insight into the number of pendant vertices of a graph. We then further our study of the spectrum by proving the recently proved Grone-Merris Conjecture which gives an upper bound on each eigenvalue of the Laplacian matrix of a graph. This is supplemented by the study of maximal or threshold graphs in which the Grone-Merris Conjecture is sharp for each eigenvalue. Such graphs have an interesting structure in that they are created by taking the successive joins and complements of complete graphs, empty graphs, and other maximal graphs. Moreover, since the upper bounds provided by the Grone-Merris Conjecture are integers, it becomes natural to study other graphs in which all eigenvalues of the Laplacian matrix are integers. In such graphs, the number of cycles comes into play.

In Chapter 5 we focus our study on the most important and most studied eigenvalue of the Laplacian matrix - the second smallest eigenvalue. This eigenvalue is known as the algebraic connectivity of a graph as it is used extensively to measure how connected a graph is. For example, the algebraic connectivity of a disconnected graph is always zero while the algebraic connectivity of a connected graph is always strictly positive. For a fixed $n$, the connected graph on $n$ vertices with the largest algebraic connectivity is the complete graph as it is clearly the "most connected" graph. The path on $n$ vertices is the connected graph on $n$ vertices with the smallest algebraic connectivity since it is seen as the "least connected" graph. Also, the algebraic connectivity is bounded above by the vertex connectivity. Hence graphs with cut vertices such as trees will never have an algebraic connectivity greater than one. Overall, graphs containing more edges are likely to be "more connected" and hence will usually have larger algebraic connectivities. Adding an edge to a graph or increasing the weight of an existing edge will cause the algebraic connectivity to monotonically increase. Additionally, graphs with larger diameters tend to have fewer edges and thus usually have lower algebraic connectivities. The same holds true for planar graphs and graphs with low genus. In Chapter 5, we prove many theorems regarding the algebraic connectivity of a graph and how it relates to the structure of a graph.

Once we have studied the interesting ideas surrounding the algebraic connectivity of a graph, it is natural to want to study the eigenvector(s) corresponding to this eigenvalue. Such an eigenvector is known as the Fiedler vector. We dedicate Chapters 6 and 7 to the study of Fiedler vectors. Since the entries in a Fiedler vector correspond to the vertices of the graph, we begin our study of Fiedler vectors by illustrating how the entries of the Fiedler vector change as we travel along various paths in a graph. This leads us to classifying graphs into one of two types depending if there is a zero entry in the Fiedler vector corresponding to a cut vertex of the graph. We spend Chapter 6 focusing on trees since there is much literature concerning the Fiedler vectors of trees. Moreover, it is helpful to understand the ideas behind Fiedler vectors of trees before generalizing these results to graphs which is done in Chapter 7. When studying trees, we take the inverse of the submatrix of Laplacian matrix created by eliminating a row and column corresponding to a given vertex $k$ of the tree. This matrix is known as the bottleneck matrix at vertex $k$. Bottleneck matrices give us much useful information about the tree. In an unweighted tree, the $(i,j)$ entry of the bottleneck matrix is the number of edges that lie simultaneously on the path from $i$ to $k$ and on the path from $j$ to $k$. An analogous result holds for weighted trees. Bottleneck matrices are also helpful in determining the algebraic connectivity of a tree as the spectral radius of bottleneck matrices and the algebraic connectivity are closely related. When generalizing these results to graphs, we gain much insight into the structure of a graph. We learn a great deal about its cut vertices, girth, and cycle structure.

Chapter 8 deals with the more modern aspects of Laplacian matrices. Since zero is an eigenvalue of the Laplacian matrix, it is singular, and hence we cannot take the inverse of such matrices. However, we can take the group generalized inverse of the Laplacian matrix and we discuss this in this chapter. Since the formula for the group inverse of the Laplacian matrix relies heavily on bottleneck matrices, we use many of the results of the previous two chapters to prove theorems concerning group inverses. We then apply these results to sharpen earlier results in this book. For example, we use the group inverse to create the Zenger function which is another upper bound on the algebraic connectivity. We also use the group inverse to investigate the rate of change of increase (the second derivative) in the algebraic connectivity when we increase the weight of an edge of a graph. The group inverse of the Laplacian matrix is interesting in its own right as its combinatorial properties give us much information about the stucture of a graph, especially trees. The distances between each pair of vertices in a tree is closely reflected in the entries of the group inverse. Moreover, within each row $k$ of the group inverse, the entries in that row decrease as you travel along any path in the tree beginning at vertex $k$.

Matrix-graph theory is a fascinating subject that ties togtether two seemingly unrealted branches of mathematics. Because it makes use of both the combinatorial properties and the numerical properties of a matrix, this area of mathematics is fertile ground for research at the undergraduate, graduate, and experienced levels. I hope this book can serve as exploratory literature for the undergraduate student who is just learning how to do mathematical reasearch, a useful "start-up" book for the graduate student begining research in matrix-graph theory, and a convenient reference for the more experienced researcher.

# Acknowledgments

The author would like to thank Dr. Stephen Kirkland and Dr. Michael Neumann for conversations that took place during the early stages of writing this book.

The author would also like to thank Sacred Heart University for providing support in the form of (i) a sabbatical leave, (ii) a University Research and Creativity Grant (URCG), and (iii) a College of Arts and Sciences release time grant.

# Notation

$\Re$ - the set of real numbers

$\Re^n$ - the space of $n$-dimensional real-valued vectors

$A[X, Y]$ - the submatrix of $A$ corresponding to the rows indexed by $X$ and the columns indexed by $Y$

$A[X] = A[X, X]$

$[\overline{X}] = \{1, \ldots, n\} \setminus X$

$\|x\|$ - the Euclidean norm of the vector $x$

$e$ - the column vector of all ones (the dimension is understood by the context)

$e^{(n)}$ - the $n$-dimensional column vector of all ones

$e_i$ - the column vector with 1 in the $i^{th}$ component and zeros elsewhere

$y_i$ - the $i^{th}$ component of the vector $y$

$I$ - the identity matrix

$J$ - the matrix of all ones

$E_{i,j}$ - the matrix with 1 in the $(i, j)$ entry and zeros elsewhere

$M_n$ - the set of all $n \times n$ matrices

$M_{m,n}$ - the set of all $m \times n$ matrices

$A \leq B$ - entries $a_{ij} \leq b_{ij}$ for all ordered pairs $(i, j)$

$A < B$ - entries $a_{ij} \leq b_{ij}$ for all ordered pairs $(i, j)$ with strict inequality for at least one $(i, j)$

$A << B$ - entries $a_{ij} < b_{ij}$ for all ordered pairs $(i,j)$.

$A^T$ - the transpose of the matrix $A$

$A^{-1}$ - the inverse of the matrix $A$

$A^{\#}$ - the group inverse of the matrix $A$

$A^+$ - the Moore-Penrose inverse of the matrix $A$

$\text{diag}(A)$ - the diagonal matrix consisting of the diagonal entries of $A$

$\det(A)$ - the determinant of the matrix $A$

$\text{Tr}(A)$ - the trace of the matrix $A$

$m_A(\lambda)$ - the multiplicity of the eigenvalue $\lambda$ of the matrix $A$

$L(\mathcal{G})$ - the Laplacian matrix of the graph $\mathcal{G}$

$m_\mathcal{G}(\lambda)$ - the multiplicity of the eigenvalue $\lambda$ of $L(\mathcal{G})$

$\rho(A)$ - the spectral radius of the matrix $A$

$\lambda_k(A)$ - the $k^{th}$ smallest eigenvalue of the matrix $A$. (Note that we will always use $\lambda_n$ to denote the largest eigenvalue of the matrix $A$.)

$\sigma(A)$ - the spectrum of $A$, i.e., the set of eigenvalues of the matrix $A$ counting multiplicity

$\sigma(\mathcal{G})$ - the set of eigenvalues, counting multiplicity, of $L(\mathcal{G})$

$\mathcal{Z}(A)$ - the Zenger of the matrix $A$

$|X|$ - the cardinality of a set $X$

$w(e)$ - the weight of the edge $e$

$|\mathcal{G}|$ - the number of vertices in the graph $\mathcal{G}$

$d_v$ or $\deg(v)$ - the degree of vertex $v$

$m_v$ - the average of the degrees of the vertices adjacent to $v$

$v \sim w$ - vertices $v$ and $w$ are adjacent

$N(v)$ - the set of vertices in $\mathcal{G}$ adjacent to the vertex $v$

$d(u,v)$ - the distance between vertices $u$ and $v$

$\tilde{d}(u,v)$ - the inverse weighted distance between vertices $u$ and $v$

$\tilde{d}_v$ - the inverse status of the vertex $v$

$\mathrm{diam}(\mathcal{G})$ - the diameter of the graph $\mathcal{G}$

$\overline{\rho}(\mathcal{G})$ - the mean distance of the graph $\mathcal{G}$

$V(\mathcal{G})$ - the vertex set of the graph $\mathcal{G}$

$E(\mathcal{G})$ - the edge set of the graph $\mathcal{G}$

$v(\mathcal{G})$ - the vertex connectivity of the graph $\mathcal{G}$

$e(\mathcal{G})$ - the edge connectivity of the graph $\mathcal{G}$

$a(\mathcal{G})$ - the algebraic connectivity of the graph $\mathcal{G}$

$\delta(\mathcal{G})$ - the minimum vertex degree of the graph $\mathcal{G}$

$\Delta(\mathcal{G})$ - the maximum vertex degree of the graph $\mathcal{G}$

$\gamma(\mathcal{G})$ - the genus of the graph $\mathcal{G}$

$p(\mathcal{G})$ - the number of pendant vertices of the graph $\mathcal{G}$

$q(\mathcal{G})$ - the number of quasipendant vertices of the graph $\mathcal{G}$

$K_n$ - the complete graph on $n$ vertices

$K_{m,m}$ - the complete bipartite graph whose partite sets contain $m$ and $n$ vertices, respectively

$P_n$ - the path on $n$ vertices

$C_n$ - the cycle on $n$ vertices

$W_n$ - the wheel on $n+1$ vertices

$\mathcal{G}^c$ - the complement of the graph $\mathcal{G}$

$\mathcal{G}_1 + \mathcal{G}_2$ - the sum (union) of the graphs $\mathcal{G}_1$ and $\mathcal{G}_2$

$\mathcal{G}_1 \vee \mathcal{G}_2$ - the join of the graphs $\mathcal{G}_1$ and $\mathcal{G}_2$

$\mathcal{G}_1 \times \mathcal{G}_2$ - the product of the graphs $\mathcal{G}_1$ and $\mathcal{G}_2$

$\mathcal{L}(\mathcal{G})$ - the line graph of the graph $\mathcal{G}$

# Chapter 1

# Matrix Theory Preliminaries

As stated in the Preface, this book assumes an undergraduate knowledge of linear algebra. In this chapter, we study topics that are typically beyond that of an undergraduate linear algebra course, but are useful in later chapters of this book. Much of the material is taken from [6] and [41] which are two standard resources in linear algebra. We begin with a study of vector and matrix norms. Vector and matrix norms are useful in finding bounds on the spectral radius of a square matrix. We study the spectral radius of matrices more extensively in the next section which covers Perron-Frobenius theory. Perron-Frobenius theory is the study of nonnegative matrices. We will study nonnegative matrices in general, but also study interesting subsets of this class of matrices, namely positive matrices and irreducible matrices. We will see that positive matrices and irreducible matrices have many of the same properties. Nonnegative matrices will play an important role throughout this book and will be useful in understanding the theory behind M-matrices which also play an important role in later chapters. Hence we dedicate a section to M-matrices and apply the theory of nonnegative matrices to proofs of theorems involving M-matrices. Nonnegative matrices are also useful in the study of doubly stochastic matrices. Doubly stochastic matrices, which we study in the section following the section on M-matrices, are nonnegative matrices whose row sums and column sums are each one. Doubly stochastic matrices will play an important role in the study of the algebraic connectivity of graphs. Finally, we close this chapter with a section on generalized inverses of matrices. Since many of the matrices we will utilize in this book are singular, we need to familiarize ourselves with more general inverses, namely the group inverse of matrices.

## 1.1 Vector Norms, Matrix Norms, and the Spectral Radius of a Matrix

Vector and matrix norms have many uses in mathematics. In this section, we investigate vector and matrix norms and show how they give us insight into the spectral radius of a square matrix. To do this, we begin by understanding vector norms. In $\Re^n$, vectors are used to quantify length and distance. The length of a vector, or equivalently, the distance between two points in $\Re^n$, can be defined in many ways. However, for the sake of convenience, there are conditions that are often placed on the way such distances can be defined. This leads us to the formal definition of a vector norm:

**DEFINITION 1.1.1** In $\Re^n$, the function $\|\bullet\| : \Re^n \to \Re$ is a *vector norm* if for all vectors $x, y \in \Re^n$, it satisfies the following properties:

i) $\|x\| \geq 0$ and $\|x\| = 0$ if and only if $x = 0$
ii) $\|cx\| = |c|\|x\|$ for all scalars $c$
iii) $\|x + y\| \leq \|x\| + \|y\|$

**EXAMPLE 1.1.2** The most commonly used norm is the *Euclidean norm*:

$$\|x\|_2 = \left(\sum_{i=1}^{n} |x_i|^2\right)^{1/2}$$

Given two vectors $x$ and $y$ whose initial point is at the origin, we often use the Euclidean norm to find the distance between the end points of these vectors. We do this by finding $\|x - y\|_2$. In other words, the Euclidean norm is often used to find the distance between two points in $\Re^n$. For example, the set of all points in $\Re^2$ whose Euclidean distance from the origin is at most 1 is the following:

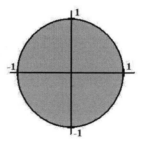

**EXAMPLE 1.1.3** We can generalize the Euclidean norm to the $\ell_p$ norm for $p \geq 1$:

$$\|x\|_p = \left(\sum_{i=1}^{n} |x_i|^p\right)^{1/p}$$

**OBSERVATION 1.1.4** The $\ell_1$ norm is often referred to as the *sum norm* since:

$$\|x\|_1 = |x|_1 + |x|_2 + \ldots + |x|_n.$$

Since norms are often used to measure distance, we can compare the manners in which distance is defined between the norms $\ell_1$ and $\ell_2$. We saw above that the set of all points whose distance from the origin in $\Re^2$ is at most 1 with respect to $\ell_2$ is the unit disc. However, the set of all points whose distance from the origin in $\Re^2$ is at most 1 with respect to $\ell_1$ is the following:

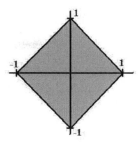

**OBSERVATION 1.1.5** The $\ell_\infty$ norm is often referred to as the *max norm* since:

$$\|x\|_\infty = \max\{|x|_1, |x|_2, \ldots, |x|_n\}.$$

Keeping with the concept of distance, the set of all points whose distance from the origin in $\Re^2$ is at most 1 with respect to $\ell_\infty$ is the following

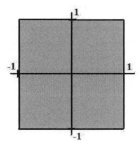

Since norms are used to quantify distance in $\Re^n$, this leads us to the concept of a sequence of vectors converging. To this end, we have the following definition:

**DEFINITION 1.1.6** Let $\{x^{(k)}\}$ be a sequence of vectors in $\Re^n$. We say that $\{x^{(k)}\}$ converges to the vector $x$ with respect to the norm $\|\bullet\|$ if $\|x^{(k)} - x\| \to 0$ as $k \to \infty$.

With the idea of convergence, we are now able to compare various vector norms in $\Re^n$. We do this in the following theorem from [41]:

**THEOREM 1.1.7** Let $\|\bullet\|_\alpha$ and $\|\bullet\|_\beta$ be any two vector norms in $\Re^n$. Then there exist finite positive constants $c_m$ and $c_M$ such that $c_m \|x\|_\alpha \leq \|x\|_\beta \leq c_M \|x\|_\alpha$ for all $x \in \Re^n$.

**Proof**: Define the function $h(x) = \|x\|_\beta / \|x\|_\alpha$ on the Euclidean unit ball $S = \{x \in \Re^n \mid \|x\|_2 = 1\}$ which is a compact set in $\Re^n$. Observe that the denominator of $h(x)$ is never zero on $S$ by (i) of Definition 1.1.1. Since vector norms are continuous functions and since the denominaror of $h(x)$ is never zero on $S$, it follows that $h(x)$ is continuous on the compact set $S$. Hence by the Weierstrass theorem, $h$ achieves a finite positive maximum $c_M$ and a positive minimum $c_m$ on $S$. Hence $c_m \|x\|_\alpha \leq \|x\|_\beta \leq c_M \|x\|_\alpha$ for all $x \in S$. Because $x / \|x\|_2 \in S$ for every nonzero vector $x \in \Re^n$, it follows that these inequalities hold for all nonzero $x \in \Re^n$.

These inequalities trivially hold for $x = 0$. This completes the proof. □

Theorem 1.1.7 suggests that given a vector $x \in \Re^n$, the values of $x$ with respect to various norms will not vary too much. This leads to the idea of equivalent norms.

**DEFINITION 1.1.8** Two norms are *equivalent* if whenever a sequence of vectors $\{x^{(k)}\}$ converges to a vector $x$ with respect to the first norm, then it converges to the same vector with respect to the second norm.

With this definition, we can now prove a corollary for Theorem 1.1.7 which is also from [41].

**COROLLARY 1.1.9** *All vector norms in $\Re^n$ are equivalent.*

**Proof**: Let $\|\bullet\|_\alpha$ and $\|\bullet\|_\beta$ be vector norms in $\Re^n$. Let $\{x^{(k)}\}$ be a sequence of vectors that converges to a vector $x$ with respect to $\|\bullet\|_\alpha$. By Theorem 1.1.7, there exist constants $c_M \geq c_m > 0$ such that

$$c_m \|x^{(k)} - x\|_\alpha \leq \|x^{(k)} - x\|_\beta \leq c_M \|x^{(k)} - x\|_\alpha$$

for all $k$. Therefore, it follows that $\|x^{(k)} - x\|_\alpha \to 0$ if and only if $\|x^{(k)} - x\|_\beta \to 0$ as $k \to \infty$. □

The idea of equivalent norms will be useful as we turn our attention to matrix norms. We begin with a definition of a matrix norm. Observe that this definition is of similar flavor to that of a vector norm.

**DEFINITION 1.1.10** Let $M_n$ denote the set of all $n \times n$ matrices. The function $\|\bullet\| : M_n \to \Re$ is a *matrix norm* if for all $A, B \in M_n$, it satisfies the following properties:

i) $\|A\| \geq 0$ and $\|A\| = 0$ if and only if $A = 0$
ii) $\|cA\| = |c|\|A\|$ for all complex scalars $c$
iii) $\|A + B\| \leq \|A\| + \|B\|$
iv) $\|AB\| \leq \|A\|\|B\|$

Matrix norms are often defined in terms of vector norms. For example, a commonly used matrix norm is $\|A\|_p$ which is defined as

$$\|A\|_p = \max_{\|x\|_p \neq 0} \frac{\|Ax\|_p}{\|x\|_p} = \max_{\|x\|_p = 1} \|Ax\|_p.$$

As with vector norms, letting $p = 1$ and letting $p \to \infty$ are of interest. We now present the following observations from [41] concerning $p$-norms for matrices for important values of $p$:

**OBSERVATION 1.1.11** For any $n \times n$ matrix $A$,

$$\|A\|_1 = \max_{1 \leq j \leq n} \sum_{i=1}^{n} |a_{i,j}|.$$

In other words, the 1-norm of a matrix is the maximum of the 1-norm of the column vectors of the matrix.

**OBSERVATION 1.1.12** For any $n \times n$ matrix $A$,

$$\|A\|_\infty = \max_{1 \leq i \leq n} \sum_{j=1}^{n} |a_{i,j}|.$$

In other words, the $\infty$-norm of a matrix is the maximum of the 1-norm of the row vectors of the matrix.

Matrix norms are very useful in finding bounds on the eigenvalues of a square matrix. The following theorem from [41] shows that the spectral radius of a matrix is always bounded above by any norm of a matrix:

**THEOREM 1.1.13** If $\|\bullet\|$ is any matrix norm and if $A \in M_n$, then $\rho(A) \leq \|A\|$.

**Proof**: Let $\lambda$ be an eigenvalue of $A$ such that $|\lambda| = \rho(A)$. Let $x$ be a corresponding eigenvector. Using the properties of matrix norms, we have

$$|\lambda|\|x\| = \|\lambda x\| = \|Ax\| \leq \|A\|\|x\|.$$

Since $\|x\| > 0$, dividing through by $\|x\|$ gives us $\rho(A) = |\lambda| \leq \|A\|$. □

We can use Observations 1.1.11 and 1.1.12 to obtain the following corollary from [41] which gives conditions as to when $\rho(A)$ and $\|A\|$ can be equal.

**COROLLARY 1.1.14** Let $A \in M_n$ and suppose that $A$ is nonnegative. If the row sums of $A$ are constant, then $\rho(A) = \|A\|_\infty$. If the column sums are constant, then $\rho(A) = \|A\|_1$.

**Proof**: We know from Theorem 1.1.13 that $\rho(A) \leq \|A\|$ for any matrix norm $\|\bullet\|$. However, if the row sums are constant, then $e$ is an eigenvector of $A$ with eigenvalue $\|A\|_\infty$, and so $\rho(A) = \|A\|_\infty$. The statement for column sums follows from applying the same argument to $A^T$. □

The goal for the remainder of this section is to prove a theorem which gives us a formula for the spectral radius in terms of matrix norms. To this end, we begin with an important lemma from [41].

**LEMMA 1.1.15** Let $A \in M_n$ and $\epsilon > 0$ be given. Then there is a matrix norm $\|\bullet\|$ such that $\rho(A) \leq \|A\| \leq \rho(A) + \epsilon$.

**Proof**: By the Schur triangularization theorem (see [41]), there is a unitary matrix $U$ and an upper triangular matrix $V$ such that $A = U^T V U$. Let $D_t = \text{diag}(t, t^2, \ldots, t^n)$ and observe

$$D_t V D_t^{-1} = \begin{bmatrix} \lambda_1 & t^{-1}d_{12} & t^{-2}d_{13} & \ldots & t^{-n+1}d_{1n} \\ 0 & \lambda_2 & t^{-1}d_{23} & \ldots & t^{-n+2}d_{2n} \\ 0 & 0 & \lambda_3 & \ldots & t^{-n+3}d_{3n} \\ \ldots & \ldots & \ldots & \ldots & \ldots \\ 0 & 0 & 0 & \ldots & t^{-1}d_{n-1,n} \\ 0 & 0 & 0 & 0 & \lambda_n \end{bmatrix},$$

where $\lambda_1, \ldots, \lambda_n$ are the eigenvalues of $A$. For $t > 0$ large enough, the sum of the off-diagonal entries of $D_t V D_t^{-1}$ are less that $\epsilon$. In particular, by Observation 1.1.11 we have $\|D_t V D_t^{-1}\|_1 \leq \rho(A) + \epsilon$ for large enough $t$. Hence if we define the matrix norm $\|\bullet\|$ by

$$\|B\| = \|D_t UTBU D_t^{-1}\|_1 = \|(UD_t^{-1})^{-1} B (UD_t^{-1})\|_1$$

for any $B \in M_n$, and if we choose $t$ large enough, we will have constructed a matrix norm such that $\|A\| \leq \rho(A) + \epsilon$. Since by Theorem 1.1.13, we have $\rho(A) \leq \|A\|$, this lemma is proven. □

We now consider matrices whose norm is less than one for some norm. We do this with a lemma from [41].

**LEMMA 1.1.16** *Let $A \in M_n$ be a given matrix. If there is a matrix norm $\|\bullet\|$ such that $\|A\| < 1$, then $\lim_{k \to \infty} A^k = 0$; that is, all the entries of $A^k$ tend to zero as $k \to \infty$.*

**Proof**: If $\|A\| < 1$, then $\|A^k\| \leq \|A\|^k \to 0$ as $k \to \infty$. Thus $\|A^k\| \to 0$ as $k \to \infty$. But since all vector norms on the $n^2$-dimensional space $M_n$ are equivalent by Corollary 1.1.9, it must also be the case that $\|A^k\|_\infty \to 0$. The result follows. □

Intuitively, if $\lim_{k \to \infty} A^k = 0$, then the entries of $A$ must be relatively small. Hence the spectral radius should be small. In the following lemma from [41], we make this idea more precise.

**LEMMA 1.1.17** *Let $A \in M_n$. Then $\lim_{k \to \infty} A^k = 0$ if and only if $\rho(A) < 1$.*

**Proof**: If $A^k \to 0$ and if $x \neq 0$ is an eigenvector corresponding to the eigenvalue $\lambda$, then $A^k x = \lambda^k x \to 0$ if and only if $|\lambda| < 1$. Since this inequality must hold for every eigenvalue of $A$, we conclude that $\rho(A) < 1$. Conversely, if $\rho(A) < 1$, then by Lemma 1.1.15, there is some matrix norm $\|\bullet\|$ such that $\|A\| < 1$. Thus by Lemma 1.1.16, it follows that $A^k \to 0$ as $k \to \infty$. □

We now prove the main result of this section which gives us a formula for the spectral radius of a matrix. This result is from [41].

**THEOREM 1.1.18** Let $A \in M_n$. For any matrix norm $\|\bullet\|$

$$\rho(A) = \lim_{k \to \infty} \|A^k\|^{1/k}$$

**Proof**: Observe $\rho(A)^k = \rho(A^k) \leq \|A^k\|$, the last inequality follows from Theorem 1.1.13. Hence $\rho(A) \leq \|A^k\|^{1/k}$ for all natural numbers $k$. Given $\epsilon > 0$, the matrix $\hat{A} := [1/(\rho(A)+\epsilon)]A$ has a spectral radius strictly less than one and hence it follows from Lemma 1.1.17 that $\|\hat{A}^k\| \to 0$ as $k \to \infty$. Thus for a fixed $A$ and $\epsilon$, there exists $N$ (depending on $A$ and $\epsilon$) such that $\|\hat{A}^k\| < 1$ for all $k \geq N$. But this is equivalent to saying $\|A^k\| \leq (\rho(A) + \epsilon)^k$ for all $k \geq N$, or that $\|A^k\|^{1/k} \leq \rho(A) + \epsilon$ for all $k \geq N$. Since $\epsilon$ was arbitrary, it follows that $\|A^k\|^{1/k} \leq \rho(A)$ for $k \geq N$. But we saw earlier in the proof that $\rho(A) \leq \|A^k\|^{1/k}$ for all $k$. Hence $\rho(A) = \lim_{k \to \infty} \|A^k\|^{1/k}$. □

Theorem 1.1.18 will be useful to us in later sections and chapters when we need to compare the spectral radii of matrices, especially nonnegative matrices. To this end, we close this section with three corollaries from [41] which allow us to compare the spectral radii of matrices. We prove the first corollary and leave the proofs of the remaining corollaries as exercies.

**COROLLARY 1.1.19** Let $A$ and $B$ be $n \times n$ matrices. If $|A| \leq B$, then $\rho(A) \leq \rho(|A|) \leq \rho(B)$.

**Proof**: First note that for every natural number $m$ we have $|A^m| \leq |A|^m \leq B^m$. Hence

$$\|A^m\|_2 \leq \||A|^m\|_2 \leq \|B^m\|_2$$

and

$$\|A^m\|_2^{1/m} \leq \||A|^m\|_2^{1/m} \leq \|B^m\|_2^{1/m}$$

for all natural numbers $m$. Letting $m$ tend to infinity and applying Theorem 1.1.18 results in $\rho(A) \leq \rho(|A|) \leq \rho(B)$. □

**COROLLARY 1.1.20** Let $A$ and $B$ be $n \times n$ matrices. If $0 \leq A \leq B$, then $\rho(A) \leq \rho(B)$.

**COROLLARY 1.1.21** Let $A$ be an $n \times n$ matrix where $A \geq 0$. If $\tilde{A}$ is any principle submatrix of $A$, then $\rho(\tilde{A}) \leq \rho(A)$. In particular, $max_{1 \leq i \leq n} a_{i,i} \leq \rho(A)$.

# Exercises:

**1.** (See [41]) Prove that for each $p \geq 1$ that $\ell_p$ is a vector norm by verifying the properties in Definition 1.1.1.

**2.** (See [41]) Prove that $\ell_\infty$ is a vector norm by verifying the properties in Definition 1.1.1, and show that

$$\|x\|_\infty = \lim_{p \to \infty} \|x\|_p.$$

**3.** Define the Frobenius norm for a matrix $A$ as

$$\|A\|_F = \left(\sum_{i,j=1}^{n} |a_{ij}|^2\right)^{1/2}$$

Use Definition 1.1.10 to verify that this is a matrix norm.

**4.** Prove Corollary 1.1.20.

**5.** Prove Corollary 1.1.21.

## 1.2 Location of Eigenvalues

In this section, we develop theory that shows where the eigenvalues of a matrix lie and how the eigenvalues of a matrix change when the matrix is perturbed. Most of this section will focus on symmetric matrices since mainly symmetric matrices will be used throughout this book. We begin with a well-known theorem known as the Gersgorin Disc Theorem which states that all of the eigenvalues of a square matrix lie in certain discs on the complex plane.

**THEOREM 1.2.1** The Gersgorin Disc Theorem. *Let $A$ be an $n \times n$ matrix and let $\sigma$ be the set of all eigenvalues of $A$. Then*

$$\sigma \subset \bigcup_{i=1}^{n} \left\{ r \in C : |a_{i,i} - r| \leq \sum_{\substack{k=1 \\ k \neq i}}^{n} |a_{i,k}| \right\} \tag{1.2.1}$$

**Proof:** Suppose $\lambda$ is an eigenvalue of $A$ with $x$ as a corresponding eigenvector, i.e., $Ax = \lambda x$. Let $x_i$ be the entry of $x$ such that $x_i = \max_{1 \leq k \leq n} |x_k|$. Observe

$$\sum_{k=1}^{n} a_{i,k} x_k = \lambda x_i$$

and therefore

$$(\lambda - a_{i,i}) x_i = \sum_{\substack{k=1 \\ k \neq i}}^{n} a_{i,k} x_k.$$

By the triangle inequality we have

$$|\lambda - a_{i,i}|\, |x_i| \leq \sum_{\substack{k=1 \\ k \neq i}}^{n} |a_{i,k}|\, |x_k|.$$

Dividing through by $|x_i|$ and recalling that $x_i = \max_{1 \leq k \leq n} |x_k|$, we obtain

$$|\lambda - a_{i,i}| \leq \sum_{\substack{k=1 \\ k \neq i}}^{n} |a_{i,k}| \frac{|x_k|}{|x_i|} \leq \sum_{\substack{k=1 \\ k \neq i}}^{n} |a_{i,k}|.$$

Therefore, the distance from $a_{i,i}$ to $\lambda$ is at most $\sum_{\substack{k=1 \\ k \neq i}}^{n} |a_{i,k}|$ on the complex plane, i.e.,

$$\lambda \in \{r \in C : |a_{i,i} - r| \leq \sum_{\substack{k=1 \\ k \neq i}}^{n} |a_{i,k}|\}.$$

Taking all eigenvalues of $A$ into account gives us (1.2.1). □

In summary, the Gersgorin Disc Theorem states that all of the eigenvalues of a square matrix lie in the union of discs whose centers are the diagonal entries of the matrix and whose radii are the sum of the absolute values of the off-diagonal entries in the corresponding row.

**EXAMPLE 1.2.2** Consider the matrix

$$A = \begin{bmatrix} 1 + 2i & 0 & 1 \\ -1 & 3 & 1 \\ 0 & i & -i \end{bmatrix}$$

We create three discs in accordance with the Gersgorin Disc Theorem. The first disc has center $1 + 2i$ and radius 1; the second disc has center 3 and radius 2; the third disc has center $-i$ and radius 1. All eigenvalues of $A$ will lie in the union of these discs.

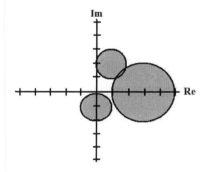

Note that the eigenvalues of $A$ are $3.1 + 0.2i$, $1.1 + 2.1i$, and $-0.2 - 1.3i$.

Since we will primarily deal with symmetric matrices in this book, we present a well-known theorem which shows that all eigenvalues of a symmetric matrix are real numbers.

**THEOREM 1.2.3** *Let $A$ be a real symmetric matrix. Then all eigenvalues of $A$ are real.*

**<u>Proof</u>**: Let $x^H$ and $A^H$ denote the conjugate transpose of the vector $x$ and matrix $A$, respectively. If $\lambda$ is a complex number such that $\lambda = a + bi$ for real numbers $a$ and $b$, note that $\lambda^H = a - bi$. We will prove this statement for the set of complex matrices $A$ such that $A = A^H$ noting that the set of real symmetric matrices is a subset of this set. Let $\lambda$ be an eigenvalue of $A$ with corresponding eigenvector $x$ normalized so that $x^H x = 1$. Then

$$\lambda = x^H A x = x^H A^H x = (x^H A x)^H = \lambda^H.$$

Since $\lambda = \lambda^H$, it follows that $\lambda$ is real. □

Since all of eigenvalues of a symmetric matrix are real, we can order the eigenvalues as follows:

$$\lambda_{min} = \lambda_1 \leq \lambda_2 \leq \ldots \leq \lambda_{n-1} \leq \lambda_n = \lambda_{max}$$

Now that we know that all of the eigenvalues of a symmetric matrix are real and the approximate location of such eigenvalues via the Gersgorin Disc Theorem, we now proceed with the goal of this section which is to gain insight into the eigenvalues of symmetric matrices with respect to unit vectors. We begin by investigating the well-known Rayleigh-Ritz equations with a theorem found in [41] which give us useful formulas for the largest and smallest eigenvalues of a symmetric matrix in terms of unit vectors.

**THEOREM 1.2.4** *Let $A \in M_n$ be symmetric. Then*

$$(i) \quad \lambda_1 x^T x \leq x^T A x \leq \lambda_n x^T x$$

*for all $x \in \Re^n$. In addition*

$$(ii) \quad \lambda_n = \max_{x \neq 0} \frac{x^T A x}{x^T x} = \max_{x^T x = 1} x^T A x$$

*and*

$$(iii) \quad \lambda_1 = \min_{x \neq 0} \frac{x^T A x}{x^T x} = \min_{x^T x = 1} x^T A x.$$

**Proof**: Since $A$ is symmetric, there exists a unitary matrix $U \in M_n$ such that $A = UDU^T$ where $D = \text{diag}(\lambda_1, \ldots, \lambda_n)$. For any vector $x \in \Re^n$, we have

$$x^T A x = x^T U D U^T x = (U^T x)^T D (U^T x) = \sum_{i=1}^{n} \lambda_i |(U^T x)_i|^2.$$

Since each term $|(U^T x)_i|^2$ is nonnegative, it follows that

$$\lambda_1 \sum_{i=1}^{n} |(U^T x)_i|^2 \leq x^T A x = \sum_{i=1}^{n} \lambda_i |(U^T x)_i|^2 \leq \lambda_n \sum_{i=1}^{n} |(U^T x)_i|^2.$$

Since $U$ is unitary, it follows that

$$\sum_{i=1}^{n} |(U^T x)_i|^2 = \sum_{i=1}^{n} |x^T U U^T x| = \sum_{i=1}^{n} |x_i|^2 = x^T x.$$

# Matrix Theory Preliminaries

Therefore
$$\lambda_1 x^T x \leq x^T A x \leq \lambda_n x^T x, \qquad (1.2.2)$$
which proves (i).

To prove (ii), we see that dividing (1.2.2) through by $x^T x$ we obtain
$$\lambda_1 \leq \frac{x^T A x}{x^T x} \leq \lambda_n$$
However, if $x$ is an eigenvector of $A$ corresponding the eigenvalue $\lambda_n$, then
$$\frac{x^T A x}{x^T x} = \frac{\lambda_n x^T x}{x^T x} = \lambda_n$$
which implies
$$\max_{x \neq 0} \frac{x^T A x}{x^T x} = \lambda_n. \qquad (1.2.3)$$
Finally, if $x \neq 0$ then
$$\frac{x^T A x}{x^T x} = \left(\frac{x}{\sqrt{x^T x}}\right)^T A \left(\frac{x}{\sqrt{x^T x}}\right) \quad \text{and} \quad \left(\frac{x}{\sqrt{x^T x}}\right)^T \left(\frac{x}{\sqrt{x^T x}}\right) = 1$$
which shows (1.2.3) is equivalent to
$$\max_{x^T x = 1} x^T A x = \lambda_n$$
This finishes the proof of (ii). The proof of (iii) is similar. □

Our goal will be to generalize the Rayleigh-Ritz equations to obtain formulas for the other eigenvalues of a symmetric matrix. This is known as the Courant-Fischer Minimax Principle. Before making such generalizations, we need a lemma from [41]:

**LEMMA 1.2.5** *Let $A \in M_n$ and let $U = [u_1, \ldots, u_n]$ be a unitary matrix such that $A = U^T D U$ where $D = diag(\lambda_1, \ldots, \lambda_n)$. Then*
$$\max_{\substack{x \neq 0 \\ x \perp u_n, u_{n-1}, \ldots, u_{n-k+1}}} \frac{x^T A x}{x^T x} = \max_{\substack{x^T x = 1 \\ x \perp u_n, u_{n-1}, \ldots, u_{n-k+1}}} x^T A x = \lambda_{n-k}$$
*where $u_1, \ldots, u_n$ are the columns of $U$.*

**Proof**: Suppose we consider only those vectors $x \in \Re^n$ that are orthogonal to $u_n, u_{n-1}, \ldots, u_{n-k+1}$. Then
$$x^T A x = \sum_{i=1}^{n} \lambda_i |(U^T x)_i|^2 = \sum_{i=1}^{n} \lambda_i |u_i^T x|^2 = \sum_{i=1}^{n-k} \lambda_i |u_i^T x|^2.$$
This is a nonnegative linear combination of $\lambda_1, \ldots, \lambda_{n-k}$. Therefore
$$x^T A x = \sum_{i=1}^{n-k} \lambda_i |u_i^T x|^2 \leq \lambda_{n-k} \sum_{i=1}^{n-k} |u_i^T x|^2 = \lambda_{n-k} \sum_{i=1}^{n} |(U^T x)_i|^2 = \lambda_{n-k} x^T x.$$
The inequality is sharp if $x = u_{n-k}$. The result now follows. □

**REMARK 1.2.6** For each $k = 1, \ldots, n$, the column vector $u_k$ of $U$ is a unit eigenvector corresponding to the eigenvalue $\lambda_k$ of $A$.

We are now ready to prove the main theorem of this section which generalizes the Rayleigh-Ritz equations. In this theorem from [41], we present the well-known Courant-Fischer Minimax Theorem.

**THEOREM 1.2.7** Let $A \in M_n$ be symmetric and let $k$ be an integer $1 \leq k \leq n$. Then

$$\lambda_k = \min_{w_1, w_2, \ldots, w_{n-k} \in \Re^n} \max_{\substack{x \neq 0 \\ x \in \Re^n \\ x \perp w_1, w_2, \ldots, w_{n-k}}} \frac{x^T A x}{x^T x} \quad (1.2.4)$$

and

$$\lambda_k = \max_{w_1, w_2, \ldots, w_{k-1} \in \Re^n} \min_{\substack{x \neq 0 \\ x \in \Re^n \\ x \perp w_1, w_2, \ldots, w_{k-1}}} \frac{x^T A x}{x^T x} \quad (1.2.5)$$

**Proof**: We will only prove (1.2.4) as the proof of (1.2.5) is similar. Writing $A = UDU^T$ as in the proof of Lemma 1.2.5 and fixing $k$ where $2 \leq k \leq n$, then if $x \neq 0$, we have

$$\frac{x^T A x}{x^T x} = \frac{(U^T x)^T D (U^T x)}{x^T x} = \frac{(U^T x)^T D (U^T x)}{(U^T x)^T (U^T x)}.$$

Since $U$ is unitary, we have

$$\{U^T x : x \in \Re^n, x \neq 0\} = \{y \in \Re^n : y \neq 0\}.$$

Therefore, if $w_1, \ldots, w_{n-k} \in \Re^n$ are given, we have

$$\sup_{\substack{x \neq 0 \\ x \perp w_1, \ldots, w_{n-k}}} \frac{x^T A x}{x^T x} = \sup_{\substack{y \neq 0 \\ y \perp U^T w_1, \ldots, U^T w_{n-k}}} \frac{y^T D y}{y^T y}$$

$$= \sup_{\substack{y^T y = 1 \\ y \perp U^T w_1, \ldots, U^T w_{n-k}}} \sum_{i=1}^n \lambda_i |y_i|^2$$

$$\geq \sup_{\substack{y^T y = 1 \\ y \perp U^T w_1, \ldots, U^T w_{n-k} \\ y_1 = y_2 = \ldots = y_{k-1} = 0}} \sum_{i=1}^n \lambda_i |y_i|^2$$

$$= \sup_{\substack{|y_k|^2 + |y_{k+1}|^2 + \ldots + |y_n|^2 = 1 \\ y \perp U^T w_1, \ldots, U^T w_{n-k}}} \sum_{i=k}^n \lambda_i |y_i|^2$$

$$\geq \lambda_k.$$

Therefore

$$\sup_{\substack{x \neq 0 \\ x \perp w_1, \ldots, w_{n-k}}} \frac{x^T A x}{x^T x} \geq \lambda_k$$

# Matrix Theory Preliminaries

for any $n-k$ vectors $w_1, \ldots, w_{n-k}$. However, Lemma 1.2.5 and Remark 1.2.6 show that equality holds for one choice of the vectors $w_i$, namely $w_i = u_{n-i+1}$. Therefore

$$\inf_{w_1,\ldots,w_{n-k}} \sup_{\substack{x \neq 0 \\ x \perp w_1,\ldots,w_{n-k}}} \frac{x^T A x}{x^T x} = \lambda_k$$

Since the extrema is achieved in all of these cases, we replace "inf" and "sup" with "min" and "max," respectively. This completes the proof. $\square$

One of the most important consequences of the Courant-Fisher Minimax Theorem are the interlacing theorems of eigenvalues. In the following theorem and corollaries from [41], we show that if we perturb a given symmetric matrix $A$ to obtain a symmetric matrx $B$, then the eigenvalues of $A$ and $B$ interlace in some fashion. In the following theorem, we investigate the eigenvalues of the matrix $A + zz^T$ where $A$ is symmetric and $z$ is any real vector.

**THEOREM 1.2.8** *Let $A \in M_n$ be symmetric and let $z \in \Re^n$ be a given vector. If the eigenvalues of $A$ and $A + zz^T$ are arranged in increasing order, then*

$(i)$ $\lambda_k(A + zz^T) \leq \lambda_{k+1}(A) \leq \lambda_{k+2}(A + zz^T)$, for $k = 1, 2, \ldots, n-2$

$(ii)$ $\lambda_k(A) \leq \lambda_{k+1}(A + zz^T) \leq \lambda_{k+2}(A)$, for $k = 1, 2, \ldots, n-2$.

**Proof**: Let $1 \leq k \leq n-2$. Then by Theorem 1.2.7 we have

$$\begin{aligned}
\lambda_{k+2}(A \pm zz^T) &= \min_{w_1,\ldots,w_{n-k-2}} \max_{\substack{x \neq 0 \\ x \perp w_1,\ldots,w_{n-k-2}}} \frac{x^T(A+zz^T)x}{x^T x} \\
&\geq \min_{w_1,\ldots,w_{n-k-2}} \max_{\substack{x \neq 0,\ x \perp z \\ x \perp w_1,\ldots,w_{n-k-2}}} \frac{x^T(A+zz^T)x}{x^T x} \\
&= \min_{\substack{w_1,\ldots,w_{n-k-2} \\ w_{n-k-1}=z}} \max_{\substack{x \neq 0 \\ x \perp w_1,\ldots,w_{n-k-1}}} \frac{x^T(A+zz^T)x}{x^T x} \\
&\geq \min_{w_1,\ldots,w_{n-k-1}} \max_{\substack{x \neq 0 \\ x \perp w_1,\ldots,w_{n-k-1}}} \frac{x^T(A+zz^T)x}{x^T x} \\
&= \lambda_{k+1}(A).
\end{aligned}$$

Similarly, for $2 \leq k \leq n-1$ we have

$$\begin{aligned}
\lambda_k(A \pm zz^T) &= \max_{w_1,\ldots,w_{k-1}} \min_{\substack{x \neq 0 \\ x \perp w_1,\ldots,w_{k-1}}} \frac{x^T(A+zz^T)x}{x^T x} \\
&\leq \max_{w_1,\ldots,w_{k-1}} \min_{\substack{x \neq 0,\, x \perp z \\ x \perp w_1,\ldots,w_{k-1}}} \frac{x^T(A+zz^T)x}{x^T x} \\
&= \max_{\substack{w_1,\ldots,w_{k-1} \\ w_k = z}} \min_{\substack{x \neq 0 \\ x \perp w_1,\ldots,w_k}} \frac{x^T(A+zz^T)x}{x^T x} \\
&\leq \max_{w_1,\ldots,w_k} \min_{\substack{x \neq 0 \\ x \perp w_1,\ldots,w_k}} \frac{x^T(A+zz^T)x}{x^T x} \\
&= \lambda_{k+1}(A).
\end{aligned}$$

Combining these inequalities proves the theorem. $\square$

We close this section with three useful corollaries (see [41]) of Theorem 1.2.8 whose proofs we leave as exercises.

**COROLLARY 1.2.9** *Let $A, B \in M_n$ be symmetric and suppose that $B$ has rank at most $r$. Then*

$(i)\ \lambda_k(A+B) \leq \lambda_{k+r}(A) \leq \lambda_{k+2r}(A+B),\ \text{for}\ k = 1, 2, \ldots, n-2r$

$(ii)\ \lambda_k(A) \leq \lambda_{k+r}(A+B) \leq \lambda_{k+2r}(A),\ \text{for}\ k = 1, 2, \ldots, n-2r.$

**COROLLARY 1.2.10** *Let $A \in M_n$ be symmetric, $z \in \Re^n$ be a vector, and $c \in \Re$. Let $\hat{A} \in M_{n+1}$ be the symmetric matrix obtained from $A$ by bordering $A$ with $z$ and $c$ as follows*

$$\hat{A} = \left[ \begin{array}{c|c} A & z \\ \hline z^T & c \end{array} \right].$$

*Then*

$\lambda_1(\hat{A}) \leq \lambda_1(A) \leq \lambda_2(\hat{A}) \leq \lambda_2(A) \leq \ldots \leq \lambda_{n-1}(A) \leq \lambda_n(\hat{A}) \leq \lambda_n(A) \leq \lambda_{n+1}(\hat{A})$

**COROLLARY 1.2.11** *Let $A, B \in M_n$ be symmetric where $B$ is positive semidefinite. Then*

$$\lambda_k(A) \leq \lambda_k(A+B)$$

*for all $k = 1, \ldots, n$.*

# Exercises:

**1**. Prove Corollary 1.2.9.

**2**. Prove Corollary 1.2.10.

**3**. Prove Corollary 1.2.11.

# 1.3 Perron-Frobenius Theory

Perron-Frobenius theory deals with the eigenvalues and eigenvectors corresponding to the spectral radius of a nonnegative matrix. Nonnegative matrices are of great importance in matrix theory and will be of special importance later in this book as we apply them extensively in graph theory. Therefore, we dedicate a section to these results. We begin with a definition:

**DEFINITION 1.3.1** A matrix $A$ is *nonnegative* if all entries of $A$ are nonnegative. In this case, we write $A \geq 0$. If all entires of $A$ are strictly positive, then we say $A$ is *positive* and write $A >> 0$.

Note that the set of positive matrices is a subset of the set of nonnegative matrices. Further if we want to denote that a nonnegative matrix $A$ has at least one positive entry, we write $A > 0$.

In this section, we will first develop Perron-Frobenius theory for positive matrices. We then relax the condition of the matrices being positive and investigate how Perron-Frobenius theory changes when dealing with nonnnegative matrices. Finally, we study a special class of nonnegative matrices known as irreducible matrices and show that they behave similarly to positive matrices. We begin with the study of positive matrices. Since the set of positive matrices is a subset of nonnegative matrices, we begin with an important preliminary lemma and three useful corollaries from [41] concerning the larger class of nonnegative matrices:

**LEMMA 1.3.2** Let $A \in M_n$ be nonnegative. Then

$$\min_{1 \leq i \leq n} \sum_{j=1}^{n} a_{i,j} \leq \rho(A) \leq \max_{1 \leq i \leq n} \sum_{j=1}^{n} a_{i,j} \qquad (1.3.1)$$

and

$$\min_{1 \leq j \leq n} \sum_{i=1}^{n} a_{i,j} \leq \rho(A) \leq \max_{1 \leq j \leq n} \sum_{i=1}^{n} a_{i,j} \qquad (1.3.2)$$

**Proof**: Let $\alpha = \min_{1 \leq i \leq n} \sum_{j=1}^{n} a_{i,j}$ and let $B \in M_n$ be such that $b_{i,j} = \alpha a_{i,j} / \sum_{j=1}^{n} a_{i,j}$. Observe $A \geq B \geq 0$. By Corollary 1.1.14 we see that $\rho(B) = \alpha$; by Corollary 1.1.20 we have $\rho(B) \leq \rho(A)$. Hence $\alpha \leq \rho(A)$ which establishes the first inequality in (1.3.1). The second inequality in (1.3.1) is established in a similar fashion. Finally, (1.3.2) is established by applying the above argument to $A^T$. □

Now that we have some preliminary bounds on the spectral radius of nonnegative matrices, we can apply this lemma to get more precise results. In our first corollary, we recall that if $S$ is an invertible matrix, then $\rho(S^{-1}AS) = \rho(A)$.

**COROLLARY 1.3.3** *Let $A \in M_n$ be nonnegative. Then for any positive vector $x \in \Re^n$ we have*

$$\min_{1 \leq i \leq n} \frac{1}{x_i} \sum_{j=1}^n a_{i,j} x_j \leq \rho(A) \leq \max_{1 \leq i \leq n} \frac{1}{x_i} \sum_{j=1}^n a_{i,j} x_j$$

*and*

$$\min_{1 \leq j \leq n} x_j \sum_{i=1}^n \frac{a_{i,j}}{x_i} \leq \rho(A) \leq \max_{1 \leq j \leq n} x_j \sum_{i=1}^n \frac{a_{i,j}}{x_i}$$

**Proof**: Let $S = diag(x_1, \ldots, x_n)$. Since $S$ is invertible, it follows that $\rho(S^{-1}AS) = \rho(A)$. Moreover, $S^{-1}AS$ is nonnegative. Thus we can apply Lemma 1.3.2 to $S^{-1}AS = [a_{i,j}x_j/x_i]$ to obtain the result. □

We now continue to sharpen our bounds on the spectral radius of nonnegative matrices found in Lemma 1.3.2 in the next corollary which helps us determine bounds on the spectral radius in terms of vectors. Observe that this corollary is somewhat reminiscent of Theorem 1.2.4(i).

**COROLLARY 1.3.4** *Let $A \in M_n$ be nonnegative and suppose $x \in \Re^n$ is a positive vector. If $\alpha, \beta \geq 0$ are such that $\alpha x \leq Ax \leq \beta x$, then $\alpha \leq \rho(A) \leq \beta$. Moreover, if $\alpha x < Ax$ then $\alpha < \rho(A)$; if $Ax < \beta x$, then $\rho(A) < \beta$.*

**Proof**: If $\alpha x \leq Ax$, then $\alpha \leq \min_{1 \leq i \leq n}(1/x_i) \sum_{j=1}^n a_{i,j} x_j$. Thus by Corollary 1.3.3, it follows that $\alpha \leq \rho(A)$. If $\alpha x < Ax$, then there exists some $\alpha' > \alpha$ such that $\alpha' x \leq Ax$. In this case, $\rho(A) \geq \alpha' > \alpha$, thus $\rho(A) > \alpha$. The upper bounds are verified in a similar fashion. □

The previous two corollaries have led up to the next corollary which will be useful when proving the first main result of this section.

**COROLLARY 1.3.5** *Let $A \in M_n$ be nonnegative. If $A$ has a positive eigenvector, then the corresponding eigenvalue is $\rho(A)$.*

**Proof**: Suppose $Ax = \lambda x$ where $x \gg 0$. Then $\lambda x \leq Ax \leq \lambda x$ by Corollary 1.3.4. Applying Corollary 1.3.4 again, we obtain $\lambda \leq \rho(A) \leq \lambda$. □

Now that we have some preliminary results concerning nonnegative matrices, we return our focus to positive matrices. The first goal of this section is to prove Perron's theorem which is a well-known theorem concerning the eigenvalues and eigenvectors of positive matrices. First we need a lemma from [41].

**LEMMA 1.3.6** *Let $A \in M_n$. Suppose that $\lambda$ is an eigenvalue of $A$ such that $|\lambda| = \rho(A)$ and that $\lambda$ is the only eigenvalue of $A$ with modulus $\rho(A)$. Suppose $x$ and $y$ are vectors such that $Ax = \lambda x$ and $A^T y = \lambda y$ where $x$ and $y$ are normalized so that $x^T y = 1$. Let $L = xy^T$. Then $\lim_{m \to \infty}[(1/\lambda(A))A]^m = L$.*

**Proof**: First, observe that (a) $L^m = L$ and (b) $A^m L = L A^m = \lambda^m L$ for all integers $m$. Then (a) and (b) imply $(A - \lambda L)^m = A^m - \lambda^m L$ for all integers $m$. Hence
$$\left(\frac{1}{\lambda} A - L\right)^m = \left[\frac{1}{\lambda}(A - \lambda L)\right]^m = \frac{1}{\lambda^m} A^m - L.$$
Therefore
$$\left(\frac{1}{\lambda} A\right)^m = L + \left(\frac{1}{\lambda} A - L\right)^m. \tag{1.3.3}$$
Since
$$\rho\left(\frac{1}{\lambda} A - L\right) = \frac{\rho(A - \lambda L)}{\rho(A)} \leq \frac{|\lambda_{n-1}(A)|}{\rho(A)} < 1,$$
the result follows from (1.3.3). □

**OBSERVATION 1.3.7** Since $L$ is the product of two vectors, it follows that the rank of $L$ is 1.

We are now ready to prove Perron's Theorem for positive matrices which is the first main result of this section. The proof is adapted from [41].

**THEOREM 1.3.8** Let $A \in M_n$ be positive. Then
(i) $\rho(A)$ is an eigenvalue of $A$,
(ii) There is a positive eigenvector corresponding to $\rho(A)$,
(iii) $|\lambda| < \rho(A)$ for every eigenvalue such that $\lambda \neq \rho(A)$,
(iv) $\rho(A)$ is a simple eigenvalue of $A$.

**Proof**: Let $x \neq 0$ be such that $Ax = \lambda x$ where $|\lambda| = \rho(A)$. Then
$$\rho(A)|x| = |\lambda||x| = |\lambda x| = |Ax| \leq |A||x| = A|x|.$$
Thus $y := A|x| - \rho(A)|x| \geq 0$. Since $|x| > 0$ and $A \gg 0$, it follows that $z := A|x| \gg 0$. If $y \neq 0$ then
$$0 < Ay = Az - \rho(A)z$$
which simplifies to $Az > \rho(A)z$. This implies that $\rho(A) > \rho(A)$ which is clearly false. Thus $y = 0$, and therefore $A|x| = \rho(A)|x|$. Hence $\rho(A)$ is a positive eigenvalue of $A$ corresponding to the positive eigenvector $|x|$, thus (i) and (ii) are proved.

To prove (iii), we will show that if $\lambda$ is an eigenvalue of $A$ where $|\lambda| = \rho(A)$, then $\lambda = \rho(A)$. Let $x$ be an eigenvector corresponding to $\lambda$. We first show that there exits an argument $0 \leq \theta < 2\pi$ such that $e^{-i\theta} x = |x| \gg 0$. To see this, observe from (i) and (ii) that
$$\rho(A)|x_k| = |\lambda||x_k| = |\lambda x_k| = \left|\sum_{p=1}^{n} a_{kp} x_p\right| \leq \sum_{p=1}^{n} |a_{kp}||x_p| = \sum_{p=1}^{n} a_{kp}|x_p| = \rho(A)|x_k|.$$

Thus equality must hold in the triangle inequality and hence the nonzero complex numbers $a_{kp} x_p$, $p = 1, \ldots, n$ must all have the same argument, say $\theta$. Since $a_{kp} > 0$

for all $p$, it follows that $e^{-i\theta}x \gg 0$. Letting $w = e^{-i\theta}x \gg 0$, we have $Aw = \lambda w$. But by Corollary 1.3.5 it follows that $\lambda = \rho(A)$.

To prove (iv), write $A = U\Delta U^T$ where $U$ is unitary and $\Delta$ is an upper triangular matrix with main diagonal entries $\rho, \ldots, \rho, \lambda_{k+1}, \ldots, \lambda_n$, where $\rho = \rho(A)$ is an eigenvalue of $A$ with algebraic multiplicity $k \geq 1$; the eigenvalues $\lambda_i$ are all such that $|\lambda_i| < \rho(A)$ for all $k+1 \leq i \leq n$ (by part (iii)). Using Lemma 1.3.6 we have

$$L = \lim_{m\to\infty} \left(\frac{1}{\rho(A)}A\right)^m = U \lim_{m\to\infty} \begin{bmatrix} 1 & & & & & * \\ & \ddots & & & & \\ & & 1 & & & \\ & & & \frac{\lambda_{k+1}}{\rho} & & \\ & 0 & & & \ddots & \\ & & & & & \frac{\lambda_n}{\rho} \end{bmatrix}^m U^T$$

$$= U \begin{bmatrix} 1 & & & & & * \\ & \ddots & & & & \\ & & 1 & & & \\ & & & 0 & & \\ & 0 & & & \ddots & \\ & & & & & 0 \end{bmatrix} U^T$$

where the diagonal entry 1 is repeated $k$ times in the last two expressions, and the diagonal entry 0 is repeated $n - k$ times. Since the upper triangular matrix in the last expression has rank at least $k$, and since $L$ has rank 1 (Observation 1.3.7), we conclude that $k > 1$ is impossible, thus proving (iv). □

**EXAMPLE 1.3.9** Consider the positive matrix

$$A = \begin{bmatrix} 8 & 8 & 8 \\ 4 & 2 & 1 \\ 4 & 12 & 4 \end{bmatrix}$$

The eigenvalues of $A$ are 16, $-1 + 3.32i$, and $-1 - 3.32i$. Note that the eigenvalue of largest modulus is 16 and that $\rho(A) = 16$. Moreover, the eigenvector corresponding to 16 is positive, namely $[3, 1, 2]^T$. Finally, 16 is the only eigenvalue with a positive eigenvector as the eigenvectors corresponding to $-1 + 3.32i$ and $-1 - 3.32i$ are $[-0.52 + 1.23i, 1, -0.93 - 1.6i]^T$ and $[-0.52 - 1.23i, 1, -0.93 + 1.6i]^T$, respectively.

Since the eigenvector corresponding to the spectral radius of a positive matrix is of special importance, we have the following definition. Note in this definition, we relax the conditions of the matrix and eigenvector corresponding to the spectral radius to be nonnnegative rather than positive. We will see in the theorem that follows that relaxing such conditions is desirable.

**DEFINITION 1.3.10** A nonnegative eigenvector of $A \geq 0$ corresponding to $\rho(A)$ is called a *Perron vector* of $A$.

# Matrix Theory Preliminaries

We now turn our attention to nonnegative matrices. Since we are relaxing the conditions of Theorem 1.3.8 by allowing our matrices to have entries of zero, we expect the conclusions of the theorem to be more relaxed in that the eigenvector corresponding to the spectral radius be allowed entries of zero. This is indeed the case as we see in the following theorem from [41].

**THEOREM 1.3.11** *Let $A$ be a nonnegative $n \times n$ matrix. Then*
*(i) $\rho(A)$ is an eigenvalue of $A$, and*
*(ii) $A$ has a nonnegative eigenvector corresponding to $\rho(A)$.*

**Proof:** For any $\epsilon > 0$, define the matrix $A(\epsilon) := [a_{i,j} + \epsilon] >> 0$. Let $x(\epsilon)$ be the positive eigenvector of $A(\epsilon)$ corresponding to $\rho(A(\epsilon))$ as per Theorem 1.3.8(i). Normalize each vector $x(\epsilon)$ so that $\sum_{i=1}^{n} x(\epsilon)_i = 1$. Since the set of vectors $\{x(\epsilon) : \epsilon > 0\}$ is contained in the compact set $\{x : x \in C^n, \|x\|_1 \leq 1\}$, there is a monotone decreasing sequence $\epsilon_1, \epsilon_2, \ldots$, with $\lim_{k \to \infty} \epsilon_k = 0$ such that $x := \lim_{k \to \infty} x(\epsilon_k)$ exists. Since $x(\epsilon_k) >> 0$ for all $k$, it follows that $x \geq 0$. However, since

$$\sum_{i=1}^{n} x_i = \lim_{k \to \infty} \sum_{i=1}^{n} x(\epsilon_k)_i = 1$$

it follows that $x \neq 0$, hence $x > 0$. By Corollary 1.1.20 it follows that $\rho(A(\epsilon_k)) \geq \rho(A(\epsilon_{k+1})) \geq \ldots \geq \rho(A)$, for any $k$. Thus the sequence of real numbers $\{\rho(A(\epsilon_k))\}_{k=1,2,\ldots}$ is a bounded monotone decreasing sequence and hence $\rho := \lim_{k \to \infty} \rho(A(\epsilon_k))$ exists and $\rho \geq \rho(A)$. However,

$$Ax = \lim_{k \to \infty} A(\epsilon_k) x(\epsilon_k)$$
$$= \lim_{k \to \infty} \rho(A(\epsilon_k)) x(\epsilon_k)$$
$$= \lim_{k \to \infty} \rho(A(\epsilon_k)) \lim_{k \to \infty} x(\epsilon_k) = \rho x.$$

Since $x \neq 0$, it follows that $\rho$ is an eigenvalue of $A$ with $x$ as the corresponding eigenvector. Therefore $\rho = \rho(A)$ and $x > 0$ is a corresponding eigenvector. □

In Theorem 1.3.8 which concerns positive matrices, i.e., nonnegative matrices which do not contain a zero entry, we see that the eigenvector corresponding to the largest eigenvalue in modulus is also positive, hence it does not contain a zero entry. Moreover, the spectral radius of such a matrix is a simple eigenvalue. However, when we relax the conditions of allowing zero entries in a nonnegative matrix as we do in Theorem 1.3.11, we see that while the spectral radius is still an eigenvalue, it need not be a simple eigenvalue. Moreover, the eigenvector corresponding to such an eigenvalue is nonnegative, hence it may have a zero entry. We now turn our attention to a specific class of nonnegative matrices known as irreducible matrices. We will see that while these matrices may have a zero entry, they will behave like positive matrices. To this end, we have a definition:

**DEFINITION 1.3.12** *A matrix $A \in M_n$ is reducible if $A$ is permutationally similar to a matrix of the form*

$$\begin{bmatrix} B & C \\ 0 & D \end{bmatrix}$$

where $B$ and $D$ are both square matrices. If $A$ is not permutationally similar to a matrix of this form, we say that $A$ is *irreducible*.

In order to be able to determine if a nonnegative matrix is irreducible, it is helpful for us to have a pictorial representation of the matrix:

**DEFINITION 1.3.13** The associated *directed graph*, $G(A)$, of a matrix $A \in M_n$ is a graph on $n$ vertices $v_1, \ldots, v_n$ where there is a directed edge from $v_i$ to $v_j$ if and only if $a_{i,j} \neq 0$.

**EXAMPLE 1.3.14** Consider the following matrices:

$$A = \begin{bmatrix} 0 & 5 & 3 \\ 0 & 0 & 1 \\ 4 & 0 & 0 \end{bmatrix} \qquad B = \begin{bmatrix} 2 & 0 & 4 & 1 \\ 0 & 0 & 0 & 3 \\ 6 & 5 & 0 & 0 \\ 0 & 7 & 0 & 2 \end{bmatrix}$$

Their associated directed graphs are

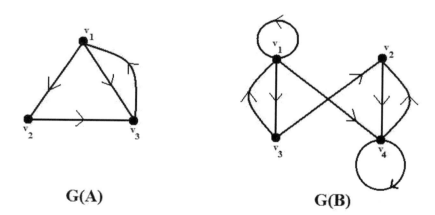

**G(A)**  **G(B)**

**DEFINITION 1.3.15** A directed graph is *strongly connected* if for any pair of vertices $v_i$ and $v_j$, it is possible to travel from $v_i$ to $v_j$ along a sequence of directed edges. We refer to such a sequence of directed edges as a *directed path*.

Observe that $G(A)$ is strongly connected. However, $G(B)$ is not strongly connected as there does not exist a directed path from $v_4$ to $v_1$ (or $v_4$ to $v_3$). The existence of directed paths between any pairs of vertices leads us to the following theorem from [6] concerning irreducible matrices:

# Matrix Theory Preliminaries

**THEOREM 1.3.16** *A matrix $A$ is irreducible if and only if $G(A)$ is strongly connected.*

**Proof**: Suppose $A$ is reducible, then there exists a permutation matrix $P$ such that
$$A = P \begin{bmatrix} B & C \\ 0 & D \end{bmatrix} P^T := P\hat{A}P^T$$
where $B \in M_r$, $D \in M_{n-r}$, $C \in M_{r,n-r}$, and $0 \in M_{n-r,r}$ for some $1 \leq r \leq n$. Let $v_1, \ldots, v_n \in V(G(A))$ and $\hat{v}_1, \ldots, \hat{v}_n \in V(G(\hat{A}))$. In $G(\hat{A})$, observe that there does not exist a directed path from $\hat{v}_i$ to $\hat{v}_j$ if $r+1 \leq i \leq n$ and $1 \leq j \leq r$. Hence $G(\hat{A})$ is not strongly connected. Since $G(A)$ and $G(\hat{A})$ are isomorphic, it follows that $G(A)$ is not strongly connected.

Now suppose $G(A)$ is not strongly connected. Then there exists nonempty sets of vertices $S_1$ and $S_2$ of $G(A)$ such that no directed path from $v_i$ to $v_j$ exists if $v_i \in S_2$ and $v_j \in S_1$. Let $|S_1| = r$ and $|S_2| = n - r$. Relabel the vertices of $G(A)$ as $\hat{v}_1, \ldots, \hat{v}_n$ where $\hat{v}_1, \ldots, \hat{v}_r \in S_1$ and $\hat{v}_{r+1}, \ldots, \hat{v}_n \in S_2$; permute the matrix $A$ in the same fashion to create $\hat{A}$. Thus graph created from relabeling the vertices of $G(A)$ is precisely $G(\hat{A})$. Since there are no directed paths in $G(\hat{A})$ from the vertices in $S_2$ to the vertices in $S_1$, it follows that $\hat{A}$ must have an $(n-r) \times r$ block of zeros in the lower left corner. Thus $\hat{A}$ is reducible. Since $A$ and $\hat{A}$ are permutationally similar, it follows that $A$ is reducible. □

**EXAMPLE 1.3.17** Revisiting the nonnegative matrices in Example 1.3.14, we see that $A$ is irreducible since the associated directed graph is strongly connected. However $B$ is reducible since its associated directed graph is not strongly connected. Partioning the matrix below to highlight reducibility, observe that $B$ is permutationally similar to

$$\left[\begin{array}{cc|cc} 2 & 4 & 0 & 1 \\ 6 & 0 & 5 & 0 \\ \hline 0 & 0 & 0 & 3 \\ 0 & 0 & 7 & 2 \end{array}\right].$$

Our goal will be to show that nonnegative irreducible matrices behave in a similar way to positive matrices. To this end, we present two lemmas from [41] and [6] which shed light on the relationship between nonnegative and positive matrices.

**LEMMA 1.3.18** *Let $A \in M_n$ and suppose $A$ is nonnegative. Then $A$ is irreducible if and only if $(I + A)^{n-1}$ is positive.*

**Proof**: Suppose first that $A$ is reducible. Then for some permutation matrix $P$ we have $A = P\hat{A}P^T$ where $\hat{A}$ is as in Theorem 1.3.16. Observe

$$(I + A)^{n-1} = (I + P\hat{A}P^T)^{n-1} = (P[I + \hat{A}]P^T)^{n-1}$$

$$= P\left[I + (n-1)\hat{A} + \binom{n-1}{2}\hat{A}^2 + \ldots + \binom{n-1}{n-1}\hat{A}^{n-1}\right]P^T.$$

By matrix multiplication, note that $\hat{A}^2, \hat{A}^3, \ldots, \hat{A}^{n-1}$ all have the same $(n-r) \times r$ block of 0's in the lower left corner as $\hat{A}$. Therefore, all of the terms in the square brackets have an $(n-r) \times r$ block of 0's in the lower left corner, and hence $(I+A)^{n-1}$ does also. Therefore $(I+A)^{n-1}$ is not positive.

Suppose now that $A$ is irreducible, then so is $I+A$. Let $Y$ be the set of all nonnegative nonzero vectors in $\Re^n$ with at least one entry of zero. Since $I+A$ is irreducible, it follows by matrix-vector multiplication that $(I+A)y$ will have fewer zero entries than $y$ for each vector $y \in Y$. Hence $(I+A)^{n-1}y$ is positive for all vectors $y \in Y$. The only way that this can hold for *every* vector $y \in Y$ is for $(I+A)^{n-1}$ to be positive. □

**LEMMA 1.3.19** *If $A \in M_n$, $A$ is nonnegative, and $A^k$ is positive for some $k \geq 1$, then $\rho(A)$ is a simple eigenvalue of $A$.*

**Proof**: If $\lambda_1, \ldots, \lambda_n$ are the eigenvalues of $A$, then $\lambda_1^k, \ldots, \lambda_n^k$ are the eigenvalues of $A^k$. By Theorem 1.3.11, $\rho(A)$ is an eigenvalue of $A$. Hence if $\rho(A)$ were a multiple eigenvalue of $A$, then $\rho(A)^k = \rho(A^k)$ would be a multiple eigenvalue of $A^k$. But this is impossible since $\rho(A^k)$ is a simple eigenvalue of $A^k$ by Theorem 1.3.8. □

We are now able to present the culminating theorem (from [41]) of this section which illustrates that the majority of Theorem 1.3.8 still holds for nonnegative matrices so long as the matrix is irreducible.

**THEOREM 1.3.20** *Let $A \in M_n$ be an irreducible nonnegative matrix. Then*
*(i) $\rho(A) > 0$*
*(ii) $\rho(A)$ is an eigenvalue of $A$*
*(iii) There is a positive eigenvector $x$ corresponding to $\rho(A)$*
*(iv) $\rho(A)$ is a simple eigenvalue.*

**Proof**: Since $A$ is nonnegative and irreducible, all row sums are positive. Thus (i) follows from Lemma 1.3.2. Statement (ii) follows from Theorem 1.3.11 and the fact that $A$ is nonnegative (irreducibility is not even required here). Theorem 1.3.11 also guarantees that there exists a nonnegative eigenvector $x$ corresponding to $\rho(A)$. To prove (iii), we need only show that $x$ does not have a zero coordinate. Since $Ax = \rho(A)x$, it follows that $(I+A)^{n-1}x = (1+\rho(A))^{n-1}x$. By Lemma 1.3.18 the matrix $(I+A)^{n-1}$ is positive, so by Theorem 1.3.8, $x$ is positive. To prove (iv), suppose $\rho(A)$ is a multiple eigenvalue of $A$. Then $1 + \rho(A) = \rho(I + A)$ is a multiple eigenvalue of $I + A$. Hence $(1 + \rho(A))^{n-1} = \rho((I+A)^{n-1})$ is a multiple eigenvalue of $(I+A)^{n-1}$. But $I + A$ is nonnegative and $(I+A)^{n-1}$ is positive by Lemma 1.3.18. Therefore by Lemma 1.3.19, $\rho((I+A)^{n-1})$ cannot be a multiple eigenvalue of $(I+A)^{n-1}$, producing a contradiction. Thus $\rho(A)$ is a simple eigenvalue of $A$. □

# Matrix Theory Preliminaries

We see from Theorems 1.3.8 and 1.3.20 that the consequences of $A$ being a positive matrix or $A$ being a nonnegative irreducible matrix are strikingly similar. In both cases, $\rho(A)$ is a simple eigenvalue and the corresponding eigenvector is positive. However, if $A$ is positive, the $\rho(A)$ is the only eigenvalue whose modulus is $\rho(A)$. This need not be the case if $A$ is nonnegative and irreducible. For example, consider

$$A = \begin{bmatrix} 0 & 1 & 0 \\ 0 & 0 & 1 \\ 1 & 0 & 0 \end{bmatrix}.$$

Observe that the eigenvalues of $A$ are $1$, $-0.5+i\sqrt{3}/2$, and $-0.5-i\sqrt{3}/2$. Note that $\rho(A) = 1$ and $1$ is an eigenvalue of $A$. However, $|-0.5+i\sqrt{3}/2| = |-0.5-i\sqrt{3}/2| = 1$. It should be noted that even though there may be more than one eigenvalue of a nonnegative irreducible matrix $A$ whose modulus is $\rho(A)$, each such eigenvalue will be simple. The proof of this idea is beyond the scope of this text but may be found in [41].

## Exercises:

**1.** (See [6]) Let $A$ and $B$ be nonnnegative matrices. Prove the following:
(i) If $0 \leq A \leq B$, then $\rho(A) \leq \rho(B)$.
(ii) If $0 \leq A \leq B$ and $A + B$ is irreducible, then $\rho(A) < \rho(B)$.
(iii) $\rho(A)$ is an eigenvalue of some proper principal submatrix of $A$ if and only if $A$ is reducible.

**2.** (See [6]) Let $A$ be nonnegative and irreducible. For every vector $x > 0$ define

$$r_x = \min_{x_i \neq 0} \frac{(Ax)_i}{x_i}.$$

Prove that $r_x$ is the largest real number $\rho$ such that $\rho x \leq Ax$.

**3.** (See [6]) Let $A$ be nonnegative and irreducible and let $r = r_z = \max_{x>0} r_x$ where $r_x$ is as in Exercise 2. Prove the following:
(i) $r > 0$
(ii) $Az = rz$
(iii) $z \gg 0$.

**4.** (See [6]) Let $A$ be nonnegative and irreducible. Prove the following minimax characteristics:

(i)
$$\rho(A) = \max_{x>0} \left( \min_{x_i>0} \frac{(Ax)_i}{x_i} \right).$$

(ii)
$$\rho(A) = \min_{x>0} \left( \max_{x_i>0} \frac{(Ax)_i}{x_i} \right).$$

**5.** (See [28]) Let $A$ be an $n \times n$ nonnegative irreducible, symmetric matrix with eigenvalues $\lambda_1 \leq \lambda_2 \leq \ldots \leq \lambda_n$. Let $u = (u_i)$ be an eigenvector corresponding to $\lambda_n$ and $v = (v_i)$ be an eigenvector corresponding to $\lambda_{n-1}$. Define

$$M_\alpha := \{i \in N | v_i + \alpha u_i \geq 0\}.$$

Prove that for any $\alpha \geq 0$, the submatrix $A(M_\alpha)$ is irreducible.

## 1.4 M-Matrices

In this section, we take the nonnegative matrices that we studied in the previous section and subtract them from a large enough multiple of the identity matrix to ensure that only the diagonal entries are nonnegative in the resulting matrix. Naturally there will be properties of nonnegative matrices that coincide with the properties of these matrices. We study a useful subset of these matrices and investigate the properties that arise from reducibility and invertibility. We begin with a definition.

**DEFINITION 1.4.1** A *Z-matrix* is an $n \times n$ matrix $A$ where $a_{i,j} \geq 0$ if $i = j$ and $a_{i,j} \leq 0$ if $i \neq j$.

Observe that a Z-matrix is of the following form:

$$\begin{bmatrix} b_{1,1} & -b_{1,2} & -b_{1,3} & \ldots \\ -b_{2,1} & b_{2,2} & -b_{2,3} & \ldots \\ -b_{3,1} & -b_{3,2} & b_{3,3} & \ldots \\ \ldots & \ldots & \ldots & \ldots \end{bmatrix}$$

where we take all $b_{i,j}$ to be nonnegative. Observe that a Z-matrix $A$ can be written as $sI - B$ for some $s > 0$ and some $B \geq 0$. Hence we can immediately see that there will be a close relationship between Z-matrices and nonnegative matrices. In fact, if a Z-matrix $A$ is irreducible and symmetric (guaranteeing real eigenvalues), we have the following two observations which are consequences of Theorem 1.3.20:

**OBSERVATION 1.4.2** Let $A$ be an irreducible symmetric Z-matrix. If $x$ is a positive eigenvector of $A$, the $x$ corresponds to the smallest eigenvalue of $A$.

**OBSERVATION 1.4.3** Let $A$ be an irreducible symmetric Z-matrix. Then the smallest eigenvalue of $A$ has multiplicity 1 and the corresponding eigenvector is positive, unique up to a scalar multiple.

We will restrict ourself to the subset of Z-matrices where $s \geq \rho(B)$. We will see in this section that the resulting matrices will have a lot of interesting properties. The properties of such matrices will have even stronger relations to those of nonnegative matrices - properties that are not as intuitive as the observations above. Since this subset of Z-matrices play an important role in graph theory, we dedicate this section to studying such matrices. We now present a formal definition of this subset of Z-matrices:

# Matrix Theory Preliminaries

**DEFINITION 1.4.4** A Z-matrix of the form $sI - B$, where $B \geq 0$, is an *M-matrix* if $s \geq \rho(B)$.

If the inequality in the above definition is strict, then the M-matrix is nonsingular. The first goal of this section is to investigate the properties of the inverse of a nonsingular M-matrix. To this end, we begin with a lemma from [6] concerning nonnegative matrices.

**LEMMA 1.4.5** *Let $T$ be a nonnegative matrix such that $\rho(T) < 1$. Then $(I - T)^{-1}$ exists and*

$$(I - T)^{-1} = \sum_{k=0}^{\infty} T^k \geq 0 \qquad (1.4.1)$$

**Proof**: If $\rho(T) < 1$ then the series in (1.4.1) converges. Moreover, since $\rho(T) < 1$, it follows that if $\lambda$ is an eigenvalue of $I - T$, then $0 < |\lambda| < 2$ by the Gersgorin Disc Theorem (Theorem 1.2.1). Hence $(I - T)^{-1}$ exists. Observe

$$(I - T)(I + T + T^2 + T^3 + \ldots + T^k) = I - T^{k+1}.$$

Since $\rho(T) < 1$, it follows that $\lim_{k \to \infty} T^{k+1} = 0$. Therefore, taking the limit of both sides as $k$ tends to infinity, we see that

$$(I - T) \sum_{k=0}^{\infty} T^k = I.$$

The result follows. □

**OBSERVATION 1.4.6** If $T$ is nonnegative and $\rho(T) < 1$, it follows that $(I - T)^{-1}$ is nonnegative.

From this lemma and observation, we see that nonnegative matrices will play a strong role in the inverses of nonsingular M-matrices. In the following theorem from [6], we see that the inverse of an M-matrix is always nonnegative.

**THEOREM 1.4.7** *If $A$ is a nonsingular M-matrix, then $A^{-1} \geq 0$. Moreover, if $A$ is irreducible, then $A^{-1} \gg 0$.*

**Proof**: Since $A$ is an M-matrix (possibly reducible), it follows that $A = sI - B$ for some nonnegative matrix $B$ and some $s > \rho(B)$. Let $C = (1/s)A = I - B/s$. Letting $T = B/s$, we see that $\rho(T) < 1$. Thus by Lemma 1.4.5 and Observation 1.4.6 we see that $C^{-1} \geq 0$. But since $A = sC$ where $s > 0$, it follows that $A^{-1} \geq 0$.

Now suppose that $A$ is irreducible. Then $B \geq 0$ is also irreducible. Let $x > 0$. Then $y := A^{-1}x > 0$. Since $(sI - B)y = Ay = x > 0$, it follows that $sy \geq By$. We claim this implies that $y \gg 0$. Suppose not. Then $y$ has a zero coordinate. Since $sy \geq By \geq 0$, it follows that $By$ has a zero coordinate in the same position as $y$. Hence $B$ would be reducible, contradicting the irreducibility of $A$. Thus $y \gg 0$.

Since $x > 0$ implies $y = A^{-1}x >> 0$, it follows by matrix-vector multiplication that $A^{-1} >> 0$. $\square$

Since irreducibility plays an important role in the theory of nonnegative matrices, it is not surprising that it plays an important role in the theory of M-matrices. If $A$ is a nonsingular M-matrix that is also irreducible, Theorem 1.4.7 shows that not only will its inverse be nonnegative, it will be strictly positive.

The theory of nonsingular M-matrices will be important when we broaden our study to singular M-matrices. To this end, we present two lemmas from [6] concerning nonsingular M-matrices that will be useful to us.

**LEMMA 1.4.8** *Let $A$ be a Z-matrix. Then $A$ is an M-matrix if and only if $A + \epsilon I$ is a nonsingular M-matrix for all scalars $\epsilon > 0$.*

**Proof**: Let $A$ be an M-matrix of the form $A = sI - B$, where $B \geq 0$ and $s \geq \rho(B)$. Then for any $\epsilon > 0$,

$$A + \epsilon I = sI - B + \epsilon I = (s+\epsilon)I - B := s'I - B. \qquad (1.4.2)$$

Since $s' > s \geq \rho(B)$, it follows that $A + \epsilon I$ is nonsingular.

Conversely, if $A + \epsilon I$ is a nonsingular M-matrix for all $\epsilon > 0$, then it follows by letting $\epsilon$ approach zero in (1.4.2) that $A$ is an M-matrix. $\square$

**LEMMA 1.4.9** *Let $A$ be a nonsingular M-matrix. Then all principal minors of $A$ are positive.*

**Proof**: We first show that there exists a diagonal matrix $D$ such that $G := D^{-1}AD$ is strictly diagonally dominant, i.e., $g_{ii} > \sum_{j=1, j\neq i}^{n} |g_{ij}|$ for each $i = 1, \ldots n$. Let $x = A^{-1}e$ and observe $x >> 0$ since $A^{-1} > 0$ by Theorem 1.4.7. Let $D = \text{diag}(x_1, \ldots, x_n)$. Observe $ADe = Ax = e >> 0$. Thus $AD$ has positive row sums. Since $AD$ is a Z-matrix, it follows that $AD$ is strictly diagonally dominant. Since multiplying $AD$ on the left by $D^{-1}$ scales the entries in the $i^{th}$ row by a factor of $1/x_i > 0$ for each $i$, it follows that $G = D^{-1}AD$ is strictly diagonally dominant.

Note that the diagonal entries of $A$ are strictly positive. This is due to the fact that (i) $A^{-1} \geq 0$ by Theorem 1.4.7, (ii) the off-diagonal entries of $A$ are nonpositive, and (iii) the dot product of the $i^{th}$ row of $A$ with the $i^{th}$ column of $A^{-1}$ is one. Since the diagonal entries of $A$ are strictly positive, so are the diagonal entries of $G$. By the fact that $G$ is strictly diagonally dominant, we apply the Gersgorin Disc Theorem (Theorem 1.2.1) to see that each eigenvalue $\lambda$ of $G$ is such that $\text{Re}\lambda > 0$. Since all entries of $G$ are real, the complex eigenvalues of $G$ come in conjugate pairs, hence the product of such pairs of eigenvalues is positive. Moreover, the real eigenvalues of $G$ are all positive. Since the determinant of a matrix is the product of the eigenvalues of the matrix, it follows that $\det G > 0$. Since $\det(G) = \det(A)$, we see that $\det(A) > 0$.

Finally, let $\hat{A}$ be a principal submatrix of $A$. Writing $A$ as $sI - B$ where $B \geq 0$ and $s > \rho(B)$, we see that $\hat{A} = sI - \hat{B}$ where $\hat{B}$ is the corresponding submatrix of $B$. Since $s > \rho(B) \geq \rho(\hat{B})$, it follows that $\hat{A}$ is also a nonsingular M-matrix. Hence by the reasoning of the previous paragraph, $\det(\hat{A}) > 0$. □

With some important lemmas in hand, we can now extend our study of M-matrices to singular M-matrices. We focus on such matrices that are irreducible and observe that the theory of irreducible nonnegative matrices will play an important role in the following theorem from [6].

**THEOREM 1.4.10** *Suppose $A$ is a singular, irreducible $n \times n$ M-matrix. Then*
*(i) Zero is a simple eigenvalue of $A$.*
*(ii) There exists a vector $x >> 0$ such that $Ax = 0$.*
*(iii) Each principal submatrix of $A$ other than $A$ itself is a nonsingular M-matrix.*

**Proof**: Since $A$ is a singular M-matrix, $A = sI - B$ for some nonnegative matrix $B$ and $s = \rho(B)$. If $A$ is irreducible, then so is $B$. By Theorem 1.3.20, $\rho(B)$ is a simple eigenvalue of $B$. Hence zero is a simple eigenvalue of $A$, proving (i). Moreover, by Theorem 1.3.20 there exists an eigenvector $x >> 0$ of $B$ such that $Bx = \rho(B)x$. Thus $Ax = (sI - B)x = sx - \rho(B)x = 0$ since $s = \rho(B)$. This proves (ii).

To prove (iii), let $\hat{A}$ be a principal submatrix of $A$ where $\hat{A} \neq A$. We saw in the proof of Lemma 1.4.9 that $\hat{A}$ is also an M-matrix. To show $\hat{A}$ to be nonsingular, let $\hat{B} \geq 0$ be the corresponding submatrix of $B$. Hence $\hat{A} = sI - \hat{B}$. Since $B$ is irreducible, it follows from Exercise 1(iii) in Section 1.3 that $s = \rho(B) > \rho(\hat{B})$. Hence $\hat{A}$ is nonsingular. □

Singular irreducible M-matrices and nonnegative matrices will both play an important role in the chapters to come when we represent graphs with Laplacian matrices. The fact that the principal submatrices of a singular irreducible M-matrix are nonsingular and that their inverses are strictly positive will be vital to us when we study the combinatorial properties of graphs.

# Exercises:

**1.** (See [25]) Let $A$ be a symmetric, irreducible M-matrix. Prove that if there exists a positive vector $x$ such that $Ax \leq \alpha x$, then $\lambda_1 \leq \alpha$ with equality if and only if $x$ is an eigenvector of $A$ corresponding to the eigenvalue $\alpha$.

**2.** (See [30]) Let $A$ be a symmetric irreducible matrix $A$ having all off-diagonal entries nonpositive and $Ax = 0$ for a real vector $x \neq 0$ that is neither positive nor negative. Prove that $A$ is not positive semidefinite.

**3.** (See [6]) Let $A$ be a symmetric Z-matrix. Prove that $A$ is a nonsingular M-matrix if and only if $A$ is positive definite.

## 1.5 Doubly Stochastic Matrices

In this section based largely on [27], we further our study of nonnegative matrices and look at a specific type of nonnegative matrix known as a doubly stochastic matrix. Such matrices are important in the study of Markov chains, probability, and statistics, and will be useful to us when we represent graphs as Laplacian matrices in subsequent chapters. When representing graphs as Laplacian matrices, we will spend considerable time studying the second smallest eigenvalue. In turn, it will be useful for us to have information on the second largest eigenvalue of doubly stochastic matrices. As we saw in the previous section, if we subtract such a matrix from a suitable multiple of the identity matrix, then any information we gather about the second largest eigenvalue of the doubly stochastic matrix will be easily transferred to the second smallest eigenvalue of the resulting matrix. We begin with a definition of a doubly stochastic matrix.

**DEFINITION 1.5.1** A *doubly stochastic matrix* is a nonnegative matrix whose row sums and column sums all equal one.

Since the row sums are all one, it follows that $e$ is an eigenvector corresponding to the eigenvalue one. Since $e$ is a nonnegative vector, it follows from Corollary 1.3.5 that one is the largest eigenvalue of any doubly stochastic matrix. The goal of this section will be to find bounds for the second largest eigenvalue of doubly stochastic matrices.

Recalling Definition 1.3.12, we can say that an $n \times n$ matrix $A = [a_{i,j}]$ is reducible if there exists a nonempty subset $M \subset N := \{1, \ldots, n\}$ such that $a_{i,j} = 0$ for all $i \in M$ and $j \in N \setminus M$. With this in mind, we have the following definition:

**DEFINITION 1.5.2** For an $n \times n$ doubly stochastic matrix $A$, the *measure of irreducibility*, $\mu(A)$ is

$$\mu(A) = \min_{\substack{M \subset N \\ M \neq \emptyset, N}} \sum_{\substack{i \in M \\ j \in N \setminus M}} a_{i,j}.$$

Definition 1.5.2 leads us to an observation from [27] which is easily verified.

**OBSERVATION 1.5.3** Let $A$ be a doubly stochastic matrix. Then $0 \leq \mu(A) \leq 1$. Moreover $A$ is reducible if and only if $\mu(A) = 0$.

**EXAMPLE 1.5.4** Consider the doubly stochastic matrix

$$A = \begin{bmatrix} 1 & 0 & 0 \\ 0 & 3/4 & 1/4 \\ 0 & 1/4 & 3/4 \end{bmatrix}.$$

Observe that $A$ is reducible. Moreover, $\mu(A) = 0$ where this value is obtained by letting $M = \{1\}$ in the language of Definition 1.5.2.

# Matrix Theory Preliminaries

**EXAMPLE 1.5.5** Consider the doubly stochastic matrix

$$B = \begin{bmatrix} 1/2 & 1/4 & 1/4 \\ 1/4 & 5/8 & 1/8 \\ 1/4 & 1/8 & 5/8 \end{bmatrix}.$$

Clearly $B$ is irreducible, hence $\mu(B) > 0$. After an exhaustive search, letting $M = \{1, 2\}$ yields $\mu(B) = 3/8$.

In the following lemma from [27], we see that if we take a convex linear combination of doubly stochastic matrices, the measure of irreducibility will be at least the sum of the measures of irreducibility of the individual matrices.

**LEMMA 1.5.6** *Let $A$ and $B$ be doubly stochastic matrices and let $\alpha, \beta \geq 0$ where $\alpha + \beta = 1$. Then $\alpha A + \beta B$ is a doubly stochastic matrix and $\mu(\alpha A + \beta B) \geq \alpha \mu(A) + \beta \mu(B)$.*

**Proof**: Clearly $\alpha A + \beta B$ is a doubly stochastic matrix. Hence there exists a nonempty proper subset $M \subset N$ such that

$$\mu(\alpha A + \beta B) = \sum_{i \in M, k \notin M} (\alpha a_{ik} + \beta b_{ik}) = \sum_{i \in M, k \notin M} \alpha a_{ik} + \sum_{i \in M, k \notin M} \beta b_{ik} \geq \alpha \mu(A) + \beta \mu(B).$$

$\square$

In the following lemma from [27], we study three important doubly stochastic matrices that will be useful to us when proving the main theorem of this section which concerns the second largest eigenvalue of doubly stochastic matrices.

**LEMMA 1.5.7** *Let $C_1$, $C_2$, and $C_3$ be $n \times n$ matrices where*

$$C_1 = \frac{1}{2} \begin{bmatrix} 1 & 1 & & & & & \\ 1 & 0 & 1 & & & & \\ & 1 & 0 & 1 & & & \\ \ldots & \ldots & \ldots & \ldots & \ldots & \ldots & \ldots \\ & & & & 1 & 0 & 1 \\ & & & & & 1 & 1 \end{bmatrix},$$

$$C_2 = \frac{1}{2} \begin{bmatrix} 0 & 1 & 1 & & & & & & & & \\ 1 & 0 & 0 & 1 & & & & & & & \\ 1 & 0 & 0 & 0 & 1 & & & & & & \\ & 1 & 0 & 0 & 0 & 1 & & & & & \\ \ldots & \ldots & \ldots & \ldots & \ldots & \ldots & \ldots & \ldots & \ldots & \ldots & \ldots \\ & & & & & & 1 & 0 & 0 & 0 & 1 \\ & & & & & & & 1 & 0 & 0 & 1 \\ & & & & & & & & 1 & 1 & 0 \end{bmatrix},$$

and $C_3 = \frac{1}{n-1}(J - I)$. Then

(a) $C_1$, $C_2$, and $C_3$ are symmetric doubly stochastic matrices that commute with each other.

(b) $\mu(C_1) = \frac{1}{2}$, $\mu(C_2) = 1$, and $\mu(C_3) = 1$.

(c) The eigenvalues $1 = \gamma_1 \geq \gamma_2 \geq \ldots \geq \gamma_n$ for $C_i$, $i = 1, 2, 3$ are $\gamma_k(C_1) = \cos((k-1)\pi/n)$, $\gamma_k(C_2) = \cos(2(k-1)\pi/n)$, and $\gamma_k(C_3) = -1/(n-1)$ for $k = 2, \ldots, n$. These eigenvalues correspond to the eigenvectors

$$z(k) := \left[ \cos\frac{(k-1)\pi}{2n}, \cos\frac{3(k-1)\pi}{2n}, \cos\frac{5(k-1)\pi}{2n}, \ldots, \cos\frac{(2n-1)(k-1)\pi}{2n} \right]^T$$

**Proof**: Parts (a) and (c) are easily verified. We leave part (b) as an exericse. □

We see that the cosine function will play an important role when considering the eigenvalues of doubly stochastic matrices. We now define a function (from [27]) that will be of great use for us. Let $n \geq 2$ be an integer. Define $\phi_n(z)$ for $0 \leq x \leq 1$ by

$$\phi_n(z) = \begin{cases} 2\left(1 - \cos\frac{\pi}{n}\right) z, & \text{if } 0 \leq z \leq \frac{1}{2}, \\ 1 - 2(1-z)\cos\frac{\pi}{n} - (2z-1)\cos\frac{2\pi}{n}, & \text{if } \frac{1}{2} \leq z \leq 1. \end{cases}$$

Observe that $\phi_n(z)$ is increasing, continuous, and convex.

In addition to introducing the function $\phi(z)$, we will introduce a matrix $B(z)$ in the following lemma from [27]. The matrix $B(z)$ makes use of the matrices $C_1$ and $C_2$ from Lemma 1.5.7.

**LEMMA 1.5.8** Let $0 \leq z \leq 1$. Let $B(z)$ be an $n \times n$ matrix defined by

$$B(z) = \begin{cases} (1-2z)I + 2zC_1, & \text{if } 0 \leq z \leq \frac{1}{2}, \\ 2(1-z)C_1 + (2z-1)C_2, & \text{if } \frac{1}{2} \leq z \leq 1. \end{cases}$$

Then:

(a) $B(z)$ is symmetric and doubly stochastic.

(b) $\mu(B(z)) = z$.

(c) $\lambda_{n-1}(B(z)) = 1 - \phi_n(z)$.

**Proof**: Part (a) is clear. To prove (b), observe that if $0 \leq z \leq \frac{1}{2}$, then $\mu(B(z)) = 2z\mu(C_1)$ since the diagonal entries of a doubly stochastic matrix have no effect on the measure of irreducibility. By Lemma 1.5.7, $2z\mu(C_1) = z$. If $\frac{1}{2} \leq z \leq 1$, it follows from Lemma 1.5.6 that

$$\mu(B(z)) \geq 2(1-z)\mu(C_1) + (2z-1)\mu(C_2) = z$$

# Matrix Theory Preliminaries

where the equality follows from Lemma 1.5.7. Letting $M = \{1\}$ as in Definition 1.5.2, we have $\mu(B(z)) \leq \sum_{i \in M, k \notin M} b_{ik} = z$. Thus $\mu(B(z)) = z$ which completes (b).

Part (c) follows from Lemma 1.5.7 and the fact that $\phi_n(z)$ is increasing and continuous. □

Before proving the main theorem of this section, we need one additional lemma from [27]. In this lemma, we define two additional matrices that will be useful to us. From this point forward, let $D_n(z)$ denote the set of symmetric doubly stochastic $n \times n$ matrices $A$ such that $\mu(A) \geq z$.

**LEMMA 1.5.9** *Let* $Z = (z_{i,j})$ *be the* $n \times n$ *matrix*

$$Z = \begin{cases} 1, & \text{if } i \leq j, \\ 0, & \text{if } i > j. \end{cases}$$

*Letting* $B(z)$ *be as in Lemma 1.5.8, then for any* $A \in D_n(z)$, *the matrix*

$$U := Z^T(B(z) - A)Z$$

*is nonnegative and all its entries in the last row and column equal zero.*

**Proof**: Let $v_j = [1, \ldots, 1, 0, \ldots, 0]^T$ be the vector with 1's as the first $j$ components and zeros elsewhere. Then for $j = 1, \ldots, n$, the $j$-th column of $Z$ is $v_j$. Therefore

$$u_{ij} = v_i^T[B(z) - A]v_j. \tag{1.5.1}$$

Define $w_j = e - v_j$. Observe for any $A \in D_n(z)$ we have

$$v_j^T A w_j = \sum_{\substack{i \in M_j \\ k \notin M_j}} a_{ik} \geq \mu(A) \geq z \tag{1.5.2}$$

where $M_j = \{1, \ldots, j\}$. Further, since for $j = 1, \ldots, n-1$ we have

$$a_{j+1,j+1} = (v_{j+1}^T - v_j^T)A(w_j - w_{j+1}) \geq 0,$$

it follows that

$$v_{j+1}^T A w_j + v_j^T A w_{j+1} \geq v_j^T A w_j + v_{j+1}^T A w_{j+1} \geq 2z \tag{1.5.3}$$

by (1.5.2). However,

$$\begin{aligned} v_{j+1}^T A w_j &= (e^T - w_{j+1}^T)A(e - v_j) \\ &= n - (n - j - 1) - j + w_{j+1}^T A v_j \\ &= 1 + v_j^T A w_{j+1}, \end{aligned}$$

the last equality following from the fact that $A$ is symmetric. Therefore,

$$v_j^T A w_{j+1} \geq z - \frac{1}{2}. \tag{1.5.4}$$

by (1.5.3). Moreover, since $A$ is nonnegative, for any $m > j$ we have

$$v_j^T A w_m \geq 0. \tag{1.5.5}$$

Recalling that $Aw_j = e - Av_j$, from (1.5.2), (1.5.4), and (1.5.5) we have

$$v_j^T A v_j \leq j - z$$

$$v_j^T A v_{j+1} \leq j - z + \tfrac{1}{2} \tag{1.5.6}$$

$$v_j^T A v_m \leq j, \quad \text{if } m > j.$$

Since $v_j^T C_1 v_j = j - \tfrac{1}{2}$ and $v_j^T C_1 v_m = j$ for $m > j$, it follows for such $0 \leq z \leq \tfrac{1}{2}$ that

$$v_j^T B(z) v_j = j - z$$
$$v_j^T B(z) v_m = j \quad \text{if } m > j. \tag{1.5.7}$$

Further, since $v_j^T C_2 v_j = j - 1$, $v_j^T C_2 v_{j+1} = j - \tfrac{1}{2}$, and $v_j^T C_2 v_m = j$ for $m > j + 1$, it follows that for $\tfrac{1}{2} \leq z \leq 1$

$$v_j^T B(z) v_j = j - z$$

$$v_j^T B(z) v_{j+1} = j - z + \tfrac{1}{2} \tag{1.5.8}$$

$$v_j^T B(z) v_m = j \quad \text{if } m > j + 1.$$

Since $0 \leq z \leq 1$, $j \geq 1$, and $U$ is symmetric, it follows from (1.5.1), (1.5.6), (1.5.7), and (1.5.8) that $u_{ij} \geq 0$. Finally, since $v_n = e$, it follows from (1.5.1) that $u_{in} = u_{ni} = 0$ for each $i = 1, \ldots, n$. $\square$

We use these lemmas to prove the main theorem of this section (see [27]) which gives us a bound on the second largest eigenvalue of a doubly stochastic matrix in terms of its measure of irreducibility.

**THEOREM 1.5.10** *Let $A$ be a symmetric doubly stochastic $n \times n$ matrix with measure of irreducibility $\mu$ and second largest eigenvalue $\lambda_{n-1}$. Then*

$$1 - \lambda_{n-1} \geq \phi_n(\mu) \geq 2\left(1 - \cos\frac{\pi}{n}\right)\mu \tag{1.5.9}$$

**Proof**: We begin with the first inequality. We will prove this by showing that for every $z$ where $0 \leq z \leq 1$ that

$$\min_{A \in D_n(z)} [1 - \lambda_{n-1}(A)] = \phi_n(z). \qquad (1.5.10)$$

If $z = 0$ then (1.5.10) simplifies to $1 - \gamma_2(A) \geq 0$ which is clearly true. Now let $z > 0$. It follows from Lemma 1.5.6 that $D_n(z)$ is nonempty, convex, closed, and bounded. Hence the minimum in (1.5.10) exists. Letting $S_n$ be the set of all unit vectors orthogonal to $e$, it follows from Theorem 1.2.7 that

$$\begin{aligned}
\min\nolimits_{A \in D_n(z)} [1 - \gamma_2(A)] &= 1 - \max\nolimits_{A \in D_n(z)} [\max\nolimits_{x \in S_n} x^T A x] \\
&= 1 - \max\nolimits_{x \in S_n} [\max\nolimits_{A \in D_n(z)} x^T A x] \\
&= 1 - \max\nolimits_{x \in S'_n} [\max\nolimits_{A \in D_n(z)} x^T A x]
\end{aligned}$$

where $S'_n$ is the set of all vectors $x = [x_1, \ldots, x_n]^T \in S_n$ such that $x_1 \geq x_2 \geq \ldots \geq x_n$. Equality in the above equations follows from the fact that if $A \in D_n(z)$ and $P$ is any permutation matrix then $PAP^T \in D_n(z)$.

We will show now that for a fixed $x \in S'_n$ that

$$\max_{A \in D_n(z)} x^T A x = x^T B(z) x. \qquad (1.5.11)$$

From (1.5.11) it will follow by Lemma 1.5.8 that

$$\min_{A \in D_n(z)} [1 - \gamma_2(A)] = 1 - \gamma_2[B(z)] = 1 - [1 - \phi_n(z)] = \phi_n(z).$$

If $x \in S'_n$ then the first $n-1$ coordinates of

$$y := [x_1 - x_2, x_2 - x_3, \ldots, x_{n-1} - x_n, x_n]^T$$

are nonnegative. Letting $U$ be as in Lemma 1.5.9, it follows from Lemma 1.5.9 that since $U$ is nonnegative and the last row and column of $U$ are all zeros, we have

$$y^T U y = \sum_{i,j=1}^{n-1} u_{ij} y_i y_j \geq 0.$$

Observing that $y = Z^{-1} x$, we obtain

$$\begin{aligned}
0 \leq y^T U y &= x^T (Z^{-1})^T [Z^T (B(z) - A) Z] Z^{-1} x \\
&= x^T [B(z) - A] x \\
&= x^T B(z) x - x^T A x.
\end{aligned}$$

This proves (1.5.11), and hence (1.5.10) since $B(z)$ is the matrix that obtains the minimum. Hence the first inequality in (1.5.9) is proven.

The second inequality in (1.5.9) follows directly from the fact that $\phi_n(z)$ is an increasing function and $0 \leq \mu \leq 1$ by definition of $\mu$. This completes the proof. $\square$

**REMARK 1.5.11** *[31]* From the proof of Theorem 1.5.10, we see that if $\mu \neq 0$ then equality holds in (1.5.9) if and only if $A$ is permutationally similar to a tridiagonal matrix.

Remark 1.5.11 will be useful in our study of Laplacian marices as the Laplacian matrix for a path is a tridiagonal matrix.

# Exercises:

**1.** Prove part (b) of Lemma 1.5.7.

**2.** Prove Remark 1.5.11.

## 1.6 Generalized Inverses

Much of linear algebra was developed in order to find efficient ways of solving systems of equations and analyzing their solutions if any exist. The most fundamental problem of this sort is if we have $m$ equations involving $n$ variables $x_1, \ldots, x_n$ and we want to solve for each variable. In such a case we represent the coefficients of the variables as the $m \times n$ matrix $A$, the right side of each equation as the vector $b$, and the variables as entries in the vector $x$. This gives us the equation $Ax = b$. If $m = n$ and $A$ is invertible, then we can solve for each variable $x_i$ since $x = A^{-1}b$. But what if $A$ is not invertible? More generally, what if $m \neq n$? Then there will be either no solution or more than one solution. Often it is desirable to get a solution that best fits the equations, even if it is not an exact fit. This leads us to the concept of *generalized inverses*. Generalized inverses have many of the same properties as inverses, and hence can often be used in place of an inverse when an inverse does not exist.

If $A$ is an $n \times n$ nonsingular matrix, then there exists a unique matrix $X$ satisfying all of the following properties:

$$
\begin{aligned}
&(1) \quad AXA = A \\
&(2) \quad XAX = X \\
&(3) \quad AX = (AX)^T \\
&(4) \quad XA = (XA)^T \\
&(5) \quad AX = XA \\
&(6) \quad A^k = XA^{k+1} \text{ for all nonnegative integers } k.
\end{aligned}
\tag{1.6.1}
$$

We typically write $A^{-1}$ for $X$ in this case. However, if a matrix $A$ is singular or rectangular, it is possible that there exists a unique matrix $X$ that satisfies some of the properties in (1.6.1). By requiring only a subset of these properties, we obtain various generalized inverses, all of which equal the traditional inverse $A^{-1}$ if it exists. In this section, we will investigate two such generalized inverses: the Moore-Penrose inverse and the group inverse.

We begin with the most common generalized inverse which is the Moore-Penrose inverse defined here:

**DEFINITION 1.6.1** Given a matrix $A$, the *Moore-Penrose inverse*, $A^+$, for $A$ is the unique matrix satisfying properties (1) through (4) in (1.6.1).

The Moore-Penrose inverse exists for every matrix, including singular and rectangular matrices. It is also unique for each matrix. We should note that for every matrix $A$, there exists a matrix $X$ satisfying properties (1) and (2). However, such a matrix is not always unique. Properties (3) and (4) guarantee the uniqueness of the Moore-Penrose inverse. To compute the Moore-Penrose inverse of an $m \times n$ matrix $A$ of rank $r$, factor $A$ into matrices $B$ and $C$ where $B$ is $m \times r$ and $C$ is $r \times n$. This can be done using the LU-factorization and letting $B$ be the first $r$ columns of $L$ and letting $C$ be the first $r$ rows of $U$. Then the formula for the Moore-Penrose inverse is (see [5])

$$A^+ = C^T(CC^T)^{-1}(B^TB)^{-1}B^T. \qquad (1.6.2)$$

**EXAMPLE 1.6.2** Consider the matrix

$$A = \begin{bmatrix} 1 & 3 & 4 \\ 2 & -1 & 5 \\ 1 & -4 & 1 \end{bmatrix}.$$

First observe that $A$ is singular, hence the traditional inverse $A^{-1}$ does not exist. However, the Moore-Penrose inverse will exist. Factoring $A$ into $BC$, we see that

$$B = \begin{bmatrix} 1 & 0 \\ 2 & 1 \\ 1 & 11 \end{bmatrix} \quad \text{and} \quad C = \begin{bmatrix} 1 & 3 & 4 \\ 0 & -7 & -3 \end{bmatrix}.$$

Applying (1.6.2) we obtain

$$A^+ = \frac{1}{1257} \begin{bmatrix} 17 & 58 & 41 \\ 135 & -57 & -192 \\ 104 & 133 & 29 \end{bmatrix}.$$

**EXAMPLE 1.6.3** Consider the matrix

$$A = \begin{bmatrix} 1 & 2 & 3 \\ 4 & 5 & 6 \end{bmatrix}.$$

Clearly $A^{-1}$ does not exist since $A$ is rectangular. However, we can still find the Moore-Penrose inverse. Factoring $A$ into $BC$, we obtain

$$B = \begin{bmatrix} 1 & 0 \\ 4 & 1 \end{bmatrix} \text{ and } C = \begin{bmatrix} 1 & 2 & 3 \\ 0 & -3 & -6 \end{bmatrix}.$$

Applying (1.6.2) we obtain

$$A^+ = \frac{1}{18} \begin{bmatrix} -17 & 8 \\ -2 & 2 \\ 13 & -4 \end{bmatrix}.$$

We now turn our attention to another type of generalized inverse known as the group inverse. The group inverse will be very useful in later chapters of this book, hence we spend some time discussing it. We begin with a definition:

**DEFINITION 1.6.4** Given a matrix $A$, the *group inverse*, $A^\#$, for $A$ is the unique matrix satisfying properties (1), (2), and (5) in (1.6.1).

By property (5), since a matrix must commute with its group inverse, it follows that the group inverse can only exist for square matrices. Moroever, if we factor $A$ into $BC$ as before, then $A^\#$ exists if and only if $CB$ is nonsingular. In such a case (see [6])

$$A^\# = B(CB)^{-2}C. \tag{1.6.3}$$

**EXAMPLE 1.6.5** Consider the matrix $A$ in Example 1.6.2. Factoring $A$ into $BC$ as we did before and applying (1.6.3), we obtain

$$A^\# = \frac{1}{81} \begin{bmatrix} -19 & -8 & -55 \\ -21 & 21 & -48 \\ -2 & 29 & 7 \end{bmatrix}.$$

Observe that $A^\# \neq A^+$. Therefore, the Moore-Penrose inverse and the group inverse for a matrix $A$ are not necessarily equal if $A^{-1}$ does not exist. If $A^{-1}$ does exist, then all generalized inverses equal $A^{-1}$.

Since the group inverse exists only for square matrices and since the group inverse itself is a square matrix, it is natural to investigate the eigenvalues of the group inverse $A^\#$ and how they relate to the eigenvalues of the original matrix $A$. Clearly if $A$ is nonsingular then the eigenvalues of $A^\#$ are the reciprocals of the eigenvalues of $A$ with the same corresponding eigenvectors. So let $A$ be singular with eigenvalues $0 = \lambda_1 \leq \lambda_2 \leq \lambda_3 \leq \ldots \leq \lambda_n$. Let $x$ be an eigenvector corresponding to the eigenvalue $\lambda = 0$. Then it follows from (1.6.1) that

$$A^\# x = A^\# A A^\# x = A^\# A^\# A x = 0.$$

# Matrix Theory Preliminaries

Hence 0 is also an eigenvalue of $A^\#$ also with corresponding eigenvector $x$. For an eigenvalue $\lambda \neq 0$ with $x$ a corresponding eigenvector, i.e., $Ax = \lambda x$, it follows from (1.6.1) that
$$\lambda x = Ax = AA^\#Ax = AA^\#\lambda x.$$
Dividing through by $\lambda$ (since $\lambda \neq 0$) and applying (1.6.1) again, we obtain
$$x = AA^\# x = A^\# Ax = \lambda A^\# x.$$
Thus $A^\# x = \frac{1}{\lambda} x$. Therefore the nonzero eigenvalues of $A^\#$ are the reciprocals of the nonzero eigenvalues of $A$. Moreover, if $x$ is an eigenvector of $A$ corresponding to $\lambda$, then $x$ is also an eigenvector of $A^\#$ corresponding to $1/\lambda$.

## Exercises:

**1.** Find the Moore-Penrose inverse and group inverse of
$$A = \begin{bmatrix} 2 & 1 & 1 \\ 4 & 0 & 3 \\ 8 & 2 & 5 \end{bmatrix}$$

**2.** Prove that for any matrix $A$, there exists a matrix $X$ satisfying properties (1) and (2) in (1.6.1). Give an example of a matrix $A$ in which the resulting matrix $X$ is not unique.

**3.** Verify that the formula for $A^+$ given in (1.6.2) satisfies properties (1) through (4) of (1.6.1).

**4.** Prove that the Moore-Penrose inverse exists for any matrix and that it is unique.

**5.** Verify that the formula for $A^\#$ given in (1.6.3) satisfies properties (1), (2), and (5) of (1.6.1).

**6.** Prove that if the group inverse exists, then it is unique.

# Chapter 2

# Graph Theory Preliminaries

We dedicate this chapter to introducing graph theory to the reader. This chapter assumes no prior knowledge of graph theory, thus we will develop all of our ideas from scratch. This chapter is not meant to cover what an entire graph theory textbook would cover, but rather to give the reader the necessary tools that will enable him or her to understand the concepts presented in the remaining chapters. In addition to exposing the reader to the graph theory concepts that will be used throughout the remainder of this book, this chapter will assist the reader in understanding how these concepts fit into the larger area of graph theory.

## 2.1 Introduction to Graphs

In 1997, Major League Baseball began Interleague Play in which teams from different leagues were scheduled to play each other. With this new initiative, the scheduling format was that each team would play every other team within their league, and every team outside their league that is in the same geographic region. Consider the following six teams which are denoted in parentheses by their league (American League or National League) and their geographic region (East, Central, or West): Yankees (AL East), Twins (AL Central), Angels (AL West), Mets (NL East), Astros (NL Central), and Padres (NL West). We can draw a picture illustrating which teams were scheduled to play each other by representing each team as a point, and then drawing a line segment between those teams that we scheduled to play each other:

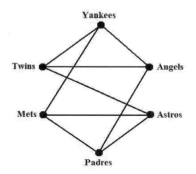

The illustration that we have just drawn is a graph. We now state a more formal definition:

**DEFINITION 2.1.1** A *graph* $\mathcal{G}$ is a set of points $V(\mathcal{G})$ called *vertices* along with a set of line segments $E(\mathcal{G})$ called *edges* joining pairs of vertices. We say that two vertices are *adjacent* if there is an edge joining them.

In the above example, we see that the vertices are adjacent if and only if the teams represented by the vertices are in the same league or in the same geographic region.

We now present some additional useful terminology:

**DEFINITION 2.1.2** The *degree* of a vertex $v$, denoted $\deg(v)$, is the number of vertices adjacent to $v$. The set of vertices adjacent to $v$ is the *neighborhood* of $v$ which we denote as $N(v)$. If $e$ is an edge joining $v$ to one of its neighbors, we say $e$ is *incident* to $v$.

**EXAMPLE 2.1.3** Consider the following graph $\mathcal{G}$:

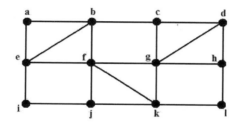

Observe that $\deg(a) = 2$ and $N(a) = \{b, e\}$. Also observe that $\deg(c) = 3$ and $N(c) = \{b, d, g\}$.

In any graph, note that each edge is incident to exactly two vertices. Thus if a graph $\mathcal{G}$ has $m$ edges, then the sum of the degrees of the vertices is $2m$. We state this more precisely as a theorem:

**THEOREM 2.1.4** *Let $v_1, \ldots, v_n$ be the vertices of a graph $\mathcal{G}$ and suppose $\mathcal{G}$ has $m$ edges, then*

$$\sum_{i=1}^{n} \deg(v_i) = 2m.$$

*Therefore, the sum of the degrees of the vertices of a graph must be even.*

Since the sum of an even number of odd numbers is even, while the sum of an odd number of odd numbers is odd, we have the following corollary:

**COROLLARY 2.1.5** *In any graph $\mathcal{G}$ there must be an even number of vertices of odd degree.*

# Graph Theory Preliminaries

Before completing our discussion on the degrees of vertices, we have one additional definition:

**DEFINITION 2.1.6** A graph is *regular* or *r-regular* if each vertex has degree $r$.

**EXAMPLE 2.1.7** Below is a 3-regular graph:

We will discuss the degrees of vertices in more depth in Section 2.5. At this time, we will focus on the neighborhood of each vertex. Naturally, neighborhoods and adjacency are the key concepts in the structure of a graph. Consider the following two graphs:

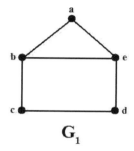
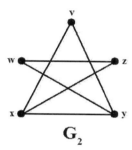

On the surface, these two graphs look rather different. However, upon further inspection we notice many similarities. For example, in each graph there are exactly two vertices of degree three and they are adjacent to each other. There are also three vertices of degree two in each graph. Moreover, in each graph, two of the vertices of degree two are adjacent to each other but neither is adjacent to the third vertex of degree two whose only neighbors are the vertices of degree three. Finally in each graph, the vertices of degree three are adjacent to exactly one common vertex. So while these graphs look different, we see that their overall structure is not so different. This leads us to the concept of an isomorphism:

**DEFINITION 2.1.8** Let $\mathcal{G}_1$ and $\mathcal{G}_2$ be two graphs with vertex sets $V(\mathcal{G}_1)$ and $V(\mathcal{G}_2)$, respectively. Then $\mathcal{G}_1$ and $\mathcal{G}_2$ are *isomorphic* if there is a one-to-one correspondence $\phi$ between $V(\mathcal{G}_1)$ and $V(\mathcal{G}_2)$ such that for any pair of vertices $v, w \in V(\mathcal{G}_1)$, we have $v$ and $w$ adjacent in $\mathcal{G}_1$ if and only if $\phi(v)$ and $\phi(w)$ are adjacent in $\mathcal{G}_2$. The function $\phi$ is an *isomorphism* from $\mathcal{G}_1$ to $\mathcal{G}_2$.

**EXAMPLE 2.1.9** In the example above, $\mathcal{G}_1$ and $\mathcal{G}_2$ are isomorphic as there exists and isomorphism $\phi$ between these two graphs, namely $\phi(a) = v$, $\phi(b) = x$, $\phi(c) = z$, $\phi(d) = w$, and $\phi(e) = y$.

At this time, we turn our attention to the edges of a graph and discuss traversing the edges in order to "travel" from one vertex to another. We begin our discussion with some definitions:

**DEFINITION 2.1.10** A *v−w walk* in a graph $\mathcal{G}$ is a sequence of vertices beginning at $v$ and ending at $w$ so that consecutive vertices in the sequence are adjacent. The number of edges encountered in a walk, including multiple appearances, is the *length* of the walk.

**EXAMPLE 2.1.11** Using the graph $\mathcal{G}$ in Example 2.1.3, the sequence $b, e, f, j, k, f, j$ is a $b - j$ walk of length 6.

**DEFINITION 2.1.12** A *closed walk* is a walk with the same starting and ending vertex. An *open walk* is a walk in which the start and end vertices differ.

**DEFINITION 2.1.13** A *trail* is a walk in which no edge is repeated.

**EXAMPLE 2.1.14** The walk in Example 2.1.11 is not a trail since edge $fj$ is repeated. Using $\mathcal{G}$ from Example 2.1.3, an example of a trail would be $d, g, h, l, k, g, f$. Observe we do not repeat an edge in this $d - f$ trail and that this $d - f$ trail is of length 6.

**DEFINITION 2.1.15** A *path* is a walk in which no vertex is repeated.

**EXAMPLE 2.1.16** The walk and trail described in Examples 2.1.11 and 2.1.14, respectively, are not paths because in each case, a vertex is repeated. Again, using $\mathcal{G}$ from Example 2.1.3, an example of a path would be $i, j, k, g, d$. Since no vertex is repeated, this is an $i - d$ path and its length is 4.

We should note that if $P$ is a path in a graph $\mathcal{G}$, then $P$ is also a trail since if no vertex is repeated, then no edge can be repeated either.

Continuing with the notion of no vertex or edge being repeated, our next definition is similar to that of a path, only the walk will be closed:

**DEFINITION 2.1.17** A *cycle* is a closed walk in which no vertex is repeated (except that the starting and ending vertices are the same).

**EXAMPLE 2.1.18** In the graph in Example 2.1.3, $abfea$ is a cycle on four vertices. The cycle $bcgkfb$ is a cycle on five vertices.

We close this section with the notion of connectivity of graphs:

**DEFINITION 2.1.19** A graph is *connected* if there exists a path between every pair of distinct vertices. A graph that is not connected is said to be *disconnected*. A connected subgraph of a disconnected graph $\mathcal{G}$ that is not a proper subgraph of any connected subgraph of $\mathcal{G}$ is called a *component* of $\mathcal{G}$.

# Graph Theory Preliminaries

**EXAMPLE 2.1.20** The following graph has three components. Note that there do not exist paths between vertices $a$, $b$, and $c$.

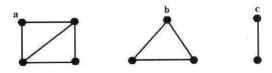

If a graph is connected, we can, in effect, "travel" between any pair of vertices. The next two definitions make the idea of travelling between vertices more precise:

**DEFINITION 2.1.21** The *distance* between two vertices $v$ and $w$ is the length of the shortest path between $v$ and $w$.

**DEFINITION 2.1.22** The *diameter* of a connected graph $\mathcal{G}$, denoted $\text{diam}(\mathcal{G})$ is the greatest distance between any two vertices of $\mathcal{G}$. In other words

$$\text{diam}(\mathcal{G}) = \max_{u,v} d(u,v)$$

**DEFINITION 2.1.23** The *mean distance* $\bar{\rho}(\mathcal{G})$ of a connected graph $\mathcal{G}$ is the average of all distances between distinct vertices of $\mathcal{G}$. Thus if $\mathcal{G}$ has $n$ vertices, then

$$\bar{\rho}(\mathcal{G}) = \frac{1}{n(n-1)} \sum_{u \in V(\mathcal{G})} \sum_{v \in V(\mathcal{G})} d(u,v)$$

**EXAMPLE 2.1.24** Consider the graph $\mathcal{G}$ below:

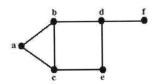

Below is a chart which gives the distances between distinct pairs of vertices.

|   | a | b | c | d | e | f |
|---|---|---|---|---|---|---|
| a | 0 | 1 | 1 | 2 | 2 | 3 |
| b | 1 | 0 | 1 | 1 | 2 | 2 |
| c | 1 | 1 | 0 | 2 | 1 | 3 |
| d | 2 | 1 | 2 | 0 | 1 | 1 |
| e | 2 | 2 | 1 | 1 | 0 | 2 |
| f | 3 | 2 | 3 | 1 | 2 | 0 |

From the chart we see that $\text{diam}(\mathcal{G}) = 3$ and $\bar{\rho}(\mathcal{G}) = 5/3$.

**OBSERVATION 2.1.25** Of all unweighted graphs on $n$ vertices with $n$ fixed, the graph with the smallest diameter and mean distance is $K_n$, both quantities equalling one. The graph with the largest diameter and mean distance is $P_n$ where $\text{diam} P_n = n - 1$ and $\bar{\rho}(P_n) = (n+1)/3$.

# Exercises:

**1.** Suppose there are ten students, numbered $1, \ldots, 10$, each of whom need to sign up for classes next semester. The classes are Math, English, Science, History, and French. The meeting times will be determined after the students sign up so that no two classes having students in common are scheduled for the same time. Supppose the class lists of students are as follows: Math: $\{1,3,4,6,7,9\}$, English: $\{2,3,4,5,8\}$, Science: $\{1,6,7,10\}$, History: $\{2,8,10\}$, French: $\{1,4,7\}$. Model this situation with a graph letting the classes be vertices and drawing an edge between vertices that share at least one common student.

**2.** Suppose that a graph $\mathcal{G}$ has one vertex of degree 6, six vertices of degree 3, and two vertices of degree 2. How many edges must $\mathcal{G}$ have?

**3.** Prove that if a graph has $n \geq 2$ vertices, then there must be at least two vertices with the same degree.

**4.** Prove that being isomorphic is an equivalence relation. In other words, letting $\mathcal{G}_1$, $\mathcal{G}_2$, and $\mathcal{G}_3$ be graphs, prove the following:

(a) If $\mathcal{G}_1$ is isomorphic to $\mathcal{G}_2$, then $\mathcal{G}_2$ isomorphic to $\mathcal{G}_1$.
(b) Any graph is isomorphic to itself.
(c) If $\mathcal{G}_1$ is isomorphic to $\mathcal{G}_2$, and $\mathcal{G}_2$ is isomorphic to $\mathcal{G}_3$, then $\mathcal{G}_1$ is isomorphic to $\mathcal{G}_3$.

**5.** In Exercise 4, we saw that being isomorphic is an equivalence relation. Thus we are able to partition sets of graphs into *isomorphism classes* in which all graphs in each class are isomorphic, but graphs belonging to different classes are not isomorphic. Partition the following sets of graphs into isomorphism classes. If the graphs are isomorphic, state the isomorphism.

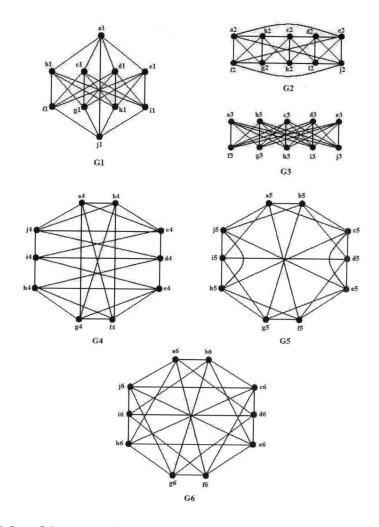

**6.** (See [12]) Let $\mathcal{G}$ be a graph on $n \geq 3$ vertices. Prove that $\mathcal{G}$ is connected if and only if $\mathcal{G}$ contains two distinct vertices $u$ and $v$ such that $\mathcal{G} - u$ and $\mathcal{G} - v$ are connected.

**7.** Consider the following graph $\mathcal{G}$:

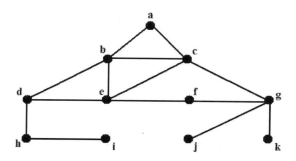

Find the following:

   (a) A $d-b$ walk of length 5.

   (b) A closed walk of length 5 beginning and ending at $g$.

   (c) An $i-e$ trail of length 6.

   (d) A $b-c$ path of length 4.

   (e) Find diam($\mathcal{G}$).

8. Prove Observation 2.1.25.

## 2.2 Operations of Graphs and Special Classes of Graphs

In this section, we discuss several useful operations of graphs. These operations will aid us in describing special classes of graphs that occur often in graph theory. We begin with the complement of a graph:

**DEFINITION 2.2.1** Given a graph $\mathcal{G}$, the *complement* $\mathcal{G}^c$ is formed from $V(\mathcal{G})$ by joining vertices $i$ and $j$ of $V(\mathcal{G})$ if and only if $i$ and $j$ are not adjacent in $\mathcal{G}$.

**EXAMPLE 2.2.2** Consider the graph $\mathcal{G}$:

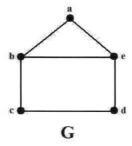

We create the complement $\mathcal{G}^c$ by adding edges joining the nonadjacent vertices of $\mathcal{G}$ and deleting the existing edges. Thus $\mathcal{G}^c$ is as follows:

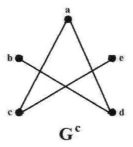

# Graph Theory Preliminaries

In addition to taking the complement of a graph, we can also take the line graph of a graph:

**DEFINITION 2.2.3** Given a graph $\mathcal{G}$, the *line graph* of $\mathcal{G}$, denoted $\mathcal{L}(\mathcal{G})$, is the graph created from $\mathcal{G}$ by letting each edge of $\mathcal{G}$ be represented by a vertex, and two vertices in $\mathcal{L}(\mathcal{G})$ are adjacent if and only if the corresponding edges in $\mathcal{G}$ are incident.

**EXAMPLE 2.2.4** For the graph $\mathcal{G}$ in Example 2.2.2, the line graph $\mathcal{L}(\mathcal{G})$ is given below. The vertices are labeled in accordance with the edges of $\mathcal{G}$.

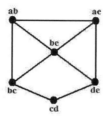

Our next operation is finding an (vertex) induced subgraph of a given graph:

**DEFINITION 2.2.5** Given a graph $\mathcal{G}$, the subgraph *induced* by a subset $U$ of the vertices of $\mathcal{G}$ is the graph consisting of the vertices in $U$ and all edges in $\mathcal{G}$ which join vertices in $U$.

**EXAMPLE 2.2.6** Consider the graph in Example 2.2.2. The subgraph induced by vertices $b$, $c$, $d$, and $e$ is the following:

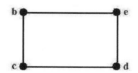

The operations discussed thus far were operations involving only one graph. The next operations are operations involving two graphs. We begin with the sum of two graphs:

**DEFINITION 2.2.7** The *sum* or *union* of two graphs $\mathcal{G}$ and $\mathcal{H}$ is the graph whose vertex set is $V(\mathcal{G}) \cup V(\mathcal{H})$ and whose edge set is $E(\mathcal{G}) \cup E(\mathcal{H})$. We denote this graph as $\mathcal{G} + \mathcal{H}$ or $\mathcal{G} \cup \mathcal{H}$.

**EXAMPLE 2.2.8** Let $\mathcal{G}$ and $\mathcal{H}$ be the graphs illustrated below.

Then $\mathcal{G} + \mathcal{H}$ is

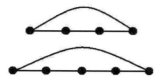

Next we consider the join of two graphs.

**DEFINITION 2.2.9** The *join* of two graphs $\mathcal{G}$ and $\mathcal{H}$, denoted $\mathcal{G} \vee \mathcal{H}$, is created by adding edges between $\mathcal{G}$ and $\mathcal{H}$ so that every vertex in $\mathcal{G}$ is adjacent to every vertex in $\mathcal{H}$.

**EXAMPLE 2.2.10** Let $\mathcal{G}$ and $\mathcal{H}$ be as in Example 2.2.7. Then $\mathcal{G} \vee \mathcal{H}$ is

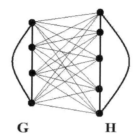

Another useful operation between two graphs is the product:

**DEFINITION 2.2.11** Suppose $V(\mathcal{G}) = \{u_1, \ldots, u_n\}$ and $V(\mathcal{H}) = \{v_1, \ldots, v_p\}$, then the *product* $\mathcal{G} \times \mathcal{H}$ is created by taking $p$ copies of $\mathcal{G}$, calling them $\mathcal{G}_1, \ldots, \mathcal{G}_p$, placing them at the location of the vertices $v_1, \ldots, v_p$ in $\mathcal{H}$, then joining the vertex $u_i$ in $\mathcal{G}_j$ to the vertex $u_i$ in $\mathcal{G}_k$ if and only if $v_j$ and $v_k$ are adjacent in $\mathcal{H}$.

**EXAMPLE 2.2.12** Consider the graphs

$$G = \triangle \qquad H = \square$$

Then $\mathcal{G} \times \mathcal{H}$ is:

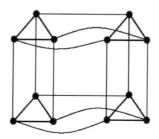

With these operations, we can create several special classes of graphs that frequently occur in graph theory. We begin with the complete graph:

**DEFINITION 2.2.13** A *complete graph* on $n$ vertices, denoted $K_n$, is the graph on $n$ vertices in which every pair of vertices is adjacent

Below we have examples of the complete graphs on $n$ vertices for $n = 1, 2, 3, 4, 5$:

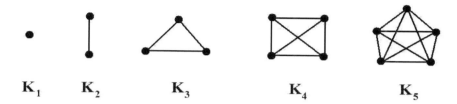

The concept of the complete graph leads us to the following definition:

**DEFINITION 2.2.14** In a graph $\mathcal{G}$, a *clique* is an induced subgraph of $\mathcal{G}$ that is complete. The *clique number*, denoted $\omega(\mathcal{G})$, is the number of vertices in the largest clique of $\mathcal{G}$.

**EXAMPLE 2.2.15** Consider the following graph $\mathcal{G}$:

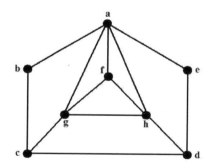

Observe the subgraph induced by vertices $a$, $f$, $g$, and $h$ is a clique because this subgraph is $K_4$. This is the largest induced complete subgraph of $\mathcal{G}$, i.e., the largest clique of $\mathcal{G}$. Hence $\omega(\mathcal{G}) = 4$.

Complete graphs have every edge possible joining pairs of vertices. On the other extreme, we have empty graphs:

**DEFINITION 2.2.16** The *empty graph*, $E_n$, on $n$ vertices is the complement of $K_n$. Thus the empty graph on $n$ vertices is the graph on $n$ vertices containing no edges.

The notions of complete graphs and empty graphs lead us to a class of graphs known as split graphs.

**DEFINITION 2.2.17** A *split graph* is a graph whose vertices can be partitioned into two sets where the induced subgraph of one set is a complete subgraph (a clique), while the induced subgraph of the other set is an empty graph (the coclique).

**EXAMPLE 2.2.18** Observe the graph $\mathcal{G}$ below is a split graph. We can partition the vertices into sets $V := \{b, d, f, g\}$ and $W := \{a, c, e\}$ where the subgraph induced by the vertices in $V$ is the complete graph $K_4$, while the subgraph induced by the vertices in $W$ is the empty graph $E_3$.

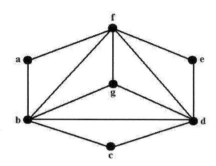

We now present two classes of graphs that were alluded to in Definitions 2.2.19 and 2.1.17. These definitions concerned paths and cycles as subgraphs of a graph. We now recast these definitions as graphs in their own right as opposed to being subgraphs of an existing graph:

**DEFINITION 2.2.19** A *path* on $n$ vertices, denoted $P_n$, is a graph which itself is a path. Likewise, a *cycle*, $C_n$, on $n$ vertices is a graph which itself is a cycle. A path or cycle is said to be odd (even) if $n$ is odd (even).

The next two classes of graphs, the wheel and the cube, make use of the operations of joins and products.

**DEFINITION 2.2.20** The *wheel*, denoted $W_n$ is created by joining a single vertex $u$ to each vertex on the cycle $C_n$. In other words $W_n = u \vee C_n$.

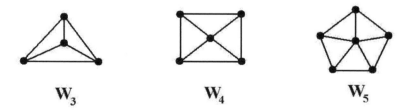

The cube is defined recursively and makes use of the product of graphs.

**DEFINITION 2.2.21** Let $Q_1 = P_2$ and define the *n-cube*, $Q_n$ as $Q_n = Q_{n-1} \times K_2$.

# Graph Theory Preliminaries

The vertices of the cube $Q_n$ are traditionally labeled as $n$-tuples of 0's and 1's where two vertices are adjacent if and only if their $n$-tuples differ in exacly one position. Below are examples of the cubes $Q_1$, $Q_2$, and $Q_3$:

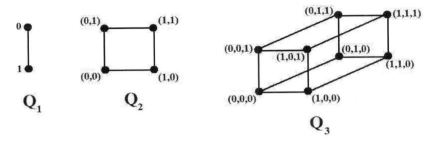

The final class of graphs we discuss are bipartite graphs. We present a formal definition:

**DEFINITION 2.2.22** A graph $\mathcal{G}$ is *bipartite* if $V(\mathcal{G})$ can be partitioned into two vertex sets $V_1$ and $V_2$ so that no vertices in $V_i$ are adjacent for $i = 1, 2$.

Often, it is straightfoward to see that a graph is bipartite. Yet sometimes, it takes more effort to determine if a graph is bipartite. Consider the following two examples.

**EXAMPLE 2.2.23** The graph below is clearly a bipartite graph:

**EXAMPLE 2.2.24** The graph $\mathcal{G}_1$ below is bipartite; however, this is difficult to see. However, we can redraw $\mathcal{G}_1$ as $\mathcal{G}_2$, with the vertices labeled accordingly, to see more clearly that $\mathcal{G}_1$ is bipartite since $\mathcal{G}_1$ and $\mathcal{G}_2$ are isomorphic.

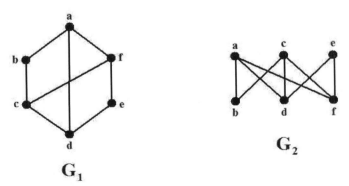

We see from the above example that it is often difficult to determine if a graph is bipartite. The following theorem from [11] and [12] assists us in determining whether graphs are bipartite by illustrating an interesting relationship between bipartite graphs and cycles:

**THEOREM 2.2.25** *A graph $\mathcal{G}$ on $n \geq 2$ vertices is bipartite if and only if it contains no odd cycles.*

**Proof**: First assume $\mathcal{G}$ is bipartite. Then $V(\mathcal{G})$ can be partitioned into two nonempty subsets $V_1$ and $V_2$ so that each edge joins a vertex in $V_1$ to a vertex in $V_2$. Let $C_p = v_1, v_2, \ldots, v_p, v_1$ be a cycle in $\mathcal{G}$. We must show that $p$ is even. Suppose $v_1 \in V_1$. Then since $\mathcal{G}$ is bipartite, it follows that $v_2 \in V_2$. By similar reasoning $v_3 \in V_1$, $v_4 \in V_2$, and so on. Since $v_p v_1$ is an edge in $\mathcal{G}$, and since $v_1 \in V_1$, it follows that $v_p \in V_2$. Hence $p$ is even.

For the converse, suppose that $\mathcal{G}$ has no odd cycles. We must show that $\mathcal{G}$ is bipartite. First assume that $\mathcal{G}$ is connected and let $u \in V(\mathcal{G})$. Then for each vertex $v \in V(\mathcal{G})$, there exists a $u - v$ path. Since there may be more than one such path for each $v$, select a shortest such path for each $v$. Let $V_1$ be the subset of $V(\mathcal{G})$ consisting of $u$ and every vertex $v$ such that a shortest $u - v$ path has even length. Define $V_2 := V(\mathcal{G}) - V_1$. Observe that every vertex adjacent to $u$ belongs to $V_2$.

We claim that every edge of $\mathcal{G}$ joins a vertex of $V_1$ to a vertex of $V_2$. Suppose that this is not true. Then some edge $e$ must join two vertices of $V_1$ or two vertices of $V_2$. Without loss of generality, let $e$ join $y$ and $z$, both of $V_2$. Since $y, z \in V_2$, it follows that each shortest $u - y$ path and each shortest $u - z$ path have odd length. Let $P = u, v_1, v_2, \ldots, v_{2s+1} = y$ be a shortest $u - y$ path and let $Q = u, w_1, w_2, \ldots, w_{2t+1} = z$ be a shortest $u - z$ path. Clearly $P$ and $Q$ have vertex $u$ in common. If $P$ and $Q$ have no other vertices in common, then $P$, $Q$, and edge $yz$ form an odd cycle which is a contradiction. Suppose $P$ and $Q$ have vertices in common other than $u$. Let $x$ be the last such vertex. Since $x \in P$, it follows $x = v_i$ for some $i \geq 1$; hence $d(u, x) = i$. Since $x \in Q$ and $w_i$ is the only vertex in $Q$ whose distance from $u$ is $i$, it follows that $x = w_i$. Thus $x = v_i = w_i$. Then $v_i, v_{i+1}, \ldots, v_{2s+1}, w_{2t+1}, w_{2t}, \ldots, w_i = v_i$ is a cycle of length $[(2s+1) - i] + [(2t+1) - i] + 1 = 2(s + t - i + 1) + 1$ is a cycle of odd length which is a contradiction.

If $\mathcal{G}$ is disconnected, we apply the above argument to show that each component of $\mathcal{G}$ is bipartite. Suppose the components of $\mathcal{G}$ are $\mathcal{G}_1, \ldots, \mathcal{G}_k$. Then we can partition the vertices of $\mathcal{G}_i$ into the vertex sets $U_i$ and $W_i$ for $i = 1, \ldots, k$ where for each $i$, no vertices of $U_i$ are adjacent and no vertices of $W_i$ are adjacent. Hence $\cup U_i$ and $\cup W_i$ form the partite sets of $V(\mathcal{G})$. □

We can extend the notion of bipartite graphs to complete bipartite graphs which make use of the operation of joining two empty graphs.

**DEFINITION 2.2.26** *A complete bipartite graph $K_{m,n}$ is the graph $E_m \vee E_n$.*

**EXAMPLE 2.2.27** Below is the graph $K_{3,4}$. Observe that every vertex in $E_3$ is adjacent to every vertex in $E_4$.

# Graph Theory Preliminaries

We can also extend the definition of bipartite graphs to *k-partite graphs* where the vertices of the graph are partitioned into $k$ subsets, $V_1, V_2, \ldots, V_k$ such that no two vertices within $V_i$ are adjacent for each $i = 1, \ldots, k$. Hence we can also define complete $k$-partite graph in a similar fashion. Below is the complete 3-partite graph $K_{1,3,2}$.

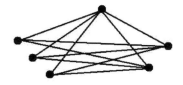

## Exercises:

1. Draw the following graphs:
   (a) $C_4 \vee C_3$
   (b) $P_2 \times W_3$
   (c) $Q_4$
   (d) $Q_3^c$
   (e) $(C_4^c \vee P_3) \times E_2$

2. Let $\mathcal{G}$ and $\mathcal{H}$ be graphs. Prove that $(\mathcal{G} \vee \mathcal{H})^c = \mathcal{G}^c + \mathcal{H}^c$.

3. Find a formula in terms of $n$ for the number of edges of $K_n$.

4. Find formulas in terms of $n$ for the number of vertices, edges, and degree of each vertex of $Q_n$.

5. Consider the following graph $\mathcal{G}$:

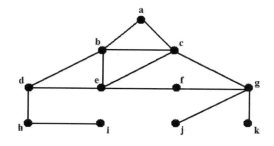

Find the subgraphs induced by the following sets of vertices:

(a) $\{a, b, d, e\}$
(b) $\{c, f, g\}$
(c) $\{f, i, j, k\}$
(d) $\{a, b, c, d, e, f, g\}$
(e) $\{d, h, i\}$

**6.** Draw the line graph for the graph in Exercise 5.

**7.** Determine if each of the following is a split graph. If it is, partition the vertex set into the subsets which form the clique and the co-clique.

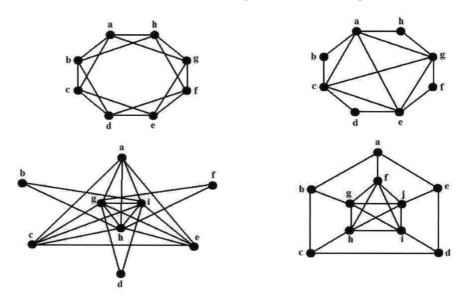

**8.** Use Theorem 2.2.25 to determine if each graph is bipartite. If the graph is bipartite, draw it in such a way that the partite sets are clear.

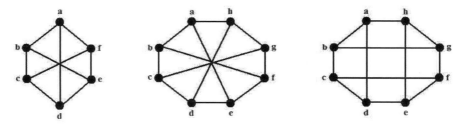

**9.** Prove that if a graph $\mathcal{G}$ on $n$ vertices is $r$-regular, then $\mathcal{G}^c$ is $n - r - 1$-regular.

**10.** (See [11]) Let $\mathcal{G}$ be a graph in which every edge joins an even vertex with an odd vertex. Prove that $\mathcal{G}$ is bipartite.

**11.** Theorem 2.2.25 can be rephrased to say that a graph $\mathcal{G}$ is bipartite if and only if all cycles are of even length. Can this theorem be generalized to say that $\mathcal{G}$ is $k$-partite if and only if all cycles have length divisible by $k$? Prove or give a counterexample.

**12.** Show that there cannot exist a disconnected graph that is isomorphic to its complement.

## 2.3 Trees

Often we wish to consider connected graphs where between any pair of vertices, there is a unique path. Such graphs are considerably easier to deal with because given two vertices, we can discuss *the* path between them rather than having to consider all paths between them. For example, in the proof of Theorem 2.2.25 we needed to consider a shortest path between two vertices. However, when there is a unique path between every pair of vertices, this would not be an issue. Because of the convenient nature of working with such graphs, there are many interesting theorems and applications.

Observe that in order for there to be a unique path between every pair of vertices, the graphs cannot contain any cycles. Hence we have a formal definition:

**DEFINITION 2.3.1** A *tree* is a connected graph that contains no cycles.

**EXAMPLE 2.3.2** Some examples of trees are as follows:

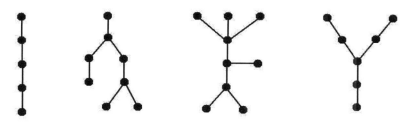

Note that the paths that we introduced in Definition 2.2.19 are trees. Another common class of trees are stars:

**DEFINITION 2.3.3** A *star* on $n$ vertices is a tree consisting of one vertex that is adjacent to the remaining $n - 1$ vertices.

**EXAMPLE 2.3.4** Examples of stars:

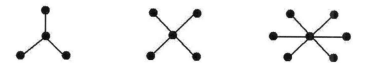

The goal of this section is to understand the structure of trees. To this end, we have the following definition:

**DEFINITION 2.3.5** A *pendant vertex* is a vertex whose degree is one.

Visually speaking, pendant vertices are often referred to as end vertices. We now prove our first lemma from [12] about trees which gives us information as to their structure with regard to pendant vertices. This lemma will be used to prove the subsequent theorem which concerns the number of edges in a tree.

**LEMMA 2.3.6** *Every tree (other than $K_1$) has at least two pendant vertices.*

**Proof**: Let $\mathcal{T}$ be a tree (other than $K_1$) and let $P$ be a path in $\mathcal{T}$ of greatest length. Let $P$ be a $u - v$ path $u = v_0, v_1, \ldots, v_k = v$. We will show that $u$ and $v$ are pendant vertices of $\mathcal{T}$. Observe that $u$ and $v$ are not adjacent to any vertex not on $P$, because otherwise a path whose length is greater than the length of $P$ will be produced. Also note that $u$ and $v$ are adjacent to $v_1$ and $v_{k-1}$, respectively. However, because $\mathcal{T}$ contains no cycles, $u$ and $v$ cannot be adjacent to any other vertices on $P$. Thus $\deg u = \deg v = 1$. Hence $u$ and $v$ are pendant vertices of $\mathcal{T}$. □

We use Lemma 2.3.6 in the proof our next theorem from [12] which derives the number of edges in a tree.

**THEOREM 2.3.7** *A tree on $n$ vertices has $n - 1$ edges.*

**Proof**: We will prove this by induction on $n$. The only tree on $n = 1$ vertex is $K_1$ which has no edges. Hence the result is true for $n = 1$. Assume now that for a fixed positive integer $k$ that all trees on $k$ vertices have $k - 1$ edges. Let $\mathcal{T}$ be a tree on $n = k + 1$ vertices. The theorem is proven if we can show $\mathcal{T}$ has $k$ edges. By Lemma 2.3.6, $\mathcal{T}$ contains at least two pendant vertices. Let $v$ be one of them. Then $\mathcal{T}' = \mathcal{T} - v$ is a tree on $k$ vertices. Hence by the inductive hypotheses, $\mathcal{T}'$ has $k - 1$ edges. Since $\mathcal{T}$ has exactly one more edge than $\mathcal{T}'$, it follows that $\mathcal{T}$ has $k$ edges. □

Observe that the proof of Theorem 2.3.7 makes direct use of Lemma 2.3.6 whose proof relies on the fact that trees have no cycles. In the beginning of this section, we pointed out that by having no cycles between any two distinct vertices of a tree there is a unique path. Hence we obtain the following theorem which summarizes the important aspects of the structures of trees.

**THEOREM 2.3.8** *Let $\mathcal{G}$ be a connected graph on $n$ vertices. The following are equivalent:*
  *(a) $\mathcal{G}$ is a tree.*
  *(b) $\mathcal{G}$ has $n - 1$ edges.*
  *(c) For every pair of distinct vertices $u$ and $v$ of $\mathcal{G}$, there exists a unique $u - v$ path.*

Given a connected graph $\mathcal{G}$ with cycles, observe that if we remove an edge $e$ from $\mathcal{G}$, then the resulting graph is connected if and only if $e$ belongs to some cycle of $\mathcal{G}$. Therefore given a graph, we can keep removing edges that belong to cycles until we are left with a tree. So by Theorem 2.3.7, it follows that a connected graph must have at least $n - 1$ edges. We state this more formally as a theorem from [12] and provide a formal proof:

**THEOREM 2.3.9** *Let $\mathcal{G}$ be a connected graph on $n$ vertices. Then $\mathcal{G}$ has at least $n - 1$ edges.*

**Proof**: This theorem is clearly true for graphs on 1, 2, or 3 vertices. Assume that this theorem is false. Then there exists a connected graph on $n$ vertices with at most $n-2$ edges. Let $\mathcal{G}$ be the smallest such graph in terms of the number of vertices $n$. We first claim that $\mathcal{G}$ has a pendant vertex. Assume not. Then the degree of every vertex of $\mathcal{G}$ is at least 2. Hence $\sum_{v \in \mathcal{G}} \deg v \geq 2n$. Since by Theorem 2.1.4 the sum of the degrees of the vertices is twice the number of edges of a graph, it follows that $\mathcal{G}$ has at least $n$ edges which contradicts our assumption that $\mathcal{G}$ has at most $n-2$ edges. Therefore, $\mathcal{G}$ contains a pendant vertex.

Let $v$ be a pendant vertex of $\mathcal{G}$. Since $\mathcal{G}$ is connected, has $n$ vertices, and has at most $n-2$ edges, it follows that $\mathcal{G} - v$ is connected (since $v$ is pendant), has $n-1$ vertices, and at most $n-3$ edges. But this contradicts the fact that $\mathcal{G}$ is a graph on the smallest number of vertices such that the number of edges is at most the number of vertices minus two. $\square$

Theorem 2.3.9 tells us that given any connected graph $\mathcal{G}$ on $n$ vertices having more than $n-1$ edges, we can find a subgraph of $\mathcal{G}$ that is a tree that contains all of the vertices of $\mathcal{G}$. This gives rise to the concept of spanning trees:

**DEFINITION 2.3.10** Given a graph $\mathcal{G}$, a *spanning tree* of $\mathcal{G}$ is a tree that is a subgraph of $\mathcal{G}$ containing all of the vertices of $\mathcal{G}$.

**EXAMPLE 2.3.11** Consider the following graph $\mathcal{G}$:

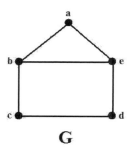

Below are examples of spanning trees of $\mathcal{G}$:

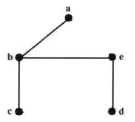 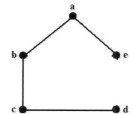 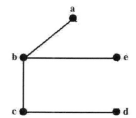

The number of spanning trees that a graph has is an interesting problem that makes surprising use of linear algebra. We will discuss this problem in depth in Section 3.2. For now, we will turn our discussion to systematic ways of creating spanning trees from a given graph. To this end, we need to discuss the concept of a rooted tree:

**DEFINITION 2.3.12** A *rooted tree* is a tree in which a vertex is designated as the *root vertex*.

Typically, when we draw rooted trees, we place the root vertex on top. The vertices adjacent to the root vertex are usually all drawn one level below, the vertices adjacent to those vertices are drawn two levels below, and so on. We say that a vertex is on level $i$ if it is of distance $i$ from the root vertex. The root vertex is said to be on level 0. Rooted trees give rise to a relationship between vertices as we see in the following definition.

**DEFINITION 2.3.13** In a rooted tree, if a vertex $v$ is adjacent to a vertex $w$ and $v$ lies one level above $w$, then we say $v$ is a *parent* of $w$ and that $w$ is a *child* of $v$.

We can generalize the above definition to vertices that are not adjacent:

**DEFINITION 2.3.14** In a rooted tree, if $w$ lies any number of levels below $v$ and the path from $w$ to $v$ contains only vertices on levels below $v$, then we say $w$ is a *descendent* of $v$ and that $v$ is an *ancestor* of $w$.

In general, the most common rooted trees are Bethe trees.

**DEFINITION 2.3.15** A *rooted $m$-ary Bethe tree* on $k$ levels is a rooted tree on $k$ levels where each vertex in levels $0, \ldots, k-1$ has $m$ children.

**EXAMPLE 2.3.16** Below is a 2-ary Bethe tree on 3 levels. Note that each non-pendant vertex has 2 children and that there are 3 levels.

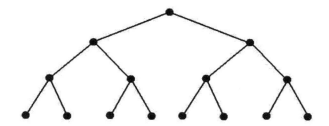

Rooted trees play an important role in algorithms that create spanning trees of graphs. There are two common algorithms that create spanning trees of graphs: the *depth-first search* and the *breadth-first search* algorithms. With the depth-first search, we first designate a vertex to be the root vertex. We then create a spanning tree by travelling from the root vertex as far as we can go until we cannot travel further without going to a vertex we already accounted for. We then backtrack through the path we created until we encounter a vertex that is adjacent to a vertex we have not accounted for. We then travel along a path accounting for new vertices until we cannot travel any further. We then backtrack as we did before until we find another vertex that is adjacent to a vertex we have not acounted for. We continue this procedure until all vertices are accounted for. It is important to note that since we have a choice in which vertex we designate as the root vertex and since we often have a choice of unaccounted vertices that are adjacent to a vertex we are at, the spanning tree we produce using the depth-first search algorithm is not unique.

**EXAMPLE 2.3.17** Consider the graph below. Let's designate vertex $a$ to be the root vertex.

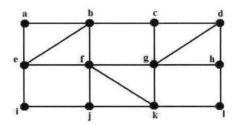

Starting at vertex $a$, we can create a path by travelling to vertices $b$, $f$, $j$, $i$, then $e$. At this point, $e$ is not adjacent to any unaccounted vertices. So we backtrack until we reach a vertex adjacent to an unaccounted vertex. The first such vertex we reach when we backtrack is $j$. From $j$ we create the path by travelling to vertices $k$, $g$, $d$, then $c$. Since $c$ is not adjacent to any accounted vertices, we backtrack to $d$. From $d$ we create the path by travelling to vertices $h$ then $l$. We have now accounted for all of the vertices of our graph. Below is the spanning tree we created using the depth-first search algorithm:

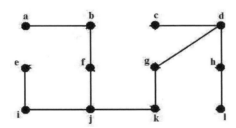

The depth-first search algorithm got its name because, at each step, we create the spanning tree by travelling as deep into the graph as possible. We now discuss the breadth-first search algorithm. In the breadth-first search algorithm, we designate a root vertex. However, instead of travelling deep into the graph, we choose all vertices that are adjacent to the root vertex. We then go one by one through these vertices and choose all of the vertices adjacent to these vertices that have not been accounted for. We continue in this fashion creating additional levels of our spanning tree until all vertices of the graph are accounted for. As with the depth-first search algorithm, since there is a choice in which vertex is designated the root vertex, and since there is often a choice in the order in which the subsequent vertices are considered, the spanning tree we produce using the breadth-first search algorithm is not unique.

**EXAMPLE 2.3.18** Consider the graph used in Example 2.3.17. Again, designate $a$ to be the root vertex. Since $a$ is adjacent to $b$ and $e$, let edges $ab$ and $ae$ be in our spanning tree. Now we consider vertices $b$ and $e$. The vertices adjacent to $b$ that we have not accounted for are $c$ and $f$, hence we let edges $bc$ and $bf$ be in our spanning tree. The only unaccounted vertex adjacent to vertex $e$ is $i$, thus the edge $ei$ will be in our tree. Now we consider the vertices $c$, $f$, and $i$ which are currently at the bottom level of our spanning tree. The unaccounted vertices adjacent to $c$ are $d$ and $g$, thus adding edges $cd$ and $dg$ to the tree; the unaccounted vertices adjacent to $f$ are $j$ and $k$, thus adding edges $fj$ and $fk$ to the tree; and there are no unaccounted vertices adjacent to $i$. Now vertices $d$, $g$, $j$, and $k$ are at the bottom level of our tree. Beginning with $d$, we see $d$ is adjacent to the unaccounted vertex $h$. Likewise, vertex $k$ is adjacent to the unaccounted vertex $l$. Thus we complete our tree by adding edges $dh$ and $kl$. Below is the spanning tree we created using the breadth-first search algorithm:

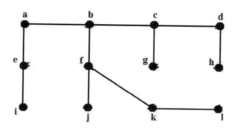

Spanning trees will be discussed in greater detail in Section 3.2 when we use matrix theory to determine the number of spanning trees in a given graph. In Chapter 7 it will be of interest to investigate spanning trees which are not connected. Thus we close this section with the following definition which we make use of in the exercises that follow.

**DEFINITION 2.3.19** A *forest* is a disconnected graph in which each connected component is a tree.

# Exercises:

1. (a) Draw all trees on 6 vertices.

   (b) Draw all forests on 6 vertices.

2. (See [11]) Find all trees $\mathcal{T}$ in which $\mathcal{T}^c$ is also a tree.

3. (See [12]) Suppose a tree $\mathcal{T}$ has 21 vertices and the vertices have only degrees 1, 3, 5, and 6. If $\mathcal{T}$ has exactly 15 pendant vertices and one vertex of degree 6, how many vertices of $\mathcal{T}$ have degree 5.

4. (See [12]) Suppose a tree $\mathcal{T}$ on 35 vertices has 25 pendant vertices, two vertices of degree 2, three vertices of degree 4, one vertex of degree 5, and two vertices

# Graph Theory Preliminaries

of degree 6. It also contains two vertices of the same unknown degree $x$. What is $x$?

**5.** Prove Theorem 2.3.8.

**6.** Let $F$ be a forest on $n$ vertices that consists of $k$ components. Prove that $F$ has $n - k$ edges.

**7.** Letting vertex $a$ be the root vertex, create a spanning tree using each of the depth-first search and breadth-first search algorithms.

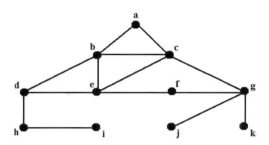

## 2.4 Connectivity of Graphs

Connectivity of graphs will play a crucial role in later chapters of this book when we study the algebraic connectivity of graphs. Therefore, it is helpful to us to have a better understanding of connectivity in general. In Section 2.1 we said that a graph $\mathcal{G}$ is connected if there exists a path between any pair of vertices of $\mathcal{G}$. In this section, we discuss ways of measuring how connected a connected graph is. We begin our discussion by investigating graphs in which the removal of a vertex along with its incident edges renders the graph disconnected. We define such vertices as follows:

**DEFINITION 2.4.1** A vertex $v$ is a *cut vertex* of a connected graph $\mathcal{G}$ if $\mathcal{G} - v$ is disconnected.

**EXAMPLE 2.4.2** Consider the following graph $\mathcal{G}$:

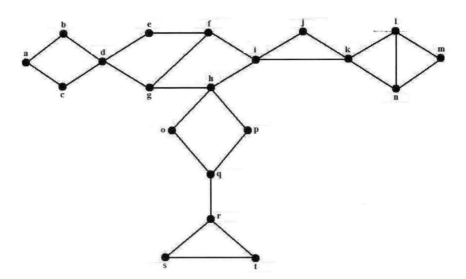

Observe that the cut vertices are $d$, $h$, $i$, $k$, $q$, and $r$, as the removal of any of one these vertices along with its incident edges will result in a disconnected graph.

To characterize cut vertices, we have the following theorem from [12]:

**THEOREM 2.4.3** *A vertex $v$ is a cut vertex in a connected graph $\mathcal{G}$ if and only if there exist vertices $u$ and $w$ distinct from $v$ such that $v$ lies on every $u - w$ path of $\mathcal{G}$.*

**Proof**: Suppose that $v$ is a cut vertex of $\mathcal{G}$. Then $\mathcal{G} - v$ is disconnected. Let $u$ and $w$ be in different components of $\mathcal{G} - v$. Since there does not exist a $u - w$ path in $\mathcal{G} - v$ but there does exist such a path in $\mathcal{G}$, it follows that every $u - w$ path in $\mathcal{G}$ must contain $v$.

Suppose now that $\mathcal{G}$ contains two vertices $u$ and $w$ such that every $u - w$ path in $\mathcal{G}$ contains $v$. Then there is no $u - w$ path in $\mathcal{G} - v$. Hence $\mathcal{G} - v$ is disconnected. Thus $v$ is a cut vertex of $\mathcal{G}$. □

Looking at Example 2.4.2, we saw that $h$ was a cut vertex. So by Theorem 2.4.3, there exists two vertices in $\mathcal{G}$ such that $h$ lies on every path between them. One example of such a pair of vertices is $e$ and $p$ as every $e - p$ path contains $h$. Conversely, we can discover cut vertices by finding pairs of vertices such that every path contains a certain vertex. For example, observe that every $d - k$ path contains vertex $i$. Hence by Theorem 2.4.3, $i$ is a cut vertex of $\mathcal{G}$.

We now investigate more deeply the structure of graphs having cut vertices. To this end, we need two definitions:

**DEFINITION 2.4.4** *A graph is nonseparable if it does not contain a cut vertex.*

**DEFINITION 2.4.5** *A block of a graph $\mathcal{G}$ is a nonseparable subgraph of $\mathcal{G}$ that is not a proper subgraph of any other nonseparable subgraph of $\mathcal{G}$.*

**EXAMPLE 2.4.6** Consider the graph $G$ in Example 2.4.2. Observe that the subgraph induced by vertices $a$, $b$, $c$, and $d$ is a block. Calling this subgraph $\mathcal{H}$, we see that $\mathcal{H}$ is nonseparable, yet any subgraph of $G$ for which $\mathcal{H}$ is a subgraph is not nonseparable for it will have $d$ as a cut vertex. We see that the other blocks of $G$ are the subgraph induced by vertices $d, e, f, g, h, i$; the subgraph induced by vertices $i, j, k$; the subgraph induced by vertices $k, l, m, n$; the subgraph induced by vertices $h, o, p, q$; the subgraph induced by vertices $q, r$; and the subgraph induced by vertices $r, s, t$.

Overall, we see that the basic structure of any graph with cut vertices is that such a graph is a set of blocks with cut vertices separating them. Note that a vertex of a graph belongs to more than one block if and only if it is a cut vertex of the graph.

At this point, we now investigate the structure of nonseparable graphs. To begin our investigation, we require two definitions:

**DEFINITION 2.4.7** Given a graph $G$, a *vertex cut* is a set of vertices $U$ such that $G - U$ is disconnected.

**DEFINITION 2.4.8** If the cardinality of vertices in a vertex cut $U$ of a graph $G$ is the minimum number of vertices required to be removed from $G$ (along with its incident edges) to render $G$ disconnected, we say that $U$ is a *minimal vertex cut*. This cardinality is known as the *vertex connectivity* of $G$ and is denoted by $v(G)$.

**EXAMPLE 2.4.9** Consider the graph $G$ below:

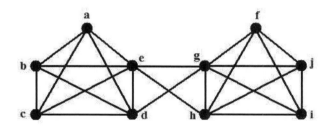

Observe that $G$ contains no cut vertices. However, there exists a vertex cut $U$ of cardinality 2, namely $U = \{e, g\}$, since $G - U$ is disconnected. Hence the vertex connectivity of $G$ is 2, i.e., $v(G) = 2$.

While the concept of the vertex connectivity is useful in measuring how connected a connected graph is, we can also imply a similar concept of connectivity regarding the edges of a graph. Our next definition concerns the edges of a graph and is analagous to the definition of cut vertex:

**DEFINITION 2.4.10** An edge $e$ is a *bridge* of a connected graph $G$ if $G - e$ is disconnected.

**EXAMPLE 2.4.11** In Example 2.4.2, the edge $qr$ is a bridge since $\mathcal{G} - qr$ is disconnected.

As with vertex cuts, edge cuts also play an important role in measuring the connectivity of a graph. With this in mind, we have the following definitions which are similar to that of a vertex cut:

**DEFINITION 2.4.12** Given a graph $\mathcal{G}$, an *edge cut* is a set of edges $X$ such that $\mathcal{G} - X$ is disconnected.

**DEFINITION 2.4.13** If the cardinality of edges in an edge cut $X$ of a graph $\mathcal{G}$ is the minimum number of edges required to be removed from $\mathcal{G}$ to render $\mathcal{G}$ disconnected, we say that $X$ is a *minimal edge cut*. This cardinality is known as the *edge connectivity* of $\mathcal{G}$ and is denoted by $e(\mathcal{G})$.

**EXAMPLE 2.4.14** Consider the graph $\mathcal{G}$ in Example 2.4.9. Observe that $\mathcal{G}$ has no bridges. Also observe that removing any two edges of $\mathcal{G}$ will not render the graph disconnected. However, there does exist an edge cut $X$ of cardinality 3, namely $X = \{dg, eg, eh\}$ since $\mathcal{G} - X$ is disconnected. Thus $e(\mathcal{G}) = 3$.

At this point, it is natural to ask if there is some relationship between the vertex connectivity and edge connectivity of a graph. We close this section with the following theorem from [12] which shows a relationship between these quantities and also the minimum vertex degree, denoted $\delta(\mathcal{G})$.

**THEOREM 2.4.15** *For every graph $\mathcal{G}$, we have $v(\mathcal{G}) \leq e(\mathcal{G}) \leq \delta(\mathcal{G})$.*

**<u>Proof</u>**: If $\mathcal{G}$ is disconnected, then $v(\mathcal{G}) = e(\mathcal{G}) = 0 \leq \delta(\mathcal{G})$. If $\mathcal{G} = K_n$ for $n \geq 2$ then $v(\mathcal{G}) = e(\mathcal{G}) = \delta(\mathcal{G}) = n - 1$. So for the remainder of the proof, we will assume that $\mathcal{G}$ is a connected noncomplete graph on $n \geq 3$ vertices. Hence $\delta(\mathcal{G}) \leq n - 2$.

Let's first show $e(\mathcal{G}) \leq \delta(\mathcal{G})$. Let $v$ be a vertex of $\mathcal{G}$ such that $\deg v = \delta(\mathcal{G})$. Since the set of $\delta(\mathcal{G})$ edges incident with $v$ is an edge cut of $\mathcal{G}$, it follows that $e(\mathcal{G}) \leq \delta(\mathcal{G}) \leq n - 2$.

To show $v(\mathcal{G}) \leq e(\mathcal{G})$, let $X$ be a minimum edge cut of $\mathcal{G}$. Then $|X| = e(\mathcal{G}) \leq n - 2$. Observe $\mathcal{G} - X$ contains exactly two components, say $\mathcal{G}_1$ and $\mathcal{G}_2$. Suppose $\mathcal{G}_1$ has $k \geq 1$ vertices; hence $\mathcal{G}_2$ has $n - k \geq 1$ vertices. Since $X$ is an edge cut, every edge in $X$ joins a vertex in $\mathcal{G}_1$ to a vertex in $\mathcal{G}_2$. We consider two cases:

*Case I*: $\mathcal{G} = \mathcal{G}_1 \vee \mathcal{G}_2$. In this case, $|X| = k(n-k)$. Since $(k-1)(n-k-1) \geq 0$ and since $(k-1)(n-k-1) = k(n-k) - n + 1$, it follows that $k(n-k) - n + 1 \geq 0$. Since $X$ is assumed to be a minimal edge cut, it follows that $e(\mathcal{G}) = |X| = k(n-k) \geq n - 1$. However, $e(\mathcal{G}) \leq n - 2$, so this case cannot occur.

*Case II*: Suppose that $\mathcal{G} \neq \mathcal{G}_1 \vee \mathcal{G}_2$. Then there exists a vertex $u \in \mathcal{G}_1$ and a vertex $v \in \mathcal{G}_2$ such that $u$ and $v$ are not adjacent in $\mathcal{G}$. We now define a set of vertices $U$. For each edge $e \in X$, we select a vertex for $U$ in the following way: If $u$

# Graph Theory Preliminaries

is incident with $e$, then choose the vertex in $\mathcal{G}_2$ incident with $e$ as a vertex belonging to $U$. Otherwise, select the vertex in $\mathcal{G}_1$ incident with $e$ as a vertex belonging to $U$. Since there may be more than one edge incident to a vertex in $U$, it follows that $|U| \leq |X|$. Since $u, v \notin U$ and there is no $u - v$ path in $\mathcal{G} - U$, it follows that $\mathcal{G} - U$ is disconnected. Thus $U$ is a vertex cut of $\mathcal{G}$. Hence $v(\mathcal{G}) \leq |U| \leq |X| = e(\mathcal{G})$. □

In Chapter 5 and subsequent chapters, we will use matrix theory to expand upon the ideas of connectivity presented in this section when we study the algebraic connectivity of a graph. The algebraic connectivity of a graph will also be a useful tool in measuring how connected a graph is. We will study relationships between the algebraic connectivity and the types of connectivity discussed in this section.

# Exercises:

**1.** Determine the cut vertices, bridges, and blocks of the graph below:

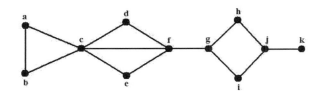

**2.** (See [12]) Let $\mathcal{G}$ be a connected graph and let $u$ be a fixed vertex in $\mathcal{G}$. Prove that if $v$ is a vertex that is farthest from $u$, then $v$ is not a cut vertex of $\mathcal{G}$.

**3.** (See [12]) Prove that if $v$ is a cut vertex of a graph $\mathcal{G}$, then $v$ is not a cut vertex of $\mathcal{G}^c$.

**4.** (See [12]) Prove that a connected graph $\mathcal{G}$ on $n \geq 3$ vertices is nonseparable if and only if any two adjacent edges of $\mathcal{G}$ lie on a common cycle.

**5.** (See [12]) Prove that if $\mathcal{G}$ is a graph on $n \geq 3$ vertices such that $\deg v \geq n/2$ for each vertex $v \in \mathcal{G}$, then $\mathcal{G}$ is nonseparable.

**6.** Suppose that a graph contains $b$ blocks and $k$ cut vertices. What are the possible values for $k$ in terms of $b$?

**7.** (See [12]) Without using Theorem 2.4.15, prove that $e(K_n) = n - 1$.

**8.** Determine $v(\mathcal{G})$ and $e(\mathcal{G})$ for each of the following: $K_{m,n}$, $C_n$, $W_n$.

**9.** (See [33]) Determine the vertex and edge connectivities of the following graph:

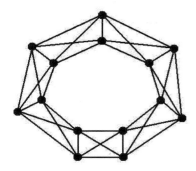

**10.** (See [12]) Prove that if $\mathcal{G}$ is a graph on $n$ vertices such that $\delta(\mathcal{G}) \geq (n-1)/2$, then $e(\mathcal{G}) = \delta(\mathcal{G})$.

## 2.5 Degree Sequences and Maximal Graphs

In this section, we explore the degrees of the individual vertices of a graph. Given a finite sequence of nonnegative integers, we determine the sequences in which there exists a graph such that the degrees of the vertices are that of the sequence. We then define a partial ordering of such sequences and investigate graphs having a degree sequence in which no graph has a degree sequence considered to be greater. Before we begin, we need to give a formal definition of a degree sequence:

**DEFINITION 2.5.1** Given a graph $\mathcal{G}$, the *degree sequence* of $\mathcal{G}$ is the sequence of the degrees of the vertices of $\mathcal{G}$ listed in nonincreasing order.

**EXAMPLE 2.5.2** Consider the graph $\mathcal{G}$:

The degree sequence of $\mathcal{G}$ is $(5, 2, 2, 1, 1, 1)$.

Most often, we will be working in the other direction. In other words, we will be given a finite nonincreasing sequence $S$ of nonnegative integers and will then want to find a graph, if any exist, whose degree sequence is $S$. This leads us to the following definition:

**DEFINITION 2.5.3** A finite nonincreasing sequence $S$ of nonnegative integers is *graphical* if there exists a graph whose degree sequence is $S$.

The first goal of this section is to determine which finite nonincreasing sequences of nonnegative integers are graphical. At this point, we already know of some such sequences which are not graphical. For example, if a graph has $n$ vertices, then a vertex can have degree at most $n-1$. This occurs when a vertex is adjacent to all of the vertices other than itself. Hence if we have a nonincreasing sequence of $n$ nonnegative integers in which at least one of the integers is at least $n$, then that sequence cannot be graphical. We also saw in Theorem 2.1.4 that the sum of the degrees in any graph must be even. Hence if the sum of the integers in a given sequence is odd, then Theorem 2.1.4 implies that the sequence is not graphical. More specifically, if a given sequence has an odd number of odd terms, then the sequence is not graphical.

While having the tools illustrated in the previous paragraph can be useful, we see that they apply only in limited cases. Hence it will be useful to have a theorem which allows us to determine which sequences are graphical and which are not. The following theorem, known as the Havel-Hakimi Theorem, is a theorem proved independently in [39] and [38]. We will use the proof found in [12]. Before presenting this proof, we will assume without loss of generality that the graph contains no isolated vertices for if the graph did, then the degree sequence would merely end with a string of zeros which we can delete in order to consider the components which are not isolated vertices. We now present the theorem:

**THEOREM 2.5.4** *A nonincreasing sequence $S : (d_1 \geq d_2 \geq \ldots \geq d_n)$ of nonnegative integers with $n \geq 2$ and $d_i \geq 1$ for each $i$ is graphical if and only if the sequence*

$$S_1 : (d_2 - 1, d_3 - 1, \ldots, d_{d_1+1} - 1, d_{d_1+2}, \ldots, d_n)$$

*is graphical.*

Before presenting the proof of this theorem, it is beneficial to discuss what this theorem is saying. The theorem says that the original sequence $S$ is graphical if and only if the sequence obtained by deleting $d_1$ and subtracting 1 from the next $d_1$ entries (i.e., $d_2$ through $d_{d_1+1}$) is graphical. With this in mind, we present the proof:

**Proof**: First assume that $S_1$ is graphical. Then there exists a graph $\mathcal{G}_1$ with vertices $v_2, \ldots, v_n$ such that

$$\deg_{\mathcal{G}} v_i = \begin{cases} d_i - 1, & \text{if } 2 \leq i \leq d_1 + 1, \\ d_i, & \text{if } d_1 + 2 \leq i \leq n. \end{cases}$$

We now construct a graph $\mathcal{G}$ from $\mathcal{G}_1$ by adding a vertex $v_1$ and adding $d_1$ edges joining $v_1$ to each of $v_2, \ldots, v_{d_1+1}$. Observe that $\deg_{\mathcal{G}} v_i = d_i$ for $i = 1, \ldots, n$. Thus $S$ is the degree sequence for $\mathcal{G}$; hence $S$ is graphical.

For the converse, assume that $S$ is graphical. We will consider two cases concerning a vertex of degree $d_1$:

*Case I*: Suppose that $\mathcal{G}$ has degree sequence $S$ and contains a vertex $u$ of degree $d_1$ such that $u$ is adjacent to vertices of degrees $d_2, \ldots d_{d_1+1}$. Then $S_1$ is the degree sequence of $\mathcal{G} - u$, showing that $S_1$ is graphical.

*Case II*: Suppose that graph $\mathcal{G}$ has degree sequence $S$ but there does not exist a vertex $u$ of degree $d_1$ which is adjacent to vertices of degrees $d_2, \ldots, d_{d_1+1}$. Among all graphs with degree sequence $S$, let $\mathcal{G}$ be the one on vertices $v_1, \ldots, v_n$ such that $\deg v_i = d_i$ for $1 \le i \le n$ and the sum of the degrees of the vertices adjacent to $v_1$ is as large as possible. Since $v_1$ is not adjacent to all of the vertices having degrees $d_2, \ldots, d_{d_1+1}$, (i.e., all of the vertices of next highest degrees down through degree $d_{d_1+1}$), it follows that $v_1$ must be adjacent to a vertex $v_s$ having smaller degree than a vertex $v_r$ to which $v_1$ is not adjacent. In other words, there exist vertices $v_r$ and $v_s$ with $d_r > d_s$ such that $v_1$ is adjacent to $v_s$ but not to $v_r$. Since $d_r > d_s$, there exists a vertex $v_t$ adjacent to $v_r$ but not to $v_s$. Create the graph $\mathcal{G}'$ from $\mathcal{G}$ by removing edges $v_1 v_s$ and $v_r v_t$ and adding edges $v_1 v_r$ and $v_s v_t$. Then $\mathcal{G}$ and $\mathcal{G}'$ each have $S$ as its degree sequence. However, the sum of the degrees of the vertices adjacent to $v_1$ in $\mathcal{G}'$ is larger than the sum of the degrees of the vertices adjacent to $v_1$ in $\mathcal{G}$, thus producing a contradiction. Hence Case II cannot hold and the theorem is proven. □

We now illustrate Theorem 2.5.4 with the following two examples. It is important to note that if the sequence we obtain from applying Theorem 2.5.4 is not nonincreasing, then we rearrange the sequence so that it is nonincreasing before we proceed.

**EXAMPLE 2.5.5** Determine if the sequence $S : (6, 5, 4, 3, 3, 3, 3, 1)$ is graphical.

**Solution**: Deleting 6 from $S$ and subtracting 1 from the next six terms, we obtain

$$S_1 : (4, 3, 2, 2, 2, 2, 1).$$

Theorem 2.5.4 states that $S$ is graphical if and only if $S_1$ is. So it suffices to determine if $S_1$ is graphical. To do this, apply Theorem 2.5.4 again by deleting 4 from $S_1$, subtracting 1 from the next four terms, then rearranging so the resulting sequence is nonincreasing to obtain

$$S_2 : (2, 2, 1, 1, 1, 1).$$

To determine if $S_2$ is graphical, we apply Theorem 2.5.4 yet again. Thus we delete the first 2 from $S_2$, subtract 1 from the next two terms, and rearrange to obtain

$$S_3 : (1, 1, 1, 1, 0).$$

We see that $S_3$ is graphical since the graph $\mathcal{G}_3$ below is a graph whose degree sequence is $S_3$.

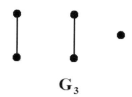

$\mathcal{G}_3$

# Graph Theory Preliminaries

We can then work backwards by adding one vertex at a time and the appropriate edges to create graphs $G_2$, $G_1$, and $G$ with the degree sequence $S_2$, $S_1$, and $S$, respectively. To see this, since $S_3$ was obtained from $S_2$ by deleting a 2 and then subtracting one each from the other 2 term and one of the 1 terms, it follows that $G_2$ is created from $G_3$ by adding a vertex $u$ of degree two which is adjacent to one vertex of degree one and the vertex of degree zero:

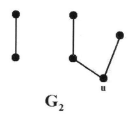

$G_2$

Since $S_2$ was obtained from $S_1$ by deleting the 4 and subtracting one each from the 3 term and three of the 2 terms, it follows that $G_1$ can be created from $G_2$ by adding a vertex $u$ of degree four which is adjacent to one vertex of degree two and three vertices of degreee one:

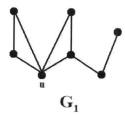

$G_1$

Finally, since $S_1$ was obtained from $S$ by deleting the 6 and subtracting one each from the 5, 4, and all of the 3 terms, it follows that $G$ can be created from $G_1$ by adding a vertex $u$ of degree six which is adjacent to the vertex of degree four, the vertex of degree three, and the four vertices of degree two:

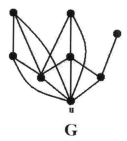

$G$

**EXAMPLE 2.5.6** Determine if the sequence $S : (8, 8, 5, 5, 5, 4, 2, 2, 1)$ is graphical.

**Solution**: Deleting 8 from $S$ and subtracting one from the next eight terms, we obtain
$$S_1 : (7, 4, 4, 4, 3, 1, 1, 0).$$

Applying Theorem 2.5.4, we delete 7 from $S_1$ and subtract one from the next seven terms to obtain
$$S_2 : (3, 3, 3, 2, 0, 0, -1)$$
Since a graph cannot have a vertex of negative degree, we see that $S_2$ is not graphical. Hence by Theorem 2.5.4, neither is $S_1$ and consequently, neither is $S$.

At this point, we will define a partial ordering of degree sequences which will be useful to us in subsequent chapters.

**DEFINITION 2.5.7** Let $(a)$ and $(b)$ be sequences of nonincreasing nonnegative integers where $(a)$ contain $r$ integers and $(b)$ contains $s$ integers. We say that $(a)$ *majorizes* $(b)$ if

$$\sum_{i=1}^{k} a_i \geq \sum_{i=1}^{k} b_i$$

for all $1 \leq k \leq \min\{r, s\}$, while

$$\sum_{i=1}^{r} a_i = \sum_{i=1}^{s} b_i.$$

If a sequence $(a)$ majorizes $(b)$, we write $(a) \succeq (b)$.

**EXAMPLE 2.5.8** Let $(a) = (5, 2, 2, 1, 1, 1)$ and $(b) = (4, 3, 1, 1, 1, 1, 1)$. Note that $r = 6$, $s = 7$, and that the sum of the entries of $(a)$ and the entries of $(b)$ are each 12. Observe that for each $k = 1, \ldots, 6$, that $\sum_{i=1}^{k} a_i \geq \sum_{i=1}^{k} b_i$. Thus $(a)$ majorizes $(b)$ and we write $(a) \succeq (b)$.

Our goal is to construct degree sequences that are not majorized by any other degree sequences. To this end, we will represent degree sequences as Ferrers-Sylvester diagrams which we now define:

**DEFINITION 2.5.9** A *Ferrers-Sylvester diagram* is a grid representing a degree sequence $(d)$ in which the $i^{th}$ row of the grid contains $d_i$ boxes.

**EXAMPLE 2.5.10** For the graph in Example 2.5.2 whose degree sequence is $d = (5, 2, 2, 1, 1, 1)$, the Ferrers-Sylvester diagram for this degree sequence is

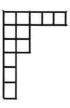

# Graph Theory Preliminaries

Ferrers-Sylvester diagrams will be useful in determining two important concepts: the conjugate and the trace of a degree sequence. Both concepts will be useful to us in achieving our goal of finding degree sequences which are not majorized by any other degree sequence. We now define both terms:

**DEFINITION 2.5.11** The *conjugate* of a degree sequence $(d)$ is the sequence $(d^*) = (d_1^*, d_2^*, \ldots, d_k^*)$ where $d_i^* = |j : d_j \geq i|$. In other words, $d_i^*$ is the number of values in $(d)$ that are greater than or equal to $i$.

Visually speaking, the value for $d_i^*$ is the number of boxes in the $i^{th}$ column of the Ferrers-Sylvester diagram.

**DEFINITION 2.5.12** The *trace*, $f(d)$, of a degree sequence $d = (d_1, d_2, \ldots, d_n)$ is $f(d) = |i : d_i \geq i|$.

In other words, the trace of a degree sequence is the largest integer $i$ such that $d_i \geq i$. Visually speaking, the trace of a degree sequence is the number of boxes on the main diagonal of its Ferrers-Sylvester diagram. Often, we will write $f(\mathcal{G})$ instead of $f(d)$.

**EXAMPLE 2.5.13** In Example 2.5.2, recall that the degree sequence is $(d) = (5, 2, 2, 1, 1, 1)$. Observe that there are six values in $(d)$ greater than or equal to 1, three values greater than or equal to 2, one value greater than or equal to 3, one value greater than or equal to 4, and one value greater than or equal to 5. Thus $(d^*) = (6, 3, 1, 1, 1)$. Note that this sequence is precisely the number of boxes in each column of the Ferrers-Sylvester diagram.

Observe the trace of this degree sequence is $f(d) = 2$ since 2 is the largest integer $i$ such that $d_i \geq i$. Note that $d_1 = 5 \geq 1$ and $d_2 = 2 \geq 2$, but $d_3 = 2 < 3$. Visually speaking, there are 2 blocks on the main diagonal of the Ferres-Sylvester diagram.

Ferrers-Sylvester diagrams, the conjugate of a degree sequence, and the trace of a degree sequence will be useful in achieving our goal of determining graphs whose degree sequence is not majorized by any other degree sequence. Before continuing, let us define such a graph:

**DEFINITION 2.5.14** A graph is a *maximal graph* or a *threshold graph* if its degree sequence is not majorized by any other degree sequence.

The following two lemmas from [73] will assist in determining precisely which graphs are maximal. These lemmas also give us further insight into which nonincreasing sequences of nonnegative integers are graphical.

**LEMMA 2.5.15** Let $(a)$ and $(b)$ be nonincreasing sequences of nonnegative integers such that $\sum a_i = \sum b_i$ and $(b) \succeq (a)$. If $(b)$ is graphical then so is $(a)$.

**Proof**: Assume that $(b)$ is graphical let $\mathcal{G}$ be a graph whose degree sequence is $(b)$. Let $(a)$ be a degree sequence satisfying the conditions of the statement of the theorem. Let $\mathcal{F}_a$ and $\mathcal{F}_b$ be the Ferrers-Sylvester diagrams for the degree sequences $(a)$ and $(b)$, respectively. Observe that $\mathcal{F}_a$ can be created from $\mathcal{F}_b$ by creating a sequence of Ferrers-Sylvester diagrams $\mathcal{F}_b = \mathcal{F}_{b_0}, \mathcal{F}_{b_1}, \ldots, \mathcal{F}_{b_p} = \mathcal{F}_a$ where for each $h = 1, \ldots, p - 1$, the diagram $\mathcal{F}_{b_{h+1}}$ is created from $\mathcal{F}_{b_h}$ by moving a box from row $i$ of $\mathcal{F}_{b_h}$ to row $j$ of $\mathcal{F}_{b_h}$ where $i$ and $j$ such that $b_{h,i} \geq b_{h,j} - 2$ with $b_{h,i}$ and $b_{h,j}$ denoting the number of boxes in rows $i$ and $j$, respectively, of $\mathcal{F}_{b_h}$. To show that $(a)$ is graphical, it suffices to show that the degree sequence represented by $\mathcal{F}_{b_1}$ is graphical as we can then conclude that $(a)$ is graphical by induction. Recalling that $\mathcal{G}$ is a graph whose degree sequence is $(b)$, since $b_i \geq b_j - 2$, it follows that there exists a vertex $v$ in $\mathcal{G}$ that is adjacent to vertex $i$ but not adjacent to vertex $j$. Let $\mathcal{H}$ be the graph created from $\mathcal{G}$ by deleting the edge joining $i$ to $v$ and replacing it with an edge joining $j$ to $v$. Observe the Ferrers-Sylvester diagram for the degree sequence of $\mathcal{H}$ is precisely $\mathcal{F}_{b_1}$, thus proving the lemma. □

**LEMMA 2.5.16** *Let $(d)$ be a degree sequence whose sum $\sum d_i$ is even. Then $(d)$ is graphical if and only if*

$$\sum_{i=1}^{k}(d_i + 1) \leq \sum_{i=1}^{k} d_i^*, \quad \text{for all } 1 \leq k \leq f(d). \tag{2.5.1}$$

**Proof**: Suppose that $(d)$ is graphical. Let $\mathcal{G}$ be a graph with degree sequence $(d)$ and let $\mathcal{F}$ be the corresponding Ferrers-Sylvester diagram. For each row $i$ of $\mathcal{F}$, label the boxes in that row with the vertices adjacent to $i$ such that the boxes get labeled with increasing values from left to right. (See Example 2.5.17 following the proof.) Then for each positive integer $k$, the labels 1 through $k$ occur in the first $k$ columns of $\mathcal{F}$. Additionally, since no row contains its own label as an index, any row with $k$ or more boxes must contain at least one label $\ell > k$ in one of the first $k$ boxes. Thus we can conlude that

$$\sum_{i=1}^{k} d_i + k \leq \sum_{i=1}^{k} d_i^* \quad \text{for all } k \leq d_k$$

which implies (2.5.1).

Now suppose (2.5.1) holds and show that $(d)$ is graphical. Suppose equality holds in (2.5.1) for all $1 \leq k \leq f(d)$. Then $d_i + 1 = d_i^*$ for all $i = 1, \ldots, f(d)$. We will create a graph that has such a degree sequence. To do this, start with vertex $v_1$ being adjacent to $d_1$ vertices. Now take a vertex adjacent to $v_1$, call it $v_2$, and add edges between $v_2$ and $d_2 - 1$ vertices that are adjacent to $v_1$. Now take a vertex adjacent to both $v_1$ and $v_2$, call it $v_3$, and add edges between $v_3$ and $d_3 - 2$ vertices that are adjacent to both $v_1$ and $v_2$. Continue this procedure until you take a vertex $v_{f(d)}$ which is adjacent to $v_1, \ldots, v_{f(d)-1}$ and add edges between $v_{f(d)}$ and $d_{f(d)} - f(d) + 1$ vertices that are adjacent to $v_1, \ldots, v_{f(d)-1}$. Because $d_i + 1 = d_i^*$ for each $1 \leq i \leq k$, it follows that this stepwise construction will lead to a unique graph whose degree

## Graph Theory Preliminaries

sequence is $(d)$. (See Example 2.5.18 following the proof.)

Now let $(d)$ be a degree sequence in which equality does not hold in (2.5.1) for all $1 \leq k \leq f(d)$. We will show that such a corresponding degree sequence is majorized by a degree sequence in which equality does hold in (2.5.1) for all $1 \leq k \leq f(d)$. Let $\mathcal{F}_0$ be the Ferrers-Sylvester diagram for $(d)$. Observe that since we don't have equality in (2.5.1) for all $1 \leq k \leq f(d)$, the first $f(d)$ columns of $\mathcal{F}_0$ must contain at least $f(d)$ boxes more than the first $f(d)$ rows of $\mathcal{F}_0$. Thus we can create a Ferrers-Sylvester diagram $\mathcal{F}_1$ from $\mathcal{F}_0$ whose degree sequence $(e)$ is such that $e_i + 1 \leq e_i^*$ for all $1 \leq i \leq f(d) = f(e)$ by successively taking boxes from columns $1, \ldots, i-1$ and adding them to column $i$. Observe that $(e) \succeq (d)$.

We now create $\mathcal{F}_2$ from $\mathcal{F}_1$ with degree sequence $(g)$ by moving exclusively boxes within the first $f(e)$ rows of $\mathcal{F}_1$ such that

$$\sum_{i=1}^{k} g_i = \begin{cases} \min\left\{\sum_{i=1}^{k}(e_i^* - 1), \sum_{i=1}^{f(e)} e_i - f(e)[f(e)-k]\right\}, & \text{for all } 1 \leq k \leq f(e), \\ \sum_{i=1}^{k} e_i, & \text{for all } k > f(e). \end{cases}$$

Observe that $(g) \succeq (e)$ and that $g_i + 1 \leq g_i^*$ for all $1 \leq i \leq f(g) = f(e) = f(d)$.

Observe that the first $f(g)$ columns of $\mathcal{F}_2$ have $f(g) + p$ boxes more than the first $f(g)$ rows of $\mathcal{F}_2$ for some $p \geq 0$. Note that $\mathcal{F}_2$ has $[f(g)]^2 + 2y + f(g) + p$ boxes for some $y \geq 0$. Since $\mathcal{F}_2$ and $\mathcal{F}_0$ have the same number of boxes and since $\mathcal{F}_0$ has an even number of boxes (since $\sum d_i$ is even) it follows that $\mathcal{F}_2$ must have an even number of boxes. Hence $p$ must be even. We will now create $\mathcal{F}_3$ by moving $p/2$ boxes from the first $f(g)$ columns of $\mathcal{F}_2$ to the first $f(g)$ rows of $\mathcal{F}_2$. Since we know $g_i + 1 \leq g_i^*$ for all $1 \leq i \leq f(g)$, let $r$ be the smallest integer such that the inequality is strict. Hence $g_r + 1 < g_r^*$, $g_{r-1} > g_r$ if $r > 1$, and $g_{f(g)}^* > f(g) + 1$. We now remove a box from column $i \leq \ell$ and add it to row $r$ where $\ell$ is the smallest integer such that $\sum_{i=1}^{\ell-1}(g_i - g_i^* + 1) > \sum_{i=1}^{f(g)}(g_i - g_i^* + 1)$. Keep doing this successively until we have equality for all $1 \leq i \leq f(g)$. (The creation of $\mathcal{F}_1, \mathcal{F}_2$, and $\mathcal{F}_3$ are illustrated in Example 2.5.19 following the proof.) Letting $(h)$ be the degree sequence for $\mathcal{F}_3$, observe that $(h) \succeq (g)$ and that equality in (2.5.1) holds for $(h)$ for all $k \leq f(h) = f(g)$. Thus by an earlier paragraph in this proof, $(h)$ is graphical. Also note that $(h) \succeq (d)$. Since both $(h)$ and $(d)$ have the same (even) number of boxes, it follows from Lemma 2.5.15 that $(d)$ is graphical. $\square$

Before continuing, we give some examples that illustrate various stages of the proof of Lemma 2.5.16.

**EXAMPLE 2.5.17** In this example, we illustrate how the boxes in the Ferrers-Sylvester diagram should be labeled in accordance with the vertices that each vertex is adjacent to. Consider the degree sequence $(d) = (5, 4, 4, 3, 3, 2, 2, 1)$ and the graph below whose degree sequence is $(d)$. The vertices are labeled so that vertex $i$ corresponds to the $i^{th}$ entry in $(d)$.

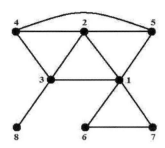

Labelling the Ferrers-Sylvester diagram in accordance with the proof of Lemma 2.5.16 we obtain:

| 1 | 2 | 3 | 5 | 6 | 7 |
|---|---|---|---|---|---|
| 2 | 1 | 3 | 4 | 5 |   |
| 3 | 1 | 2 | 4 | 8 |   |
| 4 | 2 | 3 | 5 |   |   |
| 5 | 1 | 2 | 4 |   |   |
| 6 | 1 | 7 |   |   |   |
| 7 | 1 | 6 |   |   |   |
| 8 | 3 |   |   |   |   |

**EXAMPLE 2.5.18** In this example, we will show how a graph with degree sequence $(d)$ is constructed if equality in (2.5.1) holds for all $1 \leq k \leq f(d)$. Let $(d) = (7, 5, 5, 5, 4, 4, 1)$ and note that $f(d) = 4$. Below gives the sequence of creating such a graph. Vertices $v_1, \ldots, v_4$ are labeled in accordance with the proof of Lemma 2.5.16.

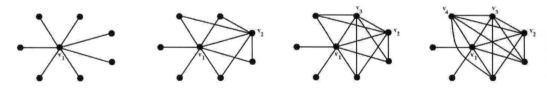

**EXAMPLE 2.5.19** Let $(d) = (6, 6, 4, 2, 2, 2, 1, 1, 1, 1, 1, 1, 1, 1)$ and note that equality in (2.5.1) does not hold for all $1 \leq k \leq f(d)$. Let $\mathcal{F}_0$ be the Ferrers-Sylvester diagram for $(d)$. We illustrate the sequence $\mathcal{F}_1, \mathcal{F}_2, \mathcal{F}_3$ described in the proof of Lemma 2.5.16. Bold lines are placed separating boxes that lie simultaneously in the first $f(d) = 3$ rows and columns.

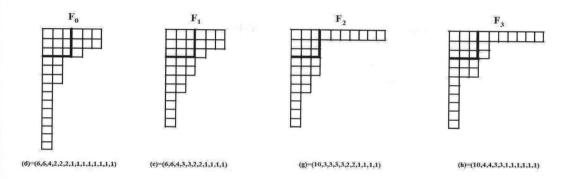

(d)=(6,6,4,2,2,2,1,1,1,1,1,1,1)  (e)=(6,6,4,3,3,2,2,1,1,1,1)  (g)=(10,3,3,3,3,2,2,1,1,1,1)  (h)=(10,4,4,3,3,1,1,1,1,1,1)

Combining Lemmas 2.5.15 and 2.5.16, we see that equality holds in (2.5.1) for all $1 \leq k \leq f(d)$ if and only if $(d)$ is the degree sequence of a maximal graph. This is equivalent to $(d)$ being the degree sequence of a maximal graph if and only if $d_i + 1 = d_i^*$ for all $1 \leq i \leq f(d)$. We state this more formally in the following theorem from [59]:

**THEOREM 2.5.20** *Let $\mathcal{G}$ be a graph with degree sequence $(d)$. Then $\mathcal{G}$ is a maximal graph if and only if $d_i + 1 = d_i^*$ for all integers $1 \leq i \leq f(d)$.*

**EXAMPLE 2.5.21** Consider the graph from Example 2.5.2. Recall that $f(d) = 2$. Thus considering $i = 1, \ldots, f(d)$, we see that $d_1 + 1 = d_1^*$ and $d_2 + 1 = d_2^*$. Since the conditions of Theorem 2.5.20 are satisfied, it follows that $\mathcal{G}$ is a maximal graph.

Theorem 2.5.20 lends itself to the following corollary from [59]:

**COROLLARY 2.5.22** *Let $\mathcal{G}$ be a maximal graph. Then $d_{i+1}(\mathcal{G}) = d_i^*(\mathcal{G})$ for all integers $i > f(\mathcal{G})$.*

**Proof**: Observe that if $\mathcal{G} = K_n$, then this corollary is vacuously true. Suppose $\mathcal{G} \neq K_n$ is a maximal graph with $\mathcal{F}$ as its Ferrers-Sylvester diagram. Let $F_1$ be the subdiagram of $\mathcal{F}$ lying in rows $f(\mathcal{G}) + 2$ through $n$ and columns 1 through $f(d)$. Let $F_2$ be the subdiagram of $\mathcal{F}$ lying in rows 1 through $f(\mathcal{G})$ and columns $f(\mathcal{G}) + 1$ through $n - 1$. The geometric interpretation of Theorem 2.5.20 is that $F_1$ and $F_2$ are transposes of each other. The result now follows. □

Now that we have an efficient way of determining if a graph is a maximal graph, we are now interested in a convenient way to construct such graphs. To do this, we prove a useful lemma from [59].

**LEMMA 2.5.23** *Let $\mathcal{G} \neq K_n$ be a maximal graph on $n$ vertices. Let $\mathcal{H}$ be the subgraph of $\mathcal{G}^c$ that remains after the isolated vertices have been removed. Then $H$ is maximal.*

**Proof**: By Theorem 2.5.20, we need to show that

$$d_i(\mathcal{H}) + 1 = d_i^*(\mathcal{H}) \tag{2.5.2}$$

for $1 \leq i \leq f(\mathcal{H})$. In the proof, we let $F(\mathcal{G})$ and $F(\mathcal{H})$ denote the Ferrers-Sylvester diagrams for $\mathcal{G}$ and $\mathcal{H}$, respectively. Let $R_1$ be the rectangular array of $n$ rows, each consisting of $n-1$ boxes, that is obtained by adding $n-1-d_i(\mathcal{G})$ boxes to row $i$ of $F(\mathcal{G})$, $1 \leq i \leq n$. Then $F(\mathcal{G})$ occupies the upper left corner of $R_1$.

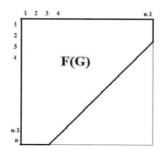

After rotating $R_1$ through 180 degrees, $F(\mathcal{G})$ can be found lying upside down and backwards in the lower right corner. Moreover, $F(\mathcal{H}) = R_1 \setminus F(\mathcal{G})$ which is the diagram that is left after the boxes of $F(\mathcal{G})$ have been removed from the rotated $R_1$.

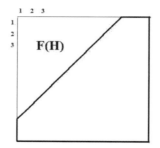

Using the appropriate indices from each drawing, we can observe the following:

$$d_i(\mathcal{H}) = n - 1 - d_{n+1-i}(\mathcal{G}) \qquad (2.5.3)$$

and

$$d_i^*(\mathcal{H}) = n - d_{n-i}^*(\mathcal{G}) \qquad (2.5.4)$$

whenever the left-hand side exists. Because $\mathcal{G}$ is maximal, $d_{f(\mathcal{G})+1} = f(\mathcal{G})$ by definition of $f(\mathcal{G})$. Thus $f(\mathcal{H}) = n - 1 - f(\mathcal{G})$ (observe the lower right-hand block is a square, the lengths of each side being $f(\mathcal{H})$):

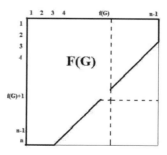

If $i \leq f(\mathcal{H})$, then $f(\mathcal{G}) + 1 = n - f(\mathcal{H}) \leq n - i$. Therefore $n - i > f(\mathcal{G})$ hence by Corollary 2.5.22 we obtain $d^*_{n-i}(\mathcal{G}) = d_{n+1-i}(\mathcal{G})$. Thus by (2.5.3) and (2.5.4), we obtain (2.5.2). □

We now show how all maximal graphs can be constructed. This is done with a technique discovered by Merris (see [59]) which classifies all maximal graphs into disjoint families. This technique is described in the form of a theorem from [59]:

**THEOREM 2.5.24** *Let* $\Gamma_1 = \{K_n : n \geq 1\}$, *i.e., the family of complete graphs. Let* $\Gamma_{j+1} = \{\mathcal{G}^c \vee K_r : \mathcal{G} \in \Gamma_j \text{ and } r \geq 1\}$. *Then* $\Gamma := \bigcup_{j \geq 1} \Gamma_j$ *is equal to the set of all maximal graphs.*

Before proving this lemma, let's use this technique to construct examples of maximal graphs. Since $\Gamma_1$ is merely the set of complete graphs, let's consider, say, $K_3 \in \Gamma_1$. To construct a maximal graph in $\Gamma_2$, we take the complement of any graph in $\Gamma_1$ (we'll use $K_3$) and join it to any complete graph, say $K_1$. Thus the following graph, $K_3^c \vee K_1$, is a maximal graph in $\Gamma_2$:

We can continue and create a graph in $\Gamma_3$. To do this, we take the complement of any graph in $\Gamma_2$, say $K_3^c \vee K_1$ from above, and join it to any complete graph, say $K_2$. Thus the following graph, $(K_3^c \vee K_1)^c \vee K_2$, is a maximal graph in $\Gamma_3$:

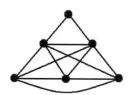

We can continue in this fashion, joining the complement of a graph in $\Gamma_j$ to any complete graph to create a maximal graph in $\Gamma_{j+1}$. Let's now prove Theorem 2.5.24:

**Proof**: We will first show that $\Gamma$ is contained in the set of maximal graphs. We will consider each set $\Gamma_j$ and do this by induction on $j$. When $j = 1$, we have the set $\Gamma_1$ which is the set of all complete graphs. Since a complete graph is a maximal graph, the case $j = 1$ is settled. Now suppose for some positive integer $k$ that if $j \leq k$ then all graphs in $\Gamma_j$ are maximal. Let $\mathcal{H} \in \Gamma_{k+1}$. We need to show that $\mathcal{H}$ is maximal. By Theorem 2.5.20, it suffices to show that

$$d_i(\mathcal{H}) + 1 = d_i^*(\mathcal{H}) \qquad (2.5.5)$$

for all integers $1 \leq i \leq f(\mathcal{H})$. Observe that $\mathcal{H} = \mathcal{G}^c \vee K_r$ for some graph $\mathcal{G} \in \Gamma_k$ on $n$ vertices and for some positive integer $r$. Let $p = n + r$. Then $\mathcal{H}$ has $p$ vertices $1, \ldots, p$ where vertices $1, \ldots, r$ correspond to the vertices in $K_r$. Since the $r$ vertices which correspond to $K_r$ are each adjacent to all other vertices in $\mathcal{H}$, it follows that $d_i(\mathcal{H}) = p - 1$ for $1 \leq i \leq r$. Similarly, since the degree of every vertex in $\mathcal{H}$ is at least $r$, it follows that $d_i^*(\mathcal{H}) = p$ for $1 \leq i \leq r$.

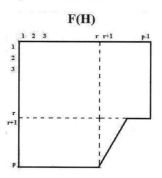

Therefore, (2.5.5) holds for $1 \leq i \leq r$. Hence it remains to establish (2.5.5) for all integers $r + 1 \leq i \leq f(\mathcal{H})$.

To establish this, observe that the Ferrers-Sylvester diagram $F(\mathcal{H})$ can be obtained from $F(\mathcal{G}^c)$ by extending each row of $F(\mathcal{G}^c)$ by $r$ boxes, then adding $r$ rows of $p - 1$ boxes to the top, and finally, for each isolated vertex of $\mathcal{G}^c$, adding a row of $r$ boxes to the bottom:

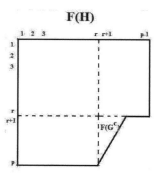

Let $R$ be the rectangular array of $p$ rows, each consisting of $p - 1$ boxes, that is obtained by adding $p - 1 - d_i(\mathcal{H})$ boxes to row $i$ of $F(\mathcal{H})$ for each $1 \leq i \leq p$. Rotating $R$ through 180 degrees, $F(\mathcal{H})$ is in the lower right corner and $F(\mathcal{G}) = R \setminus F(\mathcal{H})$.

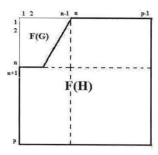

# Graph Theory Preliminaries

From this drawing and the previous drawing, we see

$$d_{r+i}(\mathcal{H}) = p - 1 - d_{n+1-i}(\mathcal{G}), \quad 1 \leq i \leq n,$$

which is equivalent to

$$d_i(\mathcal{H}) = p - 1 - d_{p+1-i}(\mathcal{G}), \quad r+1 \leq i \leq p. \tag{2.5.6}$$

Similarly,

$$d_i^*(\mathcal{H}) = p - d_{p-i}^*(\mathcal{G}), \quad r+1 \leq i \leq p-1. \tag{2.5.7}$$

By the inductive hypothesis, $\mathcal{G}$ is maximal. Using the same reasoning as in Lemma 2.5.23, we have $d_{f(\mathcal{G})+1} = f(\mathcal{G})$. Continuing with this reasoning, we see that $f(\mathcal{H}) = d_{f(\mathcal{H})+1} = p - 1 - f(\mathcal{G})$, and thus $f(\mathcal{G}) + 1 = p - f(\mathcal{H})$. If $i \leq f(\mathcal{H})$, then $f(\mathcal{G}) + 1 \leq p - i$, i.e., $p - i > f(\mathcal{G})$. Hence by Corollary 2.5.22, we get $d_{p-i}^*(\mathcal{G}) = d_{p+1-i}(\mathcal{G})$. We now use (2.5.6) and (2.5.7) to conclude (2.5.5).

We will now show that the set of maximal graphs is contained in $\Gamma$. Suppose this is false. Then there exists a graph $\mathcal{G}$, having minimum number of vertices $n$, such that $\mathcal{G} \notin \Gamma$. Then $\mathcal{G} \neq K_n$ and thus $n \neq 1$. Since $\mathcal{G}$ is maximal, we know that $d_1(\mathcal{G}) = n - 1$. Hence $\mathcal{G}^c$ has $r \geq 1$ isolated vertices. Since $\mathcal{G}$ is not complete, $r < n$. By Lemma 2.5.23, the graph $\mathcal{H}$ obtained from $\mathcal{G}^c$ by deleting the $r$ isolated vertices is maximal. Since $\mathcal{H}$ has $n - r < n$ vertices, it belongs to $\Gamma$ and hence, to $\Gamma_j$ for some $j$. But then $\mathcal{G} = H^c \vee K_r \in \Gamma_{j+1} \subset \Gamma$ which contradicts the fact that $\mathcal{G} \notin \Gamma$. Hence the set of all maximal graphs is contained in $\Gamma$. $\square$

Now that we have an efficient way to construct maximal graphs, we close this section with two theorems from [24] which help us better understand the structure of maximal graphs. The first theorem explains the conditions of when two vertices of equal degree in a maximal graph are adjacent. The second theorem explains conditions in which we can add an edge joining two nonadjacent vertices so that the resulting graph will also be a maximal graph.

**THEOREM 2.5.25** *Let $\mathcal{G}$ be a connected noncomplete maximal graph on $n$ vertices and let $(d)$ be its degree sequence. Then*

*(a) If $d_u = d_v > f(d)$, then $u$ and $v$ are adjacent.*
*(b) If $d_u = d_v = f(d)$, then $u$ and $v$ are adjacent if and only if $d_{f(d)} = f(d)$.*
*(c) If $d_u = d_v < f(d)$, then $u$ and $v$ are not adjacent.*

*Moreover, for Cases (b) and (c) with $u$ and $v$ not adjacent, it follows that $N(u) = N(v)$.*

**Proof**: Let $A$ be the adjacency matrix for $\mathcal{G}$ (see Section 3.1). Applying Theorem 2.5.20, it follows that for each $1 \leq i \leq n$, the $i^{th}$ row of $A$ has $d_i$ entries that are 1 and these entries are left-justified (maintaining zeros on the diagonal, though). Observe that if we partition $A$ into a $2 \times 2$ block matrix where the $(1,1)$ block is $f(d) \times f(d)$ and the $(2,2)$ block is $(n - f(d)) \times (n - f(d))$, then the $(2,2)$ block has all zero entries while the $(1,1)$ block is the matrix $J - I$. Since the $(2,2)$ block is the zero matrix, (c) immediately follows. Since the $(1,1)$ block is $J-I$, (a) immediately follows.

Now suppose $d_u = d_v = f(d)$. If $d_{f(d)} \neq f(d)$ then $d_{f(d)} > f(d)$. Hence $u, v > f(d)$. Then $u$ and $v$ are not adjacent since the $(2,2)$ block of $A$ is the zero matrix. If $d_{f(d)} = f(d)$, then let $K$ be the set of all vertices $x$ in $\mathcal{G}$ such that $d_x = f(d)$ and $x \leq f(d)$. Then

$$A_{x,j} = \begin{cases} 1, & \text{if } x \neq j \text{ and } x \leq f(d)+1, \\ 0, & \text{otherwise,} \end{cases}.$$

By the symmetry of $A$, it follows that all vertices in $\mathcal{G}$ whose degree is $f(d)$ are adjacent if $d_{f(d)} = f(d)$, thus proving (b).

If $d_u = d_v \leq f(d)$ and $u$ and $v$ are not adjacent, then $u, v \geq f(d) + 1$. By the left-justification of the 1 entries of $A$ as described in the first paragraph of this proof, and by the fact that the $(2,2)$ block of $A$ is the zero matrix, it follows that $N(u) = N(v)$. □

**THEOREM 2.5.26** *Let $\mathcal{G}$ be a connected maximal graph and let $(d) = (d_1, \ldots, d_n)$ be the degree sequence of $\mathcal{G}$. Let $u$ and $v$ be nonadjacent vertices of $\mathcal{G}$ with $d_u \geq d_v$. Let $\mathcal{G} + e$ be the graph created from $\mathcal{G}$ by adding an edge between $u$ and $v$. Then $\mathcal{G} + e$ is maximal if and only if $d_u \geq f(d) \geq d_v = d_{d_u+2}$.*

**Proof**: We will prove this by cases:

*Case I*: Suppose $d_u > f(d)$. Then the Ferrers-Sylvester diagram $F(\mathcal{G}+e)$ can be obtained from $F(\mathcal{G})$ by adding one box each to rows $u$ and $v$ of $F(\mathcal{G})$. By Theorem 2.5.20, $\mathcal{G} + e$ is maximal if and only if the box added to row $v$ of $F(\mathcal{G})$ occurs in column $u$. Hence $\mathcal{G} + e$ is maximal if and only if $u = d_v + 1$. In terms of adding a box to row $u$ of $F(\mathcal{G})$, we see that $u = d^*_{d_u+1} + 1$. Therefore $d_v = d^*_{d_u+1}$. Also note that $\mathcal{G} + e$ is maximal if and only if $v = d^*_u + 1 = d_u + 2$, the latter equality following from Theorem 2.5.20. Hence $d_v = d_{d_u+2}$. Finally, $d_u > f(d)$ implies that $u \leq f(d)$ which implies $d_{d^*_u+1} \leq f(d) - 1$ and hence $d_{d_u+2} \leq f(d) - 1$ by Theorem 2.5.20. Therefore $d_v = d^*_{d_u+1} = d_{d_u+2} \leq f(d) - 1 < f(d) < d_u$.

*Case II*: Suppose $d_u = f(d)$. Let $f = f(d)$ and let $f'$ denote the trace of the degree sequence of $\mathcal{G} + e$. Then the following hold:

(a) If $d^*_{f+1} = f$ then $f' = f + 1$;
(b) If $d^*_{f+1} = f - 1$ and $d_v = f$ then $f' = f + 1$;
(c) $f' = f$ otherwise.

Upon observing the diagram $F(\mathcal{G}+e)$, we see that (b) cannot happen if $\mathcal{G}+e$ is maximal. For (a) and (c), by a discussion similar to that of Case I, $\mathcal{G}+e$ is maximal if and only if $d_v = d^*_{f+1} = d_{f+2} \leq f$.

# Graph Theory Preliminaries

*Case III:* Suppose $d_u < f(d)$, and consequently $d_v < f(d)$. Then $F(\mathcal{G} + e)$ would be created from $F(\mathcal{G})$ by adding two boxes to rows below row $f(d)$. By Theorem 2.5.20, $\mathcal{G} + e$ cannot be maximal. □

## Exercises:

**1.** Use Theorem 2.5.4 to determine if the following sequences are graphical. If so, construct a graph whose degree sequence is the given sequence:

(a)  $S: (5, 4, 3, 3, 3, 3, 3, 2, 2)$

(b)  $S: (6, 6, 6, 6, 3, 3, 2)$

(c)  $S: (7, 7, 6, 5, 5, 3, 2, 1)$

(d)  $S: (5, 5, 4, 4, 3, 2, 1)$

**2.** Use Theorem 2.5.20 to verify that the following graphs are maximal graphs.

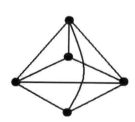
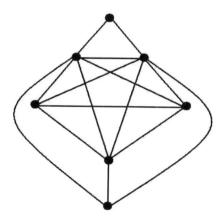

**3.** Using Theorem 2.5.26, between which pairs of nonadjacent vertices in the graphs of the previous exercise will adding an edge result in a maximal graph?

## 2.6 Planar Graphs and Graphs of Higher Genus

We begin this section immediately with a definition:

**DEFINITION 2.6.1** A graph is *embedded* on a surface if it is drawn on a surface so that no two of its edges cross.

In this section, we explore which surfaces a given graph can be embedded on. This gives rise to a second definition:

**DEFINITION 2.6.2** A graph is of *genus* $k$ if it can be embedded on a surface $S_k$ of genus $k$ but cannot be embedded on a surface $S_{k-1}$ of genus $k-1$. In such a case we write $\gamma(\mathcal{G}) = k$.

Observe that a graph can be embedded on a plane if and only if it can be embedded on a sphere. Thus such graphs have genus zero. Due to the importance of such graphs, we present a formal definition regarding them:

**DEFINITION 2.6.3** A graph is *planar* if it can be embedded on a plane (equivalently a sphere). A planar graph has genus zero.

We will spend much of this section discussing planar graphs and will then generalize these results to graphs of higher genus. Before continuing, we give an example of a planar graph:

**EXAMPLE 2.6.4** Consider the graph $\mathcal{G}$ on the left with its vertices labeled. Observe that $\mathcal{G}$ is drawn with edges crossing. However, $\mathcal{G}$ is isomorphic to the graph on the right which is drawn without edges crossing. Since $\mathcal{G}$ can be drawn without any two of its edges crossing, it is planar.

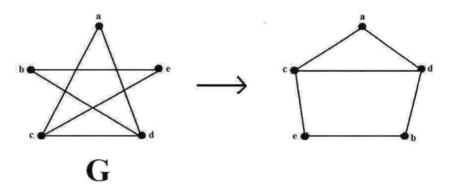

Therefore, if a graph can be drawn so that no two of its edges cross, it is considered planar even if it is being drawn with edges crossing. When a planar graph is drawn with no two edges crossing, it is referred to as a *plane graph* and the graph divides the plane into *regions*. In Example 2.6.4, the planar graph divides the plane into three regions: the region bounded by edges $ac$, $cd$, and $da$; the region bounded by edges $cd$, $db$, $be$, and $ec$; and the *exterior region* bounded by edges $ad$, $db$, $be$, $ec$, and $ca$. The number of regions that a planar graph divides the plane into is directly related to the number of edges and vertices in the graph. We now prove this relation, known as Euler's Formula, in the following theorem whose proof is taken from [11].

**THEOREM 2.6.5** *Euler's Formula: Let $\mathcal{G}$ be a connected plane graph with $n$ vertices, $m$ edges, and $r$ regions. Then $n - m + r = 2$.*

**Proof**: We will prove this by induction on $m$. If $m = 0$, the $\mathcal{G} = K_1$, so $n = 1$, $m = 0$, and $r = 1$, hence $n - m + r = 2$. Thus the result is true for $m = 0$. Assume now that the result holds for all connected plane graphs on $k-1$ edges where $k \geq 1$, and let $\mathcal{G}$ be a connected plane graph with $k$ edges. Suppose the $\mathcal{G}$ has $n$ vertices and $r$ regions. We will show that $n - k + r = 2$.

If $\mathcal{G}$ is a tree, then by Theorem 2.3.7, $n = k + 1$. Since $r = 1$ in this case, we have
$$n - k + r = (k+1) - k + 1 = 2,$$
which is our desired result. If $\mathcal{G}$ is not a tree, then some edge $e$ is on a cycle. Hence $\mathcal{G} - e$ is a connected plane graph on $n$ vertices, $k - 1$ edges, and $r - 1$ regions. By the inductive hypothesis,
$$n - (k-1) + (r-1) = 2$$
which is equivalent to $n - k + r = 2$, proving the theorem. $\square$

**EXAMPLE 2.6.6** In Example 2.6.4, note that $n = 5$, $m = 6$ and $r = 3$, which is consistent with Theorem 2.6.5.

Theorem 2.6.5 has many useful consequences which we will now prove. The first consequence basically tells us something rather intuitive: if we are able to draw a graph without any of its edges crossing, then we cannot have too many edges. The following corollary, whose proof is taken from [12] gives us an upper bound on the number of edges in a planar graph.

**COROLLARY 2.6.7** *If $\mathcal{G}$ is a planar graph on $n \geq 3$ vertices and $m$ edges, then $m \leq 3n - 6$.*

**Proof**: First suppose $\mathcal{G}$ is connected. Observe that the only connected planar graph on $n \geq 3$ vertices and $m \leq 2$ edges is $P_3$ and the inequality holds for $P_3$. So we will now assume that $\mathcal{G}$ has $m \geq 3$ edges. Drawing $\mathcal{G}$ as a plane graph, denote the regions by $R_1, \ldots, R_r$. Let $m_i$ denote the number of edges on the boundary of $R_i$ for each $1 \leq i \leq r$. Observe that $m_i \geq 3$, and therefore
$$M := \sum_{i=1}^{r} m_i \geq 3r.$$
Since each edge is on the boundary of at most two regions, it follows that $M \leq 2m$, and therefore $3r \leq M \leq 2m$. Hence $3r \leq 2m$. Applying Theorem 2.6.5 we have
$$6 = 3n - 3m + 3r \leq 3n - 3m + 2m = 3n - m.$$
Thus $m \leq 3n - 6$.

If $\mathcal{G}$ is disconnected, then edges can be added to $\mathcal{G}$ to produce a connected plane graph on $n$ vertices and $m' > m$ edges. From the preceeding paragraph we have $m' \leq 3n - 6$, and hence $m < 3n - 6$. $\square$

An immediate corollary gives a sufficient condition for a graph to be nonplanar:

**COROLLARY 2.6.8** *If $\mathcal{G}$ is a graph on $n \geq 3$ vertices with $m > 3n - 6$ edges, then $\mathcal{G}$ is nonplanar.*

We should note that Corollary 2.6.8 gives only a sufficient condition for a graph to be nonplanar; it does not give a necessary condition. By the same token, Corollary 2.6.7 gives only a necessary condition for $\mathcal{G}$ to be planar, not a sufficient condition. To understand this, consider the graph $K_{3,3}$ which has $n = 6$ vertices and $m = 9$ edges. Yet we will soon see from Theorem 2.6.15 that $K_{3,3}$ is nonplanar despite the fact that $m \leq 3n - 6$.

Another immediate consequence of Corollary 2.6.7 is the following whose proof is adapted from [12]:

**COROLLARY 2.6.9** *If $\mathcal{G}$ is a planar graph, then the minimum degree $\delta \leq 5$.*

**Proof**: Suppose that $\mathcal{G}$ is a graph in which $\delta \geq 6$. Let $\mathcal{G}$ have $n$ vertices and $m$ edges. Then by Theorem 2.1.4 we have

$$2m = \sum_{v \in \mathcal{G}} \deg v \geq 6n.$$

Hence $m \geq 3n > 3n - 6$. So by Corollary 2.6.8, $\mathcal{G}$ is nonplanar. □

Suppose $\mathcal{G}$ is a planar graph and that equality holds in Corollary 2.6.7, i.e., $m = 3n - 6$. Then if an edge is added to $\mathcal{G}$, the graph will no longer be planar. This gives rise to our next definition:

**DEFINITION 2.6.10** A graph is *maximal planar* if it is planar and the addition of an edge joining any pair of nonadjacent vertices results in a nonplanar graph.

**EXAMPLE 2.6.11** Observe the graph below is maximal planar. It is planar, yet the addition and an edge joining nonadjacent vertices will render the graph nonplanar.

**OBSERVATION 2.6.12** If $\mathcal{G}$ is a maximal planar graph drawn as a plane graph, then every region is bounded by three edges. Further, each vertex $v$ is the center of a wheel subgraph induced by $v$ and all vertices adjacent to $v$. Thus $\delta(\mathcal{G}) \geq 3$ for maximal graphs.

While maximal graphs fit into our theme of planar graphs not being allowed to have too many edges, it will be helpful to be able to easily determine precisely which graphs are planar. To this end, we require a definition:

**DEFINITION 2.6.13** A *subdivision* of a graph $\mathcal{G}$ is a graph obtained by inserting vertices into the edges of $\mathcal{G}$.

**EXAMPLE 2.6.14** Below on the left we have a graph $\mathcal{G}$. The graph on the right is a subdivision of $\mathcal{G}$.

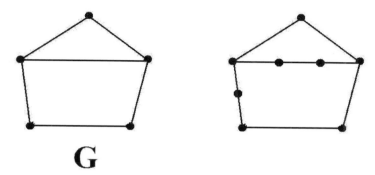

We now present Kuratowski's Theorem which characterizes all planar graphs. Because the proof is rather lengthy, we omit it but refer the reader to [53].

**THEOREM 2.6.15 Kuratowski's Theorem**: *A graph $\mathcal{G}$ is planar if and only if $\mathcal{G}$ does not contain $K_5$, $K_{3,3}$, or any subdivision of $K_5$ or $K_{3,3}$ as a subgraph.*

**EXAMPLE 2.6.16** Consider the following graph $\mathcal{G}$:

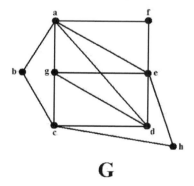

Observe that $\mathcal{G}$ is isomorphic to the graph $\mathcal{H}$ below. The vertices $a$, $b$, $c$, $d$, $e$, $g$, $h$ induce a subgraph that is a subdivision of $K_5$. Thus by Kuratowski's Theorem, $\mathcal{G}$ is nonplanar.

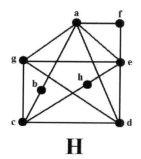

Using Kuratowski's Theorem, not only can we determine if a given graph is planar, but we can also determine if it is nonplanar. If a graph is nonplanar, then its genus is at least one. While it is sometimes difficult to determine the genus of a nonplanar graph, there are some techniques we can employ. For example, we can determine if a nonplanar graph is of genus one by embedding it on a torus, i.e., showing that it can be drawn on a torus so that no two of its edges cross. To see this, we must first show a convenient way to construct a torus. We begin with a rectangle, each of whose sides are oriented (drawings taken from [82]).

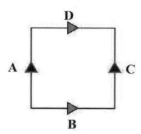

We first identify side $B$ with $D$ by folding the torus as shown below. We then identify side $A$ with $C$ to create the torus.

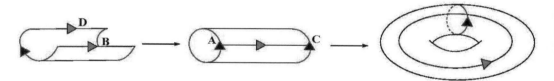

Therefore, we can draw a graph of genus one on a rectangle where the two horizontal sides and two vertical sides are each identified with each other. Thus an edge which passes through the top of the rectangle continues through the bottom of the rectangle and we label the exit and entry points the same to signify that it is the same edge. We do a similar labeling for edges which pass through the left side of the rectangle and continue through the right side. By Kuratowski's Theorem, we know that $K_5$ and $K_{3,3}$ are nonplanar. However, they are of genus one as we will see by the following two examples:

**EXAMPLE 2.6.17** The graph $K_5$ embedded on a torus:

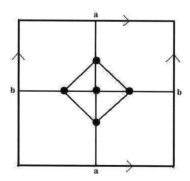

# Graph Theory Preliminaries

**EXAMPLE 2.6.18** The graph $K_{3,3}$ embedded on a torus:

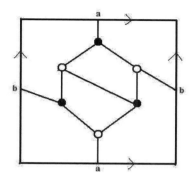

We now generalize the earlier material in this section to graphs of higher genus. To this end, we need a definition:

**DEFINITION 2.6.19** A graph is *2-cell embedded* on a surface if any closed curve on the surface not intersecting the graph can be continuously contracted to a point.

Observe that the above embeddings of $K_5$ and $K_{3,3}$ on the torus are 2-cell embeddings. Below, we have an embedding of $K_4$ on the torus that is not a 2-cell embedding because the curve $C$ cannot be continuously contracted to a point.

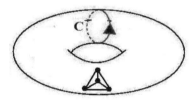

We now generalize Theorem 2.6.5 to obtain Euler's Formula for graphs of higher genus. Our proof is from [12].

**THEOREM 2.6.20 Generalized Euler's Formula**: *Let $\mathcal{G}$ be a connected graph that is 2-cell embedded on the surface $S_k$ for some $k \geq 0$. If $\mathcal{G}$ has $n$ vertices, $m$ edges, and the embedding creates $r$ regions, then $n - m + r = 2 - 2k$. More specifically, if $\mathcal{G}$ of genus $\gamma(\mathcal{G})$ embedded on a surface $S_{\gamma(\mathcal{G})}$, then $n - m + r = 2 - 2\gamma(\mathcal{G})$.*

**Proof**: We will prove this by induction on $k$. Let $\mathcal{G}$ be a graph that is 2-cell embedded on a surface of genus zero. Then more simply put, $\mathcal{G}$ is a planar graph embedded on a plane. By Theorem 2.6.5, $n - m + r = 2$. Hence the theorem holds for $k = 0$.

Assume now that every connected graph $\mathcal{G}'$ having $n'$ vertices, $m'$ edges, and $r'$ regions, when 2-cell embedded on $S_k$, is such that $n' - m' + r' = 2 - 2k$. Let $\mathcal{G}$ be a connected graph on $n$ vertices and $m$ edges that results in $r$ regions when 2-cell embedded on $S_{k+1}$. Thinking of $S_{k+1}$ as a torus with $k+1$ holes and $k+1$ handles, let $H$ be one of the handles. Assume no vertices of $\mathcal{G}$ lie on $H$. Since $\mathcal{G}$

is 2-cell embedded on $S_{k+1}$, there are edges of $\mathcal{G}$ lying on $H$. Moreover, any closed curve $C$ drawn around $H$ must intersect $t \geq 1$ edges of $\mathcal{G}$. Create a graph $\mathcal{G}_1$ by adding $t$ vertices of degree two at all points where $C$ intersects the edges of $\mathcal{G}$, letting the segments of $C$ between these added vertices become edges, and adding two additional vertices of degree two along $C$ to produce two additional edges. Let $\mathcal{G}_1$ have $n_1$ vertices, $m_1$ edges, and $r_1$ regions. Observe that $\mathcal{G}_1$ is 2-cell embedded on $S_{k+1}$ and that $n_1 = n + t + 2$, $m_1 = m + 2t + 2$. Further, since each portion of $C$ that became an edge of $\mathcal{G}_1$ is in a region of $\mathcal{G}$, the addition of each such edge divides that region into two regions. Therefore $r_1 = r + t$.

Next cut the handle $H$ along $C$ and "patch" the two resulting holes, producing two duplicate copies of the vertices and edges along $C$. Denote the resulting graph $\mathcal{G}_2$ which is 2-cell embedded on $S_k$. Let $\mathcal{G}_2$ have $n_2$ vertices, $m_2$ edges, and let the embedding of $\mathcal{G}_2$ on $S_k$ result in $r_2$ regions. Then $n_2 = n_1 + t + 2$, $m_2 = m_1 + t + 2$, and $r_2 = r_1 + 2$. Therefore

$$n_2 = n + 2t + 4, \ m_2 = m + 3t + 4, \text{ and } r_2 = r + t + 2.$$

By the induction hypothesis, $n_2 - m_2 + r_2 = 2 - 2k$ and so

$$(n + 2t + 4) - (m + 3t + 4) + (r + t + 2) = 2 - 2k.$$

Simplifying we obtain $n - m + r + 2 = 2 - 2k$, and hence $n - m + r = 2 - 2(k+1)$. $\square$

As Theorem 2.6.5 implied that a planar graph cannot have too many edges, Theorem 2.6.20 implies that for each $k \geq 1$, a graph of genus $k$ must have a limited number of edges also. Conversely, if a graph $\mathcal{G}$ is of large genus, then $\mathcal{G}$ requires some minimum number of edges to warrant the necessity of a surface of large genus for the embedding of $\mathcal{G}$ upon. Hence Theorem 2.6.20 leads us to the following corollary (see [12]) whose proof is left as an exercise.

**COROLLARY 2.6.21** Let $\mathcal{G}$ be a connected graph on $n \geq 3$ vertices and $m$ edges. Then
$$m \leq 3n - 6(1 - \gamma(\mathcal{G}))$$
and
$$\gamma(\mathcal{G}) \geq \frac{m}{6} - \frac{n}{2} + 1.$$

Loosely speaking, Corollary 2.6.21 says that graphs of large genus must have a sufficient number of edges. From this idea, it is worth noting the genus of a complete graph which we state formally as a theorem

**THEOREM 2.6.22**
$$\gamma(K_n) = \left\lceil \frac{(n-3)(n-4)}{12} \right\rceil.$$

# Graph Theory Preliminaries

The proof of Theorem 2.6.22 is beyond the scope of this book, but can be found in [72].

## Exercises:

**1.** Use Corollary 2.6.7 to prove the $K_5$ is nonplanar.

**2.** (See [12]) Without using Kuratowski's Theorem, prove that $K_{3,3}$ is nonplanar. *Hint:* Since $K_{3,3}$ is bipartite, each region must be bounded by at least 4 edges. Apply Theorem 2.6.5 to obtain a contradiction.

**3.** Determine if each of the graphs is planar. Note that the first graph is the famous Petersen graph and the third graph is the famous Headwood graph [11].

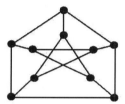  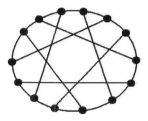

**4.** (See [11]) Prove that a graph is planar if and only if each of its blocks is planar.

**5.** Show that each of the following nonplanar graphs is of genus one by embedding them on a torus.
   (a) $K_6$
   (b) $K_7$
   (c) $K_{4,4}$
   (d) $K_{3,3,3}$.

**5.** Prove Corollary 2.6.21.

# Chapter 3

# Introduction to Laplacian Matrices

Since the focus of this book is on representing graphs as Laplacian matrices, we dedicate this chapter to the foundations of Laplacian matrices. The goal of this chapter is to provide an overview of Laplacian matrices before drilling into more specific and rigorous topics in later chapters. We begin by providing a context of Laplacian matrices in their relationship to other matrix representations of graphs. We then proceed to the Matrix Tree Theorem which is a theorem that first motivated the use of Laplacian matrices in graph theory. Since graphs are discrete objects, we provide further context of Laplacian matrices by investigating the Laplace operator which is the continuous version of the Laplacian matrix. The Laplace operator has applications in energy flow, and we adapt these applications to graphs and learn how to draw graphs which minimize energy. With this in mind, we close this section by applying Laplacian matrices to network flow. Historically, network flow has been a major motivation to studying Laplacian matrices.

## 3.1 Matrix Representations of Graphs

In graph theory, it is often desirable to represent graphs in terms of matrices. Not only is it convenient, but often the properties of the matrix give us useful information about the graph that the matrix is associated with. While this book focuses on the Laplacian matrix, we will study two additional types of matrix representations of graphs in this section: adjacency matrices and incidence matrices. Both types of matrices are useful, not only in their own right, but also in the theory of Laplacian matrices. We begin with the adjacency matrix.

**DEFINITION 3.1.1** Let $\mathcal{G}$ be a graph on $n$ vertices labelled $1, \ldots, n$. The *adjacency matrix* of $\mathcal{G}$ on $n$ vertices is the $n \times n$ matrix $A = [a_{ij}]$ where

$$a_{i,j} = \begin{cases} 1, & \text{if } i \neq j \text{ and } i \text{ and } j \text{ are adjacent,} \\ 0, & \text{if } i = j, \text{ or } i \neq j \text{ and } i \text{ is not adjacent to } j. \end{cases}$$

**EXAMPLE 3.1.2** Consider the following graph $\mathcal{G}$:

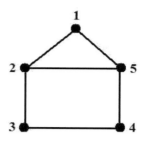

Its adjacency matrix is

$$A = \begin{bmatrix} 0 & 1 & 0 & 0 & 1 \\ 1 & 0 & 1 & 0 & 1 \\ 0 & 1 & 0 & 1 & 0 \\ 0 & 0 & 1 & 0 & 1 \\ 1 & 1 & 0 & 1 & 0 \end{bmatrix}.$$

Before we continue, we make several observations. First, since we are not allowing loops in our graphs, the diagonal entries of the adjacency matrix will always be zero. As for the off-diagonal entries, since $a_{ij} = 1$ if and only if $a_{ji} = 1$, i.e., if and only if vertices $i$ and $j$ are adjacent, it follows that the adjacency matrix is symmetric. We also note the number of 1's in each row and each column is the degree of the corresponding vertex.

The powers of the adjacency matrix give us important information as to the structure of the graph. We present this result as a theorem whose proof is adapted from [12]:

**THEOREM 3.1.3** *Let $\mathcal{G}$ be a graph with vertices $v_1, \ldots, v_n$ and adjacency matrix $A = [a_{ij}]$. Then the entry $a_{ij}^{(k)}$ in row $i$ and column $j$ of $A^k$ is the number of distinct $v_i - v_j$ walks of length $k$ in $\mathcal{G}$.*

**Proof**: We prove this by induction on $k$. For $k = 1$, we have the matrix $A^1 = A$. An entry of $a_{ij}^{(1)}$ is 1 if $v_i$ and $v_j$ are adjacent, i.e., there is a $v_i - v_j$ walk of length one in $\mathcal{G}$. Likewise, an entry of $a_{ij}^{(1)}$ is 0 if $v_i$ and $v_j$ are not adjacent, i.e., there are no $v_i - v_j$ walks of length one in $\mathcal{G}$. This proves the base step.

For the inductive hypothesis, assume that for a positive integer $k$, the number of $v_i - v_j$ walks of length $k$ in $\mathcal{G}$ is $a_{ij}^{(k)}$. By matrix multiplication, the entry $a_{ij}^{(k+1)}$ of $A^{k+1}$ is the dot product of row $i$ of $A^k$ and column $j$ of $A$. In other words

$$a_{ij}^{(k+1)} = \sum_{s=1}^{n} a_{is}^{(k)} a_{sj} = a_{i1}^{(k)} a_{1j} + a_{i2}^{(k)} a_{2j} + \ldots a_{in}^{(k)} a_{nj}. \qquad (3.1.1)$$

Every $v_i - v_j$ walk $W$ of length $k+1$ in $\mathcal{G}$ is produced by beginning with a $v_i - v_s$ walk $W'$ of length $k$ in $\mathcal{G}$ for some vertex $v_s$ adjacent to $v_j$, then following $W'$ by $v_j$. By the inductive hypothesis, the number of $v_i - v_j$ walks of length $k$ is $a_{ij}^{(k)}$, while

# Introduction to Laplacian Matrices

$a_{sj} = 1$ if and only if $v_s$ is adjacent to $v_j$. Hence by (3.1.1), $a_{ij}^{(k+1)}$ represents the number of $v_i - v_j$ walks of length $k+1$ in $\mathcal{G}$. □

**EXAMPLE 3.1.4** Consider the graph from Example 3.1.2. Observe that the second and third powers of $A$ are

$$A^2 = \begin{bmatrix} 2 & 1 & 1 & 1 & 1 \\ 1 & 3 & 0 & 2 & 1 \\ 1 & 0 & 2 & 0 & 2 \\ 1 & 2 & 0 & 2 & 0 \\ 1 & 1 & 2 & 0 & 3 \end{bmatrix} \text{ and } A^3 = \begin{bmatrix} 2 & 4 & 2 & 2 & 4 \\ 4 & 2 & 5 & 1 & 6 \\ 2 & 5 & 0 & 4 & 1 \\ 2 & 1 & 4 & 0 & 5 \\ 4 & 6 & 1 & 5 & 2 \end{bmatrix}.$$

Thus, for example, since $a_{2,4}^{(2)} = 2$, there are two $v_2 - v_4$ walks of length two, namely $v_2, v_3, v_4$ and $v_2, v_5, v_4$. Also, for example, since $a_{3,2}^{(3)} = 5$, there are five $v_3 - v_2$ walks of length three, namely $v_3, v_4, v_5, v_2$; $v_3, v_2, v_3, v_2$; $v_3, v_2, v_5, v_2$; $v_3, v_4, v_3, v_2$; and $v_3, v_2, v_1, v_2$.

Looking more closely at the powers of the adjacency matrix, we notice that the diagonal entries of $A^2$ are precisely the degrees of each vertex. This is because the walks of length two from a vertex $v_i$ to itself are the walks $v_i, v_j, v_i$ where $v_j$ is a vertex adjacent to $v_i$. Thus the number of vertices $v_j$ where such a walk is possible is precisely the number of vertices adjacent to $v_i$, i.e., the degree of vertex $i$. Similarly, a walk of length three from a vertex $v_i$ is a cycle of length three. Thus for each $i = 1, \ldots, n$, the $i^{th}$ diagonal entry of $A^3$ are the number of cycles of length three beginning and ending at $v_i$.

Another type of matrix that can be used to represent a graph is the incidence matrix. Unlike the adjacency matrix where both the rows and columns represent the vertices of a graph, in the incidence matrix the rows represent the vertices while the columns represent the edges. Hence the incidence matrix is based on which edges and vertices are incident to each other. Each edge $e$ in a graph is incident to two vertices, say $v_a$ and $v_b$. Note that $v_a$ and $v_b$ can be chosen arbitrarily, but it is traditional to regard the vertex with the lower-numbered label as $v_a$. For each vertex-edge pair $(v, e)$ in which the vertex $v$ is incident to the edge $e$, define a function $p(v, e)$ where $p(v_a, e) = 1$ and $p(v_b, e) = -1$. We now present a formal definition of the incidence matrix.

**DEFINITION 3.1.5** Let $\mathcal{G}$ be a graph on $n$ vertices labeled $v_1, \ldots, v_n$, and $m$ edges labeled $e_1, \ldots, e_m$. The *incidence matrix* $F$ is the $n \times m$ matrix such that

$$f_{i,j} = \begin{cases} p(v_i, e_j), & \text{if } v_i \text{ and } e_j \text{ are incident,} \\ 0, & \text{if } v_i \text{ and } e_j \text{ are not incident,} \end{cases} \quad (3.1.2)$$

**EXAMPLE 3.1.6** Revisiting the graph in Example 3.1.2, we show this graph again with the same vertex labellings $v_1, \ldots, v_5$, but now with the edges labeled $e_1, \ldots e_6$:

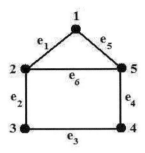

The incidence matrix $F$ is

$$F = \begin{bmatrix} 1 & 0 & 0 & 0 & 1 & 0 \\ -1 & 1 & 0 & 0 & 0 & 1 \\ 0 & -1 & 1 & 0 & 0 & 0 \\ 0 & 0 & -1 & 1 & 0 & 0 \\ 0 & 0 & 0 & -1 & -1 & -1 \end{bmatrix}.$$

**OBSERVATION 3.1.7** Each column $j$ has exactly two non-zero entries. These entries are in the rows corresponding to the two vertices that are incident to $e_j$. The row corresponding to the vertex with the lower-numbered label has an entry of 1, while the row corresponding to the vertex with the larger-numbered label has an entry of $-1$.

Given a graph $\mathcal{G}$, since both the incidence matrix and the adjacency matrix are based on the vertex-edge structure of the graph, it is plausible that these matrices are related. Their relationship is decribed in the following theorem:

**THEOREM 3.1.8** Let $\mathcal{G}$ be a graph on $n$ vertices and $m$ edges, and let $F$ be its incidence matrix. Let $A$ be the adjacency matrix for $\mathcal{L}(\mathcal{G})$, the line graph of $\mathcal{G}$. Then $A = |F^T F - 2I|$.

**Proof**: Observe that for $i, j = 1, \ldots, m$ that the $(i, j)$ entry of $F^T F$ is dot product of the $i^{th}$ column of $F$ with the $j^{th}$ column of $F$. Hence by Observation 3.1.7, it follows that the diagonal entries of $F^T F$ are each 2 and hence the diagonal entries of $F^T F - 2I$ are each zero. For the off-diagonal entries of $F^T F$ (and hence $F^T F - 2I$), the $(i, j)$ entry is zero if and only if edges $e_i$ and $e_j$ are not incident with a common vertex. Otherwise, the $(i, j)$ entry of $F^T F$ is either 1 or $-1$. The conclusion follows. □

We will see in the next section that incidence matrices will be very useful in proving a result that determines the number of spanning trees that a graph has. However, while incidence matrices are used in the proof of this result, the number of spanning trees contained in a graph is actually directly related to the Laplacian matrix which is the main focus of this book:

# Introduction to Laplacian Matrices

**DEFINITION 3.1.9** Let $\mathcal{G}$ be a graph on $n$ vertices labelled $1, \ldots, n$. The *Laplacian matrix* of $\mathcal{G}$ is the $n \times n$ matrix $L = [\ell_{i,j}]$ where

$$\ell_{i,j} = \begin{cases} -1, & \text{if } i \neq j \text{ and } i \text{ and } j \text{ are adjacent,} \\ 0, & \text{if } i \neq j \text{ and } i \text{ is not adjacent to } j, \\ d_i, & \text{if } i = j. \end{cases}$$

**EXAMPLE 3.1.10** Revisiting the graph $\mathcal{G}$ in Example 3.1.2, the Laplacian matrix for this graph is:

$$L = \begin{bmatrix} 2 & -1 & 0 & 0 & -1 \\ -1 & 3 & -1 & 0 & -1 \\ 0 & -1 & 2 & -1 & 0 \\ 0 & 0 & -1 & 2 & -1 \\ -1 & -1 & 0 & -1 & 3 \end{bmatrix}.$$

At this point, we can notice that the Laplacian matrix is related to the other two matrix representations that we have discussed so far. Most obviously, if $D$ is the diagonal matrix consisting of the degrees of the vertices of a graph $\mathcal{G}$ and $A$ is the adjacency matrix for $\mathcal{G}$, then the Laplacian matrix $L$ of $\mathcal{G}$ is $L = D - A$. The Laplacian matrix is also related to the incidence matrix which we state more formally in the following observation:

**OBSERVATION 3.1.11** Since the number of nonzero entries in row $i$ of the incidence matrix $F$ is the degree of vertex $i$, it follows that taking the dot product of row $i$ with itself will yield the degree of vertex $i$. Moreover, for $i \neq j$, taking the dot product of row $i$ and row $j$ will yield $-1$ if vertices $i$ and $j$ are adjacent and zero otherwise. Hence $L = FF^T$.

In addition to being useful in determining the number of spanning trees of a graph (which we discuss in greater detail in the next section), Laplacian matrices have interesting spectral properties. First, since Laplacian matrices are symmetric, all of the eigenvalues are real. Moreover, since the diagonal entry of each row is equal to the sum of the absolute values of the off-diagonal entries, it follows from Theorem 1.2.1 that all eigenvalues are nonnegative real numbers. Also, since the row sums of the Laplacian matrix are each zero, it follows that zero is always the smallest eigenvalue since the vector of all ones will be a corresponding eigenvector. We will see in later chapters that the second smallest eigenvalue will be another measure of the connectivity of the graph. We will also investigate in later chapters many interesting properties of the other eigenvalues of the Laplacian matrix. For now, we can relate the spectral properties of Laplacian matrices to that of incidence matrices. We do so in the following theorem whose proof we leave as an exercise.

**THEOREM 3.1.12** Let $L$ be the Laplacian matrix for a graph $\mathcal{G}$ and $F$ be its incidence matrix. Then $L = FF^T$ and $F^T F$ have the same nonzero eigenvalues.

We close this section by noting that the adjacency matrix, incidence matrix, and Laplacian matrix can each be generalized to weighted graphs. Generalizing the adjacency and Laplacian matrices to weighted graphs is straightforward in that we simply replace the off-diagonal entries with $w$ and $-w$, respectively, where $w$ denotes the weight of the corresponding edge. Further, in the Laplacian matrix of a weighted graph, each diagonal entry is the sum of the weighted of the edges incident to the corresponding vertex. Observe that with this definition of the Laplacian matrix for weighted graphs, the spectral properties discussed earlier are preserved. Generalizing the incidence matrix is less straightforward. To generalize the incidence matrix to weighted graphs, alter the definition of $p(v_i, e_j)$ by defining $p(v_a, e_j) = \sqrt{w}$ and $p(v_b, e_j) = -\sqrt{w}$ where $v_a$ and $v_b$ are as before and where $w$ is the weight of the edge incident to $v_a$ and $v_b$. Then define the incidence matrix $F$ using (3.1.2). By defining $F$ this way, Observation 3.1.11 still holds for weighted graphs.

**EXAMPLE 3.1.13** Consider the weighted graph:

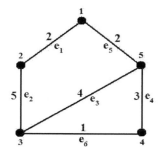

The adjacency matrix $A$, incidence matrix $F$, and Laplacian matrix $L$ are as follows:

$$A = \begin{bmatrix} 0 & 2 & 0 & 0 & 2 \\ 2 & 0 & 5 & 0 & 0 \\ 0 & 5 & 0 & 1 & 4 \\ 0 & 0 & 1 & 0 & 3 \\ 2 & 0 & 4 & 3 & 0 \end{bmatrix}, \quad F = \begin{bmatrix} \sqrt{2} & 0 & 0 & 0 & \sqrt{2} & 0 \\ -\sqrt{2} & \sqrt{5} & 0 & 0 & 0 & 0 \\ 0 & -\sqrt{5} & \sqrt{1} & 0 & 0 & \sqrt{4} \\ 0 & 0 & -\sqrt{1} & \sqrt{3} & 0 & 0 \\ 0 & 0 & 0 & -\sqrt{3} & -\sqrt{2} & -\sqrt{4} \end{bmatrix},$$

$$L = \begin{bmatrix} 4 & -2 & 0 & 0 & -2 \\ -2 & 7 & -5 & 0 & 0 \\ 0 & -5 & 10 & -1 & -4 \\ 0 & 0 & -1 & 4 & -3 \\ -2 & 0 & -4 & -3 & 9 \end{bmatrix}.$$

## Exercises:

1. For each of the following graphs, compute $A$, $A^2$, and $A^3$ without performing the matrix multiplications.

# Introduction to Laplacian Matrices

 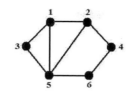 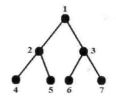 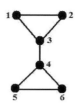

**2.** Find the incidence and Laplacian matrices for the graphs in Exercise 1.

**3.** The $i^{th}$ diagonal entry of $A^3$ represents the number of cycles of length three beginning and ending at $v_i$. Is it true that the $i^{th}$ diagonal entry of $A^4$ represents the number of cycles of length four beginning and ending at $v_i$? Why or why not?

**4.** (See [33]) Recall the trace of a matrix $A$, denoted $Tr(A)$, is the sum of the diagonal entries of $A$. Let $A$ be the adjacency matrix for a graph $\mathcal{G}$ having $m$ edges and $t$ cycles of length three. Prove the following:
(i) $Tr(A) = 0$
(ii) $Tr(A^2) = 2m$
(iii) $Tr(A^3) = 6t$.

**5.** Prove Theorem 3.1.12.

**6.** Find the adjacency, incidence, and Laplacian matrices of the weighted graph below:

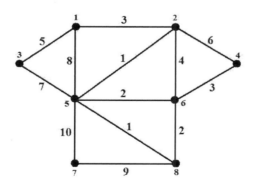

## 3.2 The Matrix Tree Theorem

In the previous section, we stated that the incidence matrix and the Laplacian matrix are useful in determining the number of spanning trees that a graph contains. In this section, we investigate these results more rigorously and prove the famous Matrix Tree Theorem. The Matrix Tree Theorem states that the number of spanning trees of a graph $\mathcal{G}$ on $n$ vertices is equal to the absolute value of the determinant of any $(n-1) \times (n-1)$ submatrix of its Laplacian. Before stating this theorem more formally and proving it, we need to introduce the Cauchy-Binet Theorem. To this end, let $I_m$ denote the set of natural numbers $\{1, 2, \ldots, m\}$. Suppose $S \subset I_m$ where

$|S| = n \leq m$. For an $n \times m$ matrix $A$, let $A_S$ denote the $n \times n$ matrix created from $A$ by using the columns of $A$ indexed by the elements in $S$. For an $m \times n$ matrix $B$, let $B_S$ denote the $n \times n$ matrix created from $B$ by using the rows of $B$ indexed by the elements in $S$. With this in mind, we now present the Cauchy-Binet Theorem whose proof can be found in [54].

**THEOREM 3.2.1** *Let $A$ be an $n \times m$ matrix and let $B$ be an $m \times n$ matrix where $m \geq n$. Then*
$$\det(AB) = \sum_S \det(A_S) \det(B_S)$$
*where the sum is taken over all subsets $S \subset I_m$ where $|S| = n$.*

First observe that if $n = m$, then this theorem merely reduces to $\det(AB) = \det(A)\det(B)$ since the only subset $S$ of $I_m$ such that $|S| = n$ would be the entire set $I_m$. Secondly, we include the condition $m \geq n$ because if $m < n$, then there would not exist any appropriate subsets $S$ of $I_m$. Thus $\det(AB) = 0$ in this case.

At this point, we are now ready to prove the Matrix Tree Theorem for unweighted graphs (see [7]).

**THEOREM 3.2.2 Matrix Tree Theorem** *Let $\mathcal{G}$ be an unweighted graph on $n$ vertices labeled $1, \ldots, n$ with Laplacian matrix $L$. Then the number of (connected) spanning trees of $\mathcal{G}$ is the absolute value of the determinant of any $(n-1) \times (n-1)$ submatrix of $L$.*

**Proof**: First suppose that $\mathcal{G}$ is disconnected. Then by Theorem 4.1.1 it follows that zero is an eigenvalue of $L$ with multiplicity of at least two. Therefore any $(n-1) \times (n-1)$ submatrix $\hat{L}$ of $L$ will be singular and thus $\det(\hat{L}) = 0$. Moreover, if $\mathcal{G}$ is disconnected, then clearly it cannot have any spanning trees.

Now assume that $\mathcal{G}$ is connected. Then zero is a simple eigenvalue of $L$. Hence the vector $e$ is a basis for the eigenspace corresponding to zero. Let $B = [b_{i,j}]$ be the adjoint of $L$ and thus $b_{i,j} = (-1)^{i+j} \det(L[\overline{j,i}])$ where $L[\overline{j,i}]$ is the $(n-1) \times (n-1)$ matrix obtained from $L$ by deleting row $j$ and column $i$. Since $L$ is an M-matrix of rank $n-1$, it follows from Lemma 1.4.9 that $(-1)^{i+j} \det(L[\overline{j,i}])$ is always positive. Thus the entries of $B$ consist of the absolute values of all of the determinants of the $(n-1) \times (n-1)$ submatrices of $L$. However, by the properties adjoint matrices, $LB = \det(L)I = 0$. Since $LB = 0$, it follows that every column of $B$ is in the eigenspace of $L$ corresponding to zero. Since $\{e\}$ is a basis for the eigenspace of $L$ corresponding to zero, it follows that all of the entries in any column of $B$ are identical. Similarly by the properties of adjoint matrices, $BL = \det(L)I = 0$. Since $L$ is symmetric, it follows that every row of $B$ is in the eigenspace of $L$ corresponding to zero, and hence all of the entries in any row of $B$ are identical. Thus all entries of $B$ are identical.

It now remains to show that this common entry of $B$ is the number of spanning trees of $\mathcal{G}$ (in absolute value). Since all entries of $B$ are identical, without loss of generality, we will consider the matrix $\hat{L}$ formed by deleting the last row and last column of $L$, i.e., $\hat{L} = L[\overline{n,n}]$. Let $F$ be the incidence matrix of $\mathcal{G}$ and let $\hat{F}$ be the matrix obtained from $F$ by deleting its last row. By Observation 3.1.11 we have $FF^T = L$ and $\hat{F}\hat{F}^T = \hat{L}$. By the Cauchy-Binet Theorem

$$\det(\hat{L}) = \det(\hat{F}\hat{F}^T) = \sum_S \det(\hat{F}_S)\det(\hat{F}_S^T) \qquad (3.2.1)$$

where the sum is taken over all subsets $S \subset I_m$ such that $|S| = n - 1$. However, since the determinant of any matrix is equal to the determinant of its transpose, we can simplify (3.2.1) to

$$\det(\hat{L}) = \sum_S \det(\hat{F}_S)^2 \qquad (3.2.2)$$

It now remains to show that for a given subset $S_k \in I_m$ where $|S_k| = n - 1$, we have $|\det(\hat{F}_{S_k})| = 1$ if the edges corresponding to the columns of $\hat{F}$ represented by $S_k$ form a spanning tree, and $\det(\hat{F}_{S_k}) = 0$ otherwise. First suppose that the set of edges $E$ in the columns corresponding to $\hat{F}_{S_k}$ do not form a spanning tree. Then there exists at least one vertex in $\mathcal{G}$ that is not in the subgraph induced by the edges in $E$. Let $v$ be such a vertex. Recalling that $\hat{L} = L[\overline{n,n}]$, there are two cases to consider: (i) if $v \neq v_n$ and (ii) if $v = v_n$. If $v \neq v_n$, then the row of $\hat{F}_{S_k}$ corresponding to $v$ will contain all zeros. Thus $\det(\hat{F}_{S_k}) = 0$. If $v = v_n$, then observe that for each edge $e \in E$, both vertices incident to $e$ are repesented by rows in $\hat{F}_{S_k}$. Thus each column has a zero sum since one entry in each column is 1, another is $-1$, and the rest are zeros (the 1 and $-1$ corresponding to the vertices incident to $e$). Hence $\hat{F}_{S_k}$ is singular and therefore $\det(\hat{F}_{S_k}) = 0$.

Now suppose that the edges in $E$ corresponding to the columns of $\hat{F}_{S_k}$ do form a spanning tree of $\mathcal{G}$. We will now show that $|\det(\hat{F}_{S_k})| = 1$. We will prove this by induction on $n$. (Recall that a spanning tree will have $n - 1$ edges.) If $n = 2$, then $\mathcal{G} = K_2$, and thus $|\det(\hat{F}_{S_k})| = 1$. For our inductive hypothesis, assume that if $\mathcal{G}$ has $n = t$ vertices and the edges in $E$ corresponding to the columns of $\hat{F}_{S_k}$ form a spanning tree of $\mathcal{G}$, then $|\det(\hat{F}_{S_k})| = 1$. Now suppose that $\mathcal{G}$ has $n = t + 1$ vertices and the edges in $E$ corresponding to the columns of $\hat{F}_{S_k}$ form a spanning tree $\mathcal{T}$ of $\mathcal{G}$. Since every tree has at least two pendant vertices, it follows that $\mathcal{T}$ has at least one pendant vertex $v \neq v_n$. Then the row in $\hat{F}_{S_k}$ corresponding to $v$ has only one nonzero entry ($\pm 1$) and that entry will be in the column of the unique edge $e$ incident to $v$ in $\mathcal{T}$. Thus $\det(\hat{F}_{S_k})$ will equal $\pm 1$ times the determinant of the submatrix $\hat{F}_{S_k}^v$ of $\hat{F}_{S_k}$ formed by eliminating the row corresponding to $v$ and the column corresponding to $e$. However, the edges corresponding to the columns in $\hat{F}_{S_k}^v$ form a spanning tree of $\mathcal{G} - v$. Since $\mathcal{G} - v$ has $t$ vertices, by the inductive hypothesis, $|\det(\hat{F}_{S_k}^v)| = 1$. Since $\det(\hat{F}_{S_k}) = \pm 1 \det(\hat{F}_{S_k}^v)$, it follows that $|\det(\hat{F}_{S_k})| = 1$. $\square$

**EXAMPLE 3.2.3** Consider the graph $\mathcal{G}$ below.

# 100 Applications of Combinatorial Matrix Theory to Laplacian Matrices of Graphs

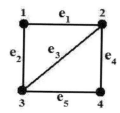

Its Laplacian matrix $L$ is:

$$L = \begin{bmatrix} 2 & -1 & -1 & 0 \\ -1 & 3 & -1 & -1 \\ -1 & -1 & 3 & -1 \\ 0 & -1 & -1 & 2 \end{bmatrix}$$

Observe that if any row and any column of $L$ is deleted, the determinant of the remaining matrix (in absolute value) is 8. For example, if we delete row 2 and column 3 of $L$ we obtain:

$$L[2,3] = \begin{bmatrix} 2 & -1 & 0 \\ -1 & -1 & -1 \\ 0 & -1 & 2 \end{bmatrix}$$

It can easily be computed that $\det(L[2,3]) = -8$ in accordance with the Matrix Tree Theorem. Thus $\mathcal{G}$ has 8 spanning trees. The 8 spanning trees are as follows:

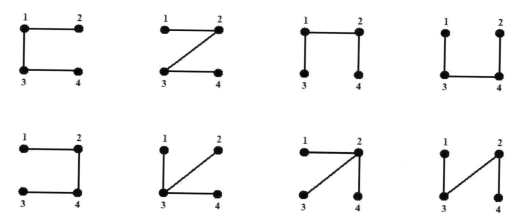

To better understand the proof of the Matrix Tree Theorem, observe that the incidence matrix $F$ for $\mathcal{G}$ is

$$F = \begin{bmatrix} 1 & 1 & 0 & 0 & 0 \\ -1 & 0 & 1 & 1 & 0 \\ 0 & -1 & -1 & 0 & 1 \\ 0 & 0 & 0 & -1 & -1 \end{bmatrix}.$$

# Introduction to Laplacian Matrices

In the proof, we created $\hat{F}$ by deleting the last row, i.e., the row corresponding to $v_4$ to obtain:

$$\hat{F} = \begin{bmatrix} 1 & 1 & 0 & 0 & 0 \\ -1 & 0 & 1 & 1 & 0 \\ 0 & -1 & -1 & 0 & 1 \end{bmatrix}.$$

Since $\mathcal{G}$ has four vertices, it follows that any spanning tree of $\mathcal{G}$ would necessarily have exactly three edges. Therefore, we consider the $3 \times 3$ submatrices $\hat{F}_{S_k}$ of $\hat{F}$ where $S_k$ denotes the columns of $\hat{F}$ (or equivalently, the edges of $\mathcal{G}$) that we use to form the submatrix. If the edges represented by $S_k$ do not form a spanning tree of $\mathcal{G}$, then $\det(\hat{F}_{S_k}) = 0$ because either one of two things holds: (i) $\hat{F}_{S_k}$ will contain a row of all zeros or (ii) all of the column sums of $\hat{F}_{S_k}$ are zero. In case (i), the induced subgraph of $\mathcal{G}$ formed by the edges represented by $S_k$ is missing at least one of the vertices $v_i \neq v_4$ of $\mathcal{G}$; hence it is not a spanning tree. For example, let $S_k = \{3, 4, 5\}$. Then row 1 of $\hat{F}_{3,4,5}$ is a row of all zeros because the subgraph of $\mathcal{G}$ induced by edges $e_3$, $e_4$, and $e_5$ does not contain $v_1$, and hence is not a spanning tree. In case (ii), the induced subgraph of $\mathcal{G}$ formed by the edges represented by $S_k$ does not contain vertex $v_4$; hence it is not a spanning tree. For example, let $S_k = \{1, 2, 3\}$. Since $v_4$ is not a vertex in the subgraph of $\mathcal{G}$ induced by $e_1$, $e_2$, and $e_3$, the two vertices that each of these edges are incident to must be from the remaining three vertices, hence causing each column of $\hat{F}_{1,2,3}$ to have a zero sum.

If the edges represented by $S_k$ do form a spanning tree of $\mathcal{G}$, then $|\det(\hat{F}_{S_k})| = 1$. For example, let $S_k = \{1, 2, 4\}$ and observe the subgraph of $\mathcal{G}$ induced by $e_1$, $e_2$, and $e_4$ is a spanning tree $\mathcal{T}$ of $\mathcal{G}$. As stated in the proof, each spanning tree will have a pendant vertex other than $v_4$. In this case, $v_3$ is such a pendant vertex of $\mathcal{T}$. Observe that the row of $\hat{F}_{1,2,4}$ corresponding to $v_3$ has only one nonzero entry, namely $-1$. Thus $\det(\hat{F}_{1,2,4})$ is equal to $-1$ times the determinant of the submatrix created from $\hat{F}_{1,2,4}$ by removing the row corresponding to $v_3$ and the column corresponding to $e_2$ which is the unique edge in $\mathcal{T}$ incident to $v_3$. Calling this submatrix $\hat{F}^3_{1,2,4}$, we see that

$$\hat{F}^3_{1,2,4} = \begin{bmatrix} 1 & 0 \\ -1 & 1 \end{bmatrix}.$$

Observe now that the edges in $\mathcal{G} - v_3$ corresponding to the columns of $\hat{F}^3_{1,2,4}$ form a spanning tree $\mathcal{T}'$ of $\mathcal{G} - v_3$. Thus by the inductive hypothesis of the proof, there exists a pendant vertex of $\mathcal{T}'$ other than $v_4$. We see that it is $v_1$. Observe that the row of $\hat{F}^3_{1,2,4}$ corresponding to $v_1$ has exactly one nonzero entry, namely 1. Thus $\det(\hat{F}^3_{1,2,4})$ is equal to 1 times the determinant of the submatrix of $\hat{F}^3_{1,2,4}$ created by removing the row corresponding to $v_1$ and the column corresponding to $e_1$ which is the unique edge in $\mathcal{T}'$ incident to $v_1$. Calling this submatrix $\hat{F}^{3,1}_{1,2,4}$, we see that

$$\hat{F}^{3,1}_{1,2,4} = \begin{bmatrix} 1 \end{bmatrix}.$$

Clearly $\det(\hat{F}^{3,1}_{1,2,4}) = 1$. Hence, it follows that

$$\det(\hat{F}_{1,2,4}) = (-1) \bullet 1 \bullet 1 = -1.$$

We can actually extend the results of the Matrix Tree Theorem to weighted graphs. To this end, we use the generalized version of the incidence matrix for weighted graphs discussed in the previous section. In our investigation of the Matrix Tree Theorem for weighted graphs, we require the concept of the weight of the entire graph:

**DEFINITION 3.2.4** If $\mathcal{G}$ is a weighted graph, we define the *weight of $\mathcal{G}$*, denoted $w(\mathcal{G})$, as the product of the weights of the edges of $\mathcal{G}$. A graph with no edges is defined to have a weight of 1.

Thus in the above example, $w(\mathcal{G}) = 240$. We now generalize Theorem 3.2.2 to weighted graphs (see [7]).

**THEOREM 3.2.5** *Let $\mathcal{G}$ be a weighted graph on $n$ vertices labeled $1, \ldots, n$ with Laplacian matrix $L$. Then the sum of the weights of the spanning trees of $\mathcal{G}$ is the absolute value of the determinant of any $(n-1) \times (n-1)$ submatrix of $L$.*

The proof of Theorem 3.2.5 follows similarly to the proof of Theorem 3.2.2. If $B$ is the adjoint matrix for $L$, then by similar reasoning to the proof of Theorem 3.2.2, all entries of $B$ will be identical and will be the absolute value of the determinant of any $(n-1) \times (n-1)$ submatrix of $L$. To show that the absolute value of this determinant is equal to the sum of the weights of the spanning trees of $\mathcal{G}$, let $\hat{L}$ and $\hat{F}$ be as in the proof of Theorem 3.2.2. We need to compute $\det(\hat{L})$. We do this using (3.2.2). The only difference in the rest of the proof is when the edges corresponding the columns of $\hat{F}_{S_k}$ do form a spanning tree $\mathcal{T}_{S_k}$ of $\mathcal{G}$. In this case, we use similar reasoning to the proof of Theorem 3.2.2 to see that $\det(\hat{F}_{S_k}) = \sqrt{w(\mathcal{T}_{S_k})}$. Thus $\det(\hat{F}\hat{F}^T) = w(\mathcal{T}_{S_k})$, and the rest of the proof follows from (3.2.1).

**EXAMPLE 3.2.6** Consider the weighted graph $\mathcal{G}$ below.

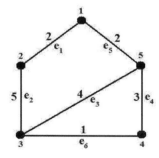

Its Laplacian matrix is

$$L = \begin{bmatrix} 4 & -2 & 0 & 0 & -2 \\ -2 & 7 & -5 & 0 & 0 \\ 0 & -5 & 10 & -1 & -4 \\ 0 & 0 & -1 & 4 & -3 \\ -2 & 0 & -4 & -3 & 9 \end{bmatrix}$$

# Introduction to Laplacian Matrices

Observe that the determinant of any $4 \times 4$ submatrix is either 536 of $-536$. Thus by Theorem 3.2.5, the sum of the weights of the spanning trees of $\mathcal{G}$ is 536. Below are the spanning trees of $\mathcal{G}$, each denoted with their weights. Observe the sum of these weights is indeed 536.

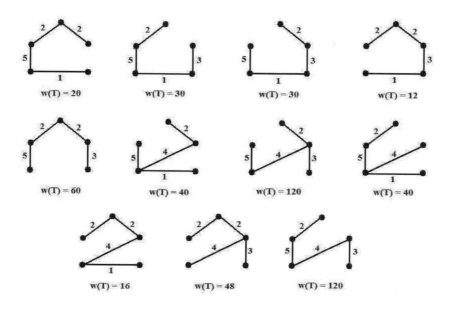

# Exercises:

**1.** Using the Matrix Tree Theorem (Theorem 3.2.2), determine the number of spanning trees in the following graph:

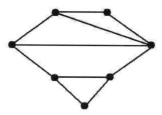

**2.** Using Theorem 3.2.5 determine the sums of the weights of all spanning trees of the following weighted graph:

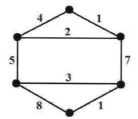

**3.** Prove Theorem 3.2.5.

**4.** Let $\mathcal{G}$ be a graph on $n$ vertices and let $0 = \lambda_1 \leq \ldots \leq \lambda_n$ be the eigenvalues of the Laplacian matrix. Prove that the number of spanning trees of $\mathcal{G}$ is $(\lambda_2 \cdots \lambda_n)/n$.

## 3.3 The Continuous Version of the Laplacian

In this section, we discuss the idea of the Laplacian matrix defined in Section 3.1 being the discrete version of the continuous Laplacian operator. We begin first by defining and motivating the use of the Laplacian operator. Starting with the one-dimensional version (see [16]), consider a rod located on the $x$-axis. We will investigate functions describing the temperature of this rod. Let $u = u(x,t)$ denote the temperature of the rod at position $x$ at time $t$. Assume that the rod has a uniform cross-section of area $A$ perpendicular to the $x$-axis and that $u$ is constant on each cross-section (think of $A$ being small). Consider a small portion of the rod from $x$ to $x + \Delta x$. Then the total heat $Q(t)$ in this portion of the rod at time $t$ can be seen as the temperature $u(x,t)$ multiplied by the area $A$ of a cross-section multiplied by the length $\Delta x$ of the portion of the rod we are considering. Since $u(x,t)$ is not constant, it follows that

$$Q(t) = \int_x^{x+\Delta x} Au(x,t)\,dx. \tag{3.3.1}$$

Therefore

$$Q'(t) = A[u_x(x+\Delta x, t) - u_x(x,t)].$$

By the mean value theorem for integrals, we also see from (3.3.1) that

$$Q'(t) = Au_t(\bar{x}, t)\Delta x$$

for some $\bar{x} \in (x, x+\Delta x)$. Hence

$$[u_x(x+\Delta x, t) - u_x(x,t)] = u_t(\bar{x}, t)\Delta x$$

for some $\bar{x} \in (x, x+\Delta x)$. Therefore

$$u_t(\bar{x}, t) = \frac{u_x(x+\Delta x, t) - u_x(x,t)}{\Delta x}.$$

Taking the limit as $\Delta x \to 0$, we see that $\bar{x} \to x$. Thus we obtain the one-dimensional heat equation [1]

$$u_t = u_{xx}. \tag{3.3.2}$$

Often in this application, we look at a finite portion of the rod from $x = x_0$ to $x = x_1$ and determine the temperature function based on (1) a fixed temperature

---

[1] In reality, $u_t$ and $u_{xx}$ are proportional, rather than equal, since we did not take into account the density of the rod. Hence $u_t = ku_{xx}$. But for our purposes and for the sake of simplicity, we will take $k = 1$.

to a the endpoints and (2) the temperature function $f(x)$ at time $t = 0$. Since all temperature functions must satisfy (3.3.2), it follows that we obtain the boundary value problem:

$$u_t = u_{xx}$$

$$u(x_0, t) = u(x_1, t) = t_0$$

$$u(x, 0) = f(x).$$

**EXAMPLE 3.3.1** (See [16]) Find the temperature equation of the rod extending from $x = 0$ to $x = \pi$ where the temperatures at $x = 0$ and $x = 1$ are both 0°C and the temperature at position $x$ at time $t = 0$ is $f(x) = 4\sin 4x \cos 2x$.

**Solution**: From (3.3.2) and the information given, we need to solve the boundary value problem

$$u_t = u_{xx}$$

$$u(0, t) = u(\pi, t) = 0$$

$$u(x, 0) = 4\sin 4x \cos 2x$$

Observe that $u(x, t) = 2e^{-4t}\sin 2x + 2e^{-36t}\sin 6x$ satisfies all three conditions.

The one-dimensional heat equation (3.3.2) can be extended to $n > 1$ dimensions as follows [16]

$$u_t = \sum_{i=1}^{n} u_{x_i, x_i}. \qquad (3.3.3)$$

In many instances we have a steady-state temperature meaning the temperature change varies only with location and not with time. Hence $u_t = 0$ in this instance. This leads us to the concept of the Laplace equation and the Laplacian of a function.

**DEFINITION 3.3.2** The *multi-dimensional Laplace equation* occurs when we take $u_t = 0$ in (3.3.3), i.e.,

$$\sum_{i=1}^{n} u_{x_i, x_i} = 0. \qquad (3.3.4)$$

**DEFINITION 3.3.3** The *Laplacian of a function* $u(x_1, \ldots, x_n)$, denoted $\Delta u$ is

$$\Delta u = \sum_{i=1}^{n} u_{x_i, x_i}.$$

Hence (3.3.4) can be rewritten as

$$\Delta u = 0.$$

In multiple dimensions, instead of having a rod we have a region $R$ bounded by a piecewise smooth manifold $C$. Our goal is to determine the temperature function

**EXAMPLE 3.3.4** (See [16]) In $\Re^2$ (the $xy$ plane), let the region $R$ be the closed rectangle $[0, \pi] \times [0, 1]$. Suppose the temperature along the $x$-axis, $0 < x < \pi$ is $1°C$ while the temperature along each of the remaining boundary line segments is $0°C$. Find the temperature function $u(x, y)$.

**Solution**: From (3.3.4) and the information given, we need to solve the boundary value problem

$$u_{xx} + u_{yy} = 0$$

$$u(0, y) = u(\pi, y) = u(x, 1) = 0$$

$$u(x, 0) = 1.$$

Observe that

$$u(x, y) = \sum_{n=1}^{\infty} \frac{2[1 - (-1)^n]}{n\pi \sinh n} \sin nx \sinh n\pi(1 - y)$$

satisfies all of the above conditions.

Given an $n$-dimensional region $R$ bounded by a smooth manifold, the heat can flow freely throughout. Hence the Laplacian operator $\Delta u$ that we have defined for functions $u : \Re^n \to \Re$ is a continuous operator. We now focus our attention on the discrete version of the Laplacian operator. Thus instead of heat flowing through an $n$-dimensional region $R$, we will investigate heat flowing through a graph on $n$ vertices. The edges will indicate paths on which heat can flow between vertices. Let $\mathcal{G}$ be a graph on $n$ vertices with vertex set $V$ and edge set $E$. Let $u$ be a real-valued function on the vertices of $\mathcal{G}$, i.e.,

$$u : V \to \Re.$$

Since this is a discrete function, we can regard $u(V)$ as a column vector $x$ where each entry $x_i$ of $x$ is the value $u(v_i)$ of the vertex $v_i \in V$ for $i = 1, \ldots, n$. Hence the discrete Laplacian operator defined for each vertex $v_i \in V$ is

$$\Delta u(v_i) = \sum_{\substack{v_j \\ v_i \sim v_j}} [u(v_i) - u(v_j)] \qquad (3.3.5)$$

where $v_i \sim v_j$ signifies $v_i$ adjacent to $v_j$. Since $u(v_i) = x_i$ for all $i = 1, \ldots, n$, it follows (3.3.5) can be written as

$$\Delta u(v_i) = \sum_{\substack{j \\ v_i \sim v_j}} [x_i - x_j]. \qquad (3.3.6)$$

# Introduction to Laplacian Matrices

Letting $d_i$ denote the degree of $v_i$, and suppose $v_i$ is adjacent to $v_{j_1}, \ldots, v_{j_d}$, then (3.3.6) can be written as

$$\Delta u(v_i) = d_i x_i - \sum_{k=1}^{d} x_{j_k}. \tag{3.3.7}$$

However, (3.3.7) is precisely $(Lx)_i$ where $L$ is the Laplacian matrix of the graph. Hence we have proven the following:

**THEOREM 3.3.5** *Let $\mathcal{G}$ be an unweighted graph on $n$ vertices with Laplacian matrix $L$ and vertex set $V = \{v_1, \ldots, v_n\}$. Let $u : V \to \Re$ be a function represented by the column vector $x = [x_i]$ where $u(v_i) = x_i$ for all $i = 1, \ldots, n$, and let $\Delta u$ be the Laplacian of $u$ as defined in (3.3.5). Then*

$$(Lx)_i = \Delta u(v_i)$$

*for $i = 1, \ldots, n$.*

**EXAMPLE 3.3.6** Consider the following graph $\mathcal{G}$:

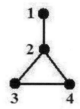

Observe that

$$L = \begin{bmatrix} 1 & -1 & 0 & 0 \\ -1 & 3 & -1 & -1 \\ 0 & -1 & 2 & -1 \\ 0 & -1 & -1 & 2 \end{bmatrix}$$

Letting $x = [x_1, x_2, x_3, x_4]^T$, we have

$$Lx = \begin{bmatrix} x_1 - x_2 \\ -x_1 + 3x_2 - x_3 - x_4 \\ -x_2 + 2x_3 - x_4 \\ -x_2 - x_3 + 2x_4 \end{bmatrix} = \begin{bmatrix} (x_1 - x_2) \\ (x_2 - x_1) + (x_2 - x_3) + (x_2 - x_4) \\ (x_3 - x_2) + (x_3 - x_4) \\ (x_4 - x_2) + (x_4 - x_3) \end{bmatrix} = \Delta u(V).$$

# Exercise:

1. Derive and prove a theorem analogous to Theorem 3.3.5 for weighted graphs.

## 3.4 Graph Representations and Energy

In this section, we continue with the theme of the Laplacian matrix of a graph being the discrete version of the continuous Laplacian operator. In the continuous case, we prove that the energy functional is minimized over all functions whose Laplacian is zero. Adapting the notion of energy to graph theory, we see that the eigenvalues and eigenvectors of the Laplacian matrix will be vital in drawing the graph in $\Re^k$ so that the energy of the graph is minimized. To further our comparison between the continuous and disrete Laplacians, we have the following definition from [17] which will be applicable to the continuous Laplacian operator.

**DEFINITION 3.4.1** *Let $U$ be an open bounded region in $\Re^k$. The energy functional $E(u)$ for a function $u : U \to \Re$ is*

$$E(u) = \frac{1}{2} \int_U \|\nabla u\|^2 \, dx$$

*where $\nabla u$ is the vector $[u_{x_1}, u_{x_2}, \ldots, u_{x_m}]$.*

In the following theorem from [17] known as Dirichlet's Principle, let $\overline{U}$ and $\partial U$ denote the closure and boundary of the region $U$, respectively, and let $C^2(U)$ denote the set of functions that are continuously twice differentiable on $U$.

**THEOREM 3.4.2** *Let $U$ be an open bounded region in $\Re^k$, let $u : U \to \Re$, and let $A = \{w \in C^2(\overline{U}) : w|_{\partial U} = 0\}$. Then $\Delta u = 0$ on $\overline{U}$ if and only if $E(u) = \min_{w \in A} E(w)$.*

**Proof**: First suppose $\Delta u = 0$ on $\overline{U}$. Then

$$0 = \int_U \Delta u (u - w) \, dx.$$

Integration by parts yields

$$0 = \int_U \nabla u \cdot \nabla (u - w) \, dx.$$

Expanding this we obtain

$$0 = \int_U \nabla u \cdot (\nabla u - \nabla w) \, dx = \int_U \|\nabla u\|^2 - (\nabla u \cdot \nabla w) \, dx.$$

Hence

$$\int_U \|\nabla u\|^2 \, dx = \int_U (\nabla u \cdot \nabla w) \, dx \le \frac{1}{2} \int_U \|\nabla u\|^2 \, dx + \frac{1}{2} \int_U \|\nabla w\|^2 \, dx,$$

where the last inequality follows from the Cauchy-Shwarz inequality. Thus

$$E(u) = \frac{1}{2} \int_U \|\nabla u\|^2 \, dx \le \frac{1}{2} \int_U \|\nabla w\|^2 \, dx = E(w)$$

# Introduction to Laplacian Matrices

as desired.

Now suppose $u$ is a function which minimizes the function $E(w)$ over all functions in $A$. Fix an arbitrary function $v \in A$ and define the function

$$g(t) := E(u + tv)$$

where $t \in \Re$. Observe

$$g(t) = \frac{1}{2}\int_U \|\nabla u + t\nabla v\|^2 \, dx = \int_U \frac{1}{2}\|\nabla u\|^2 + t\nabla u \cdot \nabla v + \frac{t^2}{2}\|\nabla v\|^2 \, dx.$$

Since $g(t)$ is minimized at $\mathcal{T} = 0$, it follows that $g'(0) = 0$ provided this derivative exists. Consequently,

$$0 = g'(0) = \int_U \nabla u \cdot \nabla v \, dx = \int_U (-\Delta u)v \, dx, \qquad (3.4.1)$$

where the last equality follows from integration by parts. Since $v$ was arbitrary, (3.4.1) holds for all $v \in A$. Hence $\Delta u = 0$ on $\overline{U}$. □

We now turn our attention to graphs. Our focus will be drawing graphs in $\Re^k$ where the energy function is minimized. Before generalizing the definition of the energy function to graphs, we need a preliminary definition:

**DEFINITION 3.4.3** *[33]* Let $\mathcal{G}$ be a graph with vertex set $V$. A *representation* of $\mathcal{G}$ in $\Re^k$ is a function $\rho : V \to \Re^k$.

Loosely speaking, $\rho$ represents the positions of the vertices when we draw $\mathcal{G}$ in $\Re^k$.

**EXAMPLE 3.4.4** Below is a representation of the wheel graph $W_4$ in $\Re^3$.

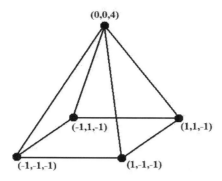

It is often convenient to represent a representation as a representation matrix $R$ where row $i$ of $R$ corresponds to $\rho(i)$, the representation of vertex $i$. Thus in Example 3.4.4, we have

$$R = \begin{bmatrix} -1 & -1 & -1 \\ 1 & -1 & -1 \\ -1 & 1 & -1 \\ 1 & 1 & -1 \\ 0 & 0 & 4 \end{bmatrix}.$$

Regarding the vertex set of a graph as a subset of $\Re^k$ in a similar way we regarded $U$ as a subset of $\Re^k$ earlier in this section, observe that $\rho$ is a discrete function from a subset of $\Re^k$ to $\Re$ as $u$ was a continuous such function. Therefore, we can define the discrete version of energy of a function with respect to a domain. However, in the discrete version, the domain is a vertex set of a graph rather than an open bounded set $U \subset \Re^k$.

**DEFINITION 3.4.5** Let $\mathcal{G}$ be a graph with vertex set $V$ represented in $\Re^k$ with representation $\rho : V \to \Re$. The *energy* $\epsilon(\rho)$ of the representation $\rho$ with respect to the graph $\mathcal{G}$ is the quantity

$$\epsilon(\rho) := \frac{1}{2} \sum_{\substack{v \in V \\ u \sim v}} \|\rho(u) - \rho(v)\|^2$$

Given an edge $ij$ of a graph, observe that in Definition 3.4.5 that each edge is being accounted for twice, once when $i = u$ and $j = v$ and again when $i = v$ and $j = u$. Hence if $E(\mathcal{G})$ is the edge set of $\mathcal{G}$, we can rewrite Definition 3.4.5 as

$$\epsilon(\rho) = \sum_{uv \in E(\mathcal{G})} \|\rho(u) - \rho(v)\|^2,$$

thus counting each edge exactly once. We now present a lemma from [33] which gives us an efficient way to calculate the energy of a representation.

**LEMMA 3.4.6** *Let $\rho$ be a representation of a graph $\mathcal{G}$ with $R$ as the representation matrix. Let $L$ be the Laplacian matrix of $\mathcal{G}$. Then*

$$\epsilon(\rho) = \mathrm{Tr}(R^T L R).$$

**Proof**: Recalling the $L = FF^T$ where $F$ is the incidence matrix of $\mathcal{G}$, let the $uv^{th}$ row of $F^T R$ represent the edge $uv \in \mathcal{G}$ as the $uv^{th}$ row of $F^T$ represents the edge $uv \in \mathcal{G}$. Observe that the $uv^{th}$ row of $F^T R$ is $\pm(\rho(u) - \rho(v))$. Therefore, the $uv^{th}$ diagonal entry of $F^T RR^T F$ is $\|\rho(u) - \rho(v)\|^2$. Hence

$$\epsilon(\rho) = \mathrm{Tr}(F^T RR^T F) = \mathrm{Tr}(R^T FF^T R) = \mathrm{Tr}(R^T L R).$$

□

**EXAMPLE 3.4.7** Let $\mathcal{G}$ and $\rho$ be the graph and representation, respectively, in Example 3.4.4. Then

$$R^T L R = \begin{bmatrix} 12 & 0 & 0 \\ 0 & 12 & 0 \\ 0 & 0 & 100 \end{bmatrix}.$$

Since $\mathrm{Tr}(R^T L R) = 124$, it follows that $\epsilon(\rho) = 124$.

Of course for a given graph $\mathcal{G}$, the energy depends on its representation. However, some representations are more convenient than others. This leads us to the following definition.

# Introduction to Laplacian Matrices

**DEFINITION 3.4.8** A representation $\rho$ of a graph $\mathcal{G}$ with vertex set $V$ is *balanced* if $\sum_{v \in V} \rho(v) = 0$.

Thus if $\rho$ is represented by $R$, then $\rho$ is balanced if and only if $e^T R = 0$. Observe that the representation of the graph $W_4$ in Example 3.4.4 is balanced.

Given a graph $\mathcal{G}$ on $n$ vertices and a positive integer $k \leq n-1$, our goal is to find a representation $\rho$ that minimizes the energy. To this end, observe that if $\mathcal{G}$ is drawn in $\Re^k$, we can translate the graph so that it balanced without changing the quantities $\|\rho(u) - \rho(v)\|^2$. Hence from this point on, we can assume that a graph $\mathcal{G}$ is drawn so that its representation $\rho$ is balanced. Letting $R$ be the $n \times k$ representation matrix, observe that if $M$ is a $k \times k$ invertible matrix, then the representation of $\mathcal{G}$ represented by the matrix $RM$ provides us with the same information about the graph $\mathcal{G}$ as $\rho$ does. From this point of view, $\rho$ is determined by the column space of $R$ (see [33]). Therefore, not only can we assume that $\rho$ is balanced, but that the columns of $R$ are orthonormal, i.e., $R^T R = I$. This gives rise to the following definition:

**DEFINITION 3.4.9** Let $\mathcal{G}$ be a graph and let $\rho$ be a balanced representation with $R$ as the corresponding representation matrix. If $R^T R = I$, then $\rho$ is an *orthogonal representation* of $\mathcal{G}$.

Before presenting the main theorem of this section, we provide a matrix-theoretic lemma from [33], which we state without proof:

**LEMMA 3.4.10** *Let $M$ be a real symmetric $n \times n$ matrix. If $R$ is an $n \times k$ matrix such that $R^T R = I$, then $\operatorname{Tr}(R^T M R)$ is greater than or equal to the sum of the $k$ smallest eigenvalues of $M$. Equality holds if and only if the column space of $R$ is spanned by the eigenvectors belonging to these eigenspaces.*

We now present the main theorem of this section which is from [33]. This theorem shows that given a connected graph, the eigenvalues of the Laplacian matrix give us insight into the minimum energy of a balanced orthogonal representation of a graph $\mathcal{G}$.

**THEOREM 3.4.11** *Let $\mathcal{G}$ be a connected graph on $n$ vertices with Laplacian matrix $L$. Let $\lambda_1 \leq \lambda_2 \leq \ldots \leq \lambda_n$ be the eigenvalues of $L$. Then the minimum energy of a balanced orthogonal representation of of $\mathcal{G}$ in $\Re^k$ equals $\sum_{i=1}^{k+1} \lambda_i$.*

Before proving this theorem, we should note that we will see in Chapter 4 that since $\mathcal{G}$ is connected that $\lambda_2 > 0$ and $e$ is the basis for the eigenspace corresponding to the eigenvalue $\lambda_1 = 0$.

**Proof**: By Lemma 3.4.6, the energy of a representation whose representation matrix is $R$ is $\operatorname{Tr}(R^T L R)$. From Lemma 3.4.10, the energy of an orthogonal representation in $\Re^k$ is bounded below by the $k$ smallest eigenvalues of $L$. We can realize this lower bound by taking the columns of $R$ to be the vectors $x_1, \ldots, x_k$ such that $Lx_i = \lambda_i x_i$ for $i = 1, \ldots, k$.

Since $x_1 = e$, it follows that the set $\{x_2, \ldots, x_k\}$ gives a balanced orthogonal representation of $\mathcal{G}$ in $\mathfrak{R}^{k-1}$ with the same energy. Conversely, we can reverse this process to obtain an orthogonal (but not necessarily balanced) representation in $\mathfrak{R}^k$ from a balanced orthogonal representation in $\mathfrak{R}^{k-1}$ such that these two representations have the same energy. Therefore, the minimum energy of a *balanced* orthogonal representation on $\mathcal{G}$ in $\mathfrak{R}^k$ equals the minimum energy of an orthogonal (not necessarily balanced) representation in $\mathfrak{R}^{k+1}$. Applying Lemma 3.4.10, we see that this mimimum equals $\sum_{i=2}^{k+1} \lambda_i$ (noting $\lambda_1 = 0$). □

**OBSERVATION 3.4.12** From Lemma 3.4.10 and the proof of Theorem 3.4.11, we see that to represent a graph in an orthogonally balanced way in $\mathfrak{R}^{k+1}$ so that its energy is minimized, we merely compute an orthonormal basis for the space spanned by the set of eigenvectors $\{x_2, \ldots, x_{k+1}\}$ of $L$ and let the vectors in this basis be the columns of the representation matrix $R$. It should be noted, as in the following example, that this balanced orthogonal representation of minimum energy is not unique because (1) we may reverse the sign of any vector in the basis and (2) it may be that $\lambda_{k+1} = \lambda_{k+2}$ in which we can take many appropriate linear combinations of the eigenvectors corresponding to $\lambda_{k+1} = \lambda_{k+2}$.

**EXAMPLE 3.4.13** Consider the graph of $W_4$ in Example 3.4.4. To represent this graph in an orthogonally balanced way in $\mathfrak{R}^3$ so that its energy is minimized, observe first that

$$L = \begin{bmatrix} 3 & -1 & -1 & 0 & -1 \\ -1 & 3 & 0 & -1 & -1 \\ -1 & 0 & 3 & -1 & -1 \\ 0 & -1 & -1 & 3 & -1 \\ -1 & -1 & -1 & -1 & 4 \end{bmatrix}.$$

It can be calculated that $\lambda_1 = 0$, $\lambda_2 = \lambda_3 = 3$, and $\lambda_4 = \lambda_5 = 5$. The corresponding eigenvectors, normalized to unit length and made orthogonal to each other by use of the Gram-Schmidt Process, are:

$$x_1 = \begin{bmatrix} 0.447 \\ 0.447 \\ 0.447 \\ 0.447 \\ 0.447 \end{bmatrix}, \ x_2 = \begin{bmatrix} -0.707 \\ 0 \\ 0 \\ 0.707 \\ 0 \end{bmatrix}, \ x_3 = \begin{bmatrix} 0 \\ -0.707 \\ 0.707 \\ 0 \\ 0 \end{bmatrix}, \ x_4 = \begin{bmatrix} 0 \\ 0.408 \\ 0.408 \\ 0 \\ -0.816 \end{bmatrix},$$

$$x_5 = \begin{bmatrix} 0.548 \\ -0.365 \\ -0.365 \\ 0.548 \\ -0.365 \end{bmatrix}.$$

# Introduction to Laplacian Matrices

Hence

$$R = \begin{bmatrix} -0.707 & 0 & 0 \\ 0 & -0.707 & 0.408 \\ 0 & 0.707 & 0.408 \\ 0.707 & 0 & 0 \\ 0 & 0 & -0.816 \end{bmatrix}$$

Recalling that the rows of $R$ represent the locations of each vertex of $W_4$, this yields the representation

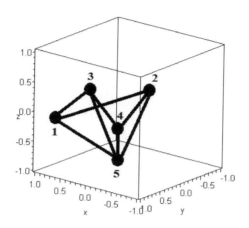

By Theorem 3.4.11, the energy is $\sum_{i=2}^{4} \lambda_i = 11$ and this is the minimum energy of any balanced orthogonal representation of $W_4$ in $\Re^3$.

## Exercises:

1. Represent the graph below in each of $\Re^2$ and $\Re^3$ as a balanced orthogonal representation of minimum energy. What is the energy in each case?

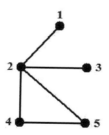

2. (See [33]) If $\mathcal{G}$ is a weighted graph where $w_{uv}$ represents weight of the edge $uv$, then the energy of a representation $\rho$ is

$$\epsilon(\rho) = \sum_{uv \in E(\mathcal{G})} w_{uv} \|\rho(u) - \rho(v)\|^2.$$

114 *Applications of Combinatorial Matrix Theory to Laplacian Matrices of Graphs*

Let $F$ be the incidence matrix of $\mathcal{G}$, $R$ be the representation matrix, and $W$ be the diagonal matrix of the weights of the edges of $\mathcal{G}$ (where the rows and columns of $W$ are indexed in correspondence to the columns of $F$). Prove that

$$\epsilon(\rho) = \text{Tr}(R^T F W F^T R).$$

**3.** Use Exercise 2 to find the energy of the following weighted graph represented in $\Re^3$:

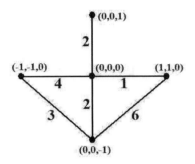

## 3.5 Laplacian Matrices and Networks

We close this chapter by investigating properties of Laplacian matrices for graphs which represent networks. Loosely speaking, we define a network to be a system of interacting entities. We represent networks as graphs where the vertices represent the entities and edges are drawn between vertices whose corresponding entities interact with each other. One example of a network is the World Wide Web. In this network, the vertices represent websites and edges are placed between vertices whose websites are linked to each other. Airline routes are another example of networks. In this network, the vertices represent cities with airports and edges join vertices whose corresponding cities have a direct flight between them. The human population (also known as "six degrees of separation") can be seen as a network where each person is represented by a vertex and an edge is placed between two vertices if the two corresponding people know each other. Citation networks are also common where each vertex represents an academic source (e.g., a paper, book, website, etc.) and an edge is placed between two sources if one source cites the other [80]. In this section, we will focus on three types of networks and investigate the eigenvalues of the Laplacian matrices of the graphs that represent these networks.

The first type of network we discuss are random graphs. We begin with a formal definition:

**DEFINITION 3.5.1** A *random graph* is a graph that is chosen with probability $p$ from all graphs on $n$ vertices and $m$ edges where $n$ and $m$ are fixed.

Oberve that there are $C_2^n = \frac{n(n-1)}{2} := N$ pairs of vertices between which to place edges. There are two common random graph models. The first is $G_1(n,m)$

where each graph on $n$ vertices and $m$ edges has equal probability of being chosen, namely $\left(C_m^N\right)^{-1}$. The second random graph model, $G_2(n,p)$, is where each of the $n(n-1)/2$ pairs of vertices has a probability $p$ of there being an edge placed between them. Given a graph $\mathcal{G}$ having $n$ vertices and $m$ edges, suppose we want to find the probability that $\mathcal{G}$ will be chosen at random from the $G_2(n,p)$ model. Then the probability that the $m$ pairs of adjacent vertices of $\mathcal{G}$ will be chosen is $p^m$ while the probability of the $N-m$ pairs of nonadjacent vertices of $\mathcal{G}$ being chosen is $(1-p)^{N-m}$. Hence the probability that $\mathcal{G}$ will be chosen is $p^m(1-p)^{N-m}$ (see [80]). In applications, it is often the case that $m \approx np$. In this case, $G_1(n,m)$ and $G_2(n,p)$ are essentially equivalent when $n$ is large [8].

On graphs with $n$ vertices where $n$ is large, it is impossible with today's computers to determine the eigenvalues of the Laplacian matrices of all possible graphs on $n$ vertices. In [81], 2000 graphs on $n = 2500$ vertices were chosen randomly using the $G_2(n,p)$ model with probabilitly $p = 0.08$. The eigenvalues of the corresponding Laplacian matrices were computed. Let $(d_1, \ldots, d_n)$ be the degree sequence of each graph (degrees listed in nonincreasing order) and let $\lambda_1 \leq \ldots \leq \lambda_n$ be the eigenvalues of the Laplacian matrix. The following trends were observed:

(a) For all graphs, $d_1 + 1 \leq \lambda_n \leq d_1 + \sqrt{d_1}$.

(b) For all graphs, $d_n - \sqrt{d_n} \leq \lambda_2 \leq d_n + \sqrt{d_n}$.

Thus $\lambda_n$ and $\lambda_2$ are strongly related to the largest and smallest vertex degrees, respectively. It was also observed that for large $n$ that $\lambda_i \approx d_{n-i+1}$. Hence in large graphs, the eigenvalues are most affected by the degrees of the vertices rather than the location of the edges connecting these vertices.

The next type of network we discuss are *small world networks*. Small world networks are characterized by having a relatively small diameter. A common small world network would be the human population and the idea of "six degrees of separation" [80]. In this network, we would represent each person as a vertex and join two vertices with an edge if the two corresponding people know each other. The theory is that this graph of approximately six billion vertices would have a diameter of only six. Small world networks are created in two ways:

**DEFINITION 3.5.2** *[80]* The *Watts-Strogatz model* (WS) starts with a cycle on $n$ vertices. In each step of the creation of the network, *replace* an edge of the existing graph with a random edge with probability $p$.

**DEFINITION 3.5.3** *[80]* The *Newman-Watts model* (NW) starts with a cycle on $n$ vertices. In each step of the creation of the network, *add* a random edge to the existing graph with probability $p$.

Below are two graphs representing small-world networks. The graph in (a) is created using the WS model while the graph in (b) is created using the NW model.

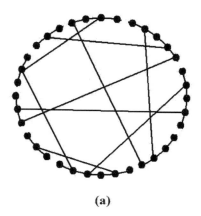 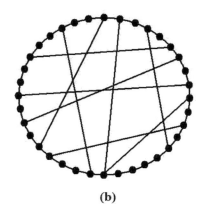

(a)                 (b)

In [81], 1000 graphs each were created using the SW and NW models. In both cases, $n = 2500$ and 50 edges were replaced/added using probabilities $p = 0.15$ and $p = 0.02$ for the SW and NW models, respectively. The following were observed in both the SW and NW models:

(a) In 80% of the graphs, $d_1 \leq \lambda_n \leq d_1 + \sqrt{d_1}$ while in the remaining 20% of the graphs, $\lambda_n > d_1 + \sqrt{d_1}$.

(b) In all graphs, $\lambda_2 < d_n - \sqrt{d_n}$.

Thus knowing only the values of $n$, $p$, and the number of edges replaced/added are needed for estimating the eigenvalues of the Laplacian matrices. The manner in which the edges are placed makes little difference in the eigenvalues of Laplacian matrices of graphs representing small world networks.

The final type of network we investigate is the *scale-free network*. In graphs representing scale-free networks, there is a higher proportion of vertices of large degree. The number of vertices of degree $k$ is proportional to $k^d$ for some exponent $d$ [80]. Scale-free networks are often used to model citation networks where the vertices represent references and edges are placed between references if one reference cites another. In academia, it is often the case that references that have been cited more often are more likely to be cited in the future [13, 14]. Hence in a graph modeling a citation network, there will be a higher proportion of vertices of large degree.

To create a scale-free network, start with a connected graph on $n_0$ vertices where $n_0$ is small. In each step $t$, introduce a new vertex and join it to $k$ existing vertices where $1 \leq k \leq n_0$. The new vertex is made adjacent to an existing vertex $i$ of degree $d_i$ with probability

$$\frac{d_i}{\sum_{j=1}^{n} d_j} \qquad (3.5.1)$$

where $n = n_0 + (t-1)$ vertices at the $(t-1)^{st}$ step [81]. A convenient way to understand which vertices a new vertex will be made adjacent to is to think of putting $d_i$ copies of each vertex in a hat, e.g., a vertex of degree 3 is placed in a hat 3 times. Then keep drawing vertices out of the hat until you have drawn $k$ distinct

# Introduction to Laplacian Matrices

vertices. Edges are then placed between the new vertex and each of the $k$ vertices drawn.

**EXAMPLE 3.5.4** Consider the graph below.

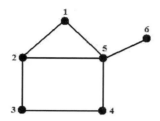

We will let $k = 2$, i.e., for each new vertex $v$ added, we will join $v$ to two existing vertices. Observe that $d_5 = 4$, $d_2 = 3$, $d_1 = d_3 = d_4 = 2$, and $d_6 = 1$. Letting $P(v_i)$ denote the probability that vertex $i$ is chosen, we see from (3.5.1) that $P(v_5) = 4/14$, $P(v_2) = 3/14$, $P(v_1) = P(v_3) = P(v_4) = 2/14$, and $P(v_6) = 1/14$. Hence the vertices with larger degree are more likely to be chosen. Suppose that vertices 4 and 5 are chosen. Then the new vertex introduced will be made adjacent to vertices 4 and 5 thus creating:

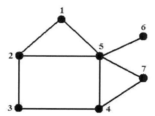

Next we choose $k = 2$ vertices for a new vertex to be made adjacent to. Observe now that $P(v_5) = 5/18$, $P(v_2) = P(v_4) = 3/18$, $P(v_1) = P(v_3) = P(v_7) = 2/18$, and $P(v_6) = 1/18$. Suppose that vertices 2 and 5 are chosen. Then the resulting graph will be:

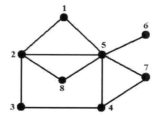

Observe that in each step, vertices with larger degree are more likely to be chosen. Hence the degrees of these vertices are likely to increase in subsequent steps, while vertices of small degree are likely to remain having a small degree.

In [81], 1000 graphs on $n = 2500$ vertices representing scale-free networks were tested with $n_0 = 5$ and $k = 4$. The results obtained were similar to the results obtained in the simulation involving random graphs. Hence the eigenvalues of the Laplacian matrices for the graphs representing scale-free networks are strongly related to the degrees of the vertices, rather than the location of the edges. For large scale-free networks, it was seen that $\lambda_i \approx d_{n-i+1}$.

We have investigated three types of networks that have numerous applications outside of mathematics. Upon investigating random graph networks, small world networks, and scale-free networks, we see that there is much interest in studying the eigenvalues of the Laplacian matrices of the graphs representing these networks, especially $\lambda_2$ and $\lambda_n$. In the remaining chapters, we will focus much attention on the eigenvalues of Laplacian matrices for graphs as they appear to provide us with a wealth of information as to the structure of the graph.

# Chapter 4

# The Spectra of Laplacian Matrices

In Chapter 3, we learned that one motivation for studying Laplacian matrices is that their spectra can give us a great deal of information about the structure of the graph. In this chapter, we investigate the overall spectra of Laplacian matrices. We do this by first investigating the effects on the spectra when we perform certain operations on graphs. We then continue our study more deeply by finding upper bounds on the largest eigenvalue of the Laplacian matrix, hence bounding the set of all eigenvalues of the Laplacian matrix. Once we do that, we are better able to study how the eigenvalues of the Laplacian matrix are distributed. Since 1 is often an eigenvalue, we investigate the properties of graphs that lead to 1 being an eigenvalue of the Laplacian matrix. We then deepen our investigation to see the distribution of eigenvalues less than 1 and eigenvalues greater than 1. By investigating the distribution of the eigenvalues and having the knowledge of an upperbound on the set of all eigenvalues of the Laplacian, we then prove the Grone-Merris Conjecture which gives upper bounds on each eigenvalue individually. This leads us to the study of maximal (threshold) graphs which are graphs in which the Grone-Merris Conjecture is sharp for each eigenvalue. Since the bounds obtained from the Grone-Merris Conjecture are integers, it becomes natural to investigate graphs which are Laplacian integral, i.e., graphs whose Laplacian eigenvalues are all integers.

## 4.1 The Spectra of Laplacian Matrices under Certain Graph Operations

In Section 3.1, we saw that Laplacian matrices are symmetric; hence, all Laplacian eigenvalues are real numbers. Moreover, since the sum of the absolute values of the off-diagonal entries in each row of a Laplacian matrix is equal to the diagonal entry of the row, it follows by the Gersgorin Disc Theorem (Theorem 1.2.1) that all eigenvalues of a Laplacian matrix lie in the right half of the complex plane. Thus it follows that all eigenvalues of a Laplacian matrix are nonnegative real numbers making it a postive semidefinite M-matrix. Since $e$, the vector of all one's, is an eigenvector for a Laplacian matrix corresponding to the eigenvalue 0, it follows that 0 is the smallest eigenvalue of a Laplacian matrix. Therefore we can order the eigenvalues of Laplacian matrices as follows:

$$0 = \lambda_1 \leq \lambda_2 \leq \lambda_3 \leq \ldots \leq \lambda_n.$$

Observe that if $\mathcal{G}$ is a disconnected graph, then we can permute its Laplacian matrix $L$ into a block diagonal matrix with each block corresponding to a component of $\mathcal{G}$. Therefore, if $\mathcal{G}$ has $k$ components, we can obtain $k$ linearly independent eigenvectors corresponding to the eigenvalue zero. We state this more precisely in the following theorem:

**THEOREM 4.1.1** *Let $\mathcal{G}$ be a weighted graph on $n$ vertices with $k$ components and Laplacian matrix $L$. Then $\lambda_1 = \ldots = \lambda_k = 0$ and $\lambda_{k+1} > 0$.*

**Proof**: Letting $L_1, \ldots, L_k$ be the Laplacian matrices for each of the $k$ components, we see that $L$ can be permuted into a block diagonal matrix with $L_1, \ldots, L_k$ as the blocks. Letting $e(i)$ be the vector consisting of ones in the entries corresponding to the $i^{th}$ component of $\mathcal{G}$ and zeros elsewhere, observe that the set of vectors $\{e(1), e(2), \ldots, e(k)\}$ is a linearly independent set of eigenvectors of $L$ corresponding to the eigenvalue zero. Since each block $L_i$ is an irreducible singular M-matrix, it follows from Theorem 1.4.10 that zero is a simple eigenvalue of $L_i$ for each $i$. Thus the set $\{e(1), e(2), \ldots, e(k)\}$ forms a basis for the eigenspace of $L$ corresponding to the eigenvalue zero. Since the algebraic and geometric multiplicities are equal for each eigenvalue of a symmetric matrix, the result follows. □

It is often of interest to observe how the eigenvalues of the Laplacian matrix change when edges are added or deleted from a graph, or if the weight of an existing edge changes. The following theorem gives us insight as to how the values of the eigenvalues change in such cases:

**THEOREM 4.1.2** *Let $\mathcal{G}$ be a graph on $n$ vertices, and let $\hat{\mathcal{G}}$ be a graph on $n$ vertices created from $\mathcal{G}$ by adding a weighted edge joining two nonadjacent vertices in $\mathcal{G}$, or increasing the weight of an existing edge in $\mathcal{G}$. Then for all $i = 1, \ldots, n$, we have $\lambda_i(L(\mathcal{G})) \leq \lambda_i(L(\hat{\mathcal{G}}))$.*

**Proof**: Observe that $L(\hat{\mathcal{G}}) = L(\mathcal{G}) + A$ for some positive semidefintie matrix $A$. The result now follows immediately from Corollary 1.2.11. □

Theorem 4.1.2 says that when we add an edge joining nonadjacent vertices, or if we increase the weight of an existing edge, then the corresponding eigenvalues will either increase or remain unchanged; they will never decrease.

**EXAMPLE 4.1.3** Consider the unweighted graph $\mathcal{G}$ on the left below. Observe the eigenvalues of the Laplacian matrix for $\mathcal{G}$ are $\lambda_1 = 0$, $\lambda_2 = 1$, $\lambda_3 = 2$, $\lambda_4 = 4$, and $\lambda_5 = 5$. Now suppose we create the graph $\hat{\mathcal{G}}$ on the right by adding an edge joining vertices 1 and 4 of $\mathcal{G}$:

# The Spectra of Laplacian Matrices

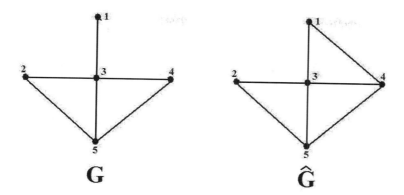

We calculate the eigenvalues of the Laplacian matrix for $\hat{\mathcal{G}}$ to be $\lambda_1 = 0$, $\lambda_2 = 1.586$, $\lambda_3 = 3$, $\lambda_4 = 4.414$, and $\lambda_5 = 5$. Observe that the values for $\lambda_2$, $\lambda_3$, and $\lambda_4$ increase when we add the edge joining vertices 1 and 4, and the values for $\lambda_1$ and $\lambda_5$ remain unchanged. (Note: By Theorem 4.1.1 we expect $\lambda_1 = 0$ in both cases.)

By Theorem 4.1.2, it follows that of all unweighted graphs on $n$ vertices (for fixed $n$), the graph with the largest eigenvalues is $K_n$, the complete graph on $n$ vertices. Observe that

$$L(K_n) = nI - J$$

where $I$ is the identity matrix and $J$ is the matrix of all one's. As usual, $e$ is an eigenvector corresponding to $\lambda_1 = 0$. As for the remaining eigenvalues, let $f(i)$ be the vector containing $n$ entries with 1 in the first position, $-1$ in the $i^{th}$ position, and zeros elsewhere. Observe that the set $\{f(i) \,|\, i = 2, \ldots, n\}$ is linearly independent. Moreover, observe for each $i = 2, \ldots, n$ that

$$(L(K_n))f(i) = (nI - J)f(i) = nf(i)$$

Hence $\lambda_2 = \lambda_3 = \ldots = \lambda_n = n$. This leads us to the following two corollaries for unweighted graphs:

**COROLLARY 4.1.4** *Let $\mathcal{G}$ be an unweighted graph on $n$ vertices with Laplacian matrix $L$. If $\lambda$ is an eigenvalue of $L$, then $\lambda \leq n$.*

**Proof**: Observe that every unweighted graph $\mathcal{G}$ on $n$ vertices is a subgraph of $K_n$. Since $n$ is the largest eigenvalue of $L(K_n)$, the result follows by Theorem 4.1.2. □

**COROLLARY 4.1.5** *Let $\mathcal{G}$ be an unweighted graph on $n$ vertices and $L$ be its Laplacian matrix with eigenvalues $\lambda_1 \leq \lambda_2 \leq \ldots \leq \lambda_n$. Then $\mathcal{G} = K_n$ if and only if $\lambda_1 = 0$ and $\lambda_i = n$ for all $2 \leq i \leq n$.*

**Proof**: We saw earlier that if $\mathcal{G} = K_n$, then $\lambda_1 = 0$ and $\lambda_i = n$ for all $2 \leq i \leq n$. For the converse, suppose $\mathcal{G} \neq K_n$. Then $\mathcal{G}$ is a subgraph of $K_n - e$ where $e$ is any edge. Without loss of generality, let $e$ be the edge joining vertices 1 and 2 of $\mathcal{G}$. Letting $\hat{L}$ be the Laplacian for $K_n - e$, we have

$$\hat{L} = \begin{bmatrix} n-2 & 0 & -1 & -1 & \cdots & -1 \\ 0 & n-2 & -1 & -1 & \cdots & -1 \\ -1 & -1 & n-1 & -1 & \cdots & -1 \\ -1 & -1 & -1 & n-1 & \cdots & -1 \\ \cdots & \cdots & \cdots & \cdots & \cdots & \cdots \\ -1 & -1 & -1 & -1 & \cdots & n-1 \end{bmatrix}$$

Clearly $\lambda_1(\hat{L}) = 0$. Observe $\lambda_2 = n - 2$ as $[1, -1, 0, 0, \ldots, 0]^T$ is a corresponding eigenvector. So by Theorem 4.1.2, $\lambda_2(L) \leq n - 2 < n$ which proves the converse. □

While $K_n$ is the only (unweighted) graph on $n$ vertices whose Laplacian matrix is such that all eigenvalues other than $\lambda_1$ are equal to $n$, we now investigate graphs where at least one eigenvalue of the Laplacian matrix equals $n$. Recalling the concept of the join of two graphs (Definition 2.2.9), we begin our investigation in the following theorem which gives us all of the eigenvalues of the Laplacian matrix for a graph which is the join of two graphs whose Laplacian spectra are known:

**THEOREM 4.1.6** Let $\mathcal{G}_1$ and $\mathcal{G}_2$ be unweighted graphs on $n_1$ and $n_2$ vertices, respectively. Let $L_1$ and $L_2$ be the Laplacian matrices for $\mathcal{G}_1$ and $\mathcal{G}_2$, respectively, and let $L$ be the Laplacian matrix for $\mathcal{G}_1 \vee \mathcal{G}_2$. If $0 = \alpha_1 \leq \alpha_2 \leq \ldots \leq \alpha_{n_1}$ and $0 = \beta_1 \leq \beta_2 \leq \ldots \leq \beta_{n_2}$ are the eigenvalues of $L_1$ and $L_2$, respectively. Then the eigenvalues of $L$ are

$$0, \quad n_2 + \alpha_2, \quad n_2 + \alpha_3, \quad \ldots, \quad n_2 + \alpha_{n_1},$$
$$n_1 + \beta_2, \quad n_1 + \beta_3, \quad \ldots, \quad n_1 + \beta_{n_2}, \quad n_1 + n_2$$

**Proof**: Observe that

$$L = \left[ \begin{array}{c|c} L_1 + n_2 I & -J \\ \hline -J & L_2 + n_1 I \end{array} \right] \tag{4.1.1}$$

where $I$ is the identity matrix and $J$ is the matrix of all ones, each matrix being of the appropriate size. Let $0^{(k)}$ and $e^{(k)}$ be the vectors consisting of $k$ entries of all zeros and ones, respectively. Observe that $e^{(n_1+n_2)}$ is an eigenvector of $L$ corresponding to 0. Similarly, $[n_2 e^{(n_1)} \,|\, n_1 e^{(n_2)}]^T$ is an eigenvector of $L$ corresponding to $n_1 + n_2$.

If $x(i)$ is an eigenvector of $L_1$ corresponding to the eigenvalue $\alpha_i$ for any $i = 2, \ldots n_1$, then since the eigenvectors of a symmetric matrix are orthogonal, it follows that $(e^{(n_1)})^T x(i) = 0$ and hence $Jx(i) = 0^{(n_1)}$. Therefore, $[x(i) \,|\, 0^{(n_2)}]^T$ is an eigenvector of $L$ corresponding to $n_2 + \alpha_i$ for $i = 2, \ldots, n_1$. By similar reasoning, if $y(j)$ is an eigenvector of $L_2$ corresponding to the eigenvalue $\beta_j$, then $[0^{(n_1)} \,|\, y(j)]^T$ is an eigenvector of $L$ corresponding to $n_1 + \beta_j$ for $j = 2, \ldots, n_2$. □

**EXAMPLE 4.1.7** Let $\mathcal{G}_1$ and $\mathcal{G}_2$ be the cycles on four and five vertices, respectively. Note that the eigenvalues of $L(\mathcal{G}_1)$ are $\alpha_1 = 0$, $\alpha_2 = 2$, $\alpha_3 = 2$, $\alpha_4 = 4$, and the eigenvalues of $L(\mathcal{G}_2)$ are $\beta_1 = 0$, $\beta_2 = 1.382$, $\beta_3 = 1.382$, $\beta_4 = 3.618$, $\beta_5 = 3.618$. Also note that $n_1 = 4$ and $n_2 = 5$. Thus by Theorem 4.1.6, the eigenvalues of $L(\mathcal{G}_1 \vee \mathcal{G}_2)$ are 0, 5.382, 5.382, 7, 7, 7.618, 7.618, 9, and 9.

Theorem 4.1.6 greatly helps us determine precisely which graphs have a Laplacian matrix where $n$ is an eigenvalue.

**THEOREM 4.1.8** *Let $\mathcal{G}$ be an unweighted connected graph on $n$ vertices with Laplacian matrix $L$. Then $n$ is an eigenvalue of $L$ if and only if $\mathcal{G}$ is the join of two graphs.*

**Proof**: If $\mathcal{G}$ is the join of two graphs, we see from Theorem 4.1.6 that $n$ is an eigenvalue of $L$. For the converse, suppose $\mathcal{G}$ is not the join of two graphs. Let $0 = \lambda_1 < \lambda_2 \leq \ldots \leq \lambda_n$ be the eigenvalues of $L$. Consider the nonnegative matrix $L + J$. Observe for $i = 2, \ldots, n$ that if $x(i)$ is an eigenvector of $L$ corresponding to $\lambda_i$, then since $Jx(i) = 0$, it follows that $x(i)$ is also an eigenvector of $L + J$ corresponding to $\lambda_i$. Also observe that $n$ is an eigenvalue of $L + J$ corresponding to $e$. Hence the eigenvalues of $L + J$ are $\lambda_2 \leq \ldots \leq \lambda_n \leq n$. Since $\mathcal{G}$ is not the join of two graphs, $L$ cannot be permuted to be of the form in (4.1.1). Therefore $L + J$ is irreducible. Thus by Theorem 1.3.20(iv), $n$ is a simple eigenvalue of $L + J$, and therefore $\lambda_n \neq n$. □

In Section 2.2, we saw that in addition to taking the join of two graphs, it is often useful to take the product of two graphs. Hence it will be beneficial to have a convenient way to find the eigenvalues of the Laplacian matrix for a graph which is the product of two graphs whose Laplacian spectra are known. The following theorem gives us the necessary information:

**THEOREM 4.1.9** *Let $L(\mathcal{G})$ and $L(\mathcal{H})$ be Laplacian matrices for $\mathcal{G}$ and $\mathcal{H}$, respectively. Let $A = \{\alpha_1, \ldots, \alpha_n\}$ and $B = \{\beta_1, \ldots, \beta_p\}$ be the eigenvalues of $L(\mathcal{G})$ and $L(\mathcal{H})$, respectively. Then the eigenvalues of $L(\mathcal{G} \times \mathcal{H})$ are $\{\alpha_i + \beta_j | \alpha_i \in A, \beta_j \in B\}$.*

**Proof**: Let $v_1, \ldots, v_p$ be vertices of $\mathcal{H}$. Observe that

$$L(\mathcal{G} \times \mathcal{H}) = \begin{bmatrix} L(\mathcal{G}) + d_1 I & F_{1,2} & \cdots & F_{1,p} \\ F_{2,1} & L(\mathcal{G}) + d_2 I & & F_{2,p} \\ \cdots & \cdots & \cdots & \cdots \\ F_{n,1} & F_{n,2} & \cdots & L(\mathcal{G}) + d_p I \end{bmatrix}$$

where $d_i$ is the degree of vertex $v_i \in \mathcal{H}$ and where $F_{i,j} = -I$ if $v_i$ and $v_j$ are adjacent in $\mathcal{H}$, and $F_{i,j} = 0$ otherwise. Suppose $[x_1, \ldots, x_n]$ is an eigenvector of $L(\mathcal{G})$ corresponding to $\alpha$; suppose $[y_1, \ldots, y_p]$ is an eigenvector of $L(\mathcal{H})$ corresponding to $\beta$. Let

$$w = [y_1 x_1 \; y_1 x_2 \; \cdots \; y_1 x_n \; | \; y_2 x_1 \; y_2 x_2 \; \cdots \; y_2 x_n \; | \; \cdots \; | \; y_p x_1 \; \cdots \; y_p x_n]^T.$$

Then by matrix-vector multiplication, $L(\mathcal{G} \times \mathcal{H})w = (\alpha + \beta)w$. □

**EXAMPLE 4.1.10** Consider the graphs $\mathcal{G}$ and $\mathcal{H}$ below:

$$G = \triangle \qquad H = \square$$

Observe that the Laplacian matrices for $\mathcal{G}$ and $\mathcal{H}$ are:

$$L(\mathcal{G}) = \begin{bmatrix} 2 & -1 & -1 \\ -1 & 2 & -1 \\ -1 & -1 & 2 \end{bmatrix}$$

and

$$L(\mathcal{H}) = \begin{bmatrix} 2 & -1 & 0 & -1 \\ -1 & 2 & -1 & 0 \\ 0 & -1 & 2 & -1 \\ -1 & 0 & -1 & 2 \end{bmatrix}.$$

Thus

$$L(\mathcal{G} \times \mathcal{H}) = \left[\begin{array}{ccc|ccc|ccc|ccc} 4 & -1 & -1 & -1 & 0 & 0 & 0 & 0 & 0 & -1 & 0 & 0 \\ -1 & 4 & -1 & 0 & -1 & 0 & 0 & 0 & 0 & 0 & -1 & 0 \\ -1 & -1 & 4 & 0 & 0 & -1 & 0 & 0 & 0 & 0 & 0 & -1 \\ \hline -1 & 0 & 0 & 4 & -1 & -1 & -1 & 0 & 0 & 0 & 0 & 0 \\ 0 & -1 & 0 & -1 & 4 & -1 & 0 & -1 & 0 & 0 & 0 & 0 \\ 0 & 0 & -1 & -1 & -1 & 4 & 0 & 0 & -1 & 0 & 0 & 0 \\ \hline 0 & 0 & 0 & -1 & 0 & 0 & 4 & -1 & -1 & -1 & 0 & 0 \\ 0 & 0 & 0 & 0 & -1 & 0 & -1 & 4 & -1 & 0 & -1 & 0 \\ 0 & 0 & 0 & 0 & 0 & -1 & -1 & -1 & 4 & 0 & 0 & -1 \\ \hline -1 & 0 & 0 & 0 & 0 & 0 & -1 & 0 & 0 & 4 & -1 & -1 \\ 0 & -1 & 0 & 0 & 0 & 0 & 0 & -1 & 0 & -1 & 4 & -1 \\ 0 & 0 & -1 & 0 & 0 & 0 & 0 & 0 & -1 & -1 & -1 & 4 \end{array}\right].$$

The eigenvalues of $L(\mathcal{G})$ are 0, 3, 3; the eigenvalues of $L(\mathcal{H})$ are 0, 2, 2, 4. From Theorem 4.1.9, we see that the eigenvalues of $L(\mathcal{G} \times \mathcal{H})$ are 0, 2, 2, 3, 3, 4, 5, 5, 5, 5, 7, 7. To find an eigenvector of $L(\mathcal{G} \times \mathcal{H})$, say corresponding to the eigenvalue 5, we find an eigenvalue of $L(\mathcal{G})$ and an eigenvalue of $L(\mathcal{H})$ that add up to 5. Observe that 3 and 2 are such eigenvalues. Note that an eigenvector of $L(\mathcal{G})$ corresponding to 3 is $[1, -1, 0]^T$; and eigenvector of $L(\mathcal{H})$ corresponding to 2 is $[1, 0, -1, 0]^T$. So according to the proof of Theorem 4.1.9, an eigenvector of $L(\mathcal{G} \times \mathcal{H})$ corresponding to 5 is $[1, -1, 0 \,|\, 0, 0, 0 \,|\, -1, 1, 0 \,|\, 0, 0, 0]^T$.

In keeping with the theme of determining the eigenvalues of Laplacian matrices of graphs which are created from the operations of other graphs, we close this section

# The Spectra of Laplacian Matrices

by investigating the eigenvalues of the Laplacian matrix of the complement of a graph (Definition 2.2.1). The spectrum of $L(\mathcal{G}^c)$ is closely related to the spectrum of $\mathcal{G}$, as we see from the following theorem:

**THEOREM 4.1.11** *Let $\mathcal{G}$ be an unweigted graph on $n$ vertices with Laplacian matrix $L$. Let $0 = \lambda_1 \leq \lambda_2 \leq \ldots \leq \lambda_n$ be the eigenvalues of $L$. Then the eigenvalues of $L(\mathcal{G}^c)$ are $0 \leq n - \lambda_n \leq n - \lambda_{n-1} \leq \ldots \leq n - \lambda_2$ with the same corresponding eigenvectors.*

**Proof**: Observe that $L(\mathcal{G}^c) = nI + J - L$. Therefore, for $i = 2, \ldots n$, if $x$ is an eigenvector of $L$ corresponding to $\lambda_i$, then $Jx = 0$. Therefore

$$L(\mathcal{G}^c)x = (nI + J - L)x = nIx + Jx - Lx = (n - \lambda_i)x.$$

Thus $(n - \lambda_i)$ is an eigenvalue with $x_i$ as a corresponding eigenvector. Finally, $e$ is an eigenvector of $L(\mathcal{G}^c)$ corresponding to 0. □

**EXAMPLE 4.1.12** Consider the following graph $\mathcal{G}$ and its complement $\mathcal{G}^c$:

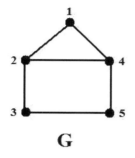
**G**

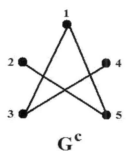
**G$^c$**

Note that

$$L(\mathcal{G}) = \begin{bmatrix} 2 & -1 & 0 & 0 & -1 \\ -1 & 3 & -1 & 0 & -1 \\ 0 & -1 & 2 & -1 & 0 \\ 0 & 0 & -1 & 2 & -1 \\ -1 & -1 & 0 & -1 & 3 \end{bmatrix}$$

and that the eigenvalues for $L(\mathcal{G})$ are 0, 1.382, 2.382, 3.618, 4.618. Since $n = 5$, we obtain the eigenvalues for $L(\mathcal{G}^c)$ by subtracting these values $\lambda_2, \ldots, \lambda_5$ from 5. Therefore, the eigenvalues of $L(\mathcal{G}^c)$ are 0, 0.382, 1.382, 2.618, 3.618.

## Exercises:

1. Using only the theorems concerning the eigenvalues of Laplacian matrices, prove that if $\mathcal{G}$ is the join of two graphs then $\mathcal{G}^c$ is disconnected.

2. Let $\mathcal{G}$ be a graph whose Laplacian eigenvalues are $0 = \lambda_1 \leq \lambda_2 \leq \ldots \leq \lambda_n$. Find the eigenvalues of the Laplacian matrices of $\mathcal{G} \vee \mathcal{G}$ and $\mathcal{G} \times \mathcal{G}$.

3. Find the eigenvalues of the Laplacian matrices of $P_2 \vee P_3$ and $P_2 \times P_3$.

## 4.2 Upper Bounds on the Set of Laplacian Eigenvalues

We saw in Corollary 4.1.4 that for an unweighted graph $\mathcal{G}$ on $n$ vertices, the set of eigenvalues of the Laplacian matrix is bounded above by $n$. Moreover, by Theorem 4.1.8 the largest eigenvalue, $\lambda_n$, will equal $n$ if and only if $\mathcal{G}$ is the join of two graphs. In this section, we derive more precise upper bounds for the set of eigenvalues of Laplacian matrices. Equivalently we will derive upper bounds on $\lambda_n$. We begin by looking at a history of such upper bounds that have been discovered and we prove the most recent discoveries. In all theorems, let $(d) = (d_1, d_2, \ldots, d_n)$ be the degree sequence of a graph where $d_1 \geq d_2 \geq \ldots \geq d_n$, and let $m_v$ denote the average of the degrees of the vertices adjacent to $v$.

We first begin with two upper bounds whose proofs are relatively simple. The first upper bound comes from [2] and dates back to 1985.

**THEOREM 4.2.1** *Let $\mathcal{G}$ be an unweighted graph on $n$ vertices with edge set $E$ and Laplacian matrix $L$. Then $\lambda_n \leq \max\{d_u + d_v : uv \in E\}$.*

**Proof**: Let $F$ be the incidence matrix for $\mathcal{G}$. Recall from Observation 3.1.11 that $L = FF^T$ and recall from Theorem 3.1.12 that $N := F^T F$ and $L$ have the same nonzero eigenvalues. Thus it suffices to show $\rho(N) \leq \max\{d_u + d_v : uv \in E\}$. Recall from Theorem 1.1.12 that $\|A\|_\infty$ is the maximum row sum of $|A|$. By Lemma 1.1.13 we have $\rho(N) \leq \|N\|_\infty$. Yet each row of $N$ corresponds to an edge $uv$ of $\mathcal{G}$ and the sum of the entries in such a row of $|N|$ is $\deg u + \deg v$. This establishes the inequality. □

An improvement of the upper bound in Theorem 4.2.1 was discovered in 1998 by Merris in [62]:

**THEOREM 4.2.2** *Let $\mathcal{G}$ be an unweighted graph on $n$ vertices with vertex set $v$. Then $\lambda_n \leq \max\{m_v + d_v : v \in V\}$.*

**Proof**: Let $L$ be the Laplacian matrix for $\mathcal{G}$ and let $D$ be the diagonal matrix consisting of the degrees of the vertices of $\mathcal{G}$. Then the eigenvalues of $L$ are the same as the eigenvalues of the matrix $C := D^{-1}LD$ which is similar to $L$. Observe that the $i^{th}$ diagonal entry $c_{i,i}$ of $C$ is $d_i$ and the off-diagonal entries $c_{i,j}$ are $-d_j/d_i$ if $i$ and $v$ are adjacent, and zero otherwise. The result now follows by applying Theorem 1.2.1 to $C$. □

Naturally, as results are improved, the proofs become more involved. In this section, we present three modern results which are improvements over Theorem 4.2.2. The first two results apply to graphs while the third result is restricted to trees. To prove our first result, we need a Lemma based on results from [57] and [75]:

**LEMMA 4.2.3** *Let $B$ be the adjacency matrix for the line graph of $\mathcal{G}$ and let $\rho$ be the largest eigenvalue of $B + 2I$. If $\lambda_n$ is the largest eigenvalue of the Laplacian matrix $L$, then $\lambda_n \leq \rho$ where equality holds if and only if $\mathcal{G}$ is bipartite.*

**Proof**: Let $F$ be the incidence matrix of $\mathcal{G}$. Recall from Theorem 3.1.12 that $|L|$ and $|F^T F|$ have the same nonzero eigenvalues. Since $L = D - A$ and $|L| = D + A$ where $D$ is the diagonal matrix of the vertex degrees of $\mathcal{G}$ and $A$ is the adjacency matrix of $\mathcal{G}$, it follows from Corollary 1.1.19 that the largest eigenvalue of $L$ cannot be more than the largest eigenvalue of $|L|$. Since $|F^T F| = B + 2I$ the inequality follows.

Let $\lambda_n(L)$ and $\lambda_n(|L|)$ denote the largest eigenvalues of $L$ and $|L|$, respectively. To complete the proof, since $|L|$ and $B + 2I$ have the same nonzero eigenvalues, it suffices to prove $\lambda_n(L) = \lambda_n(|L|)$ if and only if $\mathcal{G}$ is bipartite. First assume $\lambda_n(L) = \lambda_n(|L|)$. Let $x$ be a unit eigenvector of $L$ corresponding to $\lambda_n(L)$ and let $w$ be a unit eigenvector of $|L|$ corresponding to $\lambda_n(|L|)$. Letting "$\sim$" denote adjacency and writing $L$ as $D - A$, it follows from Theorem 1.2.4(ii) that

$$\begin{aligned}
\lambda_n(L) &= \max\nolimits_{y^T y = 1} y^T L y \\
&= \max\nolimits_{y^T y = 1} Y^T (D - A) y \\
&= \max\nolimits_{y^T y = 1} \sum\nolimits_{\substack{v_i \sim v_j \\ i < j}} (y_i - y_j)^2 \\
&= x^T (D - A) x \\
&= \sum\nolimits_{\substack{v_i \sim v_j \\ i < j}} (x_i - x_j)^2,
\end{aligned}$$

and

$$\begin{aligned}
\lambda_n(|L|) &= \max\nolimits_{y^T y = 1} y^T |L| y \\
&= \max\nolimits_{y^T y = 1} Y^T (D + A) y \\
&= \max\nolimits_{y^T y = 1} \sum\nolimits_{\substack{v_i \sim v_j \\ i < j}} (y_i + y_j)^2 \\
&= w^T (D + A) w \\
&= \sum\nolimits_{\substack{v_i \sim v_j \\ i < j}} (w_i + w_j)^2.
\end{aligned}$$

Since $\lambda_n(L) = \lambda_n(|L|)$, it follows $w = |x|$ and hence

$$\sum_{\substack{v_i \sim v_j \\ i < j}} (x_i - x_j)^2 = \sum_{\substack{v_i \sim v_j \\ i < j}} (|x_i| + |x_j|)^2. \tag{4.2.1}$$

Since for any pair of real numbers $x_i$, $x_j$ we have $(x_i - x_j)^2 \leq (|x_i| + |x_j|)^2$, (4.2.1) implies that $(x_i - x_j)^2 \leq (|x_i| + |x_j|)^2$ whenever vertices $v_i$ and $v_j$ are adjacent. Therefore, $x_i$ and $x_j$ are of opposite signs whenever $v_i$ and $v_j$ are adjacent. Let $W_1 = \{v_i | x_i > 0\}$ and $W_2 = \{v_j | x_j < 0\}$. For each edge $e = v_i v_j$, since $x_i$ and $x_j$ are of opposite signs, it follows that one of the vertices incident to $e$ lies in $W_1$ while the other vertex incident to $e$ lies in $W_2$. Thus $\mathcal{G}$ is bipartite.

For the converse, if $\mathcal{G}$ is bipartite with vertex partite sets $V_1$ and $V_2$, define the diagonal matrix $U$ such that $u_{ii} = 1$ if vertex $i$ is in $V_1$, and $u_{ii} = -1$ if vertex $i$ is in $V_2$. Then $|L| = U^{-1}LU$, thus $L$ and $|L|$ are unitarily similar. Hence if $\mathcal{G}$ is bipartite then $\lambda_n(L) = \lambda_n(|L|)$. □

At this point, we begin to see that the adjacency matrix of the line graph will play an important role in refining the upper bounds that we have obtained for the eigenvalues of Laplacian matrices. We also see that bipartite graphs are good candidates for the upper bounds being sharp. We now present our first modern upper bound on $\lambda_n$ whose proof is taken from [57].

**THEOREM 4.2.4** *Let $\mathcal{G}$ be an unweighted graph on $n$ vertices with edge set $E$ and Laplacian matrix $L$. Then*

$$\lambda_n \leq \max\left\{\frac{d_u(d_u + m_u) + d_v(d_v + m_v)}{d_u + d_v} : uv \in E\right\}. \quad (4.2.2)$$

*Equality occurs if $\mathcal{G}$ is a regular bipartite graph.*

**Proof**: Let $B$ be the adjacency matrix for the line graph of $\mathcal{G}$. If $uv$ and $xy$ are edges of $\mathcal{G}$, then $B_{uv,xy} = 1$ if $uv$ and $xy$ are adjacent, i.e., they are incident to a common vertex. Likewise, $B_{uv,xy} = 0$ if $uv$ and $xy$ are not adjacent. Let $f$ be the column vector whose entries are labeled in accordance with the labeling of $B$ and where the $uv^{th}$ entry of $f$ is $d_u + d_v$. Then the $uv^{th}$ component of $(B+2I)f$ is

$$
\begin{aligned}
((B+2I)f)_{uv} &= 2(d_u + d_v) + \sum_{xy \sim uv}(d_x + d_y) \\
&= 2(d_u + d_v) + \sum_{uy \sim uv}(d_u + d_y) + \sum_{xv \sim uv}(d_x + d_v) \\
&= 2(d_u + d_v) + d_u^2 + \sum_{y \sim u} d_y - (d_u + d_v) + d_v^2 \\
&\quad + \sum_{x \sim v} d_x - (d_u + d_v) \\
&= d_u^2 + \sum_{y \sim u} d_y + d_v^2 + \sum_{x \sim v} d_x \\
&= d_u^2 + d_u m_u + d_v^2 + d_v m_v \\
&= d_u(d_u + m_u) + d_v(d_v + m_v).
\end{aligned}
$$

Therefore

$$\frac{((B+2I)f)_{uv}}{f_{uv}} = \frac{d_u(d_u + m_u) + d_v(d_v + m_v)}{d_u + d_v}.$$

The inequality (4.2.2) now follows from Lemma 4.2.3 and Exercise 2 of Section 1.3.

Showing that equality holds when $\mathcal{G}$ is a regular bipartite graph is left as an exercise. □

# The Spectra of Laplacian Matrices

**EXAMPLE 4.2.5** Consider the following graph $\mathcal{G}$:

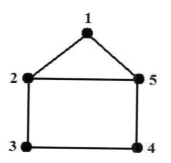

From $\mathcal{G}$ we create the following chart:

| $v$ | $d_v$ | $m_v$ | $d_v(d_v + m_v)$ |
|---|---|---|---|
| 1 | 2 | 3 | 10 |
| 2 | 3 | 2.33 | 16 |
| 3 | 2 | 2.5 | 9 |
| 4 | 2 | 2.5 | 9 |
| 5 | 3 | 2.33 | 16 |

Using this chart, we make the following chart for the edges of $\mathcal{G}$ with the bound from Theorem 4.2.4:

| edge | bound |
|---|---|
| 12 | 5.2 |
| 15 | 5.2 |
| 23 | 5 |
| 25 | 5.33 |
| 34 | 4.5 |
| 45 | 5 |

Thus if $L$ is the Laplacian for $\mathcal{G}$, then $\lambda_n(L) \leq 5.33$ by Theorem 4.2.4. (Note: By Corollary 4.1.4, we actually know that $\lambda_n \leq 5$. However, we use this example for the sake of simplicity. In subsequent examples, this bound will be better than that of Corollary 4.1.4.)

**REMARK 4.2.6** [57] to see that the bound in Theorem 4.2.4 is a better bound than the bound in Theorem 4.2.2, suppose $\max\{d_v + m_v : v \in \mathcal{G}\} = d_x + m_x$. Then

$$\frac{d_u(d_u+m_u)+d_v(d_v+m_v)}{d_u+d_v} \leq \frac{d_u(d_x+m_x)+d_v(d_x+m_x)}{d_u+d_v}$$

$$= d_x + m_x$$

$$= \max\{d_v + m_v : v \in \mathcal{G}\}.$$

While the bound in Theorem 4.2.4 is a better bound than that of Theorem 4.2.2, it should be noted that Theorem 4.2.2 is easier to apply, thus it is still relevant.

Theorem 4.2.4 relies on the edges in a graph, making it cumbersome to use. We now present a theorem from [55] which relies strictly on the degrees of the vertices and, hence, will be easier to apply.

**THEOREM 4.2.7** *Let $\mathcal{G}$ be a connected unweighted graph on $n$ vertices with Laplacian matrix $L$. Then*

$$\lambda_n \leq \max\left\{\sqrt{2d_v(d_v + m_v)} : v \in V(\mathcal{G})\right\}, \qquad (4.2.3)$$

*where equality holds if and only if $\mathcal{G}$ is a regular bipartite graph.*

**Proof**: Let $x = [x_1, x_2, \ldots, x_n]^T$ be a unit eigenvector of $L$ corresponding to $\lambda_n$. Then $Lx = \lambda_n x$. Hence for $u \in \mathcal{G}$ we have

$$\lambda_n x_u = d_u x_u - \sum_{v \in \mathcal{G}} a_{uv} x_v, \qquad (4.2.4)$$

where $A = [a_{uv}]$ is the adjacency matrix of $\mathcal{G}$. Therefore

$$\lambda_n x_u = \sum_{v \sim u} (x_u - x_v).$$

By the Cauchy-Schwarz inequality, we have

$$\lambda_n^2 x_u^2 \leq \left(\sum_{v \sim u} 1^2\right) \sum_{v \sim u} (x_u - x_v)^2$$
$$= d_u \left(\sum_{v \sim u} x_u^2 - 2x_u \sum_{v \sim u} x_v + \sum_{v \sim u} x_v^2\right).$$

Since

$$\sum_{v \sim u} (x_u + x_v)^2 = \sum_{v \sim u} x_u^2 + 2x_u \sum_{v \sim u} x_v + \sum_{v \sim u} x_v^2 \geq 0,$$

it follows that

$$-2x_u \sum_{v \sim u} x_v \leq \sum_{v \sim u} (x_u^2 + x_v^2) = d_u x_u^2 + \sum_{v \sim u} x_v^2. \qquad (4.2.5)$$

Therefore

$$\lambda_n^2 x_u^2 \leq d_u \left(\sum_{v \sim u} x_u^2 + d_u x_u^2 + 2\sum_{v \sim u} x_v^2\right)$$
$$= 2d_u^2 x_u^2 + 2d_u \sum_{v \sim u} x_v^2. \qquad (4.2.6)$$

Because $\sum x_u^2 = 1$ (since $x$ is a unit vector), it follows from (4.2.6) that

$$\lambda_n^2 = \sum_{u \in \mathcal{G}} \lambda_n^2 x_u^2 \leq 2\sum_{u \in \mathcal{G}} d_u^2 x_u^2 + 2\sum_{u \in \mathcal{G}} d_u \left(\sum_{u \sim v} x_v^2\right)$$
$$= 2\sum_{u \in \mathcal{G}} d_u^2 x_u^2 + 2\sum_{u \in \mathcal{G}} x_u^2 \left(\sum_{u \sim v} d_v\right)$$
$$= 2\sum_{u \in \mathcal{G}} (d_u^2 + m_u d_u) x_u^2$$
$$\leq \max\{2d_u(d_u + m_u) : u \in \mathcal{G}\}.$$

This proves (4.2.3).

Suppose now that equality holds in (4.2.3). Then all inequalities in the above argument must be equalities. In particular, we would have equality in (4.2.5) which would imply
$$-2x_u \sum_{v \sim u} x_v = \sum_{v \sim u} (x_u^2 + x_v^2)$$
which in turn implies
$$0 = \sum_{v \sim u} (x_u^2 + x_v^2) + 2x_u \sum_{v \sim u} x_v = \sum_{v \sim u} (x_u + x_v)^2.$$

Therefore $-x_v = x_u$ for each pair of adjacent vertices $u$ and $v$. First note that no entry $x_u$ in $x$ can be zero, for if it were then all of vertices $v$ adjacent to $u$ would be such that $x_v = 0$. Because $\mathcal{G}$ is connected, this would force $x = 0$ which cannot be an eigenvector. Since no entry of $x$ is zero, we can define the vertex sets
$$V_1 = \{u \in \mathcal{G} \mid x_u > 0\} \quad \text{and} \quad V_2 = \{u \in \mathcal{G} \mid x_u < 0\}.$$

Then $V_1, V_2$ forms a partition of the vertices of $\mathcal{G}$. Clearly there are no pairs of vertices in $V_1$ or $V_2$ that are adjacent. Thus $\mathcal{G}$ is bipartite. For each vertex $u \in \mathcal{G}$, it follows from (4.2.4) that
$$(d_u - \lambda_n)x_u = \sum_{v \sim u} x_v = d_u x_v = -d_u x_u.$$

Thus $d_u = \lambda_n / 2$ for each vertex $u$, hence $\mathcal{G}$ is regular. The converse is left as an exercise. $\square$

**EXAMPLE 4.2.8** Using the same graph and charts as in Example 4.2.5, we obtain the following chart:

| $v$ | $\sqrt{2d_v(d_v + m_v)}$ |
|---|---|
| 1 | 4.47 |
| 2 | 5.66 |
| 3 | 4.24 |
| 4 | 4.24 |
| 5 | 5.66 |

Hence by Theorem 4.2.7, $\lambda_n \leq 5.66$.

While the bounds in Theorem 4.2.4 is an improvement over that of Theorem 4.2.2, it is incomparable to bound from Theorem 4.2.7 as we see in the following example:

**EXAMPLE 4.2.9** Consider the following graphs ($\mathcal{G}_1$ is taken from [55]):

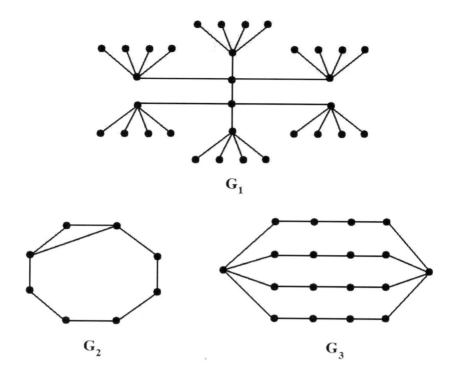

|       | $\lambda_n$ | Thm 4.2.2 | Thm 4.2.4 | Thm 4.2.7 |
|-------|-------------|-----------|-----------|-----------|
| $G_1$ | 7.092       | 8.75      | 8.75      | 8.36      |
| $G_2$ | 4.481       | 5.33      | 5.33      | 5.66      |
| $G_3$ | 5.361       | 6.00      | 5.66      | 6.92      |

Note that for $\mathcal{G}_1$, the bound from Theorem 4.2.7 is a better bound than the bound from Theorem 4.2.4; however, the bound from Theorem 4.2.4 is better than that of Theorem 4.2.7 for $\mathcal{G}_2$ and $\mathcal{G}_3$.

We now turn our attention to trees and develop an upper bound on the largest eigenvalue of the Laplacian matrix for a tree. Bethe trees will be important in the development of such an upper bound. Recall in Section 2.3 that we defined a rooted $m$-ary Bethe tree. We now recast this definition in light of our needs here. This definition is from [40] and [77].

**DEFINITION 4.2.10** A *rooted Bethe tree*, $B_{\Delta,k}$, is a rooted tree on $k$ levels where the root vertex has degree $\Delta - 1$, each vertex on levels 2 through $k-1$ has degree $\Delta$, and each vertex on level $k$ is pendant.

# The Spectra of Laplacian Matrices

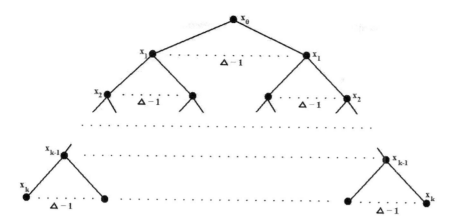

**OBSERVATION 4.2.11** If we let $b_i$ denote the number of vertices of $B_{\Delta,k}$ whose distance from the root is $i$ (i.e., on level $i-1$), then $b_0 = 1$ and
$$b_i = b_{i-1}(\Delta - 1) \quad \text{for } i = 1, \ldots, k.$$

Since our proof of an upper bound on the largest eigenvalue of the Laplacian for trees will involve the line graph of $B_{\Delta,k}$, we should note that such a line graph looks like:

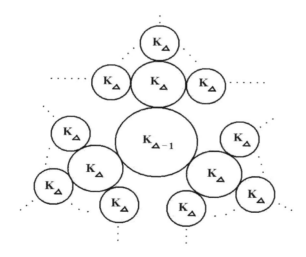

From the drawing of $B_{\Delta,k}$, we see that if $\mathcal{T}$ is a tree whose maximum vertex degree is $\Delta$ then there exists a natural number $k$ such that $\mathcal{T}$ is an induced subgraph of $B_{\Delta,k}$. To see this, let $u$ be a vertex in $\mathcal{T}$ such that $d_u < \Delta$, and declare $u$ to be the root of $\mathcal{T}$. Let $k$ be the largest distance from $u$ to any other vertex in $\mathcal{T}$. Now create $B_{\Delta,k}$ from $\mathcal{T}$ by adding the appropriate vertices and edges. Therefore, it follows from Theorem 4.1.2 that $\lambda_n(L(\mathcal{T})) \leq \lambda_n(L(B_{\Delta,k}))$. This fact will assist us in proving our next theorem from [77] which gives us an upper bound on the eigenvalues of the Laplacian matrix of a tree.

**THEOREM 4.2.12** *Let $L$ be the Laplacian matrix for a tree $\mathcal{T}$ with maximum degree $\Delta$. Then $\lambda_n(L) < \Delta + 2\sqrt{\Delta - 1}$.*

**Proof**: Let $k$ be a natural number so that $\mathcal{T}$ is an induced subgraph of $B_{\Delta,k}$, and let $B^*$ be the line graph of $B_{\Delta,k}$. Let $B$ be the adjacency matrix for $B^*$ and let $\mu$ be the largest eigenvalue of $B$. Since all trees are bipartite, it follows from Lemma 4.2.3 that
$$\lambda_n(L(B_{\Delta,k})) = \mu + 2. \qquad (4.2.7)$$

By Theorem 4.1.2, it suffices to show $\mu + 2 < \Delta + 2\sqrt{\Delta - 1}$. Let $x$ be a unit eigenvector of $B$ corresponding to the eigenvalue $\mu$.

Let $e_1$ and $e_2$ be edges of $B_{\Delta,k}$ having the same distance from the root vertex of $B_{\Delta,k}$. Then $e_1$ and $e_2$ are isomorphic in $B_{\Delta,k}$ and hence the vertices in $B^*$ corresponding to $e_1$ and $e_2$ are isomorphic. Since $B$ is nonnegative and irreducible, it follows from Theorem 1.3.20(iv) that $\mu$ is a simple eigenvalue of $B$. Hence $x_{e_1} = x_{e_2}$. Therefore, from this point on, for $i = 1, \ldots, k$, let $x_i$ denote any component of $x$ corresponding to any of the edges $e$ of $B_{\Delta,k}$ for which the distances of vertices incident with $e$ are $i-1$ and $i$ from the root vertex of $B_{\Delta,k}$. Letting $b_i$ denote the number of edges with component $x_i$ as in Observation 4.2.11, it follows from $x$ being a unit vector that
$$b_1 x_1^2 + b_2 x_2^2 + \ldots + b_k x_k^2 = 1. \qquad (4.2.8)$$

Let $E^*$ be the set of edges in $B^*$ and let $e_1$ and $e_2$ be edges in $B_{\Delta,k}$, hence vertices in $B^*$. It follows from Theorem 1.2.4(ii) that
$$\mu = 2 \sum_{(e_1,e_2) \in E^*} x_{e_1} x_{e_2}. \qquad (4.2.9)$$

Noting that the edges in $B_{\Delta,k}$ sharing a common vertex form a complete subgraph, it follows from (4.2.9) that

$$\begin{aligned}
\mu &= 2\left(\sum_{i=0}^{k-1} b_i \binom{\Delta-1}{2} x_{i+1}^2 + \sum_{i=1}^{k-1} b_i (\Delta-1) x_i x_{i+1}\right) \\
&= \sum_{i=0}^{k-1} b_i (\Delta-1)(\Delta-2) x_{i+1}^2 + 2 \sum_{i=1}^{k-1} b_i (\Delta-1) x_i x_{i+1} \\
&= (\Delta-2) \sum_{i=0}^{k-1} b_{i+1} x_{i+1}^2 + 2 \sum_{i=1}^{k-1} b_{i+1} x_i x_{i+1} \qquad (4.2.10) \\
&= (\Delta-2) \sum_{i=1}^{k} b_i x_i^2 + 2 \sum_{i=1}^{k-1} b_{i+1} x_i x_{i+1} \\
&= \Delta - 2 + 2 \sum_{i=1}^{k-1} b_{i+1} x_i x_{i+1},
\end{aligned}$$

where the third equality follows from Observation 4.2.11 and the last equality follows from (4.2.8). Turning our attention to the summation, define the vectors
$$p := \left[\sqrt{b_2} x_1, \sqrt{b_3} x_2, \ldots, \sqrt{b_k} x_{k-1}\right]$$

and
$$q := \left[\sqrt{b_2} x_2, \sqrt{b_3} x_3, \ldots, \sqrt{b_k} x_k\right].$$

# The Spectra of Laplacian Matrices

Applying the Cauchy-Schwarz inequality to $p$ and $q$ we obtain

$$\sum_{i=1}^{k-1} b_{i+1} x_i x_{i+1} \leq \sqrt{b_2 x_1^2 + b_3 x_2^2 + \ldots + b_k x_{k-1}^2} \sqrt{b_2 x_2^2 + b_3 x_3^2 + \ldots + b_k x_k^2}$$

$$= \sqrt{(\Delta - 1)(b_1 x_1^2 + b_2 x_2^2 + \ldots b_{k-1} x_{k-1}^2)} \sqrt{1 - b_1 x_1^2}$$

$$= \sqrt{(\Delta - 1)(1 - b_k x_k^2)} \sqrt{1 - b_1 x_1^2}$$

$$< \sqrt{\Delta - 1}, \tag{4.2.11}$$

where the first equality follows from Observation 4.2.11, the second equality follows from (4.2.8), and the final inequality follows from the fact that $x_1 \neq 0$. (Note: We know $x_1 \neq 0$ because if $x_1 = 0$ then by induction on (4.2.9) it would follow that $x_i = 0$ for all $i = 2, \ldots, k$.)

Plugging (4.2.11) into (4.2.10), we obtain

$$\mu < \Delta - 2 + 2\sqrt{\Delta - 1}.$$

From (4.2.7) we obtain

$$\lambda_n(L(B_{\Delta,k})) < \Delta + 2\sqrt{\Delta - 1}.$$

By Theorem 4.1.2 we have $\lambda_n(L) \leq \lambda_n(L(B_{\Delta,k}))$ which completes the proof. □

**EXAMPLE 4.2.13** Consider the tree $\mathcal{G}_1$ from Example 4.2.9. Observe that $\Delta = 5$. Thus by Theorem 4.2.12, we have that $\lambda_n(L(\mathcal{G}_1)) < 5 + 2\sqrt{5-1} = 9$. While this upper bound is slightly worse than the ones we got earlier, the bound in Theorem 4.2.12 is useful because of its simplicity to compute.

## Exercises:

**1.** Prove that equality holds in Theorem 4.2.1 if and only if $\mathcal{G}$ is bipartite where the vertices within each partite set have the same degree.

**2.** Use Theorems 4.2.4 and 4.2.7 to find upper bounds on the set of eigenvalues for the Laplacian matrices of the following graphs.

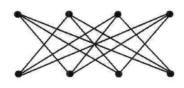

**3.** Use Theorems 4.2.4, 4.2.7, and 4.2.12 to find upper bounds on the set of eigenvalues for the Laplacian matrices of the following trees.

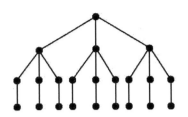
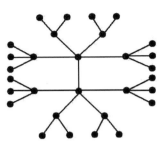

**4.** Show that equality holds in (4.2.2) when $\mathcal{G}$ is a regular bipartite graph.

**5.** Show that equality holds in Theorem 4.2.7 if $\mathcal{G}$ is a regular bipartite graph.

## 4.3 The Distribution of Eigenvalues Less than One and Greater than One

In the previous section, we investigated upper bounds on the set of all eigenvalues of the Laplacian matrix. In this section, we investigate the distribution of such eigenvalues. More specifically, we will investigate the number of eigenvalues in the interval $[0,1)$, the multiplicy of the eigenvalue $\lambda = 1$, and the number of eigenvalues in the interval $(1, \infty)$. The results we will obtain will assist us in understanding the overall structure of the graph. Since adding edges to a graph typically increases the eigenvalues according to Theorem 4.1.2, it follows that a graph having many eigenvalues in $[0,1)$ is likely to not be "as connected" as graphs in which there are more eigenvalues in $(1, \infty)$. In fact, we already saw from Theorem 4.1.1 that if zero is an eigenvalue of multiplicity greater than one, then the graph is not connected. Secondly, since we saw in Section 4.1 that graphs whose Laplacian eigenvalues are integers tend to have predicatble structures, it follows that having eigenvalues in the interval $(1, \infty)$ has the potential to be beneficial in prediciting the stucture of the graph. Finally, we will see that the multiplicity of the eigenvalue $\lambda = 1$ gives us insight into the number of pendant and quasipendant vertices of a graph. More specifically, since trees often have a proportionally high number of pendant vertices, we investigate the eigenvalues of the Laplacian matrices for trees and determine an upper and lower bound for the multiplicity of the eigenvalue $\lambda = 1$.

For a graph $\mathcal{G}$, let $p(\mathcal{G})$ denote the number of pendant vertices of $\mathcal{G}$; let $q(\mathcal{G})$ denote the number of quasipendant vertices of $\mathcal{G}$, that is, vertices that are adjacent to pendant vertices. Let $m_\mathcal{G}(\lambda)$ denote the multiplicity of the eigenvalue $\lambda$ as an eigenvalue of the Laplacian matrix of $\mathcal{G}$. Our first theorem from [26] gives a lower bound on the multiplicity of the eigenvalue 1 as an eigenvalue of the Laplacian of any graph:

**THEOREM 4.3.1** *Let $L$ be the Laplacian matrix for a graph $\mathcal{G}$. Then $m_\mathcal{G}(1) \geq p(\mathcal{G}) - q(\mathcal{G})$.*

**Proof**: Let $\mathcal{G}$ have $n$ vertices. Let $v_1, \ldots, v_p$ be the $p$ pendant vertices of $\mathcal{G}$. Let $v_{n-q+1}, v_{n-q+2}, \ldots, v_n$ be the $q$ quasipendant vertices of $\mathcal{G}$ where

$v_n$ is adjacent to $v_1, \ldots, v_{k_1}$,

$v_{n-1}$ is adjacent to $v_{k_1+1}, \ldots, v_{k_2}$,

$v_{n-2}$ is adjacent to $v_{k_2+1}, \ldots, v_{k_3}$,

$\ldots$

$v_{n-q+1}$ is adjacent to $v_{k_{q-1}+1}, \ldots v_{k_q}(=v_p)$.

Hence the Laplacian matrix $L$ for $\mathcal{G}$ is

$$\begin{bmatrix}
1 & & 0 & & & & & & & & & -1 \\
 & \ddots & & 0 & 0 & 0 & 0 & 0 & 0 & 0 & & \vdots \\
0 & & 1 & & & & & & & & & -1 \\
 & & & 1 & & 0 & & & & & -1 & \\
0 & & & & \ddots & & 0 & 0 & 0 & 0 & \vdots & 0 \\
 & & & 0 & & 1 & & & & & -1 & \\
\vdots & & \vdots & & \vdots & & \ddots & & \vdots & & & \\
 & & & & & & 1 & & 0 & & -1 & \\
0 & & 0 & & \cdots & & & \ddots & & 0 & \vdots & 0 & 0 \\
 & & & & & & 0 & & 1 & & -1 & \\
0 & & 0 & & 0 & & 0 & & * & * & * & * & * \\
0 & & 0 & & 0 & -1 & \cdots & -1 & * & d_{n-q+1} & & & * \\
0 & & 0 & & \cdots & & 0 & & * & & \cdots & & \\
0 & -1 & \cdots & -1 & 0 & & 0 & & * & & & d_{n-1} & \\
-1 & \cdots & -1 & & 0 & & 0 & & * & * & & & d_n
\end{bmatrix}$$

where the first $q$ diagonal blocks are of dimension $k_1 \times k_1, \ldots, k_q \times k_q$, and where $d_j$ is the degree of vertex $v_j$ for $j = n-q+1, \ldots, n$. Let $e_{a,b}$ be the vector on $n$ entries where the $a^{th}$ entry is 1, the $b^{th}$ entry is $-1$, and all remaining entries are 0. Letting $h_0 = 0$ and $h_i = \sum_{j=1}^i k_j$, observe that the sets $E_i := \{e_{h_i+1, h_i+2}, e_{h_i+1, h_i+3}, \ldots, e_{h_i+1, h_{i+1}}\}$ are sets of eigenvectors corresponding to the eigenvalue 1 for $i = 0, 1, \ldots, q-1$ if $k_{i+1} \geq 2$. Observe that these sets yield $p - q$ linearly independent such eigenvectors. Thus $m_\mathcal{G}(1) \geq p(\mathcal{G}) - q(\mathcal{G})$. □

We should remark that we have inequality in Theorem 4.3.1 because it is possible for there to be other linearly independent eigenvectors of $L$ corresponding to the eigenvalue 1. While this is rare, it does happen as in the following example from [26]:

# 138  Applications of Combinatorial Matrix Theory to Laplacian Matrices of Graphs

**EXAMPLE 4.3.2** Consider the graph $\mathcal{G}$ as follows:

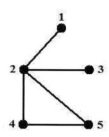

Its Laplacian matrix is:

$$L = \begin{bmatrix} 1 & -1 & 0 & 0 & 0 \\ -1 & 4 & -1 & -1 & -1 \\ 0 & -1 & 1 & 0 & 0 \\ 0 & -1 & 0 & 2 & -1 \\ 0 & -1 & 0 & -1 & 2 \end{bmatrix}$$

Note that $p(\mathcal{G}) = 2$ and $q(\mathcal{G}) = 1$, hence $p(\mathcal{G}) - q(\mathcal{G}) = 1$. However, $m_\mathcal{G}(1) = 2$. In addition to the vectors listed in the proof of Theorem 4.3.1, the vector $[-2, 0, 0, 1, 1]^T$ is also an eigenvector of $L(\mathcal{G})$ corresponding to the eigenvalue 1.

Example 4.3.2 gives rise to the question of how many eigenvalues $\lambda = 1$ are in excess of $p(\mathcal{G}) - q(\mathcal{G})$. As in [37], we define the spurious multiplicity of $\lambda = 1$ to be

$$s(\mathcal{G}) := m_\mathcal{G}(1) - [p(\mathcal{G}) - q(\mathcal{G})]. \tag{4.3.1}$$

We now prove a theorem from [37] which gives us a relevant formula for $s(\mathcal{G})$:

**THEOREM 4.3.3** *Let $\mathcal{G}$ be a graph containing $p$ pendant vertices and $q$ quasipendant vertices, and let $R$ be the set of vertices of $\mathcal{G}$ that are neither pendant nor quasipendant. Let $L$ be the Laplacian for $\mathcal{G}$. Then $s(\mathcal{G}) = m_{L[R]}(1)$.*

**Proof**: Let $v_1, \ldots, v_n$ be the vertices of $\mathcal{G}$ where $v_1, \ldots, v_r$ are the vertices of $R$, vertices $v_{r+1}, \ldots, v_{r+q}$ are the quasipendant vertices, and vertices $v_{n-p+1}, \ldots v_n$ are the pendant vertices. Assume that $v_{r+i}$ is adjacent to $v_{n-p+i}$ for each $i = 1, \ldots, q$. Then $L$ has the form

$$L = \begin{bmatrix} L[R] & X & 0 \\ X^T & Q & C \\ 0 & C^T & I \end{bmatrix}$$

where $Q$ is the $q \times q$ matrix corresponding to the quasipendant vertices and $I$ is of order $p$ corresponding to the $p$ pendant vertices. Note that the submatrix of $C$ consisting of the first $q$ columns of $C$ is $-I$. Observe that $L - I$ is row equivalent to

$$\begin{bmatrix} L[R] - I & 0 & 0 \\ 0 & 0 & B \\ 0 & B^T & 0 \end{bmatrix}$$

# The Spectra of Laplacian Matrices

where $B = [-I_q \mid 0]$. Therefore

$$m_{\mathcal{G}}(1) = \text{nullity}[L - I]$$
$$= p - q + \text{nullity}(L[R] - I).$$

From this and (4.3.1), we see that $s(\mathcal{G}) = \text{nullity}(L[R] - I) = m_{L[R]}(1)$. □

**EXAMPLE 4.3.4** Revisiting the graph $\mathcal{G}$ in Example 4.3.2, we see that $R = \{4, 5\}$. Note that

$$L[R] = \begin{bmatrix} 2 & -1 \\ -1 & 2 \end{bmatrix},$$

and that $m_{L[R]}(1) = 1$ as expected.

We now begin our discussion on the distribution of eigenvalues of the Laplacian matrix in terms of the number of eigenvalues in the interval $[0, 1)$ and the number of eigenvalues in the inverval $(1, \infty)$. To this end we supply, without proof, a necessary matrix-theoretic result from [37]:

**LEMMA 4.3.5** (i) Suppose $A$ is a $2q \times 2q$ symmetric matrix and that $\det(A_{2k}) = (-1)^k$ for all leading $2k \times 2k$ principal submatrices $A_{2k}$ for $k = 1, \ldots, q$. Then $A$ has $q$ positive and $q$ negative eigenvalues.

(ii) Suppose that $B$ is a principal submatrix of the symmetric matrix $A$. Then the number of eigenvalues of $B$ that are greater than (respectively, greater than or equal to, less than, less than or equal to) a given real number $\alpha$ is a lower bound for the number of eigenvalues of $A$ that are greater than (respectively, greater than or equal to, less than, less than or equal to) $\alpha$.

Now for the main result of this section which is also from [37]. For an interval $I$, we will denote $m_{\mathcal{G}}(I)$ as the number of eigenvalues of $L(\mathcal{G})$ in $I$, counting multiplicity.

**THEOREM 4.3.6** Let $\mathcal{G}$ be a graph and $L$ be its Laplacian. Suppose $\mathcal{G}$ has $q$ quasipendant vertices. Then $m_{\mathcal{G}}[0, 1) \geq q$ and $m_{\mathcal{G}}(1, \infty) \geq q$.

**Proof**: Number the quasipendant vertices of $\mathcal{G}$ as $1, 3, \ldots 2q - 1$. Let the vertex numbered $2k$ be a pendant vertex adjacent to vertex $2k - 1$ for $k = 1, \ldots, q$. Let $B$ be the leading $2q \times 2q$ principal submatrix of $L$. By Lemma 4.3.5(ii), it will suffice to show that $B$ has at least $q$ eigenvalues greater than one and at least $q$ eigenvalues less than one. To do this, let $C = B - I$; we will show that $C$ satisfies the hypotheses of Lemma 4.3.5(i). Observe

$$C = \begin{bmatrix} (d_1 - 1) & -1 & * & 0 & \cdots & \cdots & * & 0 \\ -1 & 0 & 0 & 0 & \cdots & \cdots & 0 & 0 \\ * & 0 & (d_3 - 1) & -1 & \cdots & \cdots & * & 0 \\ 0 & 0 & -1 & 0 & \cdots & \cdots & 0 & 0 \\ \cdots & \cdots & \cdots & \cdots & \cdots & \cdots & \cdots & \cdots \\ * & 0 & * & 0 & \cdots & \cdots & (d_{2q} - 1) & -1 \\ 0 & 0 & 0 & 0 & \cdots & \cdots & -1 & 0 \end{bmatrix}.$$

Since the even-numbered rows and columns each have a single nonzero entry, it follows that using elementary row operations and column operations involving adding even-numbered rows and columns to other rows and columns that $C$ is row/column equivalent to the direct sum of $q$ copies of $-P$ where $P = \begin{bmatrix} 0 & 1 \\ 1 & 0 \end{bmatrix}$. Hence $\det(C_{2k}) = (-1)^k$ for $k = 1, \ldots, q$. By Lemma 4.3.5(i), it follows that $C$ has at least $q$ positive and at least $q$ negative eigenvalues. Recalling that $C = B - I$ and applying Lemma 4.3.5(ii) yield the desired result. □

**EXAMPLE 4.3.7** Observe that the graph below has $q = 6$ quasipendant vertices and $p = 10$ pendant vertices:

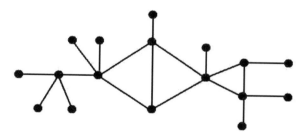

Hence by Theorem 4.3.6, there will be at least six eigenvalues of $L$ in the interval $[0, 1)$ and at least six eigenvalues of $L$ in the interval $(1, \infty)$. Moreover, by Theorem 4.3.1 it follows that $m_{\mathcal{G}}(1) \geq p - q = 4$. A computation verifies that there are exactly six eigenvalues in $[0, 1)$, the eigenvalue $\lambda = 1$ has multiplicity of four, and there are seven eigenvalues in the interval $(1, \infty)$. Finally, since $m_{\mathcal{G}}(1) = 4$, it follows that $s(\mathcal{G}) = 0$. Note that $L[R] = [3]$ and clearly $m_{L[R]}(1) = s(\mathcal{G}) = 0$ as expected from Theorem 4.3.3.

Since trees often have numerous pendant and quasipendant vertices, the spectrum of Laplacian matrices of trees is worth investigating in light of Theorem 4.3.1. For a tree $\mathcal{T}$, the quantity $p(\mathcal{T}) - q(\mathcal{T})$ is typically much greater than 1. Thus in most trees we expect 1 to appear often as an eigenvalue. However, it is often of interest to investigate the multiplicities of other eigenvalues of trees. Our next theorem from [37] discusses these ideas. Before proving this theorem, we need a matrix-theoretic lemma:

**LEMMA 4.3.8** *Let $A$ be an $n \times n$ matrix of rank $n - 1$. Then the sum of the determinants of all $(n - 1) \times (n - 1)$ principal submatrices is equal to the product of the nonzero eigenvalues.*

**Proof**: Let $\Phi(\lambda)$ be the characteristic polynomial of $A$. Since $A$ has rank $n-1$, it follows that
$$\Phi(\lambda) = \lambda f(\lambda)$$
for some polynomial $f$ where $f(0) \neq 0$. Thus
$$\Phi'(\lambda) = \lambda f'(\lambda) + f(\lambda).$$
Hence $\Phi'(0) = f(0)$ and observe that $f(0)$ is the product of the nonzero eigenvalues of $A$.

Considering $\Phi(\lambda)$ from a determinant point of view, note that
$$\Phi(\lambda) = \det(A - \lambda I).$$
By laborious calculations, we can conclude that
$$\Phi'(\lambda) = \sum_{i=1}^{n} \det(A[\bar{i}] - \lambda I)$$
where $A[\bar{i}]$ is the matrix obtained from $A$ by deleting its $i^{th}$ row and $i^{th}$ column. Hence
$$\Phi'(0) = \sum_{i=1}^{n} \det(A[\bar{i}])$$
and the lemma is proven. □

**THEOREM 4.3.9** *Suppose $\mathcal{T}$ is a tree on $n$ vertices with Laplacian matrix $L$. If $\lambda > 1$ is an integer eigenvalue of $L$ with corresponding eigenvector $x$, then*
*(i) $\lambda$ divides $n$*
*(ii) $m_{\mathcal{T}}(\lambda) = 1$*
*(iii) No coordinate of $x$ is zero.*

**Proof**: Since zero is an eigenvalue of $L$ and $L$ has all integer entries, it follows that the characteristic polynomial of $L$ is of the form $xf(x)$ where $f(x)$ is an integer monic polynomial. By the Matrix Tree Theorem (Theorem 3.2.2) the determinant of each $(n-1) \times (n-1)$ principal submatrix of $L$ is 1. Thus the sum of the determinants of all such submatrices of $L$ is $n$. Hence by Lemma 4.3.8 the product of the nonzero eigenvalues of $L$ is $n$. Therefore $f(0) = n$. Since $f(x)$ is an integer monic polynomial, the integer roots of $f(x)$ must all divide $n$ by the Rational Zero Theorem. This proves (i).

To prove (ii), we will first show that (iii) implies (ii). Using the contrapositive, suppose that $m_{\mathcal{T}}(\lambda) > 1$. Then $L$ has at least two linearly independent eigenvectors corresponding to $\lambda$. Thus we could take a linear combination of two of these eigenvectors to produce an eigenvector corresponding to $\lambda$ with a zero coordinate in any desired position. Thus (iii) implies (ii).

It now remains to show (iii). Let $x$ be an eigenvector of $L$ corresponding to $\lambda$ where the coordinates of $x$ are labeled in correspondence with the vertices $v_1, \ldots, v_n$ of $\mathcal{T}$. By way of contradiction, assume that $x$ has a zero coordinate. Without loss of generality let $x_n = 0$. Letting $d$ be the degree of $v_n$, we can write $L$ in the form

$$\begin{bmatrix} B_1 & 0 & \ldots & 0 & C_1^T \\ 0 & B_2 & \ldots & 0 & C_2^T \\ \ldots & \ldots & \ldots & \ldots & \ldots \\ 0 & 0 & \ldots & B_d & C_d^T \\ C_1 & C_2 & \ldots & C_d & d \end{bmatrix}$$

where each $B_i$ corresponds to a branch $\mathcal{T}_i$ at $v_n$. Partitioning $x$ conformally as $x = [x_1, x_2, \ldots, x_d, 0]^T$, we see that $Lx = \lambda x$ implies $B_i x_i = \lambda x_i$ for all $1 \leq i \leq d$. Since at least one of these $x_i$'s is nonzero, $\lambda$ is an eigenvalue for some $B_i$, say $B_1$ without loss of generality. Looking more closely at the branch $\mathcal{T}_1$, observe that the matrices $B_1$ and $L(\mathcal{T}_1)$ differ by exactly one entry, namely that the diagonal entry corresponding to the vertex adjacent in $\mathcal{T}$ to $v_n$ is one larger in $B_1$ than it is in $L(\mathcal{T}_1)$. Without loss of generality, let $v_1$ be the vertex adjacent in $\mathcal{T}$ to $v_n$. Thus $B_1 = L(\mathcal{T}_1) + E_{11}$ where $E_{11}$ is the matrix whose only nonzero entry is one in the $(1,1)$ position. Then $\det B_1 = \det L(\mathcal{T}_1) + \det L_{11}$ where $L_{11}$ is the submatrix of $L(\mathcal{T}_1)$ obtained by eliminating its first row and column. Since $L(\mathcal{T}_1)$ is singular, it follows that $\det L(\mathcal{T}_1) = 0$. By the Matrix Tree Theorem it follows that $\det L_{11} = 1$. Thus $\det B_1 = 1$. Hence we see that $\lambda > 1$ is an integer eigenvalue the matrix $B_1$ whose determinant is 1. Since the determinant of a matrix equals the product of its eigenvalues, it follows that if $g(x)$ is the characteristic polynomial of $B_1$, then $g(0) = 1$. Since $g(x)$ is an integer monic polynomial whose constant term in one, it follows from the Rational Zero Theorem that the only possible integer eigenvalues of $B_1$ are 1 and $-1$. Hence we have deduced a contradiction. $\square$

**REMARK 4.3.10** We should note that Theorem 4.3.9 cannot be extended to graphs in general. For example, consider $C_6$, the cycle on 6 vertices. Letting $L$ be the Laplacian for $C_6$. The eigenvalues of $L$, counting multiplicity, are $0, 1, 1, 3, 3, 4$, clearly defying (i), (ii), and (iii) of Theorem 4.3.9.

**REMARK 4.3.11** In a tree $\mathcal{T}$, $\lambda = 1$ is the only integer than can be an eigenvalue of $L$ of multiplicity greater than 1. All other integers, if they are eigenvalues of $L$, must be simple.

For the Laplacian matrix of a tree, in addition to the integer eigenvalues greater than one being simple, the next theorem from [37] shows that the largest eigenvalue of such matrices is always simple.

**THEOREM 4.3.12** *Let $\mathcal{G}$ be a connected bipartite graph with $L$ as its Laplacian. Then the maximum eigenvalue of $L$ is simple. Specifically, if $L$ is the Laplacian matrix for a tree, then the maximum eigenvalue of $L$ is simple.*

**Proof**: Since $\mathcal{G}$ is bipartite, we can partition the vertices of $\mathcal{G}$ into two subsets $V_1$ and $V_2$ so that no two vertices in $V_i$ are adjacent for $i = 1, 2$. Let $U$ be the diagonal matrix in which the $i^{th}$ diagonal entry, $u_{ii}$, is as follows:

$$u_{ii} = \begin{cases} 1, & \text{if } v_i \in V_1 \\ -1, & \text{if } v_i \in V_2. \end{cases}$$

Observe that $ULU^{-1} = |L|$. Thus $L$ and $|L|$ are unitarily similar and, hence, have the same spectrum. Since $|L|$ is nonnegative and irreducible, it follows from Theorem 1.3.8 that the maximum eigenvalue of $|L|$ is simple. The result follows for all bipartite graphs. Since all trees are bipartite, the result follows for trees. □

We close this section with a theorem from [37] that further illustrates how the multiplicity of eigenvalues for the Laplacian matrix of a tree is related to the number of pendant vertices of a tree. This theorem applies to both integer and noninteger eigenvalues.

**THEOREM 4.3.13** *Let $\mathcal{T}$ be a tree with Laplacian matrix $L(\mathcal{T})$. Let $\lambda$ be an eigenvalue of $L(\mathcal{T})$. If $\mathcal{T}$ has $p$ pendant vertices, then $m_\mathcal{T}(\lambda) \leq p - 1$.*

**Proof**: Suppose that $v_n$ is a pendant vertex of $\mathcal{T}$ and let $v_{n-1}$ be the quasipendant vertex of $\mathcal{T}$ adjacent to $v_n$. Let $x = [x_1, \ldots, x_n]^T$ be an eigenvector of $L(\mathcal{T})$ corresponding to $\lambda$. Then by the eigenvalue-eigenvector relationship we have $(1 - \lambda)x_n = x_{n-1}$. First suppose that $x_n = 0$. Then $x_{n-1} = 0$. In such a case, the vector $x' = [x_1, \ldots, x_{n-1}]^T$ is an eigenvector of $L(\mathcal{T}')$ corresponding to $\lambda$ where $\mathcal{T}'$ is the tree obtained from $\mathcal{T}$ by deleting $v_n$ and the edge incident to it. Thus if $x_r = 0$ for all pendant vertices $r$, it would follow by induction that $x = 0$, a contradiction. The same holds true if $x_r = 0$ for all but one pendant vertices $r$. On the other hand, if the eigenspace $W$ of $L(\mathcal{T})$ corresponding to $\lambda$ were to have dimension greater than $p - 1$, then via linear combinations it would be possible to create an eigenvector $w \in W$ that is zero on all but (at most) one of its coordinates corresponding to pendant vertices. □

Theorems 4.3.13 and 4.3.1 yield the following immediate corollary which provides upper and lower bounds on the algebraic multiplicity of the eigenvalue $\lambda = 1$ of a tree.

**COROLLARY 4.3.14** *Let $\mathcal{T}$ be a tree with Laplacian matrix $L$ and suppose $\mathcal{T}$ has $p$ pendant vertices and $q$ quasipendant vertices, then $p - q \leq m_\mathcal{T}(1) \leq p - 1$.*

# Exercises:

1. Consider the following graph $\mathcal{G}$:

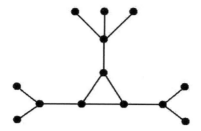

(a) Using Theorems 4.3.1 and 4.3.3, determine $m_\mathcal{G}(1)$ exactly.

(b) Determine a lower bound on the number of eigenvalues in the interval $[0, 1)$ and in the interval $(1, \infty)$.

**2.** (See [37]) Let $\mathcal{G}$ be a graph on $n$ vertices with $p$ pendant vertices and $q$ quasipendant vertices. Define $r = n - p - q$. Let $R$ be the subgraph of $\mathcal{G}$ induced by the $r$ vertices that are not pendant or quasipendant. Let $\alpha$ denote the maximum number of vertices in any independent set of $R$. Letting $s$ be the spurious multiplicity of the eigenvalue $\lambda = 1$ of $L$ as defined in (4.3.1), prove that $s \leq r - \alpha$.

**3.** (See [37]) Let $\mathcal{G}$ be a graph on $n$ vertices and let $\mathcal{H}$ be the graph obtained by making a vertex of $\mathcal{G}$ adjacent to a pendant vertex of $P_3$, the path on three vertices. Prove that $m_\mathcal{G}(1) = m_\mathcal{H}(1)$.

**4.** Consider the following tree:

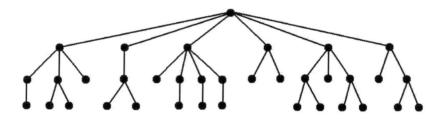

Let $L$ be its Laplacian matrix. Find a lower bound for

(a) The multiplicity of the eigenvalue $\lambda = 1$.

(b) The multiplicity of each other integer $2, \ldots, n$ as an eigenvalue of $L$. (Note that the mulitiplicity is zero if such an integer is not an eigenvalue).

(c) Any noninteger eigenvalue $\lambda$.

**5.** Use Corollary 4.3.14 to find all of the eigenvalues for the Laplacian matrix of a star on $n$ vertices.

*The Spectra of Laplacian Matrices*

## 4.4 The Grone-Merris Conjecture

The Grone-Merris conjecture gives an upper bound on each individual eigenvalue of the Laplacian matrix of a graph in terms of the degrees of the vertices. It states that the spectrum of the Laplacian matrix for any graph is majorized by the conjugate of the degree sequence of the graph (see Definition 2.5.11). The Grone-Merris Conjecture is actually a misnomer in that it is now a theorem, proven in 2010 by Hau Bai in [3]. The goal of this section, based mainly on [3] is to provide details of the proof. To begin the process of proving the Grone-Merris Conjecture we begin with a lemma from [3] concerning the Laplacian spectrum of split graphs (see Definition 2.2.17).

**LEMMA 4.4.1** *Let $\mathcal{G}$ be a split graph on $n$ vertices whose clique contains $N$ vertices. Let $L$ be the Laplacian of $\mathcal{G}$. Then*

$$\lambda_{n-N+2} \geq N \geq \tilde{\Delta}(\mathcal{G}) \geq \lambda_{n-N} \quad (4.4.1)$$

*where $\tilde{\Delta}(\mathcal{G})$ denotes the maximum degree of the vertices in the co-clique of $\mathcal{G}$.*

**Proof:** Let the co-clique of $\mathcal{G}$ contain $M$ vertices, thus $N + M = n$. Let $P$ be the $(N-1)$-dimensional subspace of $\Re^n$ consisting of vectors of the form

$$p := \begin{bmatrix} u \\ 0 \end{bmatrix}$$

where $u$ has $N$ entries, $0$ has $M$ entries, and $u^T e = 0$. Write $L$ as

$$L = \begin{bmatrix} L(K_N) + D_1 & -A \\ -A & D_2 \end{bmatrix}$$

where $D_1$ and $D_2$ are the appropriate diagonal matrices. Then for any unit vector $u$, we have

$$\begin{aligned} p^T L p &= u^T (L(K_N) + D_1) u &= u^T (NI - J + D_1) u \\ &= N u^T u - u^T J u + u^T D_1 u \\ &= N + u^T D_1 u \\ &\geq N. \end{aligned}$$

Since $p$ is a unit vector and $P$ is an $N-1$-dimensional subspace orthogonal to $e$, it follows from Theorem 1.2.7 that

$$\begin{aligned} \lambda_{n-N+2} &= \lambda_{n-(N-1)+1} \\ &= \max_{x \in P} x^T L x \\ &\geq p^T L p \\ &\geq N, \end{aligned}$$

proving the first inequality in (4.4.1). By similar reasoning, if we let $Q$ denote the $M$-dimensional subspace of $\Re^n$ consisting of vectors of the form

$$q := \begin{bmatrix} 0 \\ u \end{bmatrix}$$

where $u$ has $u$ has $M$ entries, 0 has $N$ entries, and $u^T e = 0$, then

$$\lambda_{n-N} \leq q^T L q = u^T D_2 u \leq \tilde{\Delta}(\mathcal{G}).$$

Since each vertex in the co-clique of $\mathcal{G}$ can be adjacent to at most $N$ vertices, namely the vertices in the clique of $\mathcal{G}$, it follows that $\tilde{\Delta}(\mathcal{G}) \leq N$. This completes the proof. $\square$

Observe that in the proof of Lemma 4.4.1 we labeled the eigenvalues in terms of $\lambda_{n-i}$, the $(i+1)^{st}$ largest, rather than $\lambda_i$, the $i^{th}$ smallest. This is because for purposes of majorization, we require that a sequence of numbers be listed in non-increasing order. Therefore we will denote the sequence of eigenvalues of $L(\mathcal{G})$ by $\lambda(\mathcal{G}) = (\lambda_n, \lambda_{n-1}, \ldots, \lambda_1)$.

We further our study of the spectrum of split graphs in the next lemma from [3] which states a partial result of the Grone-Merris Conjecture. We state this lemma without proof.

**LEMMA 4.4.2** *Let $\mathcal{G}$ be a split graph on $n$ vertices whose clique has $N$ vertices. Let $L$ be the Laplacian for $\mathcal{G}$. If either $\lambda_{n-N+1} > N$ or $\lambda_{n-N+1} = N > \Delta(\mathcal{G})$ (where $\Delta(\mathcal{G})$ denotes the maximum vertex degree in $\mathcal{G}$), then*

$$\sum_{i=1}^{N} \lambda_{n-i+1} \leq \sum_{i=1}^{N} d_i^*.$$

When proving the Grone-Merris conjecture, we will investigate a split graph created from a graph that is assumed to be a counterexample to the conjecture. In essence, we will be proving the Grone-Merris conjecture by way of contradiction. In the proof of the Grone-Merris conjecture, the counterexample we consider is the graph on the fewest number of edges in which there exists an integer $k$ defying Definition 2.5.7. The following lemma from [3] gives us insight as to what the structure of such a counterexample would be if such a counterexample exists.

**LEMMA 4.4.3** *Let $\mathcal{G}$ be the graph with the fewest number of edges containing a minimum value $k$ such that*

$$\sum_{i=1}^{k-1} \lambda_{n-i+1} \leq \sum_{i=1}^{k-1} d_i^* \quad \text{and} \quad \sum_{i=1}^{k} \lambda_{n-i+1} > \sum_{i=1}^{k} d_i^*.$$

*Suppose $\mathcal{G}$ has $n$ vertices and that $i$ and $j$ are vertices such that $d_i \leq k$ and $d_j \leq k$. Then $i$ and $j$ are not adjacent.*

**Proof**: We will prove this by contradiction. Assume this lemma is false, then there exists a pair of adjacent vertices $i$ and $j$ where $d_i \leq k$ and $d_j \leq k$. Let $\tilde{\mathcal{G}}$ be the graph obtained from $\mathcal{G}$ by deleting the edge $ij$. Since $\mathcal{G}$ is the graph on the fewest number of edges which satisfies the statement of the lemma, it follows that

$$\sum_{i=1}^{k} \lambda_{n-i+1}(\tilde{\mathcal{G}}) \leq \sum_{i=1}^{k} d_i^*(\tilde{\mathcal{G}}).$$

Observing that $L(\mathcal{G}) = L(\tilde{\mathcal{G}}) + E_{ij}$, it follows from the fact that $\lambda(H_1 + H_2) \preceq \lambda(H_1) + \lambda(H_2)$ for any symmetric matrices $H_1$ and $H_2$ that

$$\begin{aligned}
\sum_{i=1}^{k} \lambda_{n-i+1}(\mathcal{G}) &\leq \sum_{i=1}^{k} \lambda_{n-i+1}(\tilde{\mathcal{G}}) + \sum_{i=1}^{k} \lambda_{n-i+1}(E_{ij}) \\
&\leq \sum_{i=1}^{k} d_i^*(\tilde{\mathcal{G}}) + 2 \\
&= \left[\sum_{i=1}^{k} d_i^*(\mathcal{G}) - 2\right] + 2 \\
&= \sum_{i=1}^{k} d_i^*(\mathcal{G}).
\end{aligned}$$

However, this contradicts the minimality assumptions of $\mathcal{G}$ in the statement of the lemma, thus proving the lemma. $\square$

We are now ready to proceed with the main result of this section, the proof of the Grone-Merris Conjecture. This proof is taken from [3].

**THEOREM 4.4.4 The Grone-Merris Conjecture** *Let $\mathcal{G}$ be a graph on $n$ vertices with degree sequence $d(\mathcal{G})$ and Laplacian spectrum $\lambda(\mathcal{G})$. Then $\lambda(\mathcal{G}) \preceq d^*(\mathcal{G})$.*

**Proof**: We will prove this theorem by contradiction. Suppose that this theorem is not true. Let $\mathcal{G}$ be a counterexample to the theorem. Then there exists an integer $k$ with $1 < k < n$ such that

$$\sum_{i=1}^{k} \lambda_{n-i+1} > \sum_{i=1}^{k} d_i^*.$$

Without loss of generality, we assume that this integer $k$ is minimum over all counterexamples. Thus we have

$$\sum_{i=1}^{k-1} \lambda_{n-i+1} \leq \sum_{i=1}^{k-1} d_i^* \quad \text{and} \quad \lambda_{n-k+1} > d_k^*.$$

We can assume further without loss of generality that the number of edges in $\mathcal{G}$ is minimum over all counterexamples with the same $k$. Our goal will be to show that no such graph $\mathcal{G}$ exists, thus establishing our contradiction.

We now create a new graph $\hat{\mathcal{G}}$ by adding edges to $\mathcal{G}$ joining all pairs of nonadjacent vertices $i, j$ in $\mathcal{G}$ such that both $d_i \geq k$ and $d_j \geq k$. By Lemma 4.4.3 we see that in $\mathcal{G}$, if vertices $i$ and $j$ are such that $d_i < k$ and $d_j < k$, then $i$ and $j$ are not adjacent in $\mathcal{G}$, hence not adjacent in $\hat{\mathcal{G}}$. Therefore $\mathcal{G}$ is a split graph where the clique of $\hat{\mathcal{G}}$ consists of the vertices in $\mathcal{G}$ whose degree (in $\mathcal{G}$) is at least $k$. Thus the clique contains $N := d_k^*(\mathcal{G})$ vertices. Also observe that the co-clique of $\hat{\mathcal{G}}$ consists of the vertices in $\mathcal{G}$ whose degree (in $\mathcal{G}$) is less than $k$. Hence the maximum degree (in $\hat{\mathcal{G}}$) of the vertices in $\hat{\mathcal{G}}$ is $\Delta(\hat{\mathcal{G}}) \leq k - 1$. Note that

$$d_i^*(\hat{\mathcal{G}}) = d_i^*(\mathcal{G}) \text{ for all } i = 1, \ldots, k$$

and by Theorem 4.1.2:

$$\lambda_i(\hat{\mathcal{G}}) \geq \lambda_i(\mathcal{G}) \text{ for all } i = 1, \ldots, n.$$

Therefore, it follows that

$$\sum_{i=1}^{k} \lambda_{n-i+1}(\hat{\mathcal{G}}) > \sum_{i=1}^{k} d_i^*(\hat{\mathcal{G}}) \quad \text{and} \quad \lambda_{n-k+1}(\hat{\mathcal{G}}) > d_k^*(\hat{\mathcal{G}}) = N. \quad (4.4.2)$$

The proof by contradiction will be complete when we show that none of $N < k$, $N = k$, or $N > k$ is possible. If $N < k$ then since this is equivalent to $N + 1 \leq k$, we obtain $\lambda_{n-k+1}(\hat{\mathcal{G}}) \leq \lambda_{n-(N+1)+1} \leq N$ where the second inequality follows from Lemma 4.4.1. Simplifying we obtain $\lambda_{n-k+1}(\hat{\mathcal{G}}) \leq N$ which contradicts (4.4.2). If $N = k$ then we have that $\hat{\mathcal{G}}$ is a split graph with a clique of $N$ vertices and

$$\sum_{i=1}^{N} \lambda_{n-i+1}(\hat{\mathcal{G}}) > \sum_{i=1}^{N} d_i^*(\hat{\mathcal{G}}) \quad \text{and} \quad \lambda_{n-N+1}(\hat{\mathcal{G}}) > N$$

by (4.4.2). But this contradicts Lemma 4.4.2.

The only remaining possibility is that $N > k$. Recall that $\hat{\mathcal{G}}$ is a split graph whose clique contains $N$ vertices. Note that the minimum degree of the vertices in the clique is at least $N - 1$ while the maximum degree of the vertices in the co-clique is at most $k - 1$. Since $N > k$, it follows that

$$d_{N-1}^*(\hat{\mathcal{G}}) = \ldots = d_{k+1}^*(\hat{\mathcal{G}}) = d_k^*(\hat{\mathcal{G}}) = N.$$

Recalling from Lemma 4.4.1 that $\lambda_{n-(N-1)+1}(\hat{\mathcal{G}}) \geq N$, it follows that the inequality

$$\sum_{i=1}^{k} \lambda_{n-i+1}(\hat{\mathcal{G}}) > \sum_{i=1}^{k} d_i^*(\hat{\mathcal{G}}) \quad \text{can be extended to} \quad \sum_{i=1}^{N-1} \lambda_{n-i+1}(\hat{\mathcal{G}}) > \sum_{i=1}^{N-1} d_i^*(\hat{\mathcal{G}}).$$

We complete the case of showing that $N > k$ is not possible by considering the relationship between $\lambda_{n-N+1}(\hat{\mathcal{G}})$ and $N$ and showing that neither $\lambda_{n-N+1}(\hat{\mathcal{G}}) \geq N$ nor $\lambda_{n-N+1}(\hat{\mathcal{G}}) < N$ is possible. If $\lambda_{n-N+1}(\mathcal{G}) \geq N$, then since $N = d_{N-1}^*(\hat{\mathcal{G}}) \geq d_N^*(\hat{\mathcal{G}})$, we have that

$$\sum_{i=1}^{N} \lambda_{n-i+1}(\hat{\mathcal{G}}) > \sum_{i=1}^{N} d_i^*(\hat{\mathcal{G}}) \quad \text{and} \quad \lambda_{n-N+1}(\hat{\mathcal{G}}) > N > \Delta(\hat{\mathcal{G}})$$

# The Spectra of Laplacian Matrices

which contradicts Lemma 4.4.2. If $\lambda_{n-N+1}(\hat{\mathcal{G}}) < N$, then we consider the complement of $\hat{\mathcal{G}}$ which is also a split graph. Letting $M$ be the number of vertices in the clique of $\hat{\mathcal{G}}^c$ and noting that there are $N$ vertices in the co-clique of $\mathcal{G}^c$, it follows from Theorem 4.1.11 that

$$\lambda_{n-M+1}(\hat{\mathcal{G}}^c) = (N+M) - \lambda_{n-N+1}(\hat{\mathcal{G}}) > M.$$

Recall that if $\mathcal{G}$ is a graph on $n$ vertices with conjugate degree sequence $d^*(\mathcal{G}) = (d_1^*, \ldots, d_n^*)$ then $d^*(\mathcal{G}^c) = (n - d_{n-1}^*, \ldots, n - d_1^*, 0)$. Therefore for $k = 1, \ldots, n$, we have

$$\sum_{i=1}^{k} \lambda_{n-i+1}(\mathcal{G}) \leq \sum_{i=1}^{k} d_i^*(\mathcal{G})$$

if and only if

$$\sum_{j=1}^{n-1-k} \lambda_{n-j+1}(\mathcal{G}^c) \leq \sum_{j=1}^{n-1-k} d_j^*(\mathcal{G}^c).$$

Therefore

$$\sum_{i=1}^{M} \lambda_{n-i+1}(\hat{\mathcal{G}}) > \sum_{i=1}^{M} d_i^*(\hat{\mathcal{G}}) \quad \text{and} \quad \lambda_{n-M+1}(\hat{\mathcal{G}}^c) > M$$

which contradicts Lemma 4.4.2.

We have proven that such a graph $\hat{\mathcal{G}}$, and hence a counterexample $\mathcal{G}$, cannot exist. This proves the theorem. $\square$

Theorem 4.4.4 shows that the spectrum of the Laplacian matrix of any graph is majorized by the conjugate of the degree sequence. We now proceed with two examples.

**EXAMPLE 4.4.5** Consider the following graph $\mathcal{G}$:

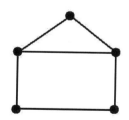

Observe that its degree sequence $(d) = (3, 3, 2, 2, 2)$. Using a Ferrers-Sylvester diagram discussed in Section 2.5, we see that $(d^*) = (5, 5, 2, 0, 0)$. Computing the eigenvalues of $L(\mathcal{G})$ to be $\lambda(\mathcal{G}) = (4.62, 3.62, 2.38, 1.38, 0)$, we obtain the following chart:

| $k$ | $\sum_{i=1}^{k} \lambda_{n-i+1}$ | $\sum_{i=1}^{k} d_i^*$ |
|---|---|---|
| 1 | 4.62 | 5 |
| 2 | 8.24 | 10 |
| 3 | 10.62 | 12 |
| 4 | 12 | 12 |
| 5 | 12 | 12 |

Observe that $\sum_{i=1}^{k} \lambda_{n-i+1} \leq \sum_{i=1}^{k} d_i^*$ for all $k = 1, \ldots 5$ and that equality holds when $k = 5$. Thus $\lambda(\mathcal{G}) \preceq d^*(\mathcal{G})$ as the Grone-Merris Conjecture states.

**EXAMPLE 4.4.6** Consider the following graph $\mathcal{G}$:

Observe that $(d) = (5, 2, 2, 1, 1, 1)$ and that $(d^*) = (6, 3, 1, 1, 1, 0)$. Computing the eigenvalues of $L(\mathcal{G})$, we see that $\lambda(\mathcal{G}) = (6, 3, 1, 1, 1, 0)$. Thus not only does the Grone-Merris Conjecture hold, but we have equality in $\sum_{i=1}^{k} \lambda_{n-i+1} \leq \sum_{i=1}^{k} d_i^*$ for all $k$. Graphs in which equality holds for all $k$ will be discussed in the next section.

# Exercises:

**1.** Verify that the Grone-Merris Conjecture holds in the following graphs:

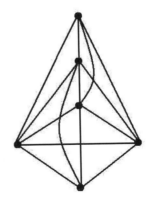

**2.** Find a graph such that $\lambda_{n-i+1} > d_i^*$ for some $1 \leq i \leq n$. This shows that while the sum of the $k$ largest eigenvalues is bounded above by the sum of the $k$ largest degree conjugates, the individual eigenvalues need not be bounded above by the individual degree conjugates.

**3.** Prove that for complete graphs, equality holds in the Grone-Merris Conjecture for all $k$.

4. Let $\mathcal{T}$ be a tree. Prove that equality holds in the Grone-Merris Conjecture for all $k$ if and only if $\mathcal{T}$ is a star.

5. Prove that split graphs of the form $K_m \vee (K_p)^c$ for some $m$ and $p$ are the only split graphs in which equality holds in the Grone-Merris Conjecture for all $k$.

## 4.5 Maximal (Threshold) Graphs and Integer Spectra

In the previous section we proved the Grone-Merris Conjecture which states that for all graphs, the spectrum is majorized by the conjugate of the degree sequence, i.e.,

$$\sum_{i=1}^{k} \lambda_{n-i+1} \leq \sum_{i=1}^{k} d_i^* \quad (4.5.1)$$

for all $k = 1, \ldots, n$. In this section, we investigate the conditions under which equality holds in (4.5.1) for each $k = 1, \ldots, n$ and determine that equality holds if and only if the graph is a maximal graph (see Definition 2.5.14). Recall that Theorem 2.5.24 says the set of all maximal graphs can be partitioned in sets $\Gamma_k$ where $\Gamma_1$ is the set of all complete graphs and $\Gamma_k = \{\mathcal{G}^c \vee K_r : \mathcal{G} \in \Gamma_{k-1}\}$ for integers $k \geq 2$. We apply this theorem to prove a theorem from [59] which shows that the sequence of eigenvalues $\lambda(\mathcal{G})$ of the Laplacian matrix for a maximal graph is precisely the conjugate $(d^*)$ of the degree sequence of $\mathcal{G}$ together with zero.

**THEOREM 4.5.1** *If $\mathcal{G}$ is a maximal graph on $n$ vertices with Laplacian matrix $L$, then all eigenvalues are integers. Moreover, if $(d) = (d_1, d_2, \ldots, d_n)$ is the degree sequence of $\mathcal{G}$, and $\lambda_1 \leq \lambda_2 \leq \ldots \leq \lambda_n$ are the eigenvalues of $L$, then $\lambda_1 = 0$ and $\lambda_{n-i+1} = d_i^*$ for $1 \leq i \leq n-1$.*

**Proof:** Clearly $\lambda_1 = 0$, by Theorem 4.1.1. By Theorem 2.5.24, it suffices to prove this theorem for $\Gamma_k$ for each integer $k \geq 1$. We will prove by induction on $k$ that the entries in $(d^*)$ are the remaining eigenvalues of $L$. For $k = 1$, we have $\Gamma_1$ which is the family of complete graphs; the theorem is clearly true for this family of graphs. Now suppose that the theorem is true for every graph in $\Gamma_k$ for a fixed $k \geq 1$. Let $\mathcal{G} \neq K_n$ be a graph on $n$ vertices such that $\mathcal{G} \in \Gamma_{k+1}$. Then $\mathcal{G} = \mathcal{H}^c \vee K_r$ for some $\mathcal{H} \in \Gamma_k$ and for some $r \geq 1$. To account for eigenvalues $\lambda_{n-i+1}$ of $L(\mathcal{G})$ where $1 \leq i \leq r$, we consider the graphs $K_r$ and $\mathcal{H}^c$ which are joined to create $\mathcal{G}$. Since $\mathcal{H}^c$ contains $n - r$ vertices, and $r$ is an eigenvalue of $L(K_r)$ with multiplicity $r - 1$, it follows from Theorem 4.1.6 that $r + (n - r) = n$ is an eigenvalue of $L(\mathcal{G})$ with a multiplicity of $r - 1$ (so far). But since $\mathcal{G}$ is the join of two graphs, it follows from Theorem 4.1.8 that $\lambda_n(L(\mathcal{G})) = n$. Hence $n$ is an eigenvalue of $L(\mathcal{G})$ with multiplicity $r$, and therefore $\lambda_{n-i+1} = n$ for all $1 \leq i \leq r$. But since $\mathcal{G} = \mathcal{H} \vee K_r$, it follows that there are exactly $r$ vertices of $\mathcal{G}$ that are adjacent to every vertex in $\mathcal{G}$. Hence $d_i^* = n$ for $1 \leq i \leq r$. Consequently $\lambda_{n-i+1} = d_i^*$ for $1 \leq i \leq r$.

We now account for eigenvalues $\lambda_{n-i+1}$ of $L(\mathcal{G})$ where $r + 1 \leq i \leq n - 1$. Since $\mathcal{H}$ consists of $n - r$ vertices and $\mathcal{H} \in \Gamma_k$, it follows by the inductive hypothesis,

$$\lambda_{n-r-i+1}(L(\mathcal{H})) = d_i^*(H) \quad (4.5.2)$$

for all $1 \leq i \leq n-r-1$. Observe from the construction of the Ferres-Sylvester diagram for $\mathcal{G} = \mathcal{H}^c \vee K_r$ that

$$d_i^*(\mathcal{H}) = n - d_{n-i}^*(\mathcal{G}) \tag{4.5.3}$$

for $1 \leq i \leq n-r-1$. Combining (4.5.2) and (4.5.3) we obtain

$$\lambda_{n-r-i+1}(L(\mathcal{H})) = n - d_{n-i}^*(\mathcal{G}) \tag{4.5.4}$$

for $1 \leq i \leq n-r-1$. Since $\mathcal{H}$ is a graph on $n-r$ vertices, it follows from Theorem 4.1.11 that

$$\lambda_i(L(\mathcal{H})) = n - r - \lambda_{n-r-i+2}(L(\mathcal{H}^c)) \tag{4.5.5}$$

for $2 \leq i \leq n-r$. Combining (4.5.4) and (4.5.5) we obtain

$$n - r - \lambda_{n-r-i+2}(L(\mathcal{H}^c)) = n - d_{r+i-1}^*(\mathcal{G}) \tag{4.5.6}$$

for $2 \leq i \leq n-r$, which simplifies to

$$r + \lambda_{n-r-i+2}(L(\mathcal{H}^c)) = d_{r+i-1}^*(\mathcal{G}). \tag{4.5.7}$$

Reindexing the subscripts, (4.5.7) can be rewritten as

$$r + \lambda_{n-i+1}(L(\mathcal{H}^c)) = d_i^*(\mathcal{G}) \tag{4.5.8}$$

for $r+1 \leq i \leq n-1$. Since $K_r$ has $r$ vertices and $\mathcal{G} = \mathcal{H}^c \vee K_r$, it follows from Theorem 4.1.6 that if $\lambda$ is an eigenvalue of $L(\mathcal{H}^c)$ then $r + \lambda$ is an eigenvalue of $L(\mathcal{G})$. Since $\lambda \leq n - r < n$, it follows that the eigenvalues of $L(\mathcal{G})$ of the form $r + \lambda$ are the eigenvalues $\lambda_i$ of $L(\mathcal{G})$ where $2 \leq i \leq n-r$. Reindexing we obtain $r + \lambda_{n-i+1}(L(\mathcal{H}^c)) = \lambda_{n-i+1}(L(\mathcal{G}))$ for $r+1 \leq i \leq n-1$. This together with (4.5.8) proves the theorem. □

**REMARK 4.5.2** While it may seem cumbersome to investigate the eigenvalues in terms of $\lambda_{n-i+1}$ instead of in terms of $\lambda_i$, we do so to be consistent with the previous section. In the previous section, we were concerned with the comparison of the sum of the $k$ largest eigenvalues with the sum of the $k$ largest values in the conjugate degree sequence. However, with maximal graphs these sums are equal for all $k$. Hence each eigenvalue is equal to its corresponding term in the conjugate degree sequence. Hence it is possible to study the eigenvalues of maximal graphs in a less cumbersome manner. This leads us to the following corollary:

**COROLLARY 4.5.3** *Let $\mathcal{G}$ be a maximal graph on $n$ vertices with degree sequence $(d) = (d_1, d_2, \ldots, d_n)$. Let $L$ be the Laplacian matrix of $\mathcal{G}$ with eigenvalues $\lambda_1 \leq \lambda_2 \leq \ldots \leq \lambda_n$. Then $\lambda_1 = 0$ and $\lambda_i = d_{n-i+1}^*$ for $2 \leq i \leq n$.*

**EXAMPLE 4.5.4** Let's reconsider the graph $\mathcal{G}$ in Example 4.4.6:

Observe that $\mathcal{G} = K_1 \vee (K_2^c \vee K_3)^c \in \Gamma_3$ and thus by Theorem 2.5.24 it follows that $\mathcal{G}$ is maximal. Recall that we had equality in the Grone-Merris Conjecture for all $k$ as Theorem 4.5.1 verifies. Since $(d^*) = (6, 3, 1, 1, 1)$, by Corollary 4.5.3 the eigenvalues of $L(\mathcal{G})$ are $\lambda_1 = 0$, $\lambda_2 = 1$, $\lambda_3 = 1$, $\lambda_4 = 1$, $\lambda_5 = 3$, $\lambda_6 = 6$.

Maximal graphs lead us to a convenient definition:

**DEFINITION 4.5.5** A graph is *Laplacian integral* if all of the eigenvalues of its Laplacian matrix are integers.

Recall from Theorem 4.1.2 that if an edge is added to a graph joining two nonadjacent vertices, each eigenvalue of the Laplacian matrix will either increase or remain unchanged. With this in mind, we now turn our attention to graphs (with a distinct focus on maximal graphs) in which adding an edge causes all eigenvalues to change by an integer (possibly zero). This leads us to another definition:

**DEFINITION 4.5.6** Let $\mathcal{G}$ be a graph and suppose we add an edge $e$ joining two nonadjacent vertices. We say that a *spectral integral variation* occurs in $k$ places if adding $e$ causes $k$ of the eigenvalues of $L(\mathcal{G})$ to increase by a positive integer.

If $\mathcal{G}$ is Laplacian integral and adding an edge $e$ to $\mathcal{G}$ causes all of the eigenvalues to increase by an integer, we have then created another Laplacian integral graph. Observe that by adding an edge $e$, we are increasing two of the diagonal entries of $L(\mathcal{G})$ each by one. Thus $Tr(L(\mathcal{G}+e)) = Tr(L(\mathcal{G})) + 2$. Hence if the eigenvalues of $L(\mathcal{G})$ increase only in integer increments then either (a) two eigenvalues of $L(\mathcal{G})$ each increase by one, or (b) one eigenvalue of $L(\mathcal{G})$ increases by two. Thus if spectral integral variation occurs by adding the edge $e$, then it can occur only in either one or two places, and all of the remaining eigenvalues remain unchanged. The next goal of this section is to determine the conditions for eigenvalues of the Laplacian matrix to remain unchanged when an edge is added to a graph. To this end, we need three matrix theoretic lemmas from [76].

**LEMMA 4.5.7** *Let $A$ be a symmetric matrix partitioned*

$$A = \begin{bmatrix} a & x^T \\ x & Z \end{bmatrix}$$

*where $x$ is an $(n-1)$-vector and $Z$ is a symmetric $(n-1) \times (n-1)$ matrix. If the spectrum of $Z$ consists of $n-1$ eigenvalues of $A$, then $x = 0$.*

**Proof**: Let $\{\alpha_1, \ldots, \alpha_n\}$ be the set of eigenvalues of $A$ and suppose the set of eigenvalues of $Z$ is $\{\alpha_1, \ldots, \alpha_{k-1}, \alpha_{k+1}, \ldots, \alpha_n\}$ for some $1 \leq k \leq n$. Then $Tr(A^2) - Tr(Z^2) = \alpha_k^2$ and $Tr(A) - Tr(Z) = \alpha_k$. But observing the partition of $A$ we see that $Tr(A^2) = a^2 + 2x^T x + Tr(Z^2)$ and $Tr(A) - Tr(Z) = a$. Consequently, $a = \alpha_k$ and therefore $x^T x = 0$, which implies $x = 0$. $\square$

**LEMMA 4.5.8** *Let $A$ be the symmetric matrix partitioned as in Lemma 4.5.7 and let $B = diag(\beta, 0, \ldots, 0)$ where $\beta \neq 0$. Let $\{\alpha_1, \ldots, \alpha_n\}$ be the eigenvalues of $A$. If the matrix $C = A + B$ has eigenvalues $\{\alpha_1, \ldots, \alpha_k + \beta, \ldots, \alpha_n\}$ for some $k$, then $x = 0$.*

**Proof**: Observe that the characteristic polynomials for $A$ and $C$, respectively, are
$$\det(\lambda I - A) = (\lambda - \alpha_1) \cdots (\lambda - \alpha_n)$$
and
$$\det(\lambda I - C) = (\lambda - \alpha_1) \cdots (\lambda - \alpha_k - \beta) \cdots (\lambda - \alpha_n).$$
Therefore
$$\det(\lambda I - C) = \det(\lambda I - A) - \beta(\lambda - \alpha_1) \cdots (\lambda - \alpha_{k-1})(\lambda - \alpha_{k+1}) \cdots (\lambda - \alpha_n).$$
From the partition of $A$, it follows that
$$\det(\lambda I - C) = \det(\lambda I - A) - \beta \det(\lambda I - Z).$$
Since $\beta \neq 0$, we have
$$\det(\lambda I - Z) = (\lambda - \alpha_1) \cdots (\lambda - \alpha_{k-1})(\lambda - \alpha_{k+1}) \cdots (\lambda - \alpha_n).$$
Therefore the spectrum of $Z$ consists of $n - 1$ eigenvalues of $A$. By Lemma 4.5.7 it follows that $x = 0$. $\square$

**LEMMA 4.5.9** *Let $A$ and $B$ be symmetric matrices with spectra $\{\alpha_1, \ldots, \alpha_n\}$ and $\{\beta, 0, \ldots, 0\}$, respectively. Then the eigenvalues of $A + B$ are $\{\alpha_1, \ldots, \alpha_k + \beta, \ldots, \alpha_n\}$ for some $k$ if and only if $AB = BA$.*

**Proof**: Without loss of generality, let $B = diag(\beta, 0, \ldots, 0)$. If $\beta = 0$ then the result is trivially true, so assume $\beta \neq 0$. Let the eigenvalues of $A + B$ be $\{\alpha_1, \ldots, \alpha_k + \beta, \ldots, \alpha_n\}$ for some $k$. By partitioning $A$ as in Lemma 4.5.7, it follows from Lemma 4.5.8 that $x = 0$. Therefore, clearly $AB = BA$. The converse is trivial. $\square$

We are now ready to investigate which eigenvalues remain unchanged when a spectral integral variation occurs by adding an edge $e$ to a graph $\mathcal{G}$. If $\mathcal{G} + e$ is created from $\mathcal{G}$ by adding the edge $e$ joining the nonadjacent vertices $i$ and $j$ of $\mathcal{G}$, then $L(\mathcal{G}+e) = L(\mathcal{G}) + E$ for the appropriate matrix $E$. Note that the spectrum of $E$ is $\{2, 0, \ldots, 0\}$. Our investigation will hinge upon the vertices in $\mathcal{G}$ to which that $i$ and $j$ are adjacent. For a vertex $v \in \mathcal{G}$, let $N(v)$ denote the set of vertices adjacent to $v$ in $\mathcal{G}$. We have the following observation from [76].

**OBSERVATION 4.5.10** *If $i$ and $j$ are vertices in $\mathcal{G}$, then $N(i) = N(j)$ if and only if $L(\mathcal{G})E = EL(\mathcal{G})$.*

The following theorem from [76] characterizes the conditions on which eigenvalues remain unchanged when an edge $e$ is added to a graph $\mathcal{G}$.

**THEOREM 4.5.11** *Suppose $\mathcal{G}+e$ is created from $\mathcal{G}$ by adding an edge $e$ between two nonadjacent vertices $i$ and $j$ of $\mathcal{G}$. Then the spectrum of $L(\mathcal{G}+e)$ overlaps the spectrum of $L(\mathcal{G})$ in $n-1$ places if and only if $N(i) = N(j)$.*

**Proof**: By Lemma 4.5.9, the spectrum of $L(\mathcal{G}+e)$ overlaps the spectrum of $L(\mathcal{G})$ in $n-1$ places if and only if $L(\mathcal{G})E = EL(\mathcal{G})$. The result now follows from Observation 4.5.10. □

We should note that if $n-1$ of the eigenvalues remain unchanged, then exactly one eigenvalue changes. Since $\text{Tr}(L(\mathcal{G}+e)) = \text{Tr}(L(\mathcal{G})) + 2$, it follows that the one eigenvalue that changes increases by two.

**EXAMPLE 4.5.12** Consider the following graph $\mathcal{G}$:

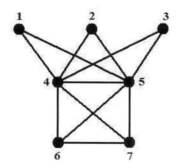

The spectrum of $L(\mathcal{G})$ is $\{0, 2, 2, 2, 4, 7, 7\}$. Observe that $N(1) = N(2) = \{4, 5\}$. Therefore, if we add an edge between vertices 1 and 2 to create $\mathcal{G}+e$, the spectrum of $L(\mathcal{G}+e)$, by Theorem 4.5.11, should differ from the spectrum of $L(\mathcal{G})$ by only one eigenvalue. Since $Tr(L(\mathcal{G}+e)) = Tr(L(\mathcal{G})) + 2$, the unique eigenvalue in $L(\mathcal{G})$ that changes should increase by two when creating $\mathcal{G}+e$. Observe that this holds as the spectrum of $L(\mathcal{G}+e)$ is $\{0, 2, 2, 4, 4, 7, 7\}$.

Since our focus is on maximal graphs, we note that Example 4.5.12 illustrates the following corollary to Theorem 4.5.11 about graphs which are Laplacian integral (see [76]):

**COROLLARY 4.5.13** *Suppose $\mathcal{G}+e$ is created from $\mathcal{G}$ by adding an edge $e$ between two nonadjacent vertices $i$ and $j$ of $\mathcal{G}$. If $N(i) = N(j)$ then $L(\mathcal{G}+e)$ is Laplacian integral if and only is $L(\mathcal{G})$ is Laplacian integral.*

The goal for the remainder of this section is to describe the conditions under which adding an edge to a maximal graph results in a maximal graph, hence also being Laplacian integral. To this end, we begin with a lemma from [23] which shows the relationship between the two eigenvalues that change when a spectral integral variation occurs in two places.

**LEMMA 4.5.14** *Let $\mathcal{G}$ be a graph on $n$ vertices. If the spectral integral variation of $\mathcal{G}$ occurs in two places by adding an edge $e$ joining nonadjacent vertices $i$ and $j$, and if the changed eigenvalues are $\lambda_k$ and $\lambda_\ell$, then*

$$\lambda_k \lambda_\ell = d_i d_j + p$$

*where $p$ is the number of vertices in $\mathcal{G}$ adjacent to both $i$ and $j$.*

**Proof**: Without loss of generality, let $i = 1$ and $j = 2$. Then the Laplacian, $L$, for $\mathcal{G}$ is of the form

$$L = \begin{bmatrix} d_1 & 0 & -e^{(r)T} & 0^{(s)T} & -e^{(p)T} & 0^{(q)T} \\ 0 & d_2 & 0^{(r)T} & -e^{(s)T} & -e^{(p)T} & 0^{(q)T} \\ -e^{(r)} & 0^{(r)} & & & & \\ 0^{(s)} & -e^{(s)} & & & & \\ -e^{(p)} & -e^{(p)} & & & & \\ 0^{(q)} & 0^{(q)} & & & & \end{bmatrix}$$

Let $L(\mathcal{G} + e) = L + E$ where $E$ is the appropriate matrix. In terms of diagonal entries, observe

$$Tr((L+E)^2) - Tr(L^2) = (d_1+1)^2 + 1 + (d_2+1)^2 + 1 - d_1^2 - d_2^2$$

$$= 2d_1 + 2d_2 + 4.$$

However, in terms of eigenvalues, observe

$$Tr((L+E)^2) - Tr(L^2) = (\lambda_k+1)^2 + (\lambda_\ell+1)^2 - \lambda_k^2 - \lambda_\ell^2$$

$$= 2\lambda_k + 2\lambda_\ell + 2.$$

Therefore
$$\lambda_k + \lambda_\ell = d_1 + d_2 + 1. \tag{4.5.9}$$

Note that

$$Tr((L+E)^3) - Tr(L^3) = Tr(ELE) + Tr(LEL) + 2Tr(LE^2) + 2Tr(L^2E) + Tr(E^3).$$

In terms of diagonal entries, observe

$$Tr((L+E)^3) - Tr(L^3) = 6(d_1 + d_2) + 3(d_1 + d_2)^2 + 3(r+s) + 8.$$

In terms of the eigenvalues, observe

$$Tr((L+E)^3) - Tr(L^3) = 3(\lambda_k^2 + \lambda_\ell^2) + 3(\lambda_k + \lambda_\ell) + 2.$$

Setting these two quantities for $Tr((L+E)^3) - Tr(L^3)$ equal and using (4.5.9), we obtain

$$2\lambda_k \lambda_\ell = 2d_1 d_2 + d_1 + d_2 - r - s.$$

Using the fact that $d_1 = r + p$ and $d_2 = s + p$ yields the desired result. □

Lemma 4.5.14 gives rise to the following corollary:

**COROLLARY 4.5.15** *Let $\mathcal{G}_1$ and $\mathcal{G}_2$ be disjoint graphs and suppose that the spectral integral variation of $L(\mathcal{G}_1 + \mathcal{G}_2)$ occurs in two places by adding an edge joining a vertex in $\mathcal{G}_1$ to a vertex in $\mathcal{G}_2$. Then either $\mathcal{G}_1$ or $\mathcal{G}_2$ is an isolated vertex.*

We now present a theorem from [23] concerning graphs in general which will be useful in our subsequent results involving maximal graphs.

**THEOREM 4.5.16** *Let $\mathcal{G}_1$ and $\mathcal{G}_2$ be connected graphs on $n_1$ and $n_2$ vertices, respectively, where $n_1 \leq n_2$. Let $u \in \mathcal{G}_1$ and $v \in \mathcal{G}_2$. Then the spectral integral variation of $L(\mathcal{G}_1 + \mathcal{G}_2)$ occurs in two places by adding the edge $uv$ if and only if $\mathcal{G}_1 = \{u\}$, $n_2 > 1$, and the degree of $v$ in $\mathcal{G}_2$ is $n_2 - 1$.*

**Proof**: Suppose the spectral integral variation occurs in two places. We know from Theorem 4.1.1 that 0 is an eigenvalue of $L(\mathcal{G}_1 + \mathcal{G}_2)$ of multiplicity 2 while 0 is an eigenvalue of $L(\mathcal{G}_1 + \mathcal{G}_2 + e)$ of multiplicity 1. Let $T$ be the set of eigenvalues in common, counting multiplicities, of $L(\mathcal{G}_1 + \mathcal{G}_2)$ and $L(\mathcal{G}_1 + \mathcal{G}_2 + e)$. Since one of the 0 eigenvalues of $L(\mathcal{G}_1 + \mathcal{G}_2)$ becomes 1 in $L(\mathcal{G}_1 + \mathcal{G}_2 + e)$ and since the spectral integral variation occurs in two places, it follows that there exists an eigenvalue $\gamma > 0$ of $L(\mathcal{G}_1 + \mathcal{G}_2)$ such that $\gamma \notin T$. Therefore $\sigma(\mathcal{G}_1 + \mathcal{G}_2 + e) = \{1, \gamma + 1\} \cup T$.

By Corollary 4.5.15, we know that $n_1 = 1$, i.e., $\mathcal{G}_1 = \{u\}$. Since the spectral variation occurs in two places, it follows that $n_2 > 1$, for otherwise $\mathcal{G}_2 = \{v\}$ and the spectral integral variation of $L(\mathcal{G}_1 + \mathcal{G}_2)$ would only occur in one place (i.e., $\{0, 0\}$ to $\{0, 2\}$). Since the number of spanning trees of $\mathcal{G}_2$ equals the number of spanning trees of $\mathcal{G}_1 + \mathcal{G}_2 + e$, it follows from Exercise 4 of Section 3.2 that

$$\frac{1}{n_2} \prod_{i=2}^{n_2} \lambda_i(L(\mathcal{G}_2)) = \frac{1}{n_2+1} \prod_{i=2}^{n_2+1} \lambda_i(L(\mathcal{G}_1 + \mathcal{G}_2 + e)).$$

Since the spectrum of $\sigma(\mathcal{G}_2)$ equals $\sigma(\mathcal{G}_1 + \mathcal{G}_2)$ less one eigenvalue of zero and since $\sigma(\mathcal{G}_1 + \mathcal{G}_2 + e) = \{1, \gamma + 1\} \cup T$, it follows that

$$\frac{1}{n_2}(\tau \gamma) = \frac{1}{n_2+1}(\tau(\gamma+1)).$$

where $\tau$ is the product of all eigenvalues in $T$. This implies $\gamma/(\gamma+1) = n_2/(n_2+1)$. Therefore $\gamma = n_2$. Since $\gamma$ is an eigenvalue of $L(\mathcal{G}_2)$ we see that $n_2$ is an eigenvalue of $L(\mathcal{G}_2)$. But since $\mathcal{G}_2$ has $n_2$ vertices, it follows from Theorem 4.1.8 that $\mathcal{G}_2$ is the join of at least two graphs. Hence $\mathcal{G}_2^c$ is disconnected. Similarly, since $\mathcal{G}_1 + \mathcal{G}_2 + e$ has $n_2 + 1$ vertices and since $n_2 + 1$ is an eigenvalue of $L(\mathcal{G}_1 + \mathcal{G}_2 + e)$, it follows that $(\mathcal{G}_1 + \mathcal{G}_2 + e)^c$ is also disconnected. Let $H_1, \ldots, H_k$ $(k \geq 2)$ be the connected components of $\mathcal{G}_2^c$ and let $v \in H_1$. Then $\mathcal{G}_2 = H_1^c \vee \ldots \vee H_k^c$ and $(\mathcal{G}_1 + \mathcal{G}_2 + e)^c = \{u\} \vee (H_1 + \ldots + H_k) - e$. Since $(\mathcal{G}_1 + \mathcal{G}_2 + e)^c$ is disconnected, $H_1$ is connected, and $u$ is adjacent in $(\mathcal{G}_1 + \mathcal{G}_2 + e)^c$ to every vertex in $H_1$ except $v$, it follows that $H_1 = \{v\}$. By the construction of $\mathcal{G}_2$, $v$ is adjacent to every other vertex in $\mathcal{G}_2$, hence the degree of $v$ in $\mathcal{G}_2$ is $n_2 - 1$.

For the converse, since the degree of the vertex $v$ in $\mathcal{G}_2$ is $n_2 - 1$, it follows that $\mathcal{G}_2^c$ is disconnected. We may assume that $\mathcal{G}_2 = H_1^c \vee \ldots \vee H_k^c$, where $H_1 = \{v\}$. Thus $\mathcal{G}_1 + \mathcal{G}_2 + e = (\{u\} + H_2^c \vee \ldots \vee H_k^c) \vee \{v\}$. Let $\sigma(H_2^c \vee \ldots \vee H_k^c) = \{0, \lambda_2, \ldots, \lambda_{n_2-1}\}$. Then $\sigma(\mathcal{G}_2) = \{0, \lambda_2 + 1, \lambda_3 + 1, \ldots, \lambda_{n_2-1} + 1, n_2\}$, and $\sigma(\mathcal{G}_1 + \mathcal{G}_2 + e) = \{0, 1, \lambda_2 + 1, \lambda_3 + 1, \ldots, \lambda_{n_2-1} + 1, n_2 + 1\}$. Since $\sigma(\mathcal{G}_1 + \mathcal{G}_2) = \{0, 0, \lambda_2 + 1, \lambda_3 + 1, \ldots, \lambda_{n_2-1} + 1, n_2\}$, the result follows. □

**EXAMPLE 4.5.17** Consider the graph $\mathcal{G}_1 + \mathcal{G}_2$ below

where $\mathcal{G}_1$ is $u$ and $\mathcal{G}_2$ is the other component. Note that $\sigma(\mathcal{G}_1 + \mathcal{G}_2)$ is $\{0, 0, 1.59, 3, 4.41, 5\}$. Observe that $v$ is adjacent to all other vertices in $\mathcal{G}_2$. Thus if we add an edge $e$ between $u$ and $v$, by Theorem 4.5.16 the spectral integral variation will occur in two places, namely at $\lambda_2 = 0$ and $\lambda_6 = 5$. As expected, $\sigma(\mathcal{G}_1 + \mathcal{G}_2 + e)$ is $\{0, 1, 1.59, 3, 4.41, 6\}$.

We now focus our attention on the spectral integral variation of maximal graphs. We begin with a lemma from [24] whose proof we leave as an exercise.

**LEMMA 4.5.18** *For a graph $\mathcal{G}$, the spectral integral variation occurs in $k$ places of $L(\mathcal{G})$ by adding an edge $e$ between two nonadjacent vertices if and only if the spectral integral variation occurs in $k$ places of $L(\mathcal{G}^c - e)$ by adding $e$.*

We now prove the first of two theorems concerning the spectral integral variation of maximal graphs. The first theorem from [24] characterizes when the spectral integral variation of a maximal graph will occur in two places.

**THEOREM 4.5.19** *Let $\mathcal{G}$ be a connected maximal graph. Then the spectral integral variation occurs in two places by adding an edge $e$ between two nonadjacent vertices $u$ and $v$ if and only if $\mathcal{G} + e$ is maximal and $d_u \neq d_v$.*

**Proof**: Suppose $\mathcal{G}$ and $\mathcal{G}+e$ are both maximal (hence Laplacian integral) and $d_u \neq d_v$. Then the spectral integral variation occurs in either one or two places by adding an edge $e$. Since $d_u \neq d_v$, it follows that $N(u) \neq N(v)$. By Theorem 4.5.11, it follows that the spectral integral variation must occur in two places.

For the converse, suppose the spectral integral variation in $\mathcal{G}$ occurs in two places by adding the edge $e$. Then by Lemma 4.5.18, the spectral integral variation also occurs in two places in $\mathcal{G}^c - e$ by adding $e$. Since $\mathcal{G}$ is maximal, then by Lemma 2.5.24 we have $\mathcal{G} := \mathcal{G}_0 = \mathcal{G}_1^c \vee K_{p_1}$ for some $p_1 \geq 1$ and some maximal graph $\mathcal{G}_1$. Observe $e \in \mathcal{G}_1$ since $e \notin \mathcal{G}$. Since $\sigma(\mathcal{G}^c) = \sigma(\mathcal{G}_1) \cup \sigma(K_{p_1}^c)$ and $\sigma(\mathcal{G}^c - e) = \sigma(\mathcal{G}_1 - e) \cup \sigma(K_{p_1}^c)$, the spectral integral variation of $\mathcal{G}_1 - e$ occurs in two places by adding $e$. By Theorem 4.5.11, $N(u) \neq N(v)$ in $\mathcal{G}_1 - e$, implying $\mathcal{G}_1$ is not complete. Thus $\mathcal{G}_1 = \mathcal{G}_2^c \vee K_{p_2}$ for some $p_2 \geq 1$ and some maximal graph $\mathcal{G}_2$. By Theorem 4.5.11, $e \notin K_{p_2}$, hence either both $u, v \in \mathcal{G}_2$, or exactly one of $u$ or $v$ is in $\mathcal{G}_2$. By Lemma 4.5.18, the spectral integral variation occurs in two places in $\mathcal{G}_1^c$ by adding $e$. Thus if both $u, v \in \mathcal{G}_2$, then $e \in \mathcal{G}_2^c$ and hence $e \notin \mathcal{G}_2$, and therefore the spectral integral variation would occur in two places in $\mathcal{G}_2$ by adding $e$. In such a case, $\mathcal{G}_2$ has the same properties as $\mathcal{G}$. Repeating the above discussion for $\mathcal{G}_2$, we then obtain graphs $\mathcal{G}_3$ and $\mathcal{G}_4$. If both $u, v \in \mathcal{G}_4$, then continue the above discussion for $\mathcal{G}_4$. Since there are only finitely many vertices in $\mathcal{G}$, this procedure must stop at the $2k^{th}$ step for some positive integer $k$. Thus in the decomposition of $\mathcal{G} = \mathcal{G}_0$, the graphs $\mathcal{G}_0, \mathcal{G}_1, \ldots, \mathcal{G}_{2k-1}$ have both of the following properties:

(a) For $i = 0, 1, \ldots, k-1$, both $u, v \in \mathcal{G}_{2i}$, $e \notin \mathcal{G}_{2i}$, and the spectral integral variation occurs in two places in $\mathcal{G}_{2i}$ by adding $e$.

(b) For $i = 0, 1, \ldots, k-1$, both $u, v \in \mathcal{G}_{2i+1}$, $e \in \mathcal{G}_{2i+1}$, and the spectral integral variation occurs in two places in $\mathcal{G}_{2i+1} - e$ by adding $e$.

Additionally, $\mathcal{G}_{2k-1} = \mathcal{G}_{2k}^c \vee K_{p_{2k}}$ where $u, v$ cannot both belong to $\mathcal{G}_{2k}$. Furthermore, $u, v$ cannot both belong to $K_{p_{2k}}$; otherwise by Theorem 4.5.11 the spectral integral variation would occur in one place in $\mathcal{G}_{2k-1} - e$ by adding $e$, contradicting (b). Without loss of generality, let $u \in K_{p_{2k}}$ and $v \in \mathcal{G}_{2k}$. By Lemma 4.5.18, the spectral integral variation occurs in two places in $\mathcal{G}_{2k-1}^c = \mathcal{G}_{2k} + K_{p_{2k}}^c$ by adding $e$. Therefore, it occurs in two places in $\mathcal{G}_{2k} + u$ by adding $e$ as

$$\sigma(\mathcal{G}_{2k-1}^c) = \sigma(\mathcal{G}_{2k} + u) \cup \sigma(K_{p_{2k}}^c - u),$$

$$\sigma(\mathcal{G}_{2k-1}^c + e) = \sigma(\mathcal{G}_{2k} + u + e) \cup \sigma(K_{p_{2k}}^c - u).$$

By Theorem 4.5.16, $v$ is adjacent to every other vertex in $\mathcal{G}_{2k}$. By Lemma 2.5.24, either $\mathcal{G}_{2k} = K_{p_{2k+1}}$ or $\mathcal{G}_{2k} = \mathcal{G}_{2k+1}^c \vee K_{p_{2k+1}}$ where $\mathcal{G}_{2k+1}$ is maximal, and hence, connected. If $v \in \mathcal{G}_{2k+1}^c$, then $v$ is adjacent to every other vertex in $\mathcal{G}_{2k+1}^c$, which means $v$ is an isolated vertex of $\mathcal{G}_{2k+1}$ rendering $\mathcal{G}_{2k+1}$ disconnected, a contradiction. Therefore $v \in K_{p_{2k+1}}$.

Observing that $\mathcal{G}_{2k-1}^c = (\mathcal{G}_{2k+1}^c \vee K_{p_{2k+1}}^c) + K_{p_{2k}}^c$, it follows that the degree of $v$ in $\mathcal{G}_{2k-1}^c$ is greater than 0 while the degree of $u$ in $\mathcal{G}_{2k-1}^c$ is 0. Since $\mathcal{G}_{2k-2}^c =$

$\mathcal{G}_{2k-1}^c \vee K_{p_{2k-1}}$, it follows that $u$ and $v$ have different degrees in $\mathcal{G}_{2k-2}^c$ (and hence in $\mathcal{G}_{2k-2}$). By similar discussion and by decomposing $\mathcal{G}$ is accordance with Lemma 2.5.24, we find that in graphs $\mathcal{G}_{2k-3}^c, \mathcal{G}_{2k-4}^c, \ldots, \mathcal{G}_1^c$, and hence in $\mathcal{G}$, that $u$ and $v$ have different degrees. Observe that

$$\begin{aligned}
\mathcal{G}_{2k-1}^c + e &= \mathcal{G}_{2k} + K_{p_{2k}}^c + e \\
&= (\mathcal{G}_{2k+1}^c \vee K_{p_{2k+1}}) + K_{p_{2k}}^c + e \\
&= (\mathcal{G}_{2k+1}^c \vee K_{p_{2k+1}-1} + u) \vee v + K_{p_{2k}-1}^c \\
&= ((\mathcal{G}_{2k+1}^c \vee K_{p_{2k+1}-1})^c \vee u)^c \vee v + K_{p_{2k}-1}^c \\
&:= H_{2k+1} + K_{p_{2k}-1}^c.
\end{aligned}$$

Therefore,

$$\begin{aligned}
\mathcal{G}_{2k-1} - e &= (H_{2k+1} + K_{p_{2k}-1}^c)^c \\
&= H_{2k+1}^c \vee K_{p_{2k}-1}.
\end{aligned}$$

By Lemma 2.5.24, $H_{2k+1}$ is maximal. Hence $\mathcal{G}_{2k-1} - e$ is maximal. Then $\mathcal{G}_{2k-2} + e = \mathcal{G}_{2k-1}^c \vee K_{p_{2k-1}} + e = (\mathcal{G}_{2k-1} - e)^c \vee K_{p_{2k-1}}$ is also maximal. Repeating this procedure, we eventually conclude that $\mathcal{G} + e$ is maximal. □

Theorem 4.5.19 tells us that the spectral integral variation of a maximal graph $\mathcal{G}$ will occur in two places if and only if $\mathcal{G} + e$ is maximal where $e$ joins nonadjacent vertices $u$ and $v$ in $\mathcal{G}$ such that $d_u \neq d_v$. In the next theorem from [24], we develop an efficient way of choosing such vertices $u$ and $v$ in $\mathcal{G}$ so that a spectral integral variation will occur in two places when an edge $e$ is added between $u$ and $v$. Equivalently, we develop an efficient way of choosing $u$ and $v$ so that $\mathcal{G} + e$ will be maximal. In this theorem, recall the definition of $f(d)$ from Definition 2.5.12.

**THEOREM 4.5.20** *Let $\mathcal{G}$ be a connected maximal graph and let $(d) = (d_1, \ldots, d_n)$ be the degree sequence of $\mathcal{G}$. Let $u$ and $v$ be nonadjacent vertices of $\mathcal{G}$ with $d_u \geq d_v$. Let $\mathcal{G} + e$ be the graph created from $\mathcal{G}$ by adding an edge between $u$ and $v$. Then the spectral integral variation of $\mathcal{G}$ occurs in two places if and only if $d_u \geq f(d) > d_v = d_{d_u+2}$.*

**Proof**: First suppose $d_u \geq f(d) > d_v = d_{d_u+2}$. Then by Theorem 2.5.26, $\mathcal{G} + e$ is maximal. Since $d_u \neq d_v$, Theorem 4.5.19 implies that the spectral integral variation of $\mathcal{G}$ occurs in two places by adding $e$.

Now suppose the spectral integral variation of $\mathcal{G}$ occurs in two places by adding $e$. Then by Theorem 4.5.19 $\mathcal{G} + e$ is maximal and $d_u \neq d_v$. By Theorem 2.5.26 we have $d_u \geq f(d) \geq d_v = d_{d_u+2}$. It remains to show that $f(d) > d_v$. We do this in two cases: where $d_u > f(d)$ and where $d_u = f(d)$.

*Case I*: Suppose $d_u > f(d)$. Then by Case I of Lemma 2.5.26, we have $d_v = d_{d_u+2} =$

# The Spectra of Laplacian Matrices

$d^*_{d_u+1} \leq f(d) - 1 < f(d)$.

*Case II:* Suppose $d_u = f(d)$. We assume $f(d) = d_v$ and deduce a contradiction. If $f(d) = d_v$ then $d_u = f(d) = d_v$. By Theorem 2.5.26, $d_v = d_{d_u+2}$. Therefore

$$f(d) = d_v = d_{d(u)+2} = d_{f(d)+2} \leq d_{f(d)}.$$

Since $u$ and $v$ are not adjacent, it follows by Theorem 2.5.25 that $d_{f(d)} \neq f(d)$. Therefore $d_{f(d)} > f(d)$. Hence $d_{d_u}, d_{d_v} > f(d)$. Therefore $d_u, d_v < f(d)$. Then by Lemma 2.5.25 it follows that $N(u) = N(v)$ since $u$ and $v$ are not adjacent. But since the spectral integral variation of $\mathcal{G}$ occurs in two places, this contradicts Theorem 4.5.11. Therefore, if $d_u = f(d)$ then $f(d) > d_v$. □

**EXAMPLE 4.5.21** Consider the maximal graph $\mathcal{G}$ below.

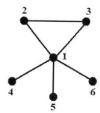

Observe that the vertices are labeled in order of nonincreasing degrees. The degree sequence $(d) = (5, 2, 2, 1, 1, 1)$ and hence by Theorem 4.5.1 we have $\sigma(\mathcal{G}) = (d^*) = (6, 3, 1, 1, 1, 0)$. Note that $f(d) = 2$. Observe that if we let $u = 2$ and $v = 4$, then the conditions of Theorem 4.5.20 are satisfied. Hence adding an edge $e$ joining vertices 2 and 4 will cause a spectral integral variation to occur in two places. Observe that $\sigma(\mathcal{G} + e) = \{0, 1, 1, 2, 4, 6\}$.

As we conclude this section, it is natural to ask if there are graphs other than maximal graphs in which the eigenvalues of Laplacian matrices are all integers. We will see in the next section that such graphs do exist.

# Exercises:

**1.** Use Theorem 4.5.1 to find all eigenvalues of the Laplacian matrices of the following maximal graphs.

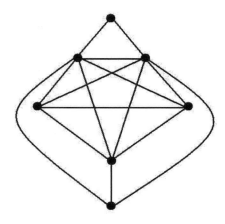

**2.** Let $\mathcal{G}$ be a maximal graph on $n$ vertices with Laplacian matrix $L$. Prove that $n$ is an eigenvalue of $L$.

**3.** Show that all of the eigenvalues of the Laplacian matrix for the graph below are integers. Then verify that this graph is *not* a maximal graph.

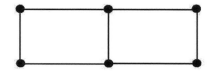

**4.** Prove Observation 4.5.10.

**5.** Prove Corollary 4.5.15.

**6.** Prove Lemma 4.5.18.

**7.** Use Theorem 4.5.20 to find all pairs of nonadjacent vertices $u$ and $v$ such that adding an edge joining $u$ and $v$ will cause a spectral integral variation to occur in two places.

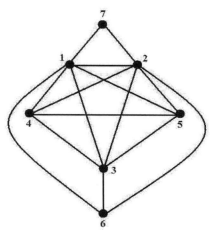

## 4.6 Graphs with Distinct Integer Spectra

We now broaden our investigation of Laplacian integral graphs. In this section based largely on [19], we will focus on graphs whose Laplacian matrices have *distinct* integer eigenvalues. Given a graph $\mathcal{G}$ on $n$ vertices with Laplacian matrix $L$, if $L$ has distinct integer eigenvalues, then $\sigma(\mathcal{G}) = \{0, 1, 2, \ldots, i-1, i+1, \ldots, n\}$ for some integer $1 \leq i \leq n$. We will denote such a set as $S_{i,n}$.

**DEFINITION 4.6.1** If there exists a graph $\mathcal{G}$ on $n$ vertices such that $\sigma(\mathcal{G}) = S_{i,n}$, we say that the set $S_{i,n}$ is *Laplacian realizable*.

**EXAMPLE 4.6.2** By an exhaustive search, there are exactly five sets $S_{i,n}$ that are Laplacian realizable for $n \leq 5$. Below are the graphs realizing each such set.

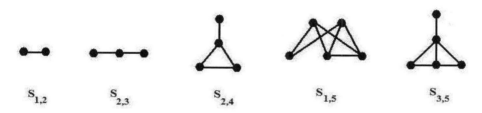

For a given $n$, our goal is to determine the values $i$ for which is $S_{i,n}$ Laplacian realizable. To do this it will help to eliminate some possibilities. First, recall that the sum of the eigenvalues (counting multiplicity) of a square matrix is equal to the sum of its diagonal entries. Since the diagonal entries of the Laplacian matrix are the degrees of the vertices of the corresponding graph, it follows that the sum of the eigenvalues of the Laplacian matrix is equal to the sum of the degrees of the vertices of the graph. Since the sum of the degrees of the vertices of any graph must be even (Theorem 2.1.4), we see that the sum of the eigenvalues must be even. This gives us the following lemma from [19]:

**LEMMA 4.6.3** Suppose for a given $n \geq 2$ and integer $1 \leq i \leq n$ that $S_{i,n}$ is Laplacian realizable. If $n \equiv 0 \bmod 4$ or $n \equiv 3 \bmod 4$, then $i$ must be even. If $n \equiv 1 \bmod 4$ or $n \equiv 2 \bmod 4$, then $i$ must be odd.

It should be noted that if a graph $\mathcal{G}$ on $n$ vertices realizes $S_{i,n}$ for some $i \neq n$, then by Theorem 4.1.8, it follows that $\mathcal{G}$ is the join of two graphs. We investigate the values for $i$ in which $S_{i,n}$ is Laplacian realizable. Our first lemma, from [19], gives us two sets which are always Laplacian realizable.

**LEMMA 4.6.4** Suppose $n \geq 2$.
(a) If $n \equiv 2 \bmod 4$ or $n \equiv 3 \bmod 4$, then $S_{n-1,n}$ is Laplacian realizable.
(b) If $n \equiv 1 \bmod n$ or $n \equiv 2 \bmod n$, then $S_{1,n}$ is Laplacian realizable.

**Proof**: (a) We use induction on $n$. Observe from Example 4.6.2 that the result holds for $n = 2, 3$. Now suppose $n$ is congruent to either 2 or 3 modulo 4. Let $\mathcal{G}$

be a graph that realizes $S_{n-1,n}$. We will show that we can create a graph from this that realizes $S_{n+3,n+4}$. Observe that

$$\sigma(\mathcal{G} + K_1) = \{0\} \cup S_{n-1,n}.$$

Therefore, by Theorem 4.1.11 it follows that $(\mathcal{G} \cup K_1)^c$ realizes $S_{2,n+1}$. Now observe that

$$\sigma(K_2 + [\mathcal{G} + K_1]^c) = \{0, 0, 1, 2, 3, \ldots, n+1\}.$$

Hence by Theorem 4.1.6, $K_1 \vee [K_2 + (\mathcal{G} + K_1)^c]$ realizes $S_{n+3,n+4}$.

(b) We use induction on $n$. Observe from Example 4.6.2 that the result holds for $n = 1, 2$. Now suppose $n$ is congruent to either 1 or 2 modulo 4. We will show that we can create a graph that realizes $S_{1,n+4}$. Note that $n + 1$ is congruent to either 2 or 3 modulo 4. Thus by part (a), there exists a graph $\mathcal{G}$ that realizes $S_{n,n+1}$. Thus

$$\sigma(\mathcal{G} + K_1) = \{0\} \cup S_{n,n+1}.$$

Hence by Theorem 4.1.6, the graph $(K_1 + K_1) \vee (\mathcal{G} + K_1)$ realizes $S_{1,n+4}$. □

Now that we have a basis of sets which are Laplacian realizable, our next lemma from [19] gives us an inductive way to create sets which are Laplacian realizable.

**LEMMA 4.6.5** *If $S_{i,n}$ is Laplacian realizable, then so is $S_{i+1,n+2}$.*

**Proof**: Let $\mathcal{G}$ be a graph on $n$ vertices that realizes $S_{i,n}$. Observe that

$$\sigma(\mathcal{G} + K_1) = \{0, 0, 1, 2, \ldots, i-1, i+1, \ldots, n\}.$$

By Theorem 4.1.6, it follows that

$$\sigma(K_1 \vee (\mathcal{G} + K_1)) = \{0, 1, 2, \ldots, i, i+2, \ldots, n+1, n+2\} = S_{i+1,n+2}$$

□

We see from the proof of Lemma 4.6.5 that we can easily create a graph whose Laplacian matrix has distinct integer eigenvalues by taking an existing such graph $\mathcal{G}$, an isolated vertex $u$, and joining an additional isolated vertex $v$ to both $u$ and $\mathcal{G}$.

**EXAMPLE 4.6.6** Consider the following graph $\mathcal{G}$. We know from Example 4.6.2 that $\sigma(\mathcal{G}) = \{0, 1, 3, 4\}$.

By Lemma 4.6.5, the Laplacian matrix for the following graph $\hat{\mathcal{G}}$ will be such that $\sigma(\hat{\mathcal{G}}) = \{0, 1, 2, 4, 5, 6\}$:

We see that we can use Lemma 4.6.5 to show which values $i$ are such that $S_{i,n}$ is Laplacian realizable. This leads us to the first main result of this section which is from [19].

**THEOREM 4.6.7** *(i) If $n \equiv 0 \bmod 4$, then for each $i = 1, 2, \ldots, \frac{n-2}{2}$, $S_{2i,n}$ is Laplacian realizable.*
*(ii) If $n \equiv 1 \bmod 4$, then for each $i = 1, 2, \ldots, \frac{n-1}{2}$, $S_{2i-1,n}$ is Laplacian realizable.*
*(iii) If $n \equiv 2 \bmod 4$, then for each $i = 1, 2, \ldots, \frac{n}{2}$, $S_{2i-1,n}$ is Laplacian realizable.*
*(iv) If $n \equiv 3 \bmod 4$, then for each $i = 1, 2, \ldots, \frac{n-1}{2}$, $S_{2i,n}$ is Laplacian realizable.*

**Proof**: We prove this using induction on $n$. Observe that the graphs in Example 4.6.2 show that the base step holds for all $n \leq 5$. For the inductive hypothesis, assume (i)–(iv) hold for all $n \leq k - 1$. We now prove each part separately for $n = k$:

(i) Suppose $k \equiv 0 \bmod 4$. Then $k - 2 \equiv 2 \bmod 4$. Thus by the inductive hypothesis, each of $S_{1,k-2}, S_{3,k-2}, \ldots S_{k-3,k-2}$ is Laplacian realizable. Hence each of $S_{2,k}, S_{4,k}, \ldots, S_{k-2,k}$ is realizable by Lemma 4.6.5.

(ii) Suppose $k \equiv 1 \bmod 4$. Then $k - 2 \equiv 3 \bmod 4$. Thus by the inductive hypothesis, each of $S_{2,k-2}, S_{4,k-2}, \ldots S_{k-3,k-2}$ is Laplacian realizable. Hence each of $S_{3,k}, S_{5,k}, \ldots, S_{k-2,k}$ is realizable by Lemma 4.6.5. Also, $S_{1,k}$ is realizable by Lemma 4.6.4(b)

The proof of cases (iii) and (iv) are similar and thus omitted. □

Between Lemma 4.6.3 and Theorem 4.6.7, we have determined exactly which values $i$ that $S_{i,n}$ is realized for a fixed $n$ with the exception of $i = n$. Thus we now need to determine if $S_{n,n}$ is realizable, and if so, under what conditions. While it is believed that $S_{n,n}$ is not realizable, this remains an open question. However, we will offer some partial results. Since $i = n$ in this case, we begin with an immediate consequence of Lemma 4.6.3 (see [19]):

**OBSERVATION 4.6.8** *If $S_{n,n}$ is Laplacian realizable, then necessarily $n \equiv 0 \bmod 4$ or $n \equiv 1 \bmod 4$.*

We now present a lemma from [19] which further narrows down the possibilities for $n$ if $S_{n,n}$ is Laplacian realizable.

**LEMMA 4.6.9** *If a graph $\mathcal{G}$ on $n$ vertices realizes $S_{n,n}$, then the number of edges in $\mathcal{G}$ is equal to $n(n-1)/4$. Further, $n|(n-1)!$ so that in particular, $n$ is not a prime number.*

**Proof**: Let $L$ be the Laplacian matrix for $\mathcal{G}$. Since the sum of the degrees of the vertices in $\mathcal{G}$ is equal to $Tr(L)$, and since in the present case we have $Tr(L) = (n-1)n/2$, we find that the number of edges in $\mathcal{G}$ must be $n(n-1)/4$. Let the eigenvalues of $L$ be given by $0, \lambda_2, \ldots, \lambda_n$. By Lemma 4.3.8 and Exercise 4 of Section 3.2 the sum of the principle minors of $L$ is $\lambda_2 \cdots \lambda_n$. According to the Matrix Tree Theorem, (Theorem 3.2.2), the number of spanning trees of $\mathcal{G}$ is equal to any of the principal minors of $L$. Hence all principal minors are equal to $\lambda_2 \cdots \lambda_n/n = (n-1)!/n$ which is the number of spanning trees of $\mathcal{G}$. As the number of spanning trees is an integer, $n$ must divide $(n-1)!$ and so, in particular, $n$ cannot be a prime number. $\square$

In the previous two sections when dealing with graphs whose Laplacian matrices had integer eigenvalues, the degree sequence of the graph was very important in our investigations. Our next two lemmas from [19] deal with the degree sequence of a graph that realizes $S_{n,n}$.

**LEMMA 4.6.10** *Let $\mathcal{G}$ be a graph on $n$ vertices that realizes $S_{n,n}$. Then for any vertex $v$, we have $2 \leq d_v \leq n-3$.*

**Proof**: Suppose to the contrary that $\mathcal{G}$ has a vertex of degree 1. Then its algebraic and vertex connectivities are equal (see Theorem 5.1.10) which implies that $\mathcal{G}$ is the join of two graphs. Therefore by Theorem 4.1.8, $\mathcal{G}$ must have $n$ as an eigenvalue, contrary to our hypothesis. Thus $d_v \geq 2$ for any vertex $v \in \mathcal{G}$. By considering the complement of $\mathcal{G}$, we also find by Theorem 4.1.11 that $d_v \leq n-3$. $\square$

**LEMMA 4.6.11** *Let $\mathcal{G}$ be a graph on $n$ vertices with degree sequence $d_1, \ldots, d_n$. If $\mathcal{G}$ realizes $S_{n,n}$, then $\sum_{i=1}^n d_i^2 = \frac{n(n-1)(n-2)}{3}$. Further, letting $c_3$ and $i_3$ denote the number of three-cycles in $\mathcal{G}$ and the number of independent triples in $\mathcal{G}$, respectively, we have $\sum_{i=1}^n d_i^3 = \frac{n(n-1)(n^2-5n+8)}{4} + 6c_3 = \frac{n(n-1)(n^2-3n-2)}{4} - 6i_3$. Finally, the number of induced subgraphs of $\mathcal{G}$ on three vertices that are neither empty nor complete is equal to $\frac{n(n-1)(n+1)}{12}$.*

**Proof**: Let $L$ be the Laplacian matrix for $\mathcal{G}$. Observe that the Frobenius norm of $L$ satisfies

$$\|L\|_F^2 = \text{Tr}(L^2) = \sum_{i=1}^{n-1} i^2 = \frac{n(n-1)(2n-1)}{6}, \qquad (4.6.1)$$

where the second equality follows from the fact that $S_{n,n}$ is the set of eigenvalues of $L$. By Lemma 4.6.9, the number of edges in $\mathcal{G}$ is $n(n-1)/4$, and hence the number of nonzero off–diagonal entries in $L$ is $n(n-1)/2$. Thus we have

$$\|L\|_F^2 = \sum_{i=1}^n d_i^2 + \frac{n(n-1)}{2}. \tag{4.6.2}$$

Equating (4.6.1) and (4.6.2) we find that

$$\sum_{i=1}^n d_i^2 = \frac{n(n-1)(2n-1)}{6} - \frac{n(n-1)}{2} = \frac{n(n-1)(n-2)}{3}. \tag{4.6.3}$$

Let $A$ be the adjacency matrix of $\mathcal{G}$, and let $D$ be the diagonal matrix of vertex degrees. Since $L = D - A$, we have $L^3 = D^3 - D^2A - AD^2 - DAD + A^2D + DA^2 + ADA - A^3$. Since the trace of a product of matrices is invariant of the order in which the product is taken, it follows that $\text{Tr}(D^2A) = \text{Tr}(AD^2) = \text{Tr}(DAD)$ and $\text{Tr}(A^2D) = \text{Tr}(DA^2) = \text{Tr}(ADA)$. However, observe that $\text{Tr}(D^2A) = 0$. Therefore

$$\text{Tr}(L^3) = \text{Tr}(D^3) + 3\text{Tr}(A^2D) - \text{Tr}(A^3) = \sum_{i=1}^n d_i^3 + 3\sum_{i=1}^n d_i^2 - 6c_3,$$

where the last equality follows from Theorem 3.1.3. Since $\sigma(\mathcal{G}) = S_{n,n}$, we have $\text{Tr}(L^3) = \frac{n^2(n-1)^2}{4}$, and it follows from (4.6.3) and Theorem 3.1.3 that

$$\sum_{i=1}^n d_i^3 = \frac{n(n-1)(n^2 - 5n + 8)}{4} + 6c_3. \tag{4.6.4}$$

Since $\mathcal{G}$ realizes $S_{n,n}$ it follows from Theorem 4.1.11 that $\mathcal{G}^c$ also realizes $S_{n,n}$. Observing that the degree sequence for $\mathcal{G}^c$ is $(n-1-d_1, \ldots, n-1-d_n)$, and letting $\bar{L}$ denote the Laplacian matrix for $\mathcal{G}^c$, we have

$$\text{Tr}(\bar{L}^3) = \sum_{i=1}^n (n-1-d_i)^3 = \frac{n(n-1)(n^2-5n+8)}{4} + 6\hat{c}_3,$$

where $\hat{c}_3$ is the number of 3-cycles in $\mathcal{G}^c$. By definition of the complement of a graph, $\hat{c}_3 = i_3$. Note that

$$\sum_{i=1}^n (n-1-d_i)^3 = \sum_{i=1}^n \left[(n-1)^3 - 3(n-1)^2 d_i + 3(n-1)d_i^2 - d_i^3\right]$$

$$= n(n-1)^3 - 3(n-1)^2 \left[\frac{n(n-1)}{2}\right] + 3(n-1)\left[\frac{n(n-1)(n-2)}{3}\right] - \sum_{i=1}^{n-1} d_i^3$$

$$= \frac{n(n-1)^2(n-3)}{2} - \sum_{i=1}^n d_i^3.$$

Thus we have

$$\frac{n(n-1)^2(n-3)}{2} - \sum_{i=1}^3 d_i^3 = \frac{n(n-1)(n^2-5n+8)}{4} + 6i_3,$$

which now yields

$$\sum_{i=1}^{n} d_i^3 = \frac{n(n-1)(n^2-3n-2)}{4} - 6i_3. \qquad (4.6.5)$$

Equating (4.6.4) and (4.6.5) it follows that $c_3 + i_3 = \frac{n(n-1)(n-5)}{12}$. Since the total number of triples of vertices in $\mathcal{G}$ is $C_{n,3} = n(n-1)(n-2)/6$, it follows that the number of induced subgraphs on three vertices which are neither empty nor complete is $\frac{n(n-1)(n+1)}{12}$. □

Lemmas 4.6.9, 4.6.10, and 4.6.11 allow us to prove our second main result of this section which shows that $S_{n,n}$ is not Laplacian realizable for $n < 12$ (see [19]).

**THEOREM 4.6.12** *If $S_{n,n}$ is Laplacian realizable, then $n \geq 12$.*

**Proof**: Suppose to the contrary that $n \leq 11$. By Lemma 4.6.9, $S_{n,n}$ is not Laplacian realizable for $n = 3, 5, 7, 11$ as they are prime. By the same lemma, $S_{4,4}$ is not Laplacian realizable as 4 is not a divisor of $(4-1)! = 6$. Finally, by Lemma 4.6.8, $S_{6,6}$ and $S_{10,10}$ are not Laplacian realizable as 6 and 10 both have a remainder of 2 when divided by 4. Thus the only remaining cases to consider are $n = 8$ and $n = 9$.

Suppose now that $n = 8$, and that $\mathcal{G}$ is a graph that realizes $S_{8,8}$. By Lemma 4.6.10, the possible degrees for $\mathcal{G}$ are 2, 3, 4, and 5. For each $i = 2, 3, 4, 5$ let $a_i$ be the number of vertices of degree $i$. By Theorem 3.1.3 we have $a_2 + a_3 + a_4 + a_5 = 8$, $2a_2 + 3a_3 + 4a_4 + 5a_5 = \sum_{i=1}^{8} d_i = n(n-1)/2 = 28$, and $4a_2 + 9a_3 + 16a_4 + 25a_5 = \sum_{i=1}^{8} d_i^2 = n(n-1)(n-2)/3 = 112$. Thus the $a_i$'s satisfy the linear system

$$\begin{bmatrix} 1 & 1 & 1 & 1 \\ 2 & 3 & 4 & 5 \\ 4 & 9 & 16 & 25 \end{bmatrix} \begin{bmatrix} a_2 \\ a_3 \\ a_4 \\ a_5 \end{bmatrix} = \begin{bmatrix} 8 \\ 28 \\ 112 \end{bmatrix}.$$

Solving, we obtain that

$$\begin{cases} a_2 = 6 - a_5, \\ a_3 = -8 + 3a_5, \\ a_4 = 10 - 3a_5. \end{cases}$$

In particular, $a_5$ must be an integer such that $10 \geq 3a_5 \geq 8$. Therefore $a_5 = 3$. Thus $a_2 = 3$, $a_3 = 1$, and $a_4 = 1$. Hence the degree sequence of $\mathcal{G}$ is $(2,2,2,3,4,5,5,5)$. From (4.6.4) and (4.6.5), we find that $c_3 = i_3 = 7$, where $c_3$ and $i_3$ are the number of three-cycles and independent triples in $\mathcal{G}$, respectively.

For a candidate graph on eight vertices with degree sequence $(2,2,2,3,4,5,5,5)$ and with $c_3 = i_3 = 7$, it can be verified that each vertex of degree five must be adjacent to a vertex with degree two. Using this observation, it follows that the subgraph induced by the vertices of degree two must be empty. At this point we are forced to consider (numerous) cases as described by:

(i) the possible neighbors of the vertices of degree two, and then

(ii) the subgraph induced by the remaining vertices.

All such nonisomorphic graphs can be constructed and their respective Laplacian eigenvalues can be obtained by a computation. In all cases (there are four such graphs) the graphs were not Laplacian integral. Thus we conclude that $S_{8,8}$ is not Laplacian realizable.

Now suppose that $n = 9$ and that $\mathcal{G}$ is a graph that realizes $S_{9,9}$. From Lemma 4.6.10, the possible degrees for the vertices of $\mathcal{G}$ are $2, 3, 4, 5$, and $6$. For each $i = 2, 3, 4, 5, 6$, let $a_i$ be the number of vertices of degree $i$. As in the case for $n = 8$, we obtain

$$\begin{bmatrix} 1 & 1 & 1 & 1 & 1 \\ 2 & 3 & 4 & 5 & 6 \\ 4 & 9 & 16 & 25 & 36 \end{bmatrix} \begin{bmatrix} a_2 \\ a_3 \\ a_4 \\ a_5 \\ a_6 \end{bmatrix} = \begin{bmatrix} 9 \\ 36 \\ 168 \end{bmatrix}.$$

Solving for $a_2, a_4, a_6$ in terms of $a_3$ and $a_5$, we find that

$$a_2 = 3 - \frac{3}{8}a_3 + \frac{1}{8}a_5$$

$$a_4 = 3 - \frac{3}{4}a_3 - \frac{3}{4}a_5$$

$$a_6 = 3 + \frac{1}{8}a_3 - \frac{3}{8}a_5$$

Considering the expression for $a_4$ we see that necessarily 4 divides $a_3 + a_5$, from which we conclude that $a_3 + a_5$ is either 0 or 4. If $a_3 + a_5 = 0$, then $a_3 = a_5 = 0$ and $a_2 = a_4 = a_6 = 3$, so that the degree sequence is $(2, 2, 2, 4, 4, 4, 6, 6, 6)$. Note that in this case we have $c_3 = i_3 = 12$ by (4.6.4) and (4.6.5). If $a_3 + a_5 = 4$, then we have $a_4 = 0$, $a_2 = (7 - a_3)/2$ and $a_6 = (3 + a_3)/2$. Necessarily, $a_3$ is odd and less than 4 so the two possible cases are $a_3 = 1$ and $a_3 = 3$. The former yields the degree sequence $(2, 2, 2, 3, 5, 5, 5, 6, 6)$, with $c_3 = 11$ and $i_3 = 13$, while the latter yields the degree sequence $(2, 2, 3, 3, 3, 5, 6, 6, 6)$, with $c_3 = 13$ and $i_3 = 11$. Observe that the last two degree sequences are complementary, i.e., if $\mathcal{G}$ has one degree sequence then its complement has the other.

Similar calculations to the ones required for the case $n = 8$ can also be used for $n = 9$ with the above degree sequences and their corresponding values of $c_3$ and $i_3$. In fact, it can be checked that, for either degree sequence, each vertex of degree six must be adjacent to a vertex with degree two, and, in addition, for the degree sequence $(2,2,2,3,5,5,5,6,6)$ at least one vertex of degree five must be adjacent to a vertex of degree two. Using this observation, again it follows that the subgraph induced by the vertices of degree two must be empty. As in the previous case we need to consider cases as described by:

(i) the possible neighbors of the vertices of degree two, and then

(ii) the subgraph induced by the remaining vertices.

All such nonisomorphic graphs can be constructed and their respective Laplacian eigenvalues can be obtained by a computation. In all cases (there are eight such graphs), the graphs were not Laplacian integral. Thus we conclude that $S_{9,9}$ is not Laplacian realizable. □

## Exercises: Note that all exercises are based on [19].

1. (a) Suppose that $n \geq 6$ and that $\mathcal{G}$ is a graph on $n$ vertices such that for $i \neq n$, $\mathcal{G}$ realizes $S_{i,n}$. Prove that either $\mathcal{G} = K_1 \vee H$ for some graph $H$ on $n-1$ vertices, or $\mathcal{G} = (K_1 + K_1) \vee (K_1 + H)$ for some graph $H$ realizing $S_{n-4,n-3}$.

   (b) Prove that if $H$ in part (a) realizes $S_{n-4,n-3}$ then necessarily $i = 1$.

2. Suppose that $\mathcal{G}$ is a graph on $n \geq 6$ vertices. Prove that $\mathcal{G}$ realizes $S_{1,n}$ if and only if $\mathcal{G}$ is formed in one of the following two ways:

   (a) $\mathcal{G} = (K_1 + K_1) \vee (K_1 + \mathcal{G}_1)$ where $\mathcal{G}_1$ is a graph on $n-3$ vertices that realizes $S_{n-4,n-3}$.

   (b) $\mathcal{G} = K_1 \vee H$ where $H$ is a graph on $n-1$ vertices that realizes $S_{n-1,n-1}$.

3. Suppose that $\mathcal{G}$ is a graph on $n \geq 6$ vertices. Prove that $\mathcal{G}$ realizes $S_{n-1,n}$ if and only if $\mathcal{G}$ is formed in one of the following two ways:

   (a) $\mathcal{G} = K_1 \vee (K_2 + \mathcal{G}_1)$ where $\mathcal{G}_1$ is a graph on $n-3$ vertices that realizes $S_{2,n-3}$.

   (b) $\mathcal{G} = K_1 \vee (K_1 + H)$ where $H$ is a graph on $n-2$ vertices that realizes $S_{n-2,n-2}$.

4. Suppose that $\mathcal{G}$ is a graph on $n \geq 6$ vertices. Prove that $\mathcal{G}$ realizes $S_{i,n}$ if and only if $\mathcal{G} = K_1 \vee (K_1 + H)$ where $H$ is a graph on $n-2$ vertices that realizes $S_{i-1,n-2}$.

5. (a) Construct the four graphs on eight vertices that are alluded to in the proof of Theorem 4.6.12.

   (b) Construct the eight graphs on nine vertices that are alluded to in the proof of Theorem 4.6.12.

6. (a) Prove that for each $n \leq 18$ that if $S_{1,n}$ is realized, then it is realized by a unique graph.

   (b) Prove that for each $n \leq 19$ that if $S_{n-1,n}$ is realized, then it is realized by a unique graph.

For Exercises 7 and 8, let $G_{k-1,k}$ be the unique graph which realizes $S_{k-1,k}$, for each $k = 2, 3, 6, 7, 10, 11, 14, 15, 19$. Similarly, let $H_{1,k}$, denote the unique graph that realizes $S_{1,k}$ for each $k = 2, 5, 6, 9, 10, 14, 18$ (and of course $G_{1,2} = H_{1,2} = K_2$). From these core graphs, we inductively construct the following classes of graphs: For each $m = 2, 5, 6, 9, 10, 14, 18$, let $A_0^m = H_{1,m}$ and for each natural number $\ell$, let $A_\ell^m = K_1 \vee (K_1 + A_{\ell-1}^m)$. For each $m = 2, 3, 6, 7, 10, 11, 14, 15, 19$, let $B_0^m = G_{m-1,m}$ and for each natural number $\ell$, let $B_\ell^m = K_1 \vee (K_1 + B_{\ell-1}^m)$.

**7.** Let $n$ be a positive integer. Prove the following:

(a) If $n$ is even, $i \in \{0, 2, 4, 6, 8\}$, and $n \geq 2i + 2$, then $S_{\frac{n}{2}-i,n}$ is uniquely realized by $A_{\frac{n}{2}-i-1}^{2i+2}$.

(b) If $n$ is even, $i \in \{2, 4\}$, and $n \geq 2i + 2$, then $S_{\frac{n}{2}+i,n}$ is uniquely realized by $B_{\frac{n}{2}-i-1}^{2i+2}$.

(c) If $n$ is odd, $i \in \{2, 4\}$, and $n \geq 2i + 1$, then $S_{\frac{n+1}{2}-i,n}$ is uniquely realized by $A_{\frac{n-1}{2}-i}^{2i+1}$.

(d) If $n$ is odd, $i \in \{0, 2, 4, 6, 8\}$, and $n \geq 2i + 3$, then $S_{\frac{n+1}{2}+i,n}$ is uniquely realized by $B_{\frac{n-3}{2}-i}^{2i+3}$.

**8.** In light of the previous exercises, why is it possible for $S_{1,22}$ not to be uniquely realizable?

# Chapter 5

# The Algebraic Connectivity

Probably the most important property of a graph that we can obtain from the Laplacian matrix is its algebraic connectivity. The algebraic connectivity of a graph is simply the second smallest eigenvalue of the Laplacian matrix. We learned in Chapter 1 that if we add a positive semidefinite matrix to an existing positive semidefinite matrix, the eigenvalues will monotonically increase. Since adding edges to an existing graph results in a Laplacian matrix obtained by adding a positive semidefinite matrix to the existing Laplacian matrix, we saw from Theorem 4.1.2 that the second smallest eigenvalue (in fact all eigenvalues) monotonically increases. Moreover, in light of Theorem 4.1.1, the second smallest eigenvalue of the Laplacian matrix is positive if and only if the graph is connected, and is zero otherwise. Hence the second smallest eigenvalue of the Laplacian matrix is a useful way of measuring the connectivity of the graph. Section 5.1 gives us an introduction to the algebraic connectivity of a graph. Here, we learn which graphs have maximum and minimum algebraic connectivity, and also develop an upper bound on the algebraic connectivity in terms of the vertex connectivity of a graph. Recalling that adding an edge to a graph causes the algebraic connectivity to monotonically increase, this motivates us to explore the effects that varying the edge weight has on the algebraic connectivity. This we do in Section 5.2 where we learn that increasing the weight of an edge causes the algebraic connectivity to increase in a concave downward fashion. In Section 5.3 we develop upper and lower bounds on the algebraic connectivity of graphs in terms of a graph's diameter and mean distance. Since graphs with large diameter and mean distance tend to have less edges, they are "less connected" and thus have lower algebraic connectivity. Section 5.4 focuses on using the edge density of a graph to give an upper bound on its algebraic connectivity. This leads us to the concept the isoperimetric number of a graph which is helpful to us in determining upper bounds on the algebraic connectivity in terms of the graph's topological properties. Section 5.5 focuses on planar graphs while Section 5.6 focuses on graphs of higher genus.

## 5.1 Introduction to the Algebraic Connectivity of Graphs

Given the Laplacian matrix for a graph $\mathcal{G}$ on $n$ vertices with eigenvalues $0 = \lambda_1 \leq \lambda_2 \leq \ldots \lambda_n$, we saw in Theorem 4.1.1 that $\lambda_2 = 0$ if and only if $\mathcal{G}$ is disconnected. Thus if $\mathcal{G}$ is a connected graph then $\lambda_2 > 0$. We also saw in Theorem 4.1.2 that $\lambda_2$ monotonically increases as we add edges to an existing graph. Hence $\lambda_2$ gives a measure of the connectivity of $\mathcal{G}$. This leads us to an important definition:

**DEFINITION 5.1.1** The *algebraic connectivity of* $\mathcal{G}$ is the second smallest eigenvalue, $\lambda_2$, of the Laplacian matrix $L(\mathcal{G})$. We denote the algebraic connectivity of $\mathcal{G}$ by $a(\mathcal{G})$.

This term was coined by Fiedler in [28]. In light of the concept of algebraic connectivity, we now present two corollaries to Theorem 4.1.2:

**COROLLARY 5.1.2** *Suppose we take a graph $\mathcal{G}$ and create a graph $\hat{\mathcal{G}}$ by either adding an edge joining a pair of nonadjacent vertices of $\mathcal{G}$ or increasing the weight of an existing adge of $\mathcal{G}$, then $a(\hat{\mathcal{G}}) \geq a(\mathcal{G})$.*

If we focus our attention on unweighted graphs, we can use Corollary 5.1.2 together with Corollary 4.1.5 to obtain another important corollary:

**COROLLARY 5.1.3** *Let $n \geq 2$ be fixed. Then $a(K_n) = n$. Moreover, of all unweighted graphs on $n$ vertices, $K_n$ is the unique graph with the largest algebraic connectivity.*

We should note that we define $a(K_1) = 1$.

**OBSERVATION 5.1.4** If we delete an edge $e$ from the graph $K_n$, then the algebraic connectivity of the resulting graph is $n - 2$. Thus if $\mathcal{G}$ is a noncomplete graph on $n$ vertices, then $a(\mathcal{G}) \leq n - 2$.

Since the algebraic connectivity of all disconnected graphs is zero while the algebraic connectivity of all connected graphs is positive, Theorem 5.1.3 begs the question: Of all connected graphs on $n$ vertices where $n$ is fixed, which graph(s) have the smallest algebraic connectivity? In our efforts to answer this question, we now present an important lemma from [31] which relates the edge connectivity (see Definition 2.4.13) to the structure of Laplacian matrices:

**LEMMA 5.1.5** *Let $\mathcal{G}$ be an unweighted connected graph on $n$ vertices with vertex set $V$. Let $L$ be the Laplacian matrix for $\mathcal{G}$. Then*

$$e(\mathcal{G}) = \min_{\substack{W \subset V \\ W \neq \emptyset, V}} \sum_{\substack{i \in W \\ j \in V \setminus W}} |\ell_{i,j}|.$$

# The Algebraic Connectivity

**Proof**: Fix a vertex set $W$ such that $W \neq \emptyset, V$. Let $E_W$ be the set of edges in $\mathcal{G}$ that are incident with one vertex in $W$ and one vertex in $V \setminus W$. Observe that $L(\mathcal{G} - E_W)$ is permutationally similar to a block diagonal matrix with the vertices of $W$ corresponding to one block and the $p_W := \sum_{\substack{i \in W \\ j \in V \setminus W}} |\ell_{i,j}|$ vertices of $V \setminus W$ corresponding to the other block. Hence $\mathcal{G} - E_W$ is disconnected. Observe the number of edges in $E_W$ is $p_W$. The result now follows. $\square$

We now use Lemma 5.1.5 to provide a lower bound on the algebraic connectivity of unweighted graphs in terms of the edge connectivity (see [31]):

**LEMMA 5.1.6** *For any unweighted graph $\mathcal{G}$ on $n$ vertices, we have $a(\mathcal{G}) \geq 2e(\mathcal{G})(1 - \cos\frac{\pi}{n})$.*

**Proof**: Let $\mathcal{G}$ be an unweighted graph on $n$ vertices with $L$ as its Laplacian. Let $\Delta$ be the maximum degree of a vertex in $\mathcal{G}$. Define the matrix $S$ as

$$S = I - \frac{1}{\Delta}L. \qquad (5.1.1)$$

Observe that $S$ is a symmetric, nonnegative, doubly stochastic matrix. Let $1 = \gamma_1 \geq \gamma_2 \geq \ldots \gamma_n$ be the eigenvalues of $S$. Then

$$\gamma_2 = 1 - \frac{1}{\Delta}a(\mathcal{G}).$$

By Lemma 5.1.5 and Definition 1.5.2, the measure of irreducibility of $L$ is

$$\mu(L) = \frac{1}{\Delta}e(\mathcal{G}). \qquad (5.1.2)$$

From (5.1.1), (5.1.2), and Theorem 1.5.10 we obtain

$$1 - \gamma_2 = \frac{1}{\Delta}a(\mathcal{G}) \geq 2\left(1 - \cos\frac{\pi}{n}\right)\frac{1}{\Delta}e(\mathcal{G}).$$

Multiplying through by $\Delta$ gives the desired result. $\square$

We are now ready to present a theorem from [28] which shows us that, for fixed $n$, the path on $n$ vertices is the unique connected graph on $n$ vertices with the smallest algebraic connectivity:

**THEOREM 5.1.7** *Let $n \geq 2$ be fixed. Then $a(P_n) = 2(1 - \cos\frac{\pi}{n})$. Moreover, of all unweighted graphs on $n$ vertices, $P_n$ is the unique graph with the smallest algebraic connectivity.*

**Proof**: Let $L$ be the Laplacian matrix for $P_n$. By matrix-vector multiplication, observe that for $k = 2, \ldots, n$ that $\lambda_k = 2(1 - \cos\frac{(k-1)\pi}{n})$ and the corresponding eigenvector is

$$\left[\cos\frac{(k-1)\pi}{2n}, \cos\frac{3(k-1)\pi}{2n}, \cos\frac{5(k-1)\pi}{2n}, \ldots, \cos\frac{(2n-1)(k-1)\pi}{2n}\right]^T$$

(see [28] Lemma 2.3). Thus $a(P_n) = \lambda_2(L(P_n)) = 2(1 - \cos\frac{\pi}{n})$. By Lemma 5.1.6 and the fact that $e(P_n) = 1$, it follows that $P_n$ is a connected graph on $n$ vertices with minimum algebraic connectivity. Uniqueness follows from Remark 1.5.11 and the fact that $P_n$ is the only connected graph $\mathcal{G}$ on $n$ vertices where $L(\mathcal{G})$ is permutationally similar to a tridiagonal matrix. $\square$

Now that we have a lower bound on the algebraic connectivity of connected unweighted graphs and know that $n$ is an upper bound on the algebraic connectivity of such graphs, our next goal will be to determine more precise upper bounds on the algebraic connectivity and to characterize all graphs in which this upper bound is attained. To this end, we will use the concept of the vertex connectivity (see Definition 2.4.8). To begin our investigation as to how the vertex connectivity gives us insight into the algebraic connectivity, we first prove the following lemma from [28]:

**LEMMA 5.1.8** *Let $\mathcal{G}$ be an unweighted graph and let $\mathcal{G}_1$ be a graph obtained from $\mathcal{G}$ by removing $k$ vertices from $\mathcal{G}$ and all incident edges. Then*

$$a(\mathcal{G}_1) \geq a(\mathcal{G}) - k \qquad (5.1.3)$$

**Proof**: Let $\mathcal{G}$ have $n$ vertices labeled $1, \ldots, n$, and let $\mathcal{G}_1$ be obtained from $\mathcal{G}$ by removing one vertex, say $n$, without loss of generality. Define a new graph $\hat{\mathcal{G}}$ by adding a vertex to $\mathcal{G}_1$ that is adjacent to all other vertices in $\mathcal{G}_1$. Then

$$L(\hat{\mathcal{G}}) = \begin{bmatrix} L(\mathcal{G}_1) + I & -e^T \\ -e & n-1 \end{bmatrix}.$$

Let $x \in \Re^{n-1}$ be an eigenvector of $L(\mathcal{G}_1)$ corresponding to the eigenvalue $a(\mathcal{G}_1)$, and let $\hat{x}$ be the vector in $\Re^n$ created from $x$ by placing zero in the $n^{th}$ position (of $\hat{x}$). Then $L(\hat{\mathcal{G}})\hat{x} = [a(\mathcal{G}_1) + 1]\hat{x}$, thus $a(\mathcal{G}_1) + 1 \neq 0$ is an eigenvalue of $L(\hat{\mathcal{G}})$. Therefore $a(\hat{\mathcal{G}}) \leq a(\mathcal{G}_1) + 1$, and thus $a(\mathcal{G}_1) \geq a(\hat{\mathcal{G}}) - 1$. But since $a(\hat{\mathcal{G}}) \geq a(\mathcal{G})$ by Theorem 5.1.2, it follows that $a(\mathcal{G}_1) \geq a(\mathcal{G}) - 1$. Thus the theorem is true for $k = 1$. For $k > 1$, the theorem follows by induction. $\square$

Lemma 5.1.8 gives us an immediate corollary from [28]:

**COROLLARY 5.1.9** *Let $\mathcal{G}$ be an unweighted graph with vertex set $V$. Let $V = V_1 \cup V_2$ where $V_1$ and $V_2$ are disjoint. For $i = 1, 2$, let $\mathcal{G}_i$ be the subgraph of $\mathcal{G}$ induced by $V_i$. Then*

$$a(\mathcal{G}) \leq \min\{a(\mathcal{G}_1) + |V_2|, a(\mathcal{G}_2) + |V_1|\}.$$

We are now ready to prove an important theorem from [28] and [46] (primarily [46]) which shows that the vertex connectivity is an upper bound on the algebraic connectivity. Moreover, we classify the graphs in which the vertex and algebraic connectivities are equal.

# The Algebraic Connectivity

**THEOREM 5.1.10** *Let $\mathcal{G}$ be a noncomplete, connected, unweighted graph on $n$ vertices. Then $a(\mathcal{G}) \leq v(\mathcal{G})$. Moreover, equality holds if and only if $\mathcal{G}$ can be written as $\mathcal{G}_1 \vee \mathcal{G}_2$, where $\mathcal{G}_1$ is a disconnected graph on $n - v(\mathcal{G})$ vertices and $\mathcal{G}_2$ is a graph on $v(\mathcal{G})$ vertices with $a(\mathcal{G}_2) \geq 2v(\mathcal{G}) - n$.*

**Proof**: Let $V$ be a set of $v(\mathcal{G})$ vertices such that $\mathcal{G} - V$ is disconnected. Then by Corollary 5.1.9 we have

$$a(\mathcal{G}) \leq a(\mathcal{G} - V) + |V| = v(\mathcal{G}),$$

where equality follows from the fact that $a(\mathcal{G} - V) = 0$ since $\mathcal{G} - V$ is disconnected. To prove the remainder of the theorem, first suppose that $a(\mathcal{G}) = v(\mathcal{G})$. Let $V$ be as above. Letting $\mathcal{G}_1 = \mathcal{G} - V$ and $\mathcal{G}_2$ be the subgraph of $\mathcal{G}$ induced by the vertices of $V$, we see that by simultaneously permuting rows and columns, the Laplacian matrix of $\mathcal{G}$ can be written as

$$L(\mathcal{G}) = \begin{bmatrix} L(\mathcal{H}_1) + D_1 & 0 & -X \\ 0 & L(\mathcal{H}_2) + D_2 & -Y \\ -X^T & -Y^T & L(\mathcal{G}_2) + D_3 \end{bmatrix},$$

where the $(1,1)$, $(1,2)$, $(2,1)$, and $(2,2)$ blocks, together, form an $(n - v(\mathcal{G})) \times (n - v(\mathcal{G}))$ matrix; the $(3,3)$ block is a $v(\mathcal{G}) \times v(\mathcal{G})$ matrix; and where $\mathcal{H}_1 + \mathcal{H}_2 = \mathcal{G}_1$, $L(\mathcal{H}_1)$ and $L(\mathcal{H}_2)$ are the corresponding Laplacian matrices, and $D_1$ and $D_2$ are suitable diagonal matrices. Suppose that $\mathcal{H}_1$ is a graph on $n_1$ vertices and that $\mathcal{H}_2$ is a graph on $n_2$ vertices with $n_1 + n_2 = n - v(\mathcal{G})$. Let

$$w^T = [\, n_2(e^{(n_1)})^T \mid -n_1(e^{(n_2)})^T \mid 0^T \,]$$

and note that $w^T e = 0$. Now

$$w^T L(\mathcal{G}) w = n_2^2 (e^{(n_1)})^T (L(\mathcal{H}_1) + D_1) e^{(n_1)} + n_1^2 (e^{(n_2)})^T (L(\mathcal{H}_2) + D_2) e^{(n_2)} \quad (5.1.4)$$
$$= n_2^2 (e^{(n_1)})^T D_1 e^{(n_1)} + n_1^2 (e^{(n_2)})^T D_2 e^{(n_2)},$$

where the last inequality follows from the fact that $L(\mathcal{H}_1)$ and $L(\mathcal{H}_2)$ have zero row sums. Notice that each of $D_1$ and $D_2$ is a diagonal matrix with diagonal entries at most $v(\mathcal{G})$. Thus we have that

$$w^T L(\mathcal{G}) w \leq n_2^2 n_1 v(\mathcal{G}) + n_1^2 n_2 v(\mathcal{G}) = v(\mathcal{G}) n_1 n_2 (n_1 + n_2) = v(\mathcal{G}) w^T w, \quad (5.1.5)$$

with equality if and only if $D_1 = v(\mathcal{G}) I_{n_1}$ and $D_2 = v(\mathcal{G}) I_{n_2}$. Now, since $w^T e = 0$, it follows from Theorem 1.2.7 that $a(\mathcal{G}) w^T w \leq w^T L(\mathcal{G}) w$. Hence, together with (5.1.5) and recalling our assumption that $a(\mathcal{G}) = v(\mathcal{G})$, we can write that

$$v(\mathcal{G}) w^T w = a(\mathcal{G}) w^T w \leq w^T L(\mathcal{G}) w \leq v(\mathcal{G}) w^T w.$$

Therefore $a(\mathcal{G})w^T w = w^T L(\mathcal{G})w$ and so from (5.1.4) we must have that $D_1 = v(\mathcal{G})I_{n_1}$ and $D_2 = v(\mathcal{G})I_{n_2}$. Thus $\mathcal{G} = \mathcal{G}_1 \vee \mathcal{G}_2$, where $\mathcal{G}_1 = \mathcal{H}_1 + \mathcal{H}_2$ is a disconnected graph. As a result,

$$L(\mathcal{G}) = \left[\begin{array}{c|c} L(\mathcal{G}_1) + v(\mathcal{G})I & -J \\ \hline -J & L(\mathcal{G}_2) + (n - v(\mathcal{G}))I \end{array}\right]. \quad (5.1.6)$$

Now we will show that if $a(\mathcal{G}) = v(\mathcal{G})$, then $a(\mathcal{G}_2) \geq 2v(\mathcal{G}) - n$. Suppose that $\lambda$ is an eigenvalue of $L(\mathcal{G}_1)$ having an eigenvector $x$ such that $x^T e = 0$. Then $[\,x^T \mid 0\,]^T$ is an eigenvector for $L(\mathcal{G})$ corresponding to $\lambda + v(\mathcal{G})$. Similarly, if $\gamma$ is an eigenvalue of $L(\mathcal{G}_2)$ having an eigenvector $y$ such that $y^T e = 0$, then $[\,0 \mid y^T\,]^T$ is an eigenvector for $L(\mathcal{G})$ corresponding to $\gamma + n - v(\mathcal{G})$. Therefore, it follows that $n - 2$ of the eigenvalues of $L(\mathcal{G})$ are of the form $\lambda + v(\mathcal{G})$ where $\lambda$ is an eigenvalue of $L(\mathcal{G}_1)$ with a corresponding eigenvector orthogonal to $e$, or of the form $\gamma + n - v(\mathcal{G})$ where $\gamma$ is an eigenvalue of $L(\mathcal{G}_2)$ with a corresponding eigenvector orthogonal to $e$. Observe that $0$ is also eigenvalue of $L(\mathcal{G})$, as is $n$, the latter with eigenvector

$$[\,v(\mathcal{G})(e^{(n-v(\mathcal{G}))})^T \mid -(n-v(\mathcal{G}))(e^{(v(\mathcal{G}))})^T\,]^T.$$

Consequently,

$$a(\mathcal{G}) = \min\{v(\mathcal{G}) + a(\mathcal{G}_1), n - v(\mathcal{G}) + a(\mathcal{G}_2)\} = \min\{v(\mathcal{G}), n - v(\mathcal{G}) + a(\mathcal{G}_2)\}$$

(since $\mathcal{G}_1$ is disconnected). Since $a(\mathcal{G}) = v(\mathcal{G})$, it follows that $n - v(\mathcal{G}) + a(\mathcal{G}_2) \geq v(\mathcal{G})$. We thus conclude that $a(\mathcal{G}_2) \geq 2v(\mathcal{G}) - n$.

Conversely, suppose that $\mathcal{G} = \mathcal{G}_1 \vee \mathcal{G}_2$, where $\mathcal{G}_1$ is disconnected graph on $n - v(\mathcal{G})$ vertices and $\mathcal{G}_2$ is a graph on $v(\mathcal{G})$ vertices with $a(\mathcal{G}_2) \geq 2v(\mathcal{G}) - n$. Then $L(\mathcal{G})$ has the form as in (5.1.6). Therefore, by the preceding analysis, we see that

$$a(\mathcal{G}) = \min\{v(\mathcal{G}) + a(\mathcal{G}_1), n - v(\mathcal{G}) + a(\mathcal{G}_2)\} = \min\{v(\mathcal{G}), n - v(\mathcal{G}) + a(\mathcal{G}_2)\}$$

(since $\mathcal{G}_1$ is disconnected). Since $a(\mathcal{G}_2) \geq 2v(\mathcal{G}) - n$, it follows that $a(\mathcal{G}) = v(\mathcal{G})$. □

**EXAMPLE 5.1.11** Consider the following graphs $\mathcal{G}_1$ and $\mathcal{G}_2$.

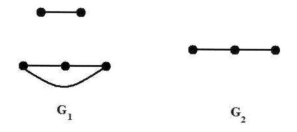

Observe the graph $\mathcal{G} := \mathcal{G}_1 \vee \mathcal{G}_2$ below:

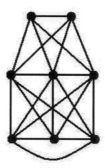

Note that $n = 8$ and that $v(\mathcal{G}) = 3$. Since $\mathcal{G}_1$ is a disconnected graph on $n - v(\mathcal{G}) = n - 3$ vertices and $\mathcal{G}_2$ is a graph on $v(\mathcal{G}) = 3$ vertices where $1 = a(\mathcal{G}_2) \geq 2v(\mathcal{G}) - n$, it follows by Theorem 5.1.10 that $a(\mathcal{G}) = v(\mathcal{G}) = 3$.

Letting $\delta(\mathcal{G})$ denote the minimum degree of a vertex in $\mathcal{G}$, recall from Theorem 2.4.15 that $v(\mathcal{G}) \leq \delta(\mathcal{G})$. Therefore we obtain the following corollary to Theorem 5.1.10.

**COROLLARY 5.1.12** *If $\mathcal{G}$ is a noncomplete unweighted graph on $n$ vertices, then*

$$a(\mathcal{G}) \leq v(\mathcal{G}) \leq \delta(\mathcal{G}) \leq n - 1.$$

Since trees are a class of graphs that will be important to us in subsequent chapters, we conclude this section with a corollary to Theorem 5.1.10 concerning the algebraic connectivity of trees. Observing that $v(\mathcal{T}) = 1$ for any tree $\mathcal{T}$ on at least three vertices, we have the following corollary to Theorem 5.1.10 whose proof is left as an exercise:

**COROLLARY 5.1.13** *Let $\mathcal{T}$ be a tree on $n \geq 3$ vertices. Then $a(\mathcal{T}) \leq 1$. Equality holds if and only if $\mathcal{T}$ is a star.*

In summary, for fixed $n$, we have seen that the unique unweighted graph on $n$ vertices with largest algebraic connectivity is the complete graph $K_n$ and that $a(K_n) = n$. We have also seen that the unique unweighted connected graph on $n$ vertices with smallest algebraic connectivity is the path $P_n$ and that $a(P_n) = 2(1 - \cos\frac{\pi}{n})$. Intuitively speaking, complete graphs are the most difficult graphs to "break apart," hence they have the largest algebraic connectivity. Likewise, paths are the easiest of the connected graphs to break apart; therefore, they have the smallest algebraic connectivity. Further continuing with this idea, since trees do not contain any cycles, intuitively they are easy to break apart, thus trees also have a low algebraic connectivity. Using the fact that the algebraic connectivity is bounded above by the vertex connectivity, it follows that the algebraic connectivity of a tree is at most one. Hence the overarching idea of this section was that as we add edges to an unweighted graph, the algebraic connectivity monotonically increases. In the next section, we will investigate the rate at which such an increase occurs.

## Exercises:

1. Prove Observation 5.1.4.

2. (See [28]) Prove $a(\mathcal{G}_1 \times \mathcal{G}_2) = \min\{a(\mathcal{G}_1), a(\mathcal{G}_2)\}$.

3. Let $K_{p,q}$ be the complete bipartite graph where the two partite sets contain $p$ and $q$ vertices, respectively. Prove $a(K_{p,q}) = \min\{p, q\}$.

4. Generalize the previous result by finding $a(K_{p_1, p_2, \ldots, p_k})$.

5. Let $C_n$ be a cycle on $n$ vertices. Prove $a(C_n) = 2(1 - \cos \frac{2\pi}{n})$.

6. Let $W_n$ be the wheel on $n$ vertices. Prove that $a(W_n) = 1 + 2(1 - \cos \frac{2\pi}{n-1})$.

7. Prove Corollary 5.1.13.

## 5.2 The Algebraic Connectivity as a Function of Edge Weight

We saw from Corollary 5.1.2 that if we take a graph and increase the weight of an existing edge, then the resulting graph will have an algebraic connectivity greater than or equal to the algebraic connectivity of the original graph. In this section, we investigate the rate at which the algebraic connectivity increases as we increase the weight of a fixed edge. We begin with a matrix theoretic result from [48]. In this theorem, and in the subsequent theorems, we order the eigenvalues $\lambda_1 \leq \lambda_2 \leq \ldots \leq \lambda_n$.

**LEMMA 5.2.1** *Let $A$ and $E$ be $n \times n$ symmetric matrices and suppose that $\lambda_2(A)$ is a simple eigenvalue of $A$. Then there exists $t_0 > 0$ such that for each $t \in [0, t_0)$:*

*(i) The matrix $A_t := A + tE$ has $\lambda_2(A_t)$ as a simple eigenvalue.*

*(ii) There is an eigenvector $v_t$ corresponding to $\lambda_2(A_t)$ such that $\|v_t\|_2 = 1$, and with the property that $v_t$ is analytic in $t$ on $[0, t_0)$.*

*(iii)*
$$\frac{d\lambda_2(A_t)}{dt} \leq \lambda_n(E)$$
*with equality holding if and only if $v_t$ is also an eigenvector for $E$ corresponding to $\lambda_n(E)$.*

**Proof**: To prove (i) (see [79], Chapter 2), denote the characteristic equation of $A$ by

$$\det(\lambda I - A) := \lambda^n + c_{n-1}\lambda^{n-1} + c_{n-2}\lambda^{n-2} + \ldots + c_0 = 0. \quad (5.2.1)$$

Then the characteristic equation of $A + tE$ is

$$\det(\lambda I - A - tE) := \lambda^n + c_{n-1}(t)\lambda^{n-1} + c_{n-2}(t)\lambda^{n-2} + \ldots + c_0(t) = 0, \quad (5.2.2)$$

where for $r = 1, \ldots, n-1$, $c_r(t)$ is a polynomial in $t$ of degree $(n-r)$ such that $c_r(0) = c_r$. Since $\lambda_2(A)$ is a simple root of (5.2.1), it follows that for sufficiently small $t$ that there is a simple root $\lambda_2(A_t)$ of (5.2.2) given by the convergent power series $\lambda_2(A_t) = \lambda_2(A) + k_1 t + k_2 t^2 + \ldots$.

To prove (ii), we use similar reasoning as in the proof of (i) to note that if $v$ is an eignvector of $A$ corresponding to the simple eigenvalue $\lambda_2(A)$, then $v_t$ can be written as the convergent power series

$$v_t = v + tz_1 + t^2 z_2 + \ldots$$

for appropriate vectors $z_i$. Thus $v_t$ is analytic for sufficiently small $t > 0$.

To prove (iii), we note that since $\lambda_2(A)$ is a simple eigenvalue of $A$, there is an open interval in $\Re$ centered at $t = 0$ on which $\lambda_2(A_t)$ is an analytic function of $t$. Similarly, there is an interval centered at $t = 0$ on which the corresponding eigenvector $v_t$ is also analytic in $t$. Hence there exists $t_0 > 0$ such that for all $t \in [0, t_0)$, the derivatives of $\lambda_2(A_t)$ and of $v_t$ of all orders exist. We claim that

$$\frac{d\lambda_2(A_t)}{dt} = \frac{v_t^T E v_t}{v_t^T v_t} \quad (5.2.3)$$

for all $t \in [0, t_0)$. To verify this claim (see [34], p. 323), let $v(t) := v_t$ and $\lambda(t) := \lambda_2(A_t)$ for notational simplicity. By the eigenvalue-eigenvector relationship, we have

$$(A + tE)v(t) = \lambda(t)v(t)$$

which we express as

$$Av(t) + tEv(t) = \lambda(t)v(t).$$

Differentiating both sides with respect to $t$, we obtain

$$Av'(t) + E[tv'(t) + v(t)] = \lambda'(t)v(t) + \lambda(t)v'(t).$$

At $t = 0$, we have

$$Av'(0) + Ev(0) = \lambda'(0)v(0) + \lambda(0)v'(0).$$

Since $Av'(0) = \lambda(0)v'(0)$ by the eigenvector-eigenvalue relationship, we simplify the above expression as

$$Ev(0) = \lambda'(0)v(0).$$

Since $v(t)$ and $\lambda(t)$ are analytic for all $t \in [0, t_0)$, we can express the above as

$$Ev(t) = \lambda'(t)v(t).$$

Multiplying both sides of the equality on the left by $v^T(t)$ and dividing through by $v^T(t)v(t)$ gives us (5.2.3). By Theorem 1.2.4, the expression on the right of (5.2.3) is at most $\lambda_n(E)$, thus proving the theorem. □

At this point, we know that the derivative of a simple eigenvalue with respect to $t$ is bounded above. We may now apply this result to Laplacian matrices. Let $E^{(i,j)}$ be the $n \times n$ matrix whose $(i,i)$ and $(j,j)$ entries are 1 and whose $(i,j)$ and $(j,i)$ entries are $-1$. Thus if $L$ is the Laplacian matrix for $\mathcal{G}$, then $L_t := L + tE$ is the Laplacian matrix for the graph obtained from $\mathcal{G}$ by increasing the weight of the edge joining vertices $i$ and $j$ by $t$ if $i$ and $j$ are adjacent, or by adding an edge of weight $t$ joining vertices $i$ and $j$ if $i$ and $j$ are not adjacent. Thus applying our results to Laplacian matrices, we can rewrite Lemma 5.2.1 as follows (see [48]):

**THEOREM 5.2.2** *Let $\mathcal{G}$ be a connected weighted graph on $n$ vertices with Laplacian matrix $L$. Let $= a(\mathcal{G})$, the algebraic connectivity of $\mathcal{G}$, be a simple eigenvalue of $L$. Fix a pair of distinct vertices $i$ and $j$ and put $L_t := L + tE^{(i,j)}$ with $t \geq 0$. Then there exists $t_0 > 0$ such that for all $t \in [0, t_0)$:*

$$\frac{da(\mathcal{G}_t)}{dt} = ((v_t)_i - (v_t)_j)^2 \leq 2 \tag{5.2.4}$$

*where $v_t$ is an analytic eigenvector corresponding to $a(\mathcal{G}_t)$ such that $\|v_t\| = 1$. Moreover, equality holds if and only if $v_t = \pm(e_i - e_j)/\sqrt{2}$.*

At this point, it is natural to ask which graphs on $n$ vertices have the property that their algebraic connectivity is a simple eigenvalue and $e_i - e_j$ is the corresponding eigenvector. We answer this in the following theorem from [48]:

**THEOREM 5.2.3** *Let $\mathcal{G}$ be an unweighted connected graph on $n \geq 3$ vertices with Laplacian matrix $L$. Then $\mathcal{G}$ has the property that $a(\mathcal{G})$ is a simple eigenvalue of $L$ with corresponding eigenvector $e_i - e_j$ for some $i \neq j$ if and only if $\mathcal{G}$ is the graph obtained from the complete graph by deleting the edge joining vertices $i$ and $j$.*

**Proof**: First note that if $\mathcal{G}$ is formed by removing the edge between vertices $i$ and $j$ from an unweighted complete graph on at least three vertices, then the eigenvalues of $L$ are 0, $n-2$, and $n$ with the eigenvalue $n$ having multiplicity $n-2$. Then $a(\mathcal{G}) = n - 2$ is a simple eigenvalue. Moreover $e_i - e_j$ is a corresponding eigenvector of $L$.

Now suppose that $\mathcal{G}$ is a connected graph on $n \geq 3$ vertices having $a(\mathcal{G})$ as a simple eigenvalue and corresponding eigenvector $e_i - e_j$. Without loss of generality, assume $i = 1$ and $j = 2$. We claim that $\mathcal{G}$ is the graph formed by removing the edge joining vertices 1 and 2 in the complete graph. This claim is clearly true if $\mathcal{G}$ has $n = 3$ vertices. Thus we will assume that $\mathcal{G}$ has $n \geq 4$ vertices. Since $a(\mathcal{G})$ is a simple eigenvalue of $L$ with $e_1 - e_2$ as the corresponding eigenvector, then if $k \neq 1, 2$, it follows by the eigenvalue-eigenvector relationship that either $k$ is adjacent to both vertices 1 and 2, or $k$ is adjacent to neither vertices 1 or 2. Suppose that there are $d$ vertices (distinct from 1 and 2) that are adjacent to both vertices 1 and 2, and that there are $m \geq 1$ vertices that are adjacent to neither vertices 1 and 2. Then the Laplacian $L$ can be written

$$\begin{bmatrix} L_0 + dI & -J & 0 \\ -J & L_1 + D_1 & -X \\ 0 & -X^T & L_2 + D_2 \end{bmatrix}$$

where the $(1,1)$ block is $2 \times 2$, the $(2,2)$ block is $d \times d$, and the $(3,3)$ block is $m \times m$; where $L_0$, $L_1$, and $L_2$ are the Laplacian matrices for the appropriate induced subgraphs $\mathcal{G}_0$, $\mathcal{G}_1$, and $\mathcal{G}_2$, respectively; and where $D_1$ and $D_2$ are appropriate diagonal matrices. At this point we note:

(1) Since vertices in $\mathcal{G}_2$ can be adjacent to at most $d$ vertices outside of $\mathcal{G}_2$, namely the vertices in $\mathcal{G}_1$, it follows that $D_2 \leq dI$, and

(2) Since $e_1 - e_2$ is an eigenvector corresponding to $a(\mathcal{G})$, it follows from the eigenvalue-eigenvector relationship that $a(\mathcal{G}) = d + 2$ if vertices 1 and 2 are adjacent, and $a(\mathcal{G}) = d$ if vertices 1 and 2 are not adjacent.

We now consider the vector

$$x = \begin{bmatrix} m/2 \\ m/2 \\ 0^{(d)} \\ -e^{(m)} \end{bmatrix}.$$

Note that $x^T e = 0$ and $x^T x = \frac{m^2}{2} + m$. Observe

$$x^T L x = \frac{dm^2}{2} + e^T D_2 e m$$

$$\leq \frac{dm^2}{2} + dm$$

$$= dx^T x$$

where the inequality follows from (1) above. Hence

$$\frac{x^T L x}{x^T x} \leq d.$$

Recall by Theorem 1.2.7 that

$$a(\mathcal{G}) = \min_{\substack{y \neq 0 \\ y^T e = 0}} \frac{y^T L y}{y^T y}.$$

Thus it follows that $a(\mathcal{G}) \leq d$. Combining this with (2) above, we conclude that $a(\mathcal{G}) = d$ and thus $x$ must be a corresponding eigenvector. Hence there are two linearly independent eigenvectors corresponding to $a(\mathcal{G})$, namely $x$ and $e_1 - e_2$. This contradicts the hypothesis that $a(\mathcal{G})$ is simple, hence our assumption that $m \geq 1$ is false. Therefore $m = 0$ and $L$ can be written

$$L = \begin{bmatrix} L_0 + dI & -J \\ -J & L_1 + 2I \end{bmatrix}$$

where the $(1,1)$ block is $2 \times 2$ and the $(2,2)$ block is $d \times d$. As noted in (2) above, $a(\mathcal{G}) = d$ or $a(\mathcal{G}) = d+2$. Since $\mathcal{G}$ has $d+2$ vertices, if $a(\mathcal{G}) = d+2$ then $\mathcal{G}$ is complete which violates our hypothesis. Thus $a(\mathcal{G}) = d$ which implies vertices 1 and 2 are not adjacent.

Write $\mathcal{G}$ as $E_2 \vee \mathcal{G}_1$ where $E_2$ is the graph consisting of two isolated vertices. The proof will be complete once we show that $\mathcal{G}_1$ is complete, or equivalently $a(\mathcal{G}_1) = d$. Let $a(\mathcal{G}_1) = c$ and let $w$ be the corresponding eigenvector of $\mathcal{G}_1$. From the structure of $L$, we see that $c+2$ is an eigenvalue of $L$ with $[0_2^T \ w^T]^T$ as a corresponding eigenvector. Since $a(\mathcal{G}) = d$ is a simple eigenvector of $L$, it follows that $c+2 > d$, i.e., $c > d-2$. Recalling Observation 5.1.4, since the algebraic connectivity of a noncomplete graph on $d$ vertices is never more than $d-2$, it follows that $c = d$ and thus $\mathcal{G}_1$ is complete, hence the structure of $\mathcal{G}$ is as claimed. □

**EXAMPLE 5.2.4** Consider the graph $\mathcal{G}_t$ below for various values of $t$ and note that $\mathcal{G}_0 = K_5 - e$. Thus by Theorems 5.2.2 and 5.2.3, we expect $da(\mathcal{G}_t)/dt = 2$. Observe that as $t$ increases by 1, the value of $a(\mathcal{G}_t)$ increases by 2.

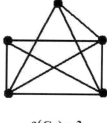
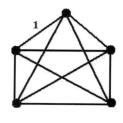
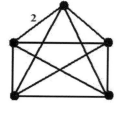

a($G_0$) = 3      a($G_1$) = 5      a($G_2$) = 7

**EXAMPLE 5.2.5** Consider the graph $\mathcal{G}_t$ below for various values of $t$ and note that $\mathcal{G}_t$ is not of the form $K_n - e$. Thus by Theorem 5.2.2, we expect $da(\mathcal{G}_t)/dt < 2$. In the graph below of $a(\mathcal{G}_t)$ as a function of $t$, observe that as $t$ increases by 1 the value of $a(\mathcal{G}_t)$ increases by less than 2.

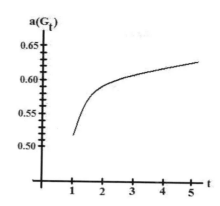

Thus far, we have obtained an upper bound of 2 on the derivative of the algebraic connectivity as a function of edge weight. Naturally, by Theorem 5.1.2 the lower bound on the derivative is zero. In the following theorem and corollaries from [48] we show conditions for when the lower bound is achieved.

**THEOREM 5.2.6** *Let $A$ and $E$ be $n \times n$ positive semidefinite matrices with a common null vector and suppose that $\lambda_2(A) > 0$. For each $t \geq 0$, let $A_t = A + tE$. If for some $t^* > 0$, $u_{t^*}$ is an eigenvector of $A_{t^*}$ corresponding to $\lambda_2(A_{t^*})$ and $Eu_{t^*} = 0$, then $\lambda_2(A_t) = \lambda_2(A_{t^*})$ for all $t \geq t^*$.*

**Proof**: Since $E$ is positive semidefinite, we see that for any $t \geq t^* \geq 0$, it follows that $\lambda_2(A_t) \geq \lambda_2(A_{t^*})$. Additionally, since $A$ and $E$ have a common null vector, it follows that $\lambda_1(A_t) = 0$ for all $t \geq 0$. Since $Eu_{t^*} = 0$, we find that for all $t \geq t^*$, we have

$$\lambda_2(A_{t^*})u_{t^*} = A_{t^*}u_{t^*} = [A_t + (t^* - t)E]u_{t^*} = A_t u_{t^*} + (t^* - t)Eu_{t^*} = A_t u_{t^*}.$$

Hence $\lambda_2(A_{t^*})$ is a positive eigenvalue of $A_t$ for all $t \geq t^*$. Hence $\lambda_2(A_{t^*}) \geq \lambda_2(A_t)$. We now conclude that $\lambda_2(A_{t^*}) = \lambda_2(A_t)$ for all $t \geq t^*$. □

In light of Laplacian matrices, observe that $E^{(i,j)}u_{t^*} = 0$ if and only if $(u_{t^*})_i = (u_{t^*})_j$. This yields the following immediate corollaries:

**COROLLARY 5.2.7** *Suppose that $\mathcal{G}$ is a weighted graph on $n \geq 3$ vertices with Laplacian matrix $L$. Suppose that for some $t^* \geq 0$, $u_{t^*}$ is an eigenvector of $L_{t^*} := L + t^*E^{(i,j)}$ corresponding to $a(\mathcal{G})$ such that $(u_{t^*})_i = (u_{t^*})_j$. Then $a(\mathcal{G}_t) = a(\mathcal{G})$ for all $t \geq t^*$.*

**COROLLARY 5.2.8** *Suppose that $\mathcal{G}$ is a weighted graph on $n \geq 3$ vertices with Laplacian matrix $L$ and that $a(\mathcal{G})$ is a multiple eigenvalue of $L$. Then for each pair of distinct indices $i$ and $j$ and for any $t \geq 0$, it follows that $a(\mathcal{G}_t) = a(\mathcal{G})$.*

**Proof**: Since $a(\mathcal{G})$ is a multiple eigenvalue of $L$, it follows that $L$ has at least two linearly independent eigenvectors corresponding to $a(\mathcal{G})$, say $x$ and $y$. Form the eigenvector $w = (y_i - y_j)x + (x_j - x_i)y$. Observe that $w_i = -x_i y_j + x_j y_i = w_j$. The conclusion now follows from Corollary 5.2.7. □

**EXAMPLE 5.2.9** Consider a star $S_n$ on $n \geq 4$ vertices. Since $a(S_n) = 1$ is a multiple eigenvalue of $L(S_n)$, it follows from Corollary 5.2.8 that adding an edge of any positive weight joining two nonadjacent vertices of $S_n$ will not increase the algebraic connectivity.

We now turn our attention to the second derivative of the algebraic connectivity of a graph. Since the first derivative is bounded above, this seems to suggest that as $t$ increases, the algebraic connectivity is increasing in a concave down manner. The following theorem and corollary from [48] verify this:

**THEOREM 5.2.10** Let $A$ and $E$ be $n \times n$ positive semidefinite matrices with a common null vector. Then for $t \geq 0$

$$\frac{d^2 \lambda_2(A_t)}{dt^2} \leq 0.$$

That is, $\lambda_2(A + tE)$ is a concave down function of $t$.

**Proof**: Let $w \neq 0$ be a common null vector to both $A$ and $E$ and let $U$ be the orthogonal complement of the subspace spanned by $w$. For any $t \geq 0$, we have

$$\begin{aligned}
\lambda_2(A + tE) &= \min_{\substack{x \in U \\ x^T x = 1}} x^T(A + tE)x \\
&= \min_{\substack{x \in U \\ x^T x = 1}} x^T[(1-t)A + t(A+E)]x \\
&\geq (1-t) \min_{\substack{x \in U \\ x^T x = 1}} x^T A x + t \min_{\substack{x \in U \\ x^T x = 1}} x^T(A+E)x \\
&= (1-t)\lambda_2(A) + t\lambda_2(A + E).
\end{aligned}$$

The result follows. □

**COROLLARY 5.2.11** Let $\mathcal{G}$ be a connected weighted graph on $n$ vertices with Laplacian matrix $L$. Let $a(\mathcal{G})$, the algebraic connectivity of $\mathcal{G}$, be a simple eigenvalue of $L$. Fix a pair of distinct vertices $i$ and $j$ and put $L_t := L + tE^{(i,j)}$ with $t \geq 0$. Then $a(\mathcal{G}_t)$ is a concave down function of $t$.

**EXAMPLE 5.2.12** Observe that in the graph $\mathcal{G}_t$ in Example 5.2.5 that the algebraic connectivity increases in a concave down fashion as expected by Corollary 5.2.11.

In Chapter 8, we introduce the concept of the group inverse of the Laplacian matrix. Once this concept is introduced, we will be able to refine our results on the second derivative of the algebraic connectivity. This we do in Section 8.4.

# The Algebraic Connectivity

## Exercise:

1. (See [48]) Let $L_1$ and $L_2$ be Laplacian matrices for graphs $\mathcal{G}_1$ and $\mathcal{G}_2$ on $k$ and $m$ vertices, respectively. Suppose that $d$ is a number such that $k > d > 0$. Also suppose that $a(\mathcal{G}_1) > d - m - 2d/k$. Consider the weighted graph $\mathcal{G}$ on $k + m + 2$ vertices whose Laplacian matrix is

$$\begin{bmatrix} dI & -(d/k)J & 0 \\ -(d/k)J & L_1 + (m + 2d/k)I & -J \\ 0 & -J & L_2 + kI \end{bmatrix}$$

where the $(1,1)$ block is $2 \times 2$, the $(2,2)$ block is $d \times d$, and the $(3,3)$ block is $m \times m$.

(a) Show that $d$ is an eigenvalue of $L$ with correspondoing eigenvector $e_1 - e_2$.

(b) Show that $a(\mathcal{G}) = d$ and that $d$ is simple.

## 5.3 The Algebraic Connectivity with Regard to Distances and Diameters

In the previous section, we saw the effect of edge weight on the algebraic connectivity of a graph. In this section, we consider the effects of distance on the algebraic connectivity of unweighted graphs. Since graphs having great distances between vertices tend to have less edges, we will ultimately be investigating the effect the number of edges has on the algebraic connectivity of a graph. Recalling the definitions and examples of diameter and mean distance from Section 2.1, we see that both the diameter and the mean distance of a graph will tend to be smaller for graphs with more edges. Hence by Theorem 5.1.2, it stands to reason that as the diameter and the mean distance decrease, the algebraic connectivity increases. In this section, we will prove specific results to this effect by deriving lower bounds for the algebraic connectivity in terms of the diameter and mean distance of a graph. To this end, we begin with a graph theoretic lemma from [64]. In this lemma, let $E(\mathcal{G})$ and $V(\mathcal{G})$ denote the edge set and vertex set, respectively, of $\mathcal{G}$.

**LEMMA 5.3.1** *Let $\mathcal{G}$ be a connected unweighted graph on $n$ vertices. For each pair of distinct vertices $u, v \in \mathcal{G}$, let $P_{uv}$ denote a shortest path from $u$ to $v$. Then any edge $e \in E(\mathcal{G})$ belongs to at most $n^2/4$ of such paths $P_{uv}$.*

**Proof**: For a fixed edge $e = xy \in E(\mathcal{G})$, define the graph $\Gamma_e$ as follows: $V(\Gamma_e) = V(\mathcal{G})$ and vertices $u$ and $v$ are adjacent in $\Gamma_e$ if and only if $e$ lies on $P_{uv}$. We need to show that $|E(\Gamma_e)| \leq n^2/4$.

First we prove that $\Gamma_e$ has no triangles. Assume that $e$ lies on $P_{uw}$ and $P_{vw}$, i.e., $uw, vw \in E(\Gamma_e)$. We need to show that $u$ and $v$ are not adjacent in $\Gamma_e$. Without loss of generality, we can orient $P_{uw}$ and $P_{vw}$ so that $d(u, y) = d(u, x) + 1$ and $d(v, y) = d(v, x) + 1$.

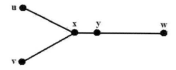

From this and from the triangle inequality we have

$$d(u,v) \le d(u,x) + d(x,v) < d(u,x) + d(y,v)$$

and similarly

$$d(u,v) \le d(u,x) + d(x,v) < d(u,y) + d(x,v).$$

Either way, the shortest path in $\mathcal{G}$ from $u$ to $v$ cannot use the edge $e = xy$. Hence $u$ and $v$ are not adjacent in $\Gamma_e$. Thus $\Gamma_e$ has no triangles.

The proof will be complete when we show that any triangle free graph $\Gamma$ on order $n$ has at most $n^2/4$ edges. We prove this by induction on $n$. This is clearly true for $n \le 3$. For the inductive hypothesis, assume that this is true for all triangle free graphs $\Gamma$ on $n \le k$ vertices for some $k \ge 3$. Now suppose $\Gamma$ has $n = k+1$ vertices and is triangle free. Take any two adjacent vertices $u$ and $v$ and observe that since $u$ and $v$ have no common neighbors that $\deg_\Gamma(u) + \deg_\Gamma(v) \le k+1$. By the induction hypothesis

$$|E(\Gamma)| = |E(\Gamma - u - v)| + \deg_\Gamma(u) + \deg_\Gamma(v) - 1 \le \frac{(k-1)^2}{4} + k = \frac{(k+1)^2}{4} = \frac{n^2}{4}.$$

□

Our next lemma from [28] is a matrix theoretic result which makes use of the eigenvector $x$ of the Laplacian matrix $L$ corresponding to the algebraic connectivity of the graph. Labelling the vertices of the graph $1, \ldots, n$ and letting $x = [x_1, \ldots, x_n]^T$, we have a natural correspondence between the vertices of $\mathcal{G}$ and the entries of $x$.

**LEMMA 5.3.2** *Let $\mathcal{G}$ be a graph with Laplacian matrix $L$, algebraic connectivity $a$, and corresponding eigenvector $x$. Then*

$$\sum_{uv \in E(\mathcal{G})} (x_u - x_v)^2 = a \sum_{v \in V(\mathcal{G})} x_v^2.$$

## The Algebraic Connectivity

**Proof**:

$$\begin{aligned}
a\sum_{v\in V(\mathcal{G})} x_v^2 &= \sum_{v\in V(\mathcal{G})}(ax_v)x_v \\
&= \sum_{v\in V(\mathcal{G})}(e_v^T Lx)x_v \\
&= \sum_{v\in V(\mathcal{G})}\left(x_v\deg(v) - \sum_{uv\in E(\mathcal{G})} x_u\right)x_v \\
&= \sum_{v\in V(\mathcal{G})}\left(x_v^2\deg(v) - \sum_{uv\in E(\mathcal{G})} x_u x_v\right) \\
&= \sum_{v\in V(\mathcal{G})}\sum_{uv\in E(\mathcal{G})}(x_v^2 - x_u x_v) \\
&= \sum_{uv\in E(\mathcal{G})}(x_v^2 - 2x_u x_v + x_u^2) \\
&= \sum_{uv\in E(\mathcal{G})}(x_v - x_u)^2.
\end{aligned}$$

□

We now present a theorem from [64] which gives lower bounds on the diameter and the mean distance, $\bar{\rho}$, of $\mathcal{G}$ in terms of the algebraic connectivity. We then present an immediate corollary which gives upper bounds on the algebraic connectivity.

**THEOREM 5.3.3** *Let $\mathcal{G}$ be an unweighted connected graph on $n$ vertices with algebraic connectivity $a$. Then*

$$\text{diam}(\mathcal{G}) \geq \frac{4}{an} \qquad (5.3.1)$$

*and*

$$(n-1)\bar{\rho}(\mathcal{G}) \geq \frac{2}{a} + \frac{n-2}{2}. \qquad (5.3.2)$$

**Proof**: For each pair of vertices $u, v \in \mathcal{G}$, choose a shortest path $P_{u,v}$ from $u$ to $v$. Let $x$ be an eigenvector corresponding to $a$. Then

$$\begin{aligned}
a\sum_{u\in V(\mathcal{G})}\sum_{v\in V(\mathcal{G})}(x_u - x_v)^2 &= a\sum_{u\in V(\mathcal{G})}\sum_{v\in V(\mathcal{G})}(x_u^2 - 2x_u x_v + x_v^2) \\
&= a(\sum_{u\in V(\mathcal{G})}\sum_{v\in V(\mathcal{G})} x_u^2 - 2a\sum_{u\in V(\mathcal{G})}\sum_{v\in V(\mathcal{G})} x_u x_v \\
&\quad + \sum_{u\in V(\mathcal{G})}\sum_{v\in V(\mathcal{G})} x_v^2) \\
&= a\left(2n\left(\tfrac{1}{a}\right)\sum_{uv\in E(\mathcal{G})}(x_u - x_v)^2 \right. \\
&\quad \left. -2\sum_{u\in V(\mathcal{G})}\sum_{v\in V(\mathcal{G})} x_u x_v\right) \\
&= 2n\sum_{uv\in E(\mathcal{G})}(x_u - x_v)^2 \\
&\quad -2a\sum_{u\in V(\mathcal{G})} x_u\left(\sum_{v\in V(\mathcal{G})} x_v\right) \\
&= 2n\sum_{uv\in E(\mathcal{G})}(x_u - x_v)^2
\end{aligned}$$

$$(5.3.3)$$

where the third equality follows from Lemma 5.3.2 and last equality follows from the fact that $\sum_{v \in V(\mathcal{G})} x_v = 0$ (since $x^T e = 0$). Letting $P_{uv} = u, v_1, v_2, \ldots, v_{k-1}, v$ (thus $d(u,v) = k$), defining $f(e) := x_a - x_b$ where the edge $e = ab$, and recalling that $(\sum_{i=1}^k a_i)^2 \leq k \sum_{i=1}^k a_i^2$ for any real numbers $a_1, \ldots, a_k$, we see that

$$(x_u - x_v)^2 = [(x_u - x_{v_1}) + (x_{v_1} - x_{v_2}) + \ldots + (x_{v_{k-1}} - x_v)]^2 \leq d(u,v) \sum_{e \in P_{uv}} f^2(e). \tag{5.3.4}$$

Let $\chi_{uv} : E(\mathcal{G}) \to \{0, 1\}$ be the characteristic function of $P_{uv}$, i.e.,

$$\chi_{uv}(e) = \begin{cases} 1, & \text{if } e \in P_{u,v} \\ 0, & \text{otherwise} \end{cases}.$$

Thus by (5.3.4)

$$\sum_{v \in V(\mathcal{G})} \sum_{u \in V(\mathcal{G})} (x_u - x_v)^2 \leq \sum_{v \in V(\mathcal{G})} \sum_{u \in V(\mathcal{G})} \left( d(u,v) \sum_{e \in E(\mathcal{G})} f^2(e) \chi_{uv}(e) \right)$$

$$= \sum_{e \in E(\mathcal{G})} \left( f^2(e) \sum_{v \in V(\mathcal{G})} \sum_{u \in V(\mathcal{G})} d(u,v) \chi_{uv}(e) \right). \tag{5.3.5}$$

To show (5.3.1), we see from Lemma 5.3.1 that since $e$ cannot belong to more than $n^2/4$ paths $P_{uv}$, it follows that

$$\sum_{v \in V(\mathcal{G})} \sum_{u \in V(\mathcal{G})} \chi_{uv}(e) \leq 2\left(\frac{n^2}{4}\right) = \frac{n^2}{2}. \tag{5.3.6}$$

Combining this, (5.3.3), (5.3.5), and using the fact that $d(u,v) \leq \operatorname{diam}(\mathcal{G})$ for any pair of vertices $u, v$, we obtain

$$2n \sum_{e \in E(\mathcal{G})} f^2(e) = 2n \sum_{uv \in E(\mathcal{G})} (x_u - x_v)^2$$

$$= a \sum_{v \in V(\mathcal{G})} \sum_{u \in V(\mathcal{G})} (x_u - x_v)^2$$

$$\leq a \sum_{e \in E(\mathcal{G})} \left( f^2(e) \sum_{v \in V(\mathcal{G})} \sum_{u \in V(\mathcal{G})} d(u,v) \chi_{uv}(e) \right)$$

$$\leq a \operatorname{diam}(\mathcal{G}) \frac{n^2}{2} \sum_{e \in E(\mathcal{G})} f^2(e).$$

Dividing through by $\frac{an}{2} \sum_{e \in E(\mathcal{G})} f^2(e)$ yields (5.3.1).

To prove (5.3.2) observe that for each edge $e \in E(\mathcal{G})$ we have

$$\sum_{u \in V(\mathcal{G})} \sum_{v \in V(\mathcal{G})} d(u,v) \chi_{uv}(e) \leq \sum_{u \in V(\mathcal{G})} \sum_{v \in V(\mathcal{G})} d(u,v)$$

$$- \sum_{v \in V(\mathcal{G})} \sum_{u \in V(\mathcal{G}) \setminus \{v\}} (1 - \chi_{uv}(e))$$

$$\leq n(n-1)\bar{\rho}(\mathcal{G}) - 2\binom{n}{2} + \frac{n^2}{2}$$

$$= n(n-1)\bar{\rho}(\mathcal{G}) - \frac{n(n-2)}{2}$$

(5.3.7)

where the last inequality follows from (5.3.6). Applying (5.3.3), (5.3.5), and (5.3.7), we obtain

$$2n \sum_{e \in E(\mathcal{G})} f^2(e) = a \sum_{v \in V(\mathcal{G})} \sum_{u \in V(\mathcal{G})} (x_u - x_v)^2$$

$$\leq a \sum_{e \in E(\mathcal{G})} f^2(e) \sum_{v \in V(\mathcal{G})} \sum_{u \in V(\mathcal{G})} d(u,v) \chi_{uv}(e)$$

$$\leq a \left( n(n-1)\overline{\rho}(\mathcal{G}) - \frac{n(n-2)}{2} \right) \sum_{e \in E(\mathcal{G})} f^2(e).$$

Dividing through by $n \sum_{e \in E(\mathcal{G})} f^2(e)$ and rearranging yields (5.3.2). □

Theorem 5.3.3 yields an immediate corollary which gives lower bounds on the algebraic connectivity:

**COROLLARY 5.3.4** *Let $\mathcal{G}$ be an unweighted connected graph on $n$ vertices with algebraic connectivity $a$. Then*

$$a \geq \frac{4}{n \operatorname{diam}(\mathcal{G})}$$

*and*

$$a \geq \frac{4}{2(n-1)\overline{\rho} - (n-2)}$$

Corollary 5.3.4 provides lower bounds for the algebraic connectivity in terms of $\operatorname{diam}(\mathcal{G})$ and $\overline{\rho}(\mathcal{G})$. Recall that if a graph has more edges, then $\operatorname{diam}(\mathcal{G})$ and $\overline{\rho}(\mathcal{G})$ are likely to be smaller. In such a case, since Corollary 5.3.4 shows that the lower bounds for $a(\mathcal{G})$ will be larger, hence yielding a larger algebraic connectivity as expected. Similarly, a graph containing fewer edges will likely have a larger diameter and larger mean distance. In this case, Corollary 5.3.4 shows that the lower bounds for $a(\mathcal{G})$ will be smaller as expected.

# Exercise:

**1.** Use Corollary 5.3.4 to find lower bounds on the algebraic connectivity of the following graphs:

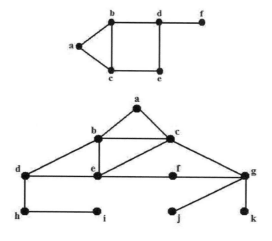

## 5.4 The Algebraic Connectivity in Terms of Edge Density and the Isoperimetric Number

In the previous two sections, we investigated how the number of edges and weight of the edges affect the algebraic connectivity of a graph. In this section, we consider how the location of edges in a graph affects the algebraic connectivity. To this end, we begin with a definition.

**DEFINITION 5.4.1** *Given a connected graph $\mathcal{G}$, a set of edges $F \subset E(\mathcal{G})$ is called an edge-cut (or cut) if there exists a set of vertices $X$ such that $F$ is precisely all edges of $\mathcal{G}$ that have one end in $X$ and the other in $X^c$.*

Given a set of vertices $X$, we let $\partial X$ denote the edge cut that it induces.

**DEFINITION 5.4.2** *Let $\mathcal{G}$ be a graph on $n$ vertices. The edge density of a cut $\partial X$ is defined as*

$$\rho(X) := \frac{n|\partial X|}{|X||X^c|}$$

**EXAMPLE 5.4.3** *Consider the graph $\mathcal{G}$ below:*

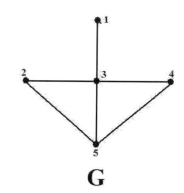

**G**

Let $X$ be the set consisting of vertices 1 and 5. Then $\partial X = \{13, 14, 25, 35, 45\}$. Hence $|X| = 2$, $|X^c| = 3$, $|\partial X| = 5$ and $n = 5$. Thus $\rho(X) = 25/6$.

We now present a theorem from [65] which gives an upper bound on the algebraic connectivity in terms of edge densities.

**THEOREM 5.4.4** *Let $\mathcal{G}$ be an unweighted connected graph on $n$ vertices with algebraic connectivity $a$. Let $L$ be its Laplacian matrix with $\lambda_n$ as its largest eigenvalue. Then for any set of vertices $X$ such that $X \neq \emptyset$ and $X \neq V(\mathcal{G})$, we have*

$$a \leq \rho(X) \leq \lambda_n \tag{5.4.1}$$

**Proof**: For a fixed set $X \subset V$, let $x \in \Re^n$ be the vector whose entries are

$$x_v = \begin{cases} |X^c| & \text{if } v \in X, \\ -|X| & \text{if } v \in X^c, \end{cases}$$

Note that $x^T e = 0$. Recalling that $e$ is an eigenvector corresponding to the eigenvalue $\lambda_1 = 0$ of $L$, it follows from Theorem 1.2.7 that

$$a \leq \frac{x^T L x}{x^T x} \leq \lambda_n(\mathcal{G}) \qquad (5.4.2)$$

for any such vector $x$. For a vertex $v$, let $\operatorname{bdeg}(v)$ denote the number of edges in $\partial X$ incident to $v$. Then

$$(Lx)_v = \begin{cases} n \operatorname{bdeg}(v) & \text{if } v \in X, \\ -n \operatorname{bdeg}(v) & \text{if } v \in X^c, \end{cases}$$

and hence

$$x^T L x = n|X^c||\partial X| + n|X||\partial X| = n|\partial X|(|X^c|+|X|) = n^2 |\partial X|. \qquad (5.4.3)$$

Also note that

$$x^T x = |X||X^c|^2 + |X|^2|X^c| = |X||X^c|(|X^c|+|X|) = |X||X^c|n. \qquad (5.4.4)$$

Substituting (5.4.3) and (5.4.4) into (5.4.2) yields (5.4.1). $\square$

**REMARK 5.4.5** Theorem 5.4.4 is valid if the graph $\mathcal{G}$ is weighted. In this case, $|\partial X|$ would merely denote the sum of the weights of the edges in $\partial X$.

Focusing our attention on the algebraic connectivity, we wish to determine what characteristics a graph $\mathcal{G}$ must possess in order for $a(\mathcal{G}) = \rho(X)$ for some subset $X \subset V$. We should first note that $\rho(X)$ is not necessarily an eigenvalue of $L(\mathcal{G})$. However, if $\rho(X)$ is indeed an eigenvalue and, moreover, if $\rho(X) = a(\mathcal{G})$, then $a(\mathcal{G})$ would necessarily be an integer. This is due to the fact that $a(\mathcal{G})$ would be a rational root of the characteristic polynomial of $L(\mathcal{G})$ which is a monic polynomial. Consequently, there would exist a corresponding eigenvector $y$ whose entries are all integers. Hence if there exists a set of vertices $X \subset V(\mathcal{G})$ inducing a cut such that $\rho(X) = a(\mathcal{G})$, then we have

$$L(\mathcal{G})y = \left[ \begin{array}{c|c} L_1 + D_1 & -A \\ \hline -A^T & L_2 + D_2 \end{array} \right] \left[ \begin{array}{c} |X^c|e^{(|X|)} \\ -|X|e^{(|X^c|)} \end{array} \right] = a(\mathcal{G})y \qquad (5.4.5)$$

where $L_1$ and $L_2$ are the Laplacian matrices for the subgraphs induced by the vertices in $X$ and $X^c$, respectively. Since $e$ is an eigenvector corresponding to the eigenvalue $\lambda_1 = 0$ of any Laplacian matrix, it follows that $L_1 e = L_2 e = 0$. Since

$L(\mathcal{G})e = 0$, it follows that $(L_1 + D_1)e^{(|X|)} - Ae^{(|X^c|)} = 0$, thus forcing $D_1 e = Ae$. Similarly, $D_2 e = A^T e$. Thus (5.4.5) simplifies to

$$\begin{bmatrix} nD_1 e^{(|X|)} \\ -nD_2 e^{(|X^c|)} \end{bmatrix} = a(\mathcal{G}) \begin{bmatrix} |X^c| e^{(|X|)} \\ -|X| e^{(|X^c|)} \end{bmatrix}. \tag{5.4.6}$$

Since the vector on the left of (5.4.6) is merely an integral multiple of the vector on the right of (5.4.6), it follows that since $D_1$ and $D_2$ are diagonal matrices, the diagonal entries in each of these matrices must be constant. Hence if $a(\mathcal{G}) = \rho(X)$ for some $X \subset V(\mathcal{G})$, the each vertex in $X$ is adjacent to the same number of vertices, say $d_1$, in $X^c$; and each vertex in $X^c$ is adjacent to the same number of vertices, say $d_2$, in $X$. This is referred to (see [22]) as *regularity across the cut*.

Conversely, suppose that every vertex in $X$ is adjacent to $d_1$ vertices in $X^c$ and every vertex in $X^c$ is adjacent to $d_2$ vertices in $X$. Then $\rho(X)$ is a nonzero eigenvalue of $L(\mathcal{G})$ (not necessarily corresponding to $a(\mathcal{G})$, though) with eigenvector

$$\begin{bmatrix} |X^c| e^{(|X|)} \\ -|X| e^{(|X^c|)} \end{bmatrix}.$$

Moreover,
$$|\partial X| = |X| d_1 = |X^c| d_2. \tag{5.4.7}$$

However, should $a(\mathcal{G}) = \rho(X)$, then

$$\begin{aligned} a(\mathcal{G}) = \rho(X) &= \frac{n|\partial X|}{|X||X^c|} \\ &= \frac{n|X|d_1}{|X||X^c|} \\ &= \frac{(|X|+|X^c|)d_1}{|X^c|} \\ &= \frac{|X|d_1 + |X^c|d_1}{|X^c|} \\ &= \frac{|X^c|(d_2+d_1)}{|X^c|} \\ &= d_1 + d_2 \end{aligned}$$

We summarize these findings (mainly from [22]) in the following theorem:

**THEOREM 5.4.6** *Let $\mathcal{G}$ be a connected unweighted graph on $n$ vertices with algebraic connectivity $a(\mathcal{G})$. If there exists a set of vertices $X \subset V(\mathcal{G})$ such that $a(\mathcal{G}) = \rho(X)$, then the following necessarily hold:*

*(i) Each vertex in $X$ is adjacent to $d_1$ vertices in $X^c$, and each vertex in $X^c$ is adjacent to $d_2$ vertices in $X$, and*

*(ii) $|X|d_1 = |X|d_2$.*

*Moreover, $a(\mathcal{G}) = d_1 + d_2$.*

# The Algebraic Connectivity

We now turn our attention to a related concept, the isoperimetric number of a graph:

**DEFINITION 5.4.7** The quantity

$$i(\mathcal{G}) = \min\left\{\frac{|\partial X|}{|X|} : X \subseteq V, 1 \leq |X| \leq \frac{n}{2}\right\}$$

is known as the *isoperimetric number* of $\mathcal{G}$.

**EXAMPLE 5.4.8** Consider the following graph $\mathcal{G}$:

Since $n = 4$, we consider all sets of vertices $X$ such that $|X| = 1$ and $|X| = 2$:

| $X$ | $|X|$ | $|\partial X|$ | $|\partial X|/|X|$ |
|---|---|---|---|
| $\{a\}$ | 1 | 2 | 2 |
| $\{b\}$ | 1 | 3 | 3 |
| $\{c\}$ | 1 | 3 | 3 |
| $\{d\}$ | 1 | 2 | 2 |
| $\{a,b\}$ | 2 | 3 | 1.5 |
| $\{a,c\}$ | 2 | 3 | 1.5 |
| $\{b,d\}$ | 2 | 3 | 1.5 |
| $\{c,d\}$ | 2 | 3 | 1.5 |
| $\{a,d\}$ | 2 | 4 | 2 |
| $\{b,c\}$ | 2 | 4 | 2 |

From the chart, we see that $i(\mathcal{G}) = 1.5$.

At this point, it seems reasonable that the concepts edge density and the isoperimetric number are related. To better understand this relation, we define

$$i_k(\mathcal{G}) := \min\left\{\frac{|\partial X|}{k} : X \subseteq V, |X| = k\right\}. \tag{5.4.8}$$

**OBSERVATION 5.4.9**

$$i(\mathcal{G}) = \min\left\{i_k(\mathcal{G}) : 1 \leq k \leq \left\lceil\frac{n}{2}\right\rceil\right\}$$

From (5.4.8) and Theorem 5.4.4 we have

$$a(\mathcal{G}) \leq \min \frac{n|\partial X|}{|X||X^c|}$$

where the minimum is taken over all nonempty sets $X \subset V$ such that $|X| \leq \lceil n/2 \rceil$. Consequently, if $|X| = k$ (hence $|X^c| = n - k$) then

$$a(\mathcal{G}) \leq \frac{n}{n-k} i_k(\mathcal{G}).$$

Hence, we obtain an important upper bound from [63] on the algebraic connectivity with respect to the isoperimetric number of a graph:

**THEOREM 5.4.10** *Let $\mathcal{G}$ be a connected graph on $n$ vertices with isoperimetric number $i(\mathcal{G})$. Then for every $k$ where $1 \leq k \leq \lceil n/2 \rceil$, we have*

$$a(\mathcal{G}) \leq \frac{n}{n-k} i_k(\mathcal{G}).$$

*Consequently $a \leq 2i(\mathcal{G})$.*

Since the isoperimetric number is a minimum over all vertex sets $|X|$ such that $1 \leq |X| \leq \lceil n/2 \rceil$, we obtain a weaker, but sometimes more useful, corollary to Theorem 5.4.10.

**COROLLARY 5.4.11** *Let $\mathcal{G}$ be a connected graph on $n$ vertices. Then for every vertex set $|X|$ such that $1 \leq |X| \leq \lceil n/2 \rceil$, we have*

$$a(\mathcal{G}) \leq \frac{2|\partial X|}{|X|}.$$

This corollary will be useful in the next two sections when we investigate how the genus of a graph affects the algebraic connectivity.

# Exercises:

**1.** (See [22]) Let $\mathcal{G}$ be a graph with vertex set $V$ where there exists a set $X \subset V$ such that (1) the subgraphs induced by the vertices in $X$ and the vertices in $X^c$ are each complete subgraphs, and (2) conditions (a) and (b) of Theorem 5.4.6 are satisfied. Show that $a(\mathcal{G}) = d_1 + d_2$; in particular $a(\mathcal{G}) = \rho(X)$.

**2.** (See [63]) Show the following:
    (i). $i(K_n) = \lceil n/2 \rceil$
    (ii). $i(C_n) = 2/\lfloor n/2 \rfloor$
    (iii). $i(P_n) = 1/\lfloor n/2 \rfloor$
    (iv). $i(K_{m,n}) = \lceil mn/2 \rceil / \lfloor (m+n)/2 \rfloor$.

**3.** Use Theorem 5.4.10 to find an upper bound on the algebraic connectivity of the following graph:

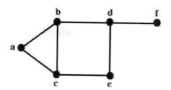

## 5.5 The Algebraic Connectivity of Planar Graphs

In this section and the next section, based largely on [67], we find upper bounds on the algebraic connectivity in terms of topological properties of a graph. Finding such bounds will make use of Theorem 5.4.10 and Corollary 5.4.11. In this section, we deal with planar graphs. Recall from Section 2.6 that a graph is planar if it can be drawn on a plane (equivalently a sphere) so that no two of its edges cross. We saw in Theorem 2.6.7 that a planar graph cannot have too many edges. Thus by Theorem 5.1.2, it follows that planar graphs have small algebraic connectivity. But how small? In this section, our goal is to prove the following:

**THEOREM 5.5.1** *If $\mathcal{G}$ is a planar graph, then $a(\mathcal{G}) \leq 4$. Moreover, equality holds if and only if $\mathcal{G} \cong K_4$ or $\mathcal{G} \cong K_{2,2,2}$.*

Since our goal is to determine an upper bound on the algebraic connectivity of planar graphs, by Theorem 5.1.2, we want to consider such planar graphs with the maximum number of edges. Thus we will restrict our attention to maximal planar graphs (see Definition 2.6.10). In order to prove Theorem 5.5.1, we recall from Observation 2.6.12 that if $\mathcal{G}$ is a maximal planar graph on $n \geq 4$ vertices, then $\delta(\mathcal{G}) \geq 3$. Also recall from Corollary 2.6.9 that for any planar graph $\mathcal{G}$ we have $\delta(\mathcal{G}) \leq 5$. Therefore, in order to determine the upper bound on the algebraic connectivity of planar graphs, we need only consider three cases: $\delta(\mathcal{G}) = 3$, $\delta(\mathcal{G}) = 4$, and $\delta(\mathcal{G}) = 5$ where $\mathcal{G}$ is maximal planar. However, note that the complete graph on the largest number of vertices that is planar is $K_4$ and that $a(K_4) = 4$. Hence by Corollary 5.1.12, we need only consider noncomplete planar graphs such that $\delta(\mathcal{G}) = 4$ and $\delta(\mathcal{G}) = 5$. Thus we will prove Theorem 5.5.1 as two claims. Claim 5.5.2 will concern the case where $\delta(\mathcal{G}) = 4$, while Claim 5.5.8 will concern the case where $\delta(\mathcal{G}) = 5$. We will first begin with Claim 5.5.2 from [67].

**CLAIM 5.5.2** *Let $\mathcal{G}$ be a noncomplete planar graph on $n$ vertices such that $\delta(\mathcal{G}) = 4$, then $a(\mathcal{G}) \leq 4$ and equality holds if and only if $\mathcal{G} \cong K_{2,2,2}$.*

**Proof**: Suppose $\mathcal{G}$ is a noncomplete planar graph such that $\delta(\mathcal{G}) = 4$. Then by Corollary 5.1.12, $a(\mathcal{G}) \leq 4$. If $a(\mathcal{G}) = 4$ then Corollary 5.1.12 implies $v(\mathcal{G}) = 4$ also. Thus by Theorem 5.1.10, $\mathcal{G} = \mathcal{G}_1 \vee \mathcal{G}_2$ where $\mathcal{G}_1$ is a disconnected graph $n - 4$ vertices, while $\mathcal{G}_2$ is a graph on four vertices with $a(\mathcal{G}_2) \geq n - 8$. Since $\mathcal{G}$ is planar, by Theorem 2.6.15 it follows that $\mathcal{G}$ cannot contain $K_{3,3}$ as a subgraph. Therefore, since $\mathcal{G} = \mathcal{G}_1 \vee \mathcal{G}_2$ and since $\mathcal{G}_2$ has four vertices, it follows that $\mathcal{G}_1$ has either one vertex or two vertices. If $\mathcal{G}_1$ has one vertex, then $\mathcal{G}_1 = K_1$. But, $a(K_1 \vee \mathcal{G}_2) < 4$

for all graphs $\mathcal{G}_2$ on four vertices in which $K_1 \vee \mathcal{G}_2$ is planar. If $\mathcal{G}_1$ has two vertices, then since $\mathcal{G}_1$ must be disconnected, it follows that $\mathcal{G}_1 = E_2$, where $E_2$ is the empty graph on two vertices. However, $a(E_2 \vee \mathcal{G}_2) \leq 4$ for all graphs $\mathcal{G}_2$ on four vertices in which $E_2 \vee \mathcal{G}_2$ is planar, and equality holds if and only if $\mathcal{G}_2 = C_4$ where $C_4$ is the cycle on four vertices. But $E_2 \vee C_4 = K_{2,2,2}$ (which is planar). Since $a(K_{2,2,2}) = 4$, the claim is proven. □

To prove Theorem 5.5.1, we need to now show that $a(\mathcal{G}) < 4$ for all planar graphs $\mathcal{G}$ in which $\delta(\mathcal{G}) = 5$. Recalling from Section 2.6 that planar graphs, when drawn as a plane graph, divide the plane into regions, we present the following definition that will help us prove Theorem 5.5.1:

**DEFINITION 5.5.3** *A planar graph $\mathcal{G}$ is* outerplanar *if it can be drawn with all vertices of $\mathcal{G}$ placed on a circle such that all edges are drawn passing through the interior of the circle with no two edges crossing.*

The following graph theoretic lemma concerning outerplanar graphs is well-known (see [78], p. 240), thus we state it without proof:

**LEMMA 5.5.4** *If $\mathcal{G}$ is outerplanar and there exists a cycle in $\mathcal{G}$ containing every vertex of $\mathcal{G}$, then there exists at least two nonadjacent vertices of degree two.*

We now state another graph theoretic lemma from [67] for which we do provide a proof:

**LEMMA 5.5.5** *Let $\mathcal{G}$ be a planar graph drawn in the plane. Let $R$ be a region bounded by a set of vertices forming a cycle $C$ such that either*

*(a) For each vertex $v \in C$ we have $2 \leq \deg(v) \leq 4$; or*

*(b) There exists exactly two adjacent vertices $w, z \in C$ such that $\deg(w) \geq \deg(z) \geq 5$, while for all remaining vertices $v \in C$ we have $2 \leq \deg(v) \leq 4$.*

*Then it is impossible to create a graph $\mathcal{H}$ from $\mathcal{G}$ by adding edges joining nonadjacent vertices of $C$, each of which passes through $R$, so that $\mathcal{H}$ is planar and all vertices in $C$ have degree five or greater in $\mathcal{H}$.*

**Proof**: Let $\mathcal{G}_c$ be the graph formed by $C$ and the edges being added that pass through $R$. Observe that $\mathcal{G}_c$ is outerplanar. Thus by Lemma 5.5.4 there exists at least two nonadjacent vertices in $\mathcal{G}_c$ with degree two in $\mathcal{G}_c$. If (a) holds, then such vertices will have degree at most four in $\mathcal{H}$. If (b) holds then since $w$ and $z$ are adjacent, it follows from Lemma 5.5.4 that at least one vertex $x$ in $C$ other than $w$ or $z$ will have degree two in $\mathcal{G}_c$. Thus $x \in C$ will have degree at most four in $\mathcal{H}$. □

Recall from Observation 2.6.12 that if a maximal planar graph is drawn so that no edges cross, then every region created by the graph is bounded by three edges.

# The Algebraic Connectivity

Thus we say that every region of a maximal planar graph is a *triangle* and that each vertex is the center of a wheel subgraph. With this in mind, we now prove the next important lemma from [67] which deals explicitly with graphs $\mathcal{G}$ where $\delta(\mathcal{G}) = 5$ and $\mathcal{G}$ is maximal planar. In this lemma, let $Adj(v) := v \cup N(v)$.

**THEOREM 5.5.6** *Let $\mathcal{G}$ be a maximal planar graph on $n$ vertices such that $\delta(\mathcal{G}) = 5$. Then $|Adj(v)| \leq \lceil n/2 \rceil$ for all $v \in \mathcal{G}$. Moreover, for each vertex $v \in \mathcal{G}$, there exists a vertex $w \in \mathcal{G}$ such that $d(v,w) \geq 3$.*

**Proof**: Consider $v \in \mathcal{G}$ and let $\deg v = k$. We need to show that $k+1 \leq \lceil n/2 \rceil$. Since $\mathcal{G}$ is maximal planar, $v$ is the center of some wheel. Therefore let $v$ be adjacent to $v_1, \ldots, v_k$ where $v_i$ is adjacent to $v_{i+1}$ for all $1 = 1, \ldots, k-1$ and where $v_k$ is adjacent to $v_1$. Let $Z$ be the set of vertices outside $Adj(v)$ that are adjacent to at least one of $v_1, \ldots, v_k$. Since each edge in a maximal planar graph lies on the boundary of two triangle regions, and since each vertex has degree at least five, it follows that for each $i = 1, \ldots, k-1$, there exists a unique vertex $w_i \in Z$ such that $w_i$ is adjacent to both $v_i$ and $v_{i+1}$ (otherwise there would be a vertex of degree four). Similarly, there exists a vertex $w_k \in Z$ adjacent to both $v_k$ and $v_1$. There may also be vertices in $Z$ adjacent to exactly one vertex in $Adj(v)$. Hence $|Z| \geq k$. Let $\hat{\mathcal{G}}$ be the subgraph of $\mathcal{G}$ that is depicted in Figure 1 (where, in this case, $k = 6$ and vertices $v_1, v_2, v_3$, and $v_5$ have other vertices in $Z$ adjacent to them). The vertices of $\hat{\mathcal{G}}$ are precisely the vertices in $Adj(v) \cup Z$. (Note that we cannot conclude that $\hat{\mathcal{G}}$ is the subgraph induced by the vertices in $Adj(v) \cup Z$ because there may be additional edges joining pairs of these vertices.)

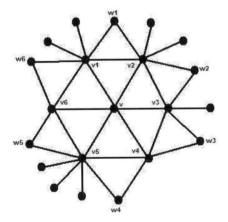

Figure 1

Therefore, $|\hat{\mathcal{G}}| \geq 2k + 1$. The theorem is proven if there exists a vertex in $\mathcal{G}$ that is not in $\hat{\mathcal{G}}$ for that would imply $n \geq 2k + 2$, and hence $k + 1 \leq \lceil n/2 \rceil$.

We first must show that the vertices in $Z$ induce a cycle. Let's suppose not and deduce a contradiction. If the vertices of $Z$ do not induce a cycle, then there are two cases to consider:

*Case (i)*: There exist two vertices in $Adj(v)$ that are adjacent in $\mathcal{G}$ but the edge joining them is not an edge in $\hat{\mathcal{G}}$.

*Case (ii)*: There exist two nonadjacent vertices in $Adj(v)$ that are adjacent to a common vertex in $x \in Z$.

First we consider Case (i). Let $v_p$ and $v_q$ where $p < q$ be vertices in $Adj(v)$ satisfying the statement of Case (i) such that none of the vertices $v_i$ where $p+1 \leq i \leq q-1$ satisfy Case (i) or Case (ii). Let $Z'$ be the set of vertices in $Z$ that are adjacent to such vertices $v_i$ for $p \leq i \leq q$. Then since $\mathcal{G}$ is maximal planar, the vertices in $Z'$ form a path $P$ in $\mathcal{G}$. Let $\mathcal{H}$ be the subgraph of $\mathcal{G}$ formed from $\hat{\mathcal{G}}$ by adding the edges of $P$ (see Figure 2.) Observe that in $\mathcal{H}$, the vertex $v_p$ followed by the vertices in $Z'$ followed by $v_q$ followed by $v_p$ forms a cycle $C$. Let $R$ denote the region that is the interior of $C$.

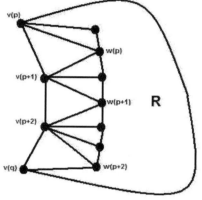

Figure 2

Thus to create $\mathcal{G}$ from $\mathcal{H}$, edges would need to be added which pass through $R$. Observe that in $\mathcal{H}$, that $2 \leq \deg(z) \leq 4$ for all $z \in Z'$. Thus the only two vertices in $C$ which can possibly have a degree at least five in $\mathcal{G}$ are $v_p$ and $v_q$ which are adjacent in $C$. Thus by Lemma 5.5.5(b), it is impossible to add edges passing through $R$ to create $\mathcal{G}$ such that all vertices in $C$ will have degree at least five in $\mathcal{G}$. Hence Case (i) cannot hold.

Now we consider Case (ii). Let $v_p$ and $v_q$ where $p < q$ be vertices in $Adj(v)$ satisfying the statement of Case (ii) such that none of the vertices $v_i$ where $p+1 \leq i \leq q-1$ satisfy Case (i) or Case (ii). Let $Z'$ be as in Case (i) and recall that the vertices in $Z'$ form a path $P$ in $\mathcal{G}$. Let $\mathcal{H}$ be the subgraph of $\mathcal{G}$ formed from $\hat{\mathcal{G}}$ by adding the edges of $P$. Observe that in $\mathcal{H}$, the vertex $v_p$ followed by the vertices in $Z'$ followed by $v_q$ followed by $x$ followed by $v_p$ forms a cycle $C$. Let $R$ be the interior of $C$ and let $u$ and $y$ be the vertices besides $x$ that are adjacent to $v_p$ and $v_q$, respectively in $C$.

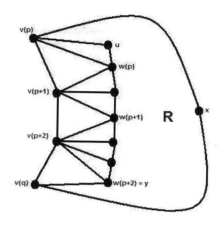

Figure 3

Thus to create $\mathcal{G}$ from $\mathcal{H}$, edges would need to be added which pass through $R$. Observe that in $\mathcal{H}$, that $2 \leq \deg(z) \leq 4$ for all $z \in Z'$. Also note in $\mathcal{H}$ that $\deg(x) = 2$, and $\deg(u), \deg(y) \leq 3$. Thus the only two vertices in $C$ which can possibly have a degree of at least five in $\mathcal{G}$ are $v_p$ and $v_q$. We now have three subcases to consider.

*Subcase (a)*: Vertices $v_p$ and $v_q$ are adjacent in $\mathcal{G}$. Then the vertex $v_p$ followed by the vertices in $P$ followed by $v_q$ and then $v_p$ forms a cycle in $\mathcal{G}$. Let $R'$ be the region bounded by this cycle. Hence by the reasoning in Case (i), edges passing through $R'$ cannot be added so that every vertex in the resulting graph has degree five or greater.

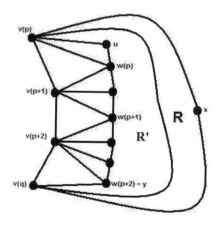

Figure 3a

*Subcase (b)*: Vertex $v_p$ or $v_q$, say $v_p$, is adjacent to a vertex $s \in Z'$. Then $R$ is divided into two regions. Consider the region $R'$ that is not bounded by $x$ and observe that $v_p$ and $s$ are the only vertices whose degree can possibly be at least five. Thus by the reasoning of Case (i), adding edges passing through this region cannot result in a graph in which every vertex is of degree five or greater.

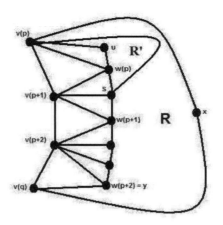

Figure 3b

*Subcase (c)*: Neither vertices $v_p$ nor $v_q$ are adjacent to other vertices in $Z'$ (except for those on $C$). Then since $\mathcal{G}$ is maximal planar and thus every region in $\mathcal{G}$ is a triangle, $x$ is adjacent to both $u$ and $y$ in $\mathcal{G}$. Since $\deg(u), \deg(y) \leq 3$ in $\mathcal{H}$, adding such edges increases the degrees of $u$ and $y$ to at most four. Thus we have created a cycle in which each vertex has degree at most four. Thus by Lemma 5.5.5(a), we cannot add edges through $R'$ such that the resulting graph has all vertices of degree five or greater.

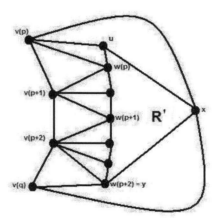

Figure 3c

Subcases (a) through (c) show that Case (ii) cannot hold. Therefore in $\mathcal{G}$, the vertices of $Z$ induce a cycle where $3 \leq \deg(z) \leq 4$ for all vertices $z \in Z$. Suppose that all vertices of $\mathcal{G}$ are also vertices of $\hat{\mathcal{G}}$. Since all the vertices of $\mathcal{G}$ have minimum degree 5 and are vertices in $\hat{\mathcal{G}}$, in order to create $\mathcal{G}$ from $\hat{\mathcal{G}}$, edges joining nonadjacent vertices of $Z$ would have to be added. By the construction of $\hat{\mathcal{G}}$, all such edges would have to pass through the same region. By Lemma 5.5.5(a), it is impossible to add such edges so that the resulting graph would be both planar and have minimum degree 5. Therefore there must exist a vertex in $\mathcal{G}$ that is not a vertex in $\hat{\mathcal{G}}$. Hence

$n \geq 2k+2$ and therefore $k+1 \leq \lceil n/2 \rceil$. Since $|Adj(v)| \leq \lceil n/2 \rceil$ for all vertices $v \in \mathcal{G}$, then for each vertex $v$, there exists a vertex $w \notin Adj(v) \cup Z$. Thus $d(v,w) \geq 3$. □

If $\mathcal{G}$ is a maximal planar graph on $n \geq 4$ vertices, then every vertex of $\mathcal{G}$ is the center of a wheel subgraph. Thus Lemma 5.5.6 says that if $\mathcal{G}$ is a maximal planar graph such that $\delta(\mathcal{G}) = 5$, then there exist two vertices $w_1$ and $w_2$ that are the centers of wheels $W_1$ and $W_2$, respectively, where $W_1$ and $W_2$ do not share any common vertices or edges and such that one wheel consists of no more than $\lceil n/2 \rceil$ vertices while the other wheel consists of no more than $\lfloor n/2 \rfloor$ vertices. Observe that the vertices adjacent to $w_1$ and $w_2$ form disjoint cycles, $C_1$ and $C_2$, respectively.

Suppose $|W_1 \cup W_2| < n$. Then there exists a vertex $v \notin W_1 \cup W_2$ that is adjacent to vertices in $C_1$ or $C_2$. Since $\deg(v) \geq 5$, it follows that $v$ is adjacent to at least two vertices of $C_1$ or two vertices of $C_2$. Without loss of generality, let $v$ be adjacent to at least two vertices of $C_1$, say $v_1$ and $v_2$. Let $v_0$ and $v_3$ be the vertices on $C_1$ not adjacent to $v$ such that $v_1$ is adjacent to $v_0$, and $v_2$ is adjacent to $v_3$. Hence we can extend $C_1$ to form the cycle $\hat{C}_1$ by beginning at $v_0$, then traveling to $v_1, v, v_2, v_3$ and then traveling the remaining vertices of $C_1$ returning to $v_0$.

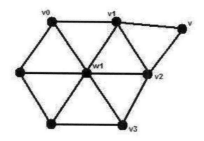

Figure 4

We can continue extending $\hat{C}_1$ and $\hat{C}_2$ in this manner until we have two disjoint sets $X$ and $Y$ such that $|X| = \lfloor n/2 \rfloor$, $|Y| = \lceil n/2 \rceil$, and the vertices in each set that are adjacent to the vertices in the other set each forms a cycle. We state this more formally as an observation from [67]:

**OBSERVATION 5.5.7** *If $\mathcal{G}$ is a maximal planar graph on $n$ vertices in which $\delta(\mathcal{G}) = 5$, then there exist disjoint sets of vertices $X$ and $Y$ in which $|X| = \lfloor n/2 \rfloor$, $|Y| = \lceil n/2 \rceil$ such that all of the following hold:*
*(i) The vertices in $X$ that are adjacent to vertices in $Y$ form a cycle.*
*(ii) The vertices in $Y$ that are adjacent to vertices in $X$ form a cycle.*
*(iii) There exists a vertex $x \in X$ that is not adjacent to any vertex in $Y$.*
*(iv) There exists a vertex $y \in Y$ that is not adjacent to any vertex in $X$.*

We can now prove Claim 5.5.8 which deals with the case of planar graphs where $\delta(\mathcal{G}) = 5$. Before doing so, we introduce notation in order to make the proof more transparent. In parts (i) and (ii), of Observation 5.5.7, we will let $X_c$ and $Y_c$ denote the respective cycles formed. We let $x_1, \ldots, x_k$ be the vertices of $X_c$, and let

$y_1, \ldots, y_m$ be the vertices of $Y_c$. We will orient each cycle so that $x_i$ is the $i^{th}$ vertex of $X_c$, and $y_i$ is the $i^{th}$ vertex of $Y_c$. This allows us to use the idea of vertices being consecutive. Thus, in $X_c$ ($Y_c$), we regard $x_{i+1}$ ($y_{i+1}$) as the vertex that directly follows $x_i$ ($y_i$) for $i = 1, \ldots, k-1$ ($m-1$), and $x_1$ ($y_1$) as the vertex that directly follows $x_k$ ($y_m$). Hence in a set of consecutive vertices of either $X_c$ or $Y_c$, the words "first" and "last" have meaning. We now proceed with the claim (see [67]) which makes use of the concept of the isoperimetric number studied in Section 5.4.

**CLAIM 5.5.8** *Let $\mathcal{G}$ be a maximal planar graph on $n$ vertices in which $\delta(\mathcal{G}) = 5$. Then $a(\mathcal{G}) < 4$.*

**Proof**: Let $X$ and $Y$ be sets of vertices in $\mathcal{G}$ that are in accordance with Observation 5.5.7. Let $X_c = x_1, x_2, \ldots, x_k, x_1$ be the (cycle of) vertices of $X$ that are adjacent to vertices in $Y$ and let $Y_c = y_1, y_2, \ldots, y_m, y_1$ be the (cycle of) vertices of $Y$ that are adjacent to vertices in $X$. By Observation 5.5.7, $k + m \leq n - 2$. We will show that there are at most $n - 2$ edges joining vertices in $X$ with vertices in $Y$.

Since $\mathcal{G}$ is maximal planar, every region is a triangle. Therefore, each $x_i \in X_c$ is adjacent to consecutive vertices (possibly only one vertex) in $Y_c$. Similarly, each $y_i \in Y_c$ is adjacent to consecutive vertices (possibly only one vertex) in $X_c$. We will now label the edges joining vertices in $X_c$ to vertices in $Y_c$ in correspondence with the vertices in $X_c$ and $Y_c$; we will show that a one-to-one correspondence exists. Since each vertex in $X_c$ is adjacent to consecutive vertices in $Y_c$, each vertex $x_i$ has a last vertex in $Y_c$ to which it is adjacent. Label the edge joining $x_i$ to such a vertex as $x_i$ (See Figure 5). (If a vertex $x_i$ is adjacent to all vertices in $Y_c$, then label the edge joining $x_i$ to $y_m$ as $x_i$.) Similarly, considering the vertices in $Y_c$, since each is adjacent to consecutive vertices in $X_c$, each vertex $y_i$ has a last vertex in $X_c$ to which it is adjacent. Label the edge joining the $y_i$ to such a vertex as $y_i$. (If a vertex $y_i$ is adjacent to all vertices in $X_c$, then label the edge joining $y_i$ to $x_k$ as $y_i$.) Thus we have created a one-to-one correspondence between the vertices in $X_c \bigcup Y_c$ and the edges joining the vertices in $X_c$ to vertices in $Y_c$. Hence $|\partial X| = k + m \leq n - 2$. Therefore by Corollary 5.4.11.

$$a(\mathcal{G}) \leq \frac{2|\partial X|}{|X|} \leq \frac{2(n-2)}{\lfloor n/2 \rfloor} \leq \frac{2(n-2)}{(n-1)/2} = \frac{4(n-2)}{n-1} < 4.$$

□

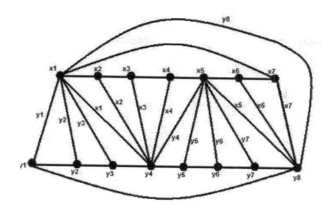

Figure 5

Claims 5.5.2 and 5.5.8 show that if $\mathcal{G}$ is a noncomplete maximal planar graph, then $a(\mathcal{G}) \leq 4$ where equality holds if and only if $\mathcal{G} = K_{2,2,2}$. Note that every planar graph $\mathcal{Q}$ is the subgraph of some maximal planar graph $\mathcal{G}$. Since $\mathcal{Q}$ is a subgraph of $\mathcal{G}$, it follows from Theorem 5.1.2 that $a(\mathcal{Q}) \leq a(\mathcal{G})$. Finally, since no subgraph $\mathcal{Q}$ of $K_{2,2,2}$ is such that $a(\mathcal{Q}) = 4$, Theorem 5.5.1 is proven.

## 5.6 The Algebraic Connectivity as a Function Genus $k$ Where $k \geq 1$

In the previous section, we found an upper bound for the algebraic connectivity of planar graphs, otherwise known as graphs of genus zero. In this section, based largely on [67], we will investigate the algebraic connectivity of graphs of genus $k \geq 1$. Since graphs of higher genus will tend to have more edges, Theorem 5.1.2 implies that the upper bound for the algebraic connectivity of such graphs should be greater. To achieve such an upper bound on the algebraic connectivity, we recall the Generalized Euler's Formula (Corollary 2.6.21) which states that if $\mathcal{G}$ is a connected graph on $n$ vertices, $m$ edges, and of genus $k$, then

$$k \geq \frac{m}{6} - \frac{n}{2} + 1. \qquad (5.6.1)$$

Recall that for all graphs, the sum of the degrees of the vertices is $2m$. Thus for all graphs of genus $k$, it makes sense to rewrite (5.6.1) as $2m \leq 6n + 12(k-1)$. If $\mathcal{G}$ has $n$ vertices, then since the degree of each vertex is at least $\delta(\mathcal{G})$, it follows that $n\delta(\mathcal{G}) \leq 2m$. Hence $n\delta(\mathcal{G}) \leq 2m \leq 6n + 12(k-1)$. Therefore, for a graph $\mathcal{G}$ of genus $k$,

$$\delta(\mathcal{G}) \leq 6 + \left\lfloor \frac{12(k-1)}{n} \right\rfloor.$$

However, since $\delta(\mathcal{G}) \leq n-1$ for all graphs on $n$ vertices, we have proven the following lemma from [67]:

**LEMMA 5.6.1** *Let $\mathcal{G}$ be a graph on $n$ vertices that is of genus $k \geq 1$. Then*

$$\delta(\mathcal{G}) \leq \min\left\{n-1,\ 6 + \left\lfloor \frac{12(k-1)}{n} \right\rfloor\right\}.$$

Since $a(\mathcal{G}) \leq \delta(\mathcal{G})$ for any noncomplete graph $\mathcal{G}$ (Corollary 5.1.12), and since our goal is to find the best upper bound for $a(\mathcal{G})$ for a given genus $k \geq 1$, then for such a given $k$ we want to find the positive integer value(s) $n$ such that equality in Lemma 5.6.1 holds. We do this in the following lemma from [67]. For the sake of simplicity, let

$$p_k := \left\lfloor \frac{7 + \sqrt{1+48k}}{2} \right\rfloor.$$

**LEMMA 5.6.2** *Let $\mathcal{G}$ be a graph of genus $k \geq 1$. Then $\delta(\mathcal{G}) \leq p_k - 1$.*

**Proof**: Observe that if

$$n - 1 = 6 + \left\lfloor \frac{12(k-1)}{n} \right\rfloor$$

then

$$n = \left\lfloor \frac{7 + \sqrt{1+48k}}{2} \right\rfloor = p_k. \qquad (5.6.2)$$

In such a case,

$$\delta(\mathcal{G}) \leq p_k - 1 \qquad (5.6.3)$$

Otherwise, since $n-1$ strictly increases in $n$ while $6 + \lfloor 12(k-1)/n \rfloor$ monotonically decreases in $n$, the result follows from Lemma 5.6.1. $\square$

We are now ready to prove the first main result of this section (see [67]) which gives upper bounds on the algebraic connectivity of graphs of genus $k \geq 1$. In this theorem, $\gamma(\mathcal{G})$ denotes the genus of the graph $\mathcal{G}$ and $S_k$ denotes a surface of genus $k$.

**THEOREM 5.6.3** *Let $\mathcal{G}$ be a graph of genus $k$ where $k \geq 1$. If $\mathcal{G}$ is the complete graph on the largest number of vertices that can be embedded on $S_k$, then $a(\mathcal{G}) = p_k$. Otherwise, $a(\mathcal{G}) \leq p_k - 1$.*

**Proof**: By Lemma 5.6.2, if $\mathcal{G}$ is a graph of genus $k \geq 1$, then $\delta(\mathcal{G}) \leq p_k - 1$. By Corollary 5.1.12, $a(\mathcal{G}) \leq \delta(\mathcal{G})$ if $\mathcal{G}$ is not complete. Thus $a(\mathcal{G}) \leq p_k - 1$ for all noncomplete graphs of genus $k$. If $\mathcal{G}$ is complete then $a(\mathcal{G}) = p_k$ if and only if $\mathcal{G} = K_{p_k}$. It remains to show that $K_{p_k}$ is the complete graph on the largest number of vertices that can be embedded on $S_k$. It is known from Theorem 2.6.22 (see [71]) that

$$\gamma(K_n) = \left\lceil \frac{(n-4)(n-3)}{12} \right\rceil. \qquad (5.6.4)$$

# The Algebraic Connectivity

Letting $n = p_k$, (5.6.4) shows that $\gamma(K_{p_k}) = k$. Thus $K_{p_k}$ is the complete graph on the largest number of vertices that can be embedded on $S_k$. This proves the theorem. □

Below is a chart giving the upper bounds for the algebraic connectivity of graphs of genus $k$. As you would expect from Theorem 5.1.2, since graphs of larger genus tend to have more edges, the upper bounds are larger for such graphs.

| genus | 0 | 1 | 2 | 3 | 4 | 5 | 6 | 7 | 8 |
|---|---|---|---|---|---|---|---|---|---|
| upper bound | 4 | 7 | 8 | 9 | 10 | 11 | 12 | 12 | 13 |

We see in Theorem 5.6.3 that if $\mathcal{G}$ is a noncomplete graph of genus $k \geq 1$, then $a(\mathcal{G}) \leq p_k - 1$. An interesting question now arises: Does there necessarily exist a noncomplete graph $\mathcal{G}$ of genus $k$ such that this upper bound is achieved, i.e., $a(\mathcal{G}) = p_k - 1$? To begin answering this question, we prove the following lemma from [67]:

**LEMMA 5.6.4** *If $k \geq 1$ is an integer such that $p_k - 7 = 12(k-1)/p_k$, then $a(\mathcal{G}) < p_k - 1$ for all noncomplete graphs $\mathcal{G}$ of genus $k$.*

**Proof**: Let $\mathcal{G}$ be a noncomplete graph on $n$ vertices of genus $k \geq 1$ where $k$ is an integer such that $p_k - 7 = 12(k-1)/p_k$. Lemma 5.6.2 dictates that $\delta(\mathcal{G}) \leq p_k - 1$. Suppose $\delta(\mathcal{G}) = p_k - 1$. This implies that $n \geq p_k$. But since, $\mathcal{G}$ is noncomplete, $n \geq p_k + 1$. Thus

$$p_k - 1 = 6 + \frac{12(k-1)}{p_k} = 6 + \left\lfloor \frac{12(k-1)}{p_k} \right\rfloor > 6 + \left\lfloor \frac{12(k-1)}{n} \right\rfloor$$

where the second equality follows from the fact that $12(k-1)/p_k$ is an integer. So by Lemma 5.6.1, if $n \geq p_k + 1$, then

$$\delta(\mathcal{G}) \leq 6 + \left\lfloor \frac{12(k-1)}{n} \right\rfloor < 6 + \left\lfloor \frac{12(k-1)}{p_k} \right\rfloor = p_k - 1 \qquad (5.6.5)$$

This contradicts $\delta(\mathcal{G}) = p_k - 1$. Hence $\delta(\mathcal{G}) < p_k - 1$, and thus it follows that $\delta(\mathcal{G}) \leq p_k - 2$. Therefore, by Corollary 5.1.12, $a(\mathcal{G}) \leq p_k - 2$. Thus there does not exist a noncomplete graph $\mathcal{G}$ of genus $k \geq 1$ such that $a(\mathcal{G}) = p_k - 1$ if $p_k - 7 = 12(k-1)/p_k$. □

**REMARK 5.6.5** If $k$ is an integer such that $p_k - 7 \neq 12(k-1)/p_k$, then $12(k-1)/p_k$ need not be an integer. Hence in this case, if $n \geq p_k + 1$, it would be possible for $\lfloor 12(k-1)/p_k \rfloor$ to equal $\lfloor 12(k-1)/n \rfloor$ in (5.6.5).

It now behooves us to determine which integers $k \geq 1$ are such that $p_k - 7 = 12(k-1)/p_k$. We determine this in the following claim from [67]:

**CLAIM 5.6.6** $p_k - 7 = 12(k-1)/p_k$ if and only if $\sqrt{1+48k}$ is an integer.

**Proof**: Observe that $1 + 48k$ is an odd integer for all integers $k \geq 1$. Thus if $\sqrt{1+48k}$ is an integer, it must also be odd. Hence $(7+\sqrt{1+48k})/2$ is an integer if and only if $\sqrt{1+48k}$ is an (odd) integer. Therefore, this claim is proven if we can show $(7+\sqrt{1+48k})/2$ is an integer if and only if $p_k - 7 = 12(k-1)/p_k$.

Suppose $(7+\sqrt{1+48k})/2$ is an integer. Then $p_k = (7+\sqrt{1+48k})/2$. Thus, rewriting this expression yields $p_k - 7 = 12(k-1)/p_k$. Now suppose $(7+\sqrt{1+48k})/2$ is not an integer. Then $p_k < (7+\sqrt{1+48k})/2$. Rewriting this expression will yield $p_k - 7 < 12(k-1)/p_k$. □

We are now ready to prove the second main result (see [67]) of this section which shows values for $k$ in which $a(\mathcal{G}) < p_k - 1$ for all graphs of genus $k$.

**THEOREM 5.6.7** *If any of the following hold for some positive integer $c$:*
*(a) $k=c(12c-1)$*
*(b) $k=c(12c+1)$*
*(c) $k=(4c-1)(3c-1)$*
*(d) $k=(4c+1)(3c+1)$*
*then $a(\mathcal{G}) < p_k - 1$ for all noncomplete graphs $\mathcal{G}$ of genus $k$.*

**Proof**: From Claim 5.6.6, we see that $p_k - 7 = 12(k-1)/p_k$ if and only if $\sqrt{1+48k}$ is an integer. But $\sqrt{1+48k}$ is an integer if and only if $1+48k$ is a perfect square, i.e., $1+48k = x^2$ for some integer $x$. This is equivalent to $(x+1)(x-1) = 48k$ for some integer $x$. Letting $y = x+1$, we see that $p_k - 7 = 12(k-1)/p_k$ if and only if there exists an integer $y$ such that $y(y-2)$ is divisible by 48. In such a case,

$$k = \frac{y(y-2)}{48}. \tag{5.6.6}$$

Observe that $y(y-2)$ is divisible by 48 if and only if any of the following hold:

(i) $y = 24c$ for some positive integer $c$
(ii) $y = 24c + 2$ for some positive integer $c$
(iii) $y = 24c - 6$ for some positive integer $c$
(iv) $y = 24c + 8$ for some positive integer $c$

Plugging (i), (ii), (iii), and (iv) into (5.6.6) yields (a), (b), (c), and (d), respectively. This together with Lemma 5.6.4 proves the theorem. □

The first nine values for $k$ in which $a(\mathcal{G}) < p_k - 1$ for all noncomplete graphs of genus $k$ that Theorem 5.6.7 produces are 6, 11, 13, 20, 35, 46, 50, 63, 88. Theorem 5.6.7 leads us to two questions. First, are there values for $k$ in which there exists a noncomplete graph $\mathcal{G}$ such that $a(\mathcal{G}) = p_k - 1$? The answer is the affirmative. For example $\gamma(K_{3,3,3}) = 1$ (see drawing below), and observe that $a(K_{3,3,3}) = p_1 - 1 = 6$. Similarly, $\gamma(K_{4,4,4}) = 3$ and note that $a(K_{4,4,4}) = p_3 - 1 = 8$.

# The Algebraic Connectivity

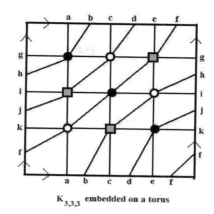

$K_{3,3,3}$ embedded on a torus

The second question worth investigating is if there are other values for $k$, besides those listed in Theorem 5.6.7, in which $a(\mathcal{G}) < p_k - 1$ for all noncomplete graphs $\mathcal{G}$ of genus $k$. Since finding the genus of a graph can be very difficult, this remains an open question.

## Exercises:

**1.** Finding an upper bound on the algebraic connectivity of graphs of genus $k \geq 1$ required extensive use of a graph's minimum degree. Explain why using the idea of a graph's minimum degree cannot be extended to prove that $a(\mathcal{G}) \leq 4$ if $\mathcal{G}$ is a planar graph.

**2.** Prove that for any $k \geq 0$ that the lower bound on the algebraic connectivity of graphs on genus $k$ is zero.

# Chapter 6

# The Fiedler Vector and Bottleneck Matrices for Trees

In this chapter we focus on the eigenvector of the Laplacian matrix corresponding to the algebraic connectivity. This eigenvector is known as a Fiedler vector. In Section 6.1 we investigate the entries of this eigenvector in relationship to the structure of the graph. We study the entries of the Fiedler vector that correspond to cut vertices of the graph and then investigate the entries that correspond to specific blocks of the graph. In a tree, since every edge is a block, we apply the results of Section 6.1 to trees and focus the remainder of this chapter on trees. In Section 6.2 we use the Fiedler vectors of trees to classify trees into two types based on the existence of zero entries in such eigenvectors. This leads us to investigating the inverses of submatrices of the Laplacian matrix obtained by deleting a row and column of the Laplacian matrix corresponding to a specific vertex of the tree, espcially vertices corresponding to a zero entry in the Fiedler vector. These inverses, known as bottleneck matrices, give us useful information about the structure of the tree and also have interesting properties that correspond to the two types of trees. Bottleneck matrices are a vital tool in Section 6.3 when we go on an excursion of studying branches at various vertices of certain unusual trees. Bottleneck matrices continue to be useful in Section 6.4 when we determine which trees of a given fixed diameter have minimum algebraic connectivity. These results help us in Section 6.5 when we consider trees having an edge of infinite weight. In Section 6.6 we generalize the results of this chapter to eigenvectors of Laplacian matrices of trees corresponding to eigenvalues other than the algebraic connectivity. Finally, in Section 6.7 we investigate the $(n-1) \times (n-1)$ submatrices of the Laplacian matrix before taking their inverses and observe how the spectral radius of these submatrices corresponding to different vertices varies.

## 6.1 The Characteristic Valuation of Vertices

Let $\mathcal{G}$ be a connected weighted graph on $n$ vertices labeled $1, \ldots, n$, with Laplacian matrix $L$. Then $a(\mathcal{G}) > 0$ by Theroem 4.1.1. In this section, we are interested in the eigenvectors corresponding to the eigenvalue $a(\mathcal{G})$. To this end, we have a definition:

**DEFINITION 6.1.1** An eigenvector $y$ corresponding to the eigenvalue $a(\mathcal{G})$ of the Laplacian matrix for a graph is called a *Fiedler vector*. This vector provides a *characteristic valuation* of $\mathcal{G}$. Labeling the entries of $y$ as $y_1, \ldots, y_n$ in correspondence to the labelings of the vertices of $\mathcal{G}$, we will denote $y_k$ and the *valuation* of vertex $k$.

In this section, we will investigate the valuation of vertices in graphs which contain cut vertices. We will prove theorems regarding the valuations of cut vertices and also how the valuations of the vertices lying in different blocks compare. We begin with two straightforward, but useful, theorems from [30].

**THEOREM 6.1.2** *Let $\mathcal{G}$ be a connected graph with Laplacian matrix $L$ and characteristic valuation $y$. Then $\sum y_i = 0$.*

**Proof**: Since $\mathcal{G}$ is connected, it follows that $0$ is a simple eigenvalue of $L$ with $e$ as its corresponding eigenvector. Since $L$ is symmetric, it follows that $y$ is orthogonal to $e$. Hence $\sum y_i = 0$. □

**THEOREM 6.1.3** *Let $\mathcal{G}$ be a weighted graph with vertex set $V = \{1, \ldots, n\}$ and edge set $E$. Let $w_{ik}$ denote the weight of edge $ik$. Letting $L$ be the Laplacian matrix for $\mathcal{G}$, if $a(\mathcal{G})$ is the algebraic connectivity of $\mathcal{G}$ and $y = (y_i)$ is the characteristic valuation, then*

$$a(\mathcal{G})y_i = \sum_{\substack{k \in V \\ (i,k) \in E}} w_{ik}(y_i - y_k).$$

**Proof**: This follows directly from matrix-vector multiplication and the fact that $Ly - a(\mathcal{G})y = 0$. □

**OBSERVATION 6.1.4** From Theorem 6.1.3, if $y_i = 0$ and $i$ is adjacent to a positively valuated vertex, then $i$ must also be adjacent to a negatively valuated vertex.

Since $a(\mathcal{G}) > 0$ if $\mathcal{G}$ is connected, we have the following immediate corollary from [30]:

**COROLLARY 6.1.5** *Let $\mathcal{G}$ be a connected weighted graph on $n$ vertices labeled $1, \ldots, n$ with Laplacian matrix $L$ and characteristic valuation $y = [y_i]$. If $y_i > 0$, then there exists a vertex $j$ adjacent to $i$ such that $y_j < y_i$.*

We are beginning to understand the relationship between the Fiedler vector and the structure of the graph. With this in mind, we further the results of Corollary 6.1.5 in the following lemma and theorem from [30]. The lemma is a matrix theoretic result which is necessary in proving the theorem. The theorem illustrates how the entries in the Fiedler vector can determine which subgraphs of a connected graph are connected.

**LEMMA 6.1.6** *Let $A$ be an $n \times n$ nonnegative irreducible, symmetric matrix. Let $u = [u_i]$ be an eigenvector corresponding to $\lambda_n$ and let $v = [v_i]$ be an eigenvector corresponding to $\lambda_{n-1}$. Then for any $\alpha \geq 0$, the submatrix $A[M_\alpha]$ is irreducible where $M_\alpha = \{i \mid v_i + \alpha u_i \geq 0\}$.*

**THEOREM 6.1.7** *Let $\mathcal{G}$ be a connected weighted graph with vertices $V = \{1,\ldots,n\}$ and Laplacian matrix $L$. Let each edge $ik$ have a positive weight $w_{ik}$. For any $r \geq 0$, let*
$$M(r) = \{i \in V \mid y_i + r \geq 0\}.$$
*Then the subgraph $\mathcal{G}(r)$ of $\mathcal{G}$ induced by the vertices of $M(r)$ is connected.*

**Proof**: Observe that for sufficiently large $s$, the matrix $B := sI - L$ is nonnegative. Moreover, the eigenvectors of $L$ are identical to those of $B$. Similarly, the second smallest eigenvalue of $L$ corresponds to the second largest eigenvalue of $B$. Then by Lemma 6.1.6, the submatrix of $B$ induced by the vertices of $M(r)$ is irreducible. Thus the corresponding submatrix of $L$ is irreducible. Hence the corresponding subgraph is connected. □

The goal of this section will be to classify connected graphs with cut vertices into two distinct classes based on their characteristic valuations. To this end, we will use Theorem 6.1.7 to further our investigation of connected subgraphs of a given connected graph $\mathcal{G}$. The subgraphs of $\mathcal{G}$ that will be important for us to study are the components of $\mathcal{G} - k$ where $k$ is a cut vertex of $\mathcal{G}$. We now prove the following lemma from [30] which deals with the valuations of the vertices in each such component of $\mathcal{G} - k$. This lemma will be crucial to proving the main theorem of this section.

**LEMMA 6.1.8** *Let $\mathcal{G}$ be a graph and $y = [y_i]$ be its characteristic valuation. Let $k$ be a cut vertex of $\mathcal{G}$ and let $\mathcal{G}_0, \mathcal{G}_1, \ldots, \mathcal{G}_r$ be components of $\mathcal{G} - k$. Then:*

*(i) If $y_k > 0$, then exactly one of the components $\mathcal{G}_i$ contains a vertex (vertices) with negative valuation. Additionally, $y_s > y_k$ for all vertices $s$ in the remaining components.*

*(ii) If $y_k = 0$ and there is a component $\mathcal{G}_i$ containing both positively and negatively valuated vertices, then there is exactly one such component. All vertices in all other components have zero valuation.*

*(iii) If $y_k = 0$ and no component contains both positively and negatively valuated vertices, then each component $\mathcal{G}_i$ contains either only positively valuated vertices, only negatively valuated vertices, or only zero valuated vertices.*

Note: We need not consider the case where $y_k < 0$ since if that does occur, we may simply multiply $y$ by $-1$ to obtain a vector which is also a characteristic valuation.

**Proof**: To prove Case (i), let $y_k > 0$. Since $\sum y_i = 0$ (Theorem 6.1.2), it follows that there exists a vertex in $\mathcal{G} - k$ with negative valuation. Without loss of generality, let $\mathcal{G}_0$ contain such a vertex. To complete the proof, we must show for all vertices $s \in \mathcal{G}_1 \cup \mathcal{G}_2 \cup \ldots \cup \mathcal{G}_r$ that $y_s > y_k$. Suppose that $y_s < y_k$ for some vertex $s \in \mathcal{G}_1 \cup \mathcal{G}_2 \cup \ldots \cup \mathcal{G}_r$. Then there exists an $\epsilon > 0$ such that $y_k - \epsilon > 0$ and $y_k - \epsilon \geq y_s$. By Theorem 6.1.7, the graph $\hat{\mathcal{G}}$ induced by the set of vertices $M = \{v \in V(\mathcal{G}) \mid y_v \leq y_k - \epsilon\}$ is connected. Since $k \notin M$, it follows that $\hat{\mathcal{G}}$ is contained in exactly one of $\mathcal{G}_0, \mathcal{G}_1, \ldots, \mathcal{G}_r$ since $k$ is a cut vertex. However, since $\mathcal{G}_0$ has a vertex of negative valuation, such a vertex must lie in $M$. Hence $\hat{\mathcal{G}}$ is contained in $\mathcal{G}_0$. But we are assuming $s \in M$ (hence $s \in \hat{\mathcal{G}}$) and that $s \in \mathcal{G}_1 \cup \mathcal{G}_2 \cup \ldots \cup \mathcal{G}_r$. Thus we have a contradiction. Now suppose that $y_s = y_k$ for some vertex $s \in \mathcal{G}_1 \cup \mathcal{G}_2 \cup \ldots \cup \mathcal{G}_r$. Since $y_s = y_k > 0$, it follows from Corollary 6.1.5 that there exists a vertex $t \in \mathcal{G}_1 \cup \mathcal{G}_2 \cup \ldots \cup \mathcal{G}_r$, $t \neq k$ where $t$ is adjacent to $s$ such that $y_t < y_s$. Thus $s$ and $t$ lie in the same component of $\mathcal{G}_1 \cup \mathcal{G}_2 \cup \ldots \cup \mathcal{G}_r$. Since $y_t < y_k$, we can use the previous argument to deduce a contradiction.

To prove Cases (ii) and (iii), assume that $y_k = 0$. Without loss of generality, let $k = n$ and that

$$L := L(\mathcal{G}) = \begin{bmatrix} L_0 & 0 & \ldots & \ldots & 0 & c_0 \\ 0 & L_1 & \ldots & \ldots & 0 & c_1 \\ \ldots & \ldots & \ldots & \ldots & \ldots & \ldots \\ 0 & 0 & \ldots & \ldots & L_r & c_r \\ c_0^T & c_1^T & \ldots & \ldots & c_r^T & \deg(k) \end{bmatrix}$$

where the rows and columns of $L_i$ correspond to the vertices in $\mathcal{G}_i$ for $i = 0, \ldots, r$. Letting $a = a(\mathcal{G})$, then $L - aI$ is singular and $(L - aI)y = 0$. Let

$$y = \begin{bmatrix} y^{(0)} \\ y^{(1)} \\ \ldots \\ y^{(r)} \\ 0 \end{bmatrix}$$

be the characteristic valuation partitioned in correspondence with the partitioning of $L$. Then

$$(L_i - aI_i)y^{(i)} = 0$$

for $i = 0, \ldots, r$ where $I_i$ are the identity matrices of the appropriate size for each $i$. We now break the proof into two subcases: Subcase (a) will prove Case (ii) and Subcase (b) will prove Case (iii):

(a) There exists a component, say $\mathcal{G}_0$, which contains both positively and negatively valuated vertices. Hence $y^{(0)}$ has both positive and negative entries. Since all of the off-diagonal entries $L_0 - aI_0$ are non-positive and since $L_0 - aI_0$ is symmetric, it follows that $L_0 - aI_0$ is not positive semidefinite (see [32]). Therefore it has at least one negative eigenvalue. Since

$$A := \begin{bmatrix} L_0 - aI_0 & & & \\ & L_1 - aI_1 & & \\ & & \cdots & \\ & & & L_r - aI_r \end{bmatrix}$$

is a principal submatrix of $L - aI$ and $L - aI$ has one negative eigenvalue, it follows from Corollary 1.2.10 that $A$ has at most one negative eigenvalue. Therefore $L_i - aI_i$ are positive semidefinite matrices for $i = 1, \ldots, r$. Assume $y^{(j)} \neq 0$ for some $j \in \{1, \ldots, r\}$. Since $L_j - aI$ is irreducible, it follows that each $y^{(j)}$ is either positive or negative. However, in the matrix $L$, we have $c_j \leq 0$. Therefore $(y^{(j)})^T c_i \neq 0$. Letting $\hat{y}^T = [0, \ldots, 0, (y^{(j)})^T, 0, \ldots, 0]^T$ and $c^T = [c_0^T, \ldots, c_r^T]$, since $(L - aI)\hat{y} = 0$ and $\hat{y}^T c \neq 0$, it follows from Lemma 1.12 of [29] that

$$s(L - aI) = \sum_{i=0}^{r} s(L_i - aI_i) + [1, 1], \qquad (6.1.1)$$

where $s(A)$ is the $1 \times 2$ row vector $[p, q]$ with $p$ denoting the number of positive eigenvalues of $A$ and $q$ denoting the number of negative eigenvalues of $A$. Since $L_0 - aI_0$ has a negative eigenvalue, it follows from (6.1.1) that $L - aI$ must have at least two negative eigenvalues. But since $a$ is the second smallest eigenvalue of $L$, it follows that $L - aI$ has one negative eigenvalue, thus a contradiction. Therefore $y^{(j)} = 0$ for all $j \in \{1, \ldots, r\}$. This proves case (ii).

(b) None of the components $\mathcal{G}_0, \mathcal{G}_1, \ldots, \mathcal{G}_r$ contain both positively and negatively valuated vertices. Since $y \neq 0$, there exists a vertex with nonzero valuation. Without loss of generality, let $\mathcal{G}_0$ contain such a vertex. Therefore $y^{(0)} \neq 0$, and hence either $y^{(0)} \geq 0$ or $y^{(0)} \leq 0$. Suppose an entry of $y^{(0)}$ is zero. Then since $(L_0 - aI_0)y^{(0)} = 0$, it would follow from Theorem 1.4.10(ii) that $L_0 - aI_0$ is not an M-matrix. Then by arguments similar to (a), we would obtain $y^{(i)} = 0$ for all $i \in \{1, \ldots, r\}$. But since $y^{(0)} \geq 0$, this contradicts Theorem 6.1.2. Thus if $y^{(0)} \geq 0$ or $y^{(0)} \leq 0$, then no entry of $y^{(0)}$ can be zero. Hence $y^{(0)} > 0$ or $y^{(0)} < 0$ which proves (iii). $\square$

We are now about ready to prove the main theorem of this section. This theorem from [30] classifies connected graphs with cut vertices into two classes depending if there exists a block with both positively and negatively valuated vertices. If such a block exists, this will be a point of interest. Otherwise, the theorem shows that there will be a zero-valuated cut vertex of interest. Moreover, we learn from this theorem that the valuations of the cut vertices as we travel away from the block / cut vertex of interest increases in absolute value. In order to be able to state this idea more precisely, we need a definition:

**DEFINITION 6.1.9** A *pure path* $P$ in a graph $\mathcal{G}$ is a path in which no block of $\mathcal{G}$ contains more than two vertices of $P$.

Now for the main theorem of this section:

**THEOREM 6.1.10** *Let $\mathcal{G}$ be a connected graph with at least one cut vertex and let y be its characteristic valuation. Then exactly one of the following cases occurs:*

*(a) There is a single block $B_0$ in $\mathcal{G}$ which contains both positively and negatively valued vertices. Each other block in $\mathcal{G}$ has either vertices with positive valuation only, vertices with negative valuation only, or vertices with zero valuation only. Every pure path $P$ starting at $B_0$ and containing just one vertex $k \in B_0$ has the property that the valuations at the cut vertices contained in $P$ form either an increasing, decreasing, or zero sequence along this path.*

*(b) No block of $\mathcal{G}$ contains both positively and negatively valued vertices. There exists a single vertex $z$ which has zero valuation and is adjacent to a vertex of non-zero valuation. Moreover, $z$ is a cut vertex. Each block contains (with the exception of $z$) either vertices with positive valuation only, vertices with negative valuation only, or vertices with zero valuation only. Every pure path $P$ starting at $z$ has the property that the valuations at the cut vertices contained in $P$ form either an increasing, decreasing, or zero sequence along this path.*

**Proof**: Clearly these cases exclude each other. The examples that follow will show that each case can occur. So now let $\mathcal{G}$ be a connected graph with $y$ as its characteristic valuation. To prove (a), let $B_0$ and $B_1$ be blocks of $\mathcal{G}$ in which there exists a cut vertex $k$ contained in $B_0$ that separates $B_0$ from $B_1$. Let $\mathcal{G}_0, \mathcal{G}_1, \ldots, \mathcal{G}_r$ be all components of the graph $\mathcal{G} - k$ where $\mathcal{G}_0$ contains the remaining vertices of $B_0$ and $\mathcal{G}_1$ contains the remaining vertices of $B_1$. If $y_k > 0$, it follows from Lemma 6.1.8 that all vertices in $\mathcal{G}_1$, and thus $B_1$ have positive valuations. If $y_k = 0$, it follows from Lemma 6.1.8 that all vertices in $\mathcal{G}_1$, and thus $B_1$ have zero valuation. If $y_k < 0$, then $-y$ is also a characteristic valuation. Thus $-y_k > 0$ and we apply the argument above. This proves the first part of (a).

To prove the second part of (a), let $P$ be a pure path and let $k$ be the only vertex of $P$ in $B_0$. Let $k, k_1, \ldots, k_s$ be cut vertices of $\mathcal{G}$ that lie on $P$ in that order as we travel away from $k$. First assume $y_k > 0$. Then by Lemma 6.1.8, $y_{k_1} > y_k$. Applying the same theorem to the cut vertex $y_{k_j}$ with $1 \leq j < s$, we obtain from Lemma 6.1.8 that $y_{k_{j+1}} > y_{k_j}$. Thus the sequence $y_k, y_{k_1}, y_{k_2}, \ldots, y_{k_s}$ increases. If $y_k < 0$, then applying this last result to $-y$ yields that this sequence decreases. If $y_k = 0$, then from (ii) of Lemma 6.1.8, we see that this sequence is the zero sequence. This completes the proof of (a).

To prove (b), suppose that no block contains both positively and negatively valuated vertices. Observe in this case that there is no edge $e$ in $\mathcal{G}$ such that one vertex incident to $e$ has positive valuation while the other vertex incident to $e$ has negative valuation. Thus if a path $P$ contains a vertex with positive valuation and a vertex with negative valuation, then $P$ must contain a vertex of zero valuation such that it is adjacent to a vertex in $P$ with non-zero valuation. Since $y \neq 0$ and since $\sum y_i = 0$, it follows that such a path $P$ does exist in $\mathcal{G}$. Therefore there exists a vertex $z$ with $y_z = 0$ where $z$ is adjacent to a vertex with non-zero valuation. By Observation 6.1.4, $z$ has neighbors with positive valuation as well as negative valuation. Since in this case these vertices cannot belong to the same block, it follows that $z$ is a cut vertex. By (ii) and (iii) in Lemma 6.1.8 it follows that $z$ is the only vertex with zero valuation that is adjacent to vertices of non-zero valuation. Hence each block contains (with the exception of $z$) either vertices with positive valuation only, vertices with negative valuation only, or vertices with zero valuation only.

To prove the final assertion of (b), let $P$ be a pure path starting at $z$. By (iii) of Lemma 6.1.8, it follows that if $P$ contains a vertex of positive (negative) valuation, then all vertices of $P$ must be of positive (negative) valuation. Since $-y$ is also an eigenvector corresponding to $a(\mathcal{G})$, we can assume without loss of generality that $P$ has vertices of positive valuation only (with the exception of $z$). Letting $z, k_1, k_2, \ldots, k_s$ be the cut vertices on $P$ in the order in which they appear, we can apply a similar argument used for (a) to obtain that $y_{k_{j+1}} > y_{k_j}$ for all $j$ such that $1 \leq j < s$. This completes the proof. □

**OBSERVATION 6.1.11** If a graph satisfies Case B, then it follows from Theorem 6.1.7 that the subgraph induced by the vertices with zero valuation forms a connected graph.

**OBSERVATION 6.1.12** If a graph $\mathcal{G}$ satisfies Case A, then $a(\mathcal{G})$ must be a simple eigenvalue. Otherwise, there would exist two linearly independent eigenvectors $g$ and $h$ corresponding to $a(\mathcal{G})$ and hence we could create a nontrivial linear combination of $g$ and $h$ having a zero coordinate.

Let's look at two examples of graphs, one satisfying Case (a) and the other satisfying Case (b):

**EXAMPLE 6.1.13** Consider the following graph:

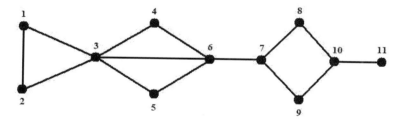

The algebraic connectivity is $a = 0.1705$ and a corresponding eigenvector to $a$ is
$[-1.2490 \ -1.2490 \ -1.0370 \ -0.7360 \ -0.7360 \ -0.3092 \ 0.5982 \ 1 \ 1 \ 1.231 \ 1.485]^T$.
Labeling each vertex of the graph with its characteristic valuation:

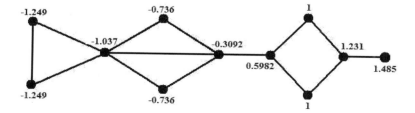

we see that this graph is an example of Case (a) since there does not exist a cut vertex whose characteristic valuation is zero. Furthermore, there is exactly one block, namely the block consisting of the edge joining vertices 6 and 7, that has both positively and negatively valuated vertices. As we travel away from that block, the valuations of the cut vertices either strictly increase or strictly decrease.

**EXAMPLE 6.1.14** Now consider the following graph:

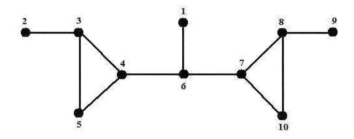

The algebraic connectivity is $a = 0.1783$ and a corresponding eigenvector to $a$ is

$$[0 \quad -1 \quad -0.8217 \quad -0.5599 \quad -0.7590 \quad 0 \quad 0.5599 \quad 0.8217 \quad 1 \quad 0.7590]^T.$$

Labeling each vertex of the graph with its characteristic valuation:

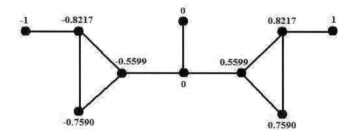

we see that this graph is an example of Case (b) since there exists a cut vertex, namely vertex 6, with valuation of zero. Notice that each block contains only vertices with positive valuations, only vertices with negative valuation, or only vertices with zero valuation. Also observe that as we travel away from vertex 6, the characteristic valuations of the cut vertices either strictly increase, strictly decrease, or are identically zero.

# Exercise:

1. Classify each graph as satisfying Case (a) or Case (b) of Theorem 6.1.10. Also, verify the statements regarding the pure paths described in each case.

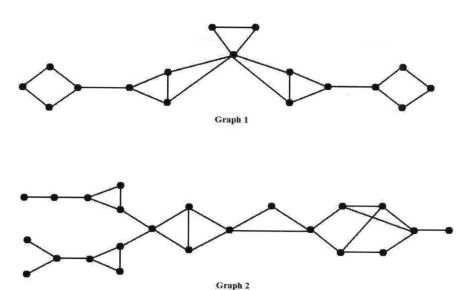

Graph 1

Graph 2

## 6.2 Bottleneck Matrices for Trees

The rest of this chapter will focus on trees. In this section, we show how the inverse of submatrices of Laplacian matrices gives us insight as to the structure of trees. These submatrices will have a direct relationship with the special vertex / block described in Theorem 6.1.10 and with the algebraic connectivity of trees. To this end, we apply the results of the previous section to trees by noting that in a tree, the blocks are merely the edges and that all vertices, except for the pendant vertices, are cut vertices. Therefore every path in a tree is a pure path. Hence, Theorem 6.1.10 yields the following corollary for trees:

**COROLLARY 6.2.1** *Let $\mathcal{T}$ be a tree and $y = [y_i]$ be its characteristic valuation. Then exactly one of two cases can occur:*

*Case (a): All values $y_i$ are different from zero. In this case, $\mathcal{T}$ contains exactly one edge pq such that $y_p > 0$ and $y_q < 0$. The valuations of the vertices on any path in $\mathcal{T}$ that begins at p and does not contain q increase. Similarly, the valuations of the vertices on any path in $\mathcal{T}$ that begins at q and does not contain p decrease.*

*Case (b): There exists exactly one vertex z where $y_z = 0$ and z is adjacent to a vertex of non-zero valuation. The valuations of the vertices along any path in $\mathcal{T}$ beginning at z are either increasing, decreasing, or identically zero.*

In [58], Merris coined the following definition:

**DEFINITION 6.2.2** *Trees that satisfy Case (b) are Type I trees; trees that satisfy Case (a) are Type II trees.*

Since the vertices $p$ and $q$ in Case (a) and the vertex $z$ in Case (b) are of extreme importance, Fiedler coined the following definition in [30]:

**DEFINITION 6.2.3** In a Type I tree, the vertex $z$ described in Case (b) is the *characteristic vertex*; in a Type II tree, the vertices $p$ and $q$ described in Case (a) are the *characteristic vertices*.

We should note that there are infinitely many vectors in the eigenspace corresponding to the algebraic connectivity of a tree. Thus before proceeding, we need to investigate if the characteristic vertex (vertices) of a given tree can differ depending on the Fiedler vector used. The next theorem from [58] shows that this will never happen.

**THEOREM 6.2.4** *Let $\mathcal{T}$ be a tree with Laplacian matrix $L$. The characteristic vertex (vertices) of $\mathcal{T}$ are independent of the eigenvector used as the characteristic valuation.*

**Proof**: Let $a$ be the algebraic connectivity of $\mathcal{T}$ and let $\xi(\mathcal{T})$ be the set of eigenvectors for $L$ corresponding to $a$. Suppose that $g$ and $h$ are eigenvectors in $\xi(\mathcal{T})$. If $a$ is simple, then $g$ is merely a nonzero multiple of $h$, thus the theorem follows immediately. So we will assume that $a$ has algebraic multiplicity of at least 2 (and that $\mathcal{T}$ is of Type I by Observation 6.1.12). Let $V_0 := \{v \in V(\mathcal{T}) \mid f_v = 0 \text{ for all } f \in \xi(\mathcal{T})\}$. We first claim that $V_0$ is not empty. To see this, since $L$ is symmetric, the geometric multiplicity of $a$ equals its algebraic multiplicity and thus is also at least 2. Thus we can create an eigenvector in $\xi(\mathcal{T})$ which is a nontrivial linear combination of $g$ and $h$ such that there exists a vertex $v$ where $f_v = 0$.

We now claim that among the vertices in $V_0$, there is exactly one vertex, say $z$, which is adjacent to a vertex not contained in $V_0$. We know there must be at least one such vertex because $\mathcal{T}$ is connected and $V_0 \neq V(\mathcal{T})$. Suppose there are two such vertices, say $z_1$ and $z_2$. Let $v_1, v_2$ be vertices not in $V_0$ such that $v_1$ is adjacent to $z_1$, and $v_2$ is adjacent to $z_2$. Then there exists $f \in \xi(\mathcal{T})$ such that $f_{v_1} \neq 0$ and $f_{v_2} \neq 0$. This contradicts Case (b) of Corollary 6.2.1.

The previous paragraph shows that for each vector $f \in \xi(\mathcal{T})$, there exists a unique vertex $z \in V_0$ such that $f_z = 0$ and $z$ is adjacent to a vertex not in $V_0$, thus making $z$ the unique characteristic vertex of $\mathcal{T}$ with respect to $f$. To complete the proof, we need to show that $z$ is the unique characteristic vertex afforded by all $f \in \xi(\mathcal{T})$. By the previous paragraph, $z$ is the only element of $V_0$ that has the chance of being such a characteristic vertex. Suppose $f^{(1)} \in \xi(\mathcal{T})$ affords characteristic vertex $x$. Since $f_z^{(1)} = 0$ and since by Corollary 6.2.1 the characteristic vertex $x$ is unique, it follows that there exists a vertex $y \in \mathcal{T}$ adjacent to $x$ such that $f_y^{(1)} \neq 0$. If $x \neq z$, then $x \notin V_0$ for all eigenvectors $f \in \xi(\mathcal{T})$. Thus there must exist $f^{(2)} \in \xi(\mathcal{T})$ such that $f_x^{(2)} \neq 0$. Since $x$ is on the unique path from $y$ to $z$, it follows from Corollary 6.2.1 that $f_y^{(2)} \neq 0$. But then there must be a nontrivial linear combination $f^{(3)}$ of $f^{(1)}$ and $f^{(2)}$ such that $f_y^{(3)} = 0$ and $f_x^{(3)} \neq 0$. This contradicts Observation 6.1.11 because $f^{(3)} \in \xi(\mathcal{T})$ and $x$ lies on the path from $y$

to $z$, yet $f_y^{(3)} = 0$, $f_x^{(3)} \neq 0$, and $f_z^{(3)} = 0$. Thus $x = z$ and our claim is established. □

Now that we know that we can choose any eigenvector corresponding to the algebraic connectivity when determining the characteristic vertex (vertices) of a tree, we can proceed with the main ideas of this section which involve the inverses of submatrices of Laplacian matrices. These inverses will be closely related to the location of the characteristic vertex (vertices). We now prove the first major theorem of this section (see [49]):

**THEOREM 6.2.5** *Suppose that $\mathcal{T}$ is a weighted tree on vertices $1, \ldots, n$ with Laplacian matrix $L$, and let $L[\bar{k}]$ be the principal submatrix of $L$ formed by deleting the $k^{th}$ row and column of $L$. Then $L[\bar{k}]$ is invertible and the $(i,j)$ entry of $(L[\bar{k}])^{-1}$ is equal to*

$$\sum_{e \in P_{i,j}} \frac{1}{w(e)}$$

*where $P_{i,j}$ is the set of edges of $\mathcal{T}$ which are simultaneously on the path from $i$ to $k$ and on the path from $j$ to $k$.*

**Proof**: Since $\mathcal{T}$ is connected, $L$ is a singular irreducible M-matrix. Thus by Theorem 1.4.10(iii), $L[\bar{k}]$ is invertible for each $k = 1, \ldots, n$. To determine the $(i,j)$ entry of $(L[\bar{k}])^{-1}$, we can let $k = n$ without loss of generality. We will proceed by induction on $n$, the number of vertices. The theorem is easily verified for $n = 2$. Suppose that the result holds for all weighted trees on $r-1$ vertices, where $r-1 \geq 2$, and that $\mathcal{T}$ is a weighted tree on $r$ vertices. To prove this result for $\mathcal{T}$, we can assume without loss of generality that vertex 1 is pendant and is adjacent to vertex 2. Let $\alpha$ be the weight of the edge joining vertices 1 and 2.

Let $L'$ be the Laplacian matrix of the weighted tree $\mathcal{T}'$ induced by the vertices $2, \ldots, r$ of $\mathcal{T}$, and let $L'[\bar{r}]$ be the principal submatrix of $L'$ formed by deleting its last row and column. Then it follows that:

$$L[\bar{r}] = \left[ \begin{array}{c|c} \alpha & -\alpha e_1^T \\ \hline -\alpha e_1 & L'[\bar{r}] + \alpha e_1 e_1^T \end{array} \right]$$

A straightforward computation yields:

$$(L[\bar{r}])^{-1} = \left[ \begin{array}{c|c} \frac{1}{\alpha} + e_1^T (L'[\bar{r}])^{-1} e_1 & e_1^T (L'[\bar{r}])^{-1} \\ \hline (L'[\bar{r}])^{-1} e_1 & (L'[\bar{r}])^{-1} \end{array} \right]$$

Observing that the $(1,1)$ block of $(L[\bar{r}])^{-1}$ is just a single entry, we see that if $i, j \geq 2$, then the formula for the $(i,j)$ entry $(L[\bar{r}])^{-1}$ is given in the $(2,2)$ block and thus follows from the induction hypothesis. Note that if $j \geq 2$, then the set $P_{1,j}$ is the same as the set $P_{2,j}$ which is the same as the corresponding set of edges in $\mathcal{T}'$. Thus the $(1,j)$ and $(j,1)$ entries of $(L[\bar{r}])^{-1}$ follow from the inductive hypothesis and from the $(1,2)$ and $(2,1)$ blocks of $(L[\bar{r}])^{-1}$, respectively. Finally, for the $(1,1)$ entry of $(L[\bar{r}])^{-1}$, since vertex 1 is pendant and adjacent to vertex 2, it follows that

$P_{1,1} = P_{2,2} \cup e_{1,2}$ where $e_{1,2}$ denotes the edge joining vertices 1 and 2. Since the weight of $e_{1,2}$ is $\alpha$, it follows that

$$\sum_{e \in P_{1,1}} \frac{1}{w(e)} = \frac{1}{\alpha} + \sum_{e \in P_{2,2}} \frac{1}{w(e)} = \frac{1}{\alpha} + e_1^T (M[\overline{r}])^{-1} e_1$$

where the equality $\sum_{e \in P_{2,2}} \frac{1}{w(e)} = e_1^T (L'[\overline{r}])^{-1} e_1$ follows from the inductive hypothesis. This is precisely the $(1,1)$ block of $(L[\overline{r}])^{-1}$, thus completing the proof. $\square$

Theorem 6.2.5 states that the $(i,j)$ entry of $(L[\overline{k}])^{-1}$ is the sum of the reciprocals of the weights of the edges that lie simultaneously on the path from $i$ to $k$ and on the path from $j$ to $k$. We should note that when numbering the rows and columns of $(L[\overline{k}])^{-1}$ in accordance with Theorem 6.2.5, we skip the number $k$.

We see from Theorem 6.2.5 (and Theorem 1.4.7) that $(L[\overline{k}])^{-1}$ is nonnegative. Since there is exactly one path between any two vertices in a tree, it follows that the $(i,j)$ entry of $(L[\overline{k}])^{-1}$ is positive if and only if vertices $i$ and $j$ belong to the same branch at vertex $k$. Otherwise $P_{i,j}$ would be empty. Therefore, $(L[\overline{k}])^{-1}$ is permutationally similar to a block diagonal matrix in which each block is positive and corresponds to a branch at $k$. This leads us to the following important definition:

**DEFINITION 6.2.6** The matrix $(L[\overline{k}])^{-1}$ is the *bottleneck matrix at vertex $k$* and each diagonal block of $(L[\overline{k}])^{-1}$ is the *bottleneck matrix for that branch at $k$*.

**OBSERVATION 6.2.7** If $\mathcal{T}$ is unweighted, then the $(i,j)$ entry of $(L[\overline{k}])^{-1}$ is the number of edges that lie simultaneously on the path from $i$ to $k$ and the path from $j$ to $k$.

**EXAMPLE 6.2.8** Consider the following tree:

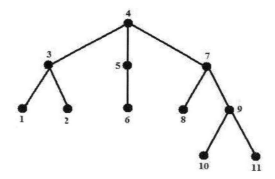

The Laplacian matrix is

$$L = \begin{bmatrix} 1 & 0 & -1 & 0 & 0 & 0 & 0 & 0 & 0 & 0 & 0 \\ 0 & 1 & -1 & 0 & 0 & 0 & 0 & 0 & 0 & 0 & 0 \\ -1 & -1 & 3 & -1 & 0 & 0 & 0 & 0 & 0 & 0 & 0 \\ 0 & 0 & -1 & 3 & -1 & 0 & -1 & 0 & 0 & 0 & 0 \\ 0 & 0 & 0 & -1 & 1 & -1 & 0 & 0 & 0 & 0 & 0 \\ 0 & 0 & 0 & 0 & -1 & 1 & 0 & 0 & 0 & 0 & 0 \\ 0 & 0 & 0 & -1 & 0 & 0 & 3 & -1 & -1 & 0 & 0 \\ 0 & 0 & 0 & 0 & 0 & 0 & -1 & 1 & 0 & 0 & 0 \\ 0 & 0 & 0 & 0 & 0 & 0 & -1 & 0 & 3 & -1 & -1 \\ 0 & 0 & 0 & 0 & 0 & 0 & 0 & 0 & -1 & 1 & 0 \\ 0 & 0 & 0 & 0 & 0 & 0 & 0 & 0 & -1 & 0 & 1 \end{bmatrix}$$

A straightforward computation shows that

$$(L[\overline{4}])^{-1} = \begin{bmatrix} 2 & 1 & 1 & 0 & 0 & 0 & 0 & 0 & 0 & 0 \\ 1 & 2 & 1 & 0 & 0 & 0 & 0 & 0 & 0 & 0 \\ 1 & 1 & 1 & 0 & 0 & 0 & 0 & 0 & 0 & 0 \\ 0 & 0 & 0 & 1 & 1 & 0 & 0 & 0 & 0 & 0 \\ 0 & 0 & 0 & 1 & 2 & 0 & 0 & 0 & 0 & 0 \\ 0 & 0 & 0 & 0 & 0 & 1 & 1 & 1 & 1 & 1 \\ 0 & 0 & 0 & 0 & 0 & 1 & 2 & 1 & 1 & 1 \\ 0 & 0 & 0 & 0 & 0 & 1 & 1 & 2 & 2 & 2 \\ 0 & 0 & 0 & 0 & 0 & 1 & 1 & 2 & 3 & 2 \\ 0 & 0 & 0 & 0 & 0 & 1 & 1 & 2 & 2 & 3 \end{bmatrix}$$

Since the degree of vertex 4 is three, we see that $(L[\overline{4}])^{-1}$ has three blocks, each of which corresponds to a branch at vertex 4. Labelling the rows and columns of $(L[\overline{4}])^{-1}$ from 1 to 11, skipping 4, observe that the $(i,j)$ entry of $(L[\overline{4}])^{-1}$ denotes the number of edges that lie simultaneously on the path from $i$ to 4 and the path from $j$ to 4. This illustrates why if vertices $i$ and $j$ belong to different branches at vertex 4, then the $(i,j)$ entry in $(L[\overline{4}])^{-1}$ is zero as there are no common edges on the paths from $i$ to 4 and from $j$ to 4. However, if vertices $i$ and $j$ belong to the same branch at vertex 4, then the $(i,j)$ entry of $(L[\overline{4}])^{-1}$ is positive. For example, consider the $(10,11)$ entry of $L_4^{-1}$. Observe that it is two and that there are two common edges lying on the path from 10 to 4 and the path from 11 to 4, namely edges 4-7 and 7-9.

**EXAMPLE 6.2.9** Consider the following weighted tree:

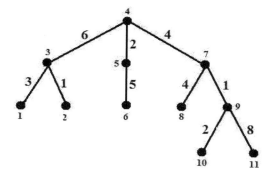

Observe that its Laplacian matrix is

$$L = \begin{bmatrix} 3 & 0 & -3 & 0 & 0 & 0 & 0 & 0 & 0 & 0 & 0 \\ 0 & 1 & -1 & 0 & 0 & 0 & 0 & 0 & 0 & 0 & 0 \\ -3 & -1 & 10 & -6 & 0 & 0 & 0 & 0 & 0 & 0 & 0 \\ 0 & 0 & -6 & 12 & -2 & 0 & -4 & 0 & 0 & 0 & 0 \\ 0 & 0 & 0 & -2 & 7 & -5 & 0 & 0 & 0 & 0 & 0 \\ 0 & 0 & 0 & 0 & -5 & 5 & 0 & 0 & 0 & 0 & 0 \\ 0 & 0 & 0 & -4 & 0 & 0 & 9 & -4 & -1 & 0 & 0 \\ 0 & 0 & 0 & 0 & 0 & 0 & -4 & 4 & 0 & 0 & 0 \\ 0 & 0 & 0 & 0 & 0 & 0 & -1 & 0 & 11 & -2 & -8 \\ 0 & 0 & 0 & 0 & 0 & 0 & 0 & 0 & -2 & 2 & 0 \\ 0 & 0 & 0 & 0 & 0 & 0 & 0 & 0 & -8 & 0 & 8 \end{bmatrix}$$

A straightforward computation shows that

$$(L[4])^{-1} = \begin{bmatrix} \frac{1}{2} & \frac{1}{6} & \frac{1}{6} & 0 & 0 & 0 & 0 & 0 & 0 & 0 \\ \frac{1}{6} & \frac{7}{6} & \frac{1}{6} & 0 & 0 & 0 & 0 & 0 & 0 & 0 \\ \frac{1}{6} & \frac{1}{6} & \frac{1}{6} & 0 & 0 & 0 & 0 & 0 & 0 & 0 \\ 0 & 0 & 0 & \frac{1}{2} & \frac{1}{2} & 0 & 0 & 0 & 0 & 0 \\ 0 & 0 & 0 & \frac{1}{2} & \frac{7}{10} & 0 & 0 & 0 & 0 & 0 \\ 0 & 0 & 0 & 0 & 0 & \frac{1}{4} & \frac{1}{4} & \frac{1}{4} & \frac{1}{4} & \frac{1}{4} \\ 0 & 0 & 0 & 0 & 0 & \frac{1}{4} & \frac{1}{2} & \frac{1}{4} & \frac{1}{4} & \frac{1}{4} \\ 0 & 0 & 0 & 0 & 0 & \frac{1}{4} & \frac{1}{2} & \frac{5}{4} & \frac{5}{4} & \frac{5}{4} \\ 0 & 0 & 0 & 0 & 0 & \frac{1}{4} & \frac{1}{2} & \frac{5}{4} & \frac{7}{4} & \frac{5}{4} \\ 0 & 0 & 0 & 0 & 0 & \frac{1}{4} & \frac{1}{2} & \frac{5}{4} & \frac{5}{4} & \frac{11}{8} \end{bmatrix}$$

As in the previous example, since vertex 4 has three branches, $(L[\overline{4}])^{-1}$ has three diagonal blocks. However, since the graph here is weighted, the $(i,j)$ entry is the sum of the reciprocals of the weights of the edges that lie simultaneously on the path from $i$ to 4 and the path from $j$ to 4. For example, consider the $(10,11)$ entry again. As in the previous example, the edges that lie on both the path from 10 to 4 and the path from 11 to 4 are 4-7 and 7-9. Since the weight of edge 4-7 is 4 and the weight of edge 7-9 is 1, it follows that the $(10,11)$ entry is $1/4+1/1 = 5/4$.

Recall from Theorem 1.3.11 that the spectral radius of a positive matrix is necessarily an eigenvalue (called the *Perron value*) of that matrix. Hence it follows that the spectral radius of $(L[\overline{k}])^{-1}$ is equal to the Perron value of the bottleneck matrix (matrices) for a branch at $k$ whose Perron value is largest. This leads us to the following definition:

**DEFINITION 6.2.10** A branch at a vertex $k$ whose bottleneck matrix has the largest Perron value is a *Perron branch at $k$*.

**EXAMPLE 6.2.11** In the tree in Example 6.2.9, observe that the bottleneck matrix for the branch at vertex 4 containing vertices 7,8,9,10,11 is the bottleneck matrix with the largest Perron value. Hence that branch is the Perron branch at vertex 4.

At this point, we can now relate the concept of Perron branches to the concept of the algebraic connectivity of a tree and the characteristic vertex (vertices). We do this in the following theorem from [49]:

**THEOREM 6.2.12** *Let $\mathcal{T}$ be a weighted tree on $n$ vertices $1,\ldots,n$ with Laplacian matrix $L$. Suppose that $i$ and $j$ are adjacent vertices of $\mathcal{T}$. Then $\mathcal{T}$ is a Type II tree with characteristic vertices $i$ and $j$ if and only if the following condition holds:*
*There exists $\gamma$ where $0 < \gamma < 1$ such that*

$$\rho\left(M_1 - \gamma\left(\frac{1}{\theta}\right)J\right) = \rho\left(M_2 - (1-\gamma)\left(\frac{1}{\theta}\right)J\right)$$

*where $\theta$ is the weight of the edge between $i$ and $j$, $M_1$ is the bottleneck matrix for the branch at $j$ containing $i$, and $M_2$ is the bottleneck matrix for the branch at $i$ containing $j$.*

*Moreover, in the case that the condition holds, it follows that*

$$\frac{1}{a(\mathcal{T})} = \rho\left(M_1 - \gamma\left(\frac{1}{\theta}\right)J\right) = \rho\left(M_2 - (1-\gamma)\left(\frac{1}{\theta}\right)J\right).$$

*Additionally, any eigenvector of $L$ corresponding to $a(\mathcal{T})$ can be permuted so it has the form $[-v_1|v_2]^T$, where $v_1$ is the Perron vector for $M_1 - \gamma(1/\theta)J$ and $v_2$ is the Perron vector for $\rho(M_2 - (1-\gamma)(1/\theta)J)$ such that $e^T v_1 = e^T v_2$.*

**Proof**: Without loss of generality, we can permute $L$ so it is of the form

$$L = \left[\begin{array}{c|c} M_1^{-1} & -\theta E_{k,1} \\ \hline -\theta E_{k,1}^T & M_2^{-1} \end{array}\right] \quad (6.2.1)$$

where $E_{k,1}$ is the $k \times (n-k)$ matrix with the $(k,1)$ entry being 1 and all other entries being 0. We should note that the partitioning is such that the last row of $M_1^{-1}$ corresponds to vertex $i$, while the first row of $M_2^{-1}$ corresponds to vertex $j$.

If $\mathcal{T}$ is a Type II tree with characteristic vertices $i$ and $j$, then it follows from Case (a) of Theorem 6.1.10 and Corollary 6.2.1 that any eigenvector of $L$ corresponding to $a := a(\mathcal{T})$ can be written as $[-v_1|v_2]^T$ where both $v_1$ and $v_2$ are positive vectors. Since by Theorem 6.1.2, the sum of the entries of $[-v_1|v_2]^T$ must be zero, we necessarily have $e^T v_1 = e^T v_2$.

Let the last entry of $v_1$ be $\beta$ and the first entry of $v_2$ be $\alpha$. From the eigenvalue-eigenvector relationship, we find that

$$-M_1^{-1} v_1 - \alpha \theta e_k = -a v_1 \quad \text{and} \quad M_2^{-1} v_2 + \beta \theta e_1 = a v_2.$$

Rewriting each of these equations we obtain

$$M_1^{-1} v_1 = a v_1 - \alpha \theta e_k \quad \text{and} \quad M_2^{-1} v_2 = a v_2 - \beta \theta e_1$$

which we further rewrite as

$$v_1 = a M_1 v_1 - \alpha \theta M_1 e_k \quad \text{and} \quad v_2 = a M_2 v_2 - \beta \theta M_2 e_1.$$

From Theorem 6.2.5, we know that the last column of $M_1$ and the first column of $M_2$ are each $(1/\theta)e$. Consequently, we have

$$v_1 = a M_1 v_1 - \alpha e \quad \text{and} \quad v_2 = a M_2 v_2 - \beta e$$

which yields

$$M_1 v_1 - \frac{\alpha}{a} e = \frac{1}{a} v_1 \quad \text{and} \quad M_2 v_2 - \frac{\beta}{a} e = \frac{1}{a} v_2. \qquad (6.2.2)$$

Looking at the last entry in the vectors of the first equation and the first entry of the vectors in the second equation, we obtain

$$e_k^T(M_1 v_1) - \frac{\alpha}{a} = \frac{\beta}{a} \quad \text{and} \quad e_1^T(M_2 v_2) - \frac{\beta}{a} = \frac{\alpha}{a}.$$

Thus

$$\frac{\alpha + \beta}{a} = e_k^T(M_1 v_1) = e_1^T(M_2 v_2)$$

which by Theorem 6.2.5, yields

$$\frac{\alpha + \beta}{a} = \frac{1}{\theta} e^T v_1 = \frac{1}{\theta} e^T v_2.$$

Hence

$$\frac{1}{a} = \frac{(1/\theta) e^T v_1}{\alpha + \beta} = \frac{(1/\theta) e^T v_2}{\alpha + \beta}.$$

Combining this with (6.2.2), we obtain

$$M_1 v_1 - \frac{\alpha}{\alpha + \beta}\left(\frac{1}{\theta}\right) J v_1 = \frac{1}{a} v_1 \quad \text{and} \quad M_2 v_2 - \frac{\beta}{\alpha + \beta}\left(\frac{1}{\theta}\right) J v_2 = \frac{1}{a} v_2. \qquad (6.2.3)$$

From Theorem 6.2.5, it follows that $M_1 \geq (1/\theta)J$ and $M_2 \geq (1/\theta)J$, since the edge joining $i$ and $j$ is on every path which we use in computing the formula (given in Theorem 6.2.5) for the entries of $M_1$ and $M_2$. Therefore, setting $\gamma = \frac{\alpha}{\alpha+\beta}$ and observing that $0 < \gamma < 1$, we see that $M_1 - \gamma(\frac{1}{\theta})J$ and $M_2 - (1-\gamma)(\frac{1}{\theta})J$ are positive matrices. By (6.2.3), we see that $v_1$ is a positive eigenvector of the positive matrix $M_1 - \gamma(\frac{1}{\theta})J$, corresponding to the eigenvalue $1/a$. Thus by Theorem 1.3.11, we have that $1/a = \rho(M_1 - \gamma(\frac{1}{\theta})J)$. Similarly from (6.2.3), we obtain $1/a = \rho(M_2 - (1-\gamma)(\frac{1}{\theta})J)$. This completes the proof of the first direction of the theorem.

Conversely, assume that there exists $\gamma$ where $0 < \gamma < 1$ such that $\rho(M_1 - \gamma(\frac{1}{\theta})J) = \rho(M_2 - (1-\gamma)(\frac{1}{\theta})J)$, and define $a'$ by

$$\frac{1}{a'} = \rho\left(M_1 - \gamma\left(\frac{1}{\theta}\right)J\right) = \rho\left(M_2 - (1-\gamma)\left(\frac{1}{\theta}\right)J\right). \quad (6.2.4)$$

Let $v_1$ and $v_2$ be Perron vectors of $M_1 - \gamma(\frac{1}{\theta})J$ and $M_2 - (1-\gamma)(\frac{1}{\theta})J$, respectively, normalized so that $e^T v_1 = e^T v_2$. By the eigenvalue-eigenvector relationship, we have that

$$M_1 v_1 - \gamma \frac{1}{\theta} J v_1 = \frac{1}{a'} v_1. \quad (6.2.5)$$

Looking at the $k^{th}$ entry of each vector, we obtain

$$e_k^T M_1 v_1 - e_k^T \gamma \frac{1}{\theta} J v_1 = \frac{1}{a'} e_k^T v_1.$$

Using the fact that $e_k^T M_1 v_1 = \frac{1}{\theta} e^T v_1$ and that $e^T v_1 = e^T v_2$, the above equation yields $(1-\gamma)(1/\theta)e^T v_2 = (1/a')e_k^T v_1$. Similarly, using the last portion of the equation in (6.2.4), we obtain that $\gamma(1/\theta)e^T v_1 = (1/a')e_1^T v_2$, i.e., $\gamma(\frac{1}{\theta})Jv_1 = \frac{1}{a'}ee_1^T v_2$. Using (6.2.5) we obtain

$$\begin{aligned}\frac{1}{a'}v_1 &= M_1 v_1 - \gamma\left(\frac{1}{\theta}\right)Jv_1 \\ &= M_1 v_1 - \frac{1}{a'}ee_1^T v_2 \\ &= M_1 v_1 - \frac{1}{a'}\theta M_1 e_k e_1^T v_2\end{aligned}$$

which yields $-M_1^{-1}v_1 - \theta E_{k,1}v_2 = -a'v_1$. Similarly, we have $M_2^{-1}v_2 + \theta E_{1,k}v_1 = a'v_2$. Using (6.2.1), this shows that $[-v_1|v_2]^T$ is an eigenvector of $L$ corresponding to the eigenvalue $a'$. It remains to show that the algebraic connectivity, $a$, of $\mathcal{T}$ is in fact equal to $a'$.

Recall that the last row of $M_1^{-1}$ corresponds to vertex $i$ and the first row of $M_2^{-1}$ corresponds to vertex $j$ (where $i$ and $j$ are adjacent). Thus if we have vertices $b$ and $c$, both in the branch of $\mathcal{T}$ at $j$ containing $i$, and if $c$ is on the path from $i$ to $b$, then from Theorem 6.2.5 we see that row $b$ of $M_1$ dominates row $c$ of $M_1$ entrywise, with strict domination in at least one position. Hence by the relationship $M_1 v_1 = a' v_1$, it follows that the entries of $v_1$ are strictly increasing along any path which starts at $i$, and similarly the entries of $v_2$ are strictly increasing along any path which starts at $j$. By multiplying all of the weights of the edges of $\mathcal{T}$ by $a'/a$, we see there is a weighting of a tree on which $\mathcal{T}$ is based, say $\mathcal{T}'$, which has algebraic connectivity $a'$ and corresponding eigenvector $[-v_1|v_2]^T$. In other words, there exists a symmetric M-matrix $L'$ with row sums of zero and the same graph as $L$ such that $L'[-v_1|v_2]^T = a'[-v_1|v_2]^T$ where $a'$ is the smallest positive eigenvalue of $L'$. To complete the proof, we need to show that $L = L'$. We do this by induction on the number of vertices. Clearly, $L = L'$ if $n = 2$. Now suppose that $n \geq 3$. Then at least one of $k$ and $n-k$ is at least 2. Without loss of generality, let $k \geq 2$. Thus the branch at $j$ containing $i$ has at least two vertices, one of which is pendant. Let $i_1$ be such a pendant vertex, adjacent to $i_2$. Let the weight of the edge joining $i_1$ to $i_2$ in $\mathcal{T}$ be $\beta$, and let the weight of such an edge in $\mathcal{T}'$ be $\beta'$. From the eigenvalue-eigenvector relationship for $L$ and $L'$, we find that since $a' e_i^T v_1 = e_{i_1}^T L[-v_1|v_2]^T = e_{i_1}^T L'[-v_1|v_2]^T$, then we must have $\beta(e_{i_1}^T - e_{i_2}^T)v_1 = \beta'(e_{i_1} - e_{i_2})v_1$. But since the entries in $v_1$ are increasing along paths starting from $i$, and since $i_1$ is pendant, we find that $(e_{i_1}^T - e_{i_2}^T)v_1 > 0$, and hence $\beta = \beta'$. By the inductive hypothesis, we necessarily have that $L = L'$. Therefore $\mathcal{T}$ has algebraic connectivity $a'$, and that $\mathcal{T}$ is a Type II tree with characteristic vertices $i$ and $j$. □

**EXAMPLE 6.2.13** Consider the tree $\mathcal{T}$ from Example 6.2.8. It can be calculated that $a(\mathcal{T}) = 0.173$ and that the corresponding Fiedler vector is

$$v = [1.48, 1.48, 0.511, 1.24, 0.827, 1, -0.619, -0.712, -1.52, -1.84, -1.84]^T.$$

Since no entry of $v$ is zero, we see that $\mathcal{T}$ is a Type II tree. Vertices 4 and 7 are the only pair of adjacent vertices whose valuations have opposite signs. Therefore these are the characteristic vertices. The bottleneck matrix $M_1$ at vertex 4 was computed in Example 6.2.8. We can compute the bottleneck matrix $M_2$ for vertex 7 in a similar fashion. The weight joining vertices 4 and 7 is $\theta = 1$. By doing some computations, we see that $\gamma = 0.341$, i.e., $\rho(M_1 - 0.341J) = \rho(M_2 - 0.659J) = 1/a(\mathcal{T})$.

We now prove a corollary (see [49]) to Theorem 6.2.12:

**COROLLARY 6.2.14** *Suppose that $\mathcal{T}$ is a weighted tree. Then $\mathcal{T}$ is a Type II tree with characteristic vertices $i$ and $j$ if and only if $i$ and $j$ are adjacent and the branch at vertex $i$ containing vertex $j$ is the unique Perron branch at $i$, while the branch at vertex $j$ containing vertex $i$ is the unique Perron branch at $j$.*

**Proof**: Using the notation of Theorem 6.2.12, $\mathcal{T}$ is a Type II tree with characteristic vertices $i$ and $j$ if and only if there exists $\gamma$ where $0 < \gamma < 1$ such that $\rho(M_1 - \gamma(1/\theta)J) = \rho(M_2 - (1-\gamma)(1/\theta)J)$. Observe that $\rho(M_1 - x(1/\theta)J)$ is a decreasing and continuous function of $x$ for $0 \leq x \leq 1$. Hence such a $\gamma$ exists if and only if $\rho(M_1) > \rho(M_2 - (1/\theta)J)$ and $\rho(M_1 - (1/\theta)J) < \rho(M_2)$. From Theorem 6.2.5, it follows that $M_2 - (1/\theta)J$ is (permutationally similar to) the block diagonal matrix whose diagonal blocks are the bottleneck matrices corresponding to the branches of $\mathcal{T}$ at $j$ which do not contain $i$. Therefore, if $\rho(M_1) > \rho(M_2 - (1/\theta)J)$, then the bottleneck matrix at the branch at vertex $j$ containing vertex $i$ has larger spectral radius than the bottleneck matrices of all other branches at vertex $j$. Hence the branch at $j$ containing $i$ is the unique Perron branch at $j$. A similar idea holds for the inequality $\rho(M_1 - (1/\theta)J) < \rho(M_2)$, thus completing the proof. □

We now shift our focus to Type I trees and prove an important theorem from [49] concerning such trees:

**THEOREM 6.2.15** *Let $\mathcal{T}$ be a weighted tree on vertices $1,\ldots,n$ with Laplacian matrix $L$. Then $\mathcal{T}$ is a Type I tree with characteristic vertex $k$ if and only if there are two or more Perron branches of $\mathcal{T}$ at $k$. Moreover, in that case, the algebraic connectivity of $\mathcal{T}$ is $1/\rho(L[\bar{k}]^{-1})$, and if $v$ is an eigenvector corresponding to the algebraic connectivity, then $v$ can be permuted and partitioned so that each of the resulting nonzero subvectors is a Perron vector for the bottleneck matrix of a Perron branch at $k$.*

**Proof**: First, suppose that no vertex of $\mathcal{T}$ has at least two Perron branches, i.e., each vertex in $\mathcal{T}$ has a unique Perron branch. We claim then that $\mathcal{T}$ is a Type II tree. To see this, select a vertex $i_1$ of $\mathcal{T}$. Let $i_2$ be the vertex adjacent to $i_1$ that is on the unique Perron branch at $i_1$. Inductively, for each $j \geq 2$, let $i_{j+1}$ be the vertex adjacent to $i_j$ that is on the unique Perron branch at $i_j$. The sequence $S: i_1, i_2, i_3, \ldots$ defines a walk in $\mathcal{T}$. Since $\mathcal{T}$ has a finite number of vertices, our sequence of vertices must contain a closed walk, hence a cycle. But since $\mathcal{T}$ is a tree, the only possible cycles are those of length 2. Therefore, for some $j$, we have $i_{j+2} = i_j$. But by our construction of $S$, it follows that the branch at $i_j$ containing $i_{j+1}$ is the unique Perron branch at $i_j$, while the branch at $i_{j+1}$ containing $i_{j+2} = i_j$ is the unique Perron branch at $i_{j+1}$. Hence by Corollary 6.2.14, it follows that $\mathcal{T}$ is a Type II tree with characteristic vertices $i_j$ and $i_{j+1}$.

Now suppose there are two Perron branches, $B_1$ and $B_2$, of $\mathcal{T}$ at vertex $k$. Let vertices $a_1 \in B_1$ and $a_2 \in B_2$ be vertices both adjacent to $k$. Since $L[\bar{k}]^{-1}$ is (permutationally similar to) a block diagonal matrix where each (positive) block is a bottleneck matrix corresponding to a branch at $k$, it follows that $L[\bar{k}]^{-1}$ has an eigenvector $w$ corresponding to $\rho(L[\bar{k}]^{-1})$ which can permuted to the form

$$w = \begin{bmatrix} -v_1 \\ v_2 \\ 0 \end{bmatrix}$$

where $v_1$ and $v_2$ are the Perron vectors of the bottleneck matrices for $B_1$ and $B_2$. Normalize $w$ so that the product of the weight of the edge between $a_1$ and $k$ with the entry in $v_1$ corresponding to $a_1$ is equal to the product of the weight of the edge between $a_2$ and $k$ with the entry in $v_2$ corresponding to $a_2$. With this normalization, it follows that by appending a 0 to $w$ in position $k$, we can generate an eigenvector of $L$ corresponding to the eigenvalue $1/\rho(L[\bar{k}]^{-1})$. It follows from Theorem 1.2.8 that $a(\mathcal{T}) \geq 1/\rho(L[\bar{j}]^{-1})$ for any vertex $j$. Furthermore, since $a(\mathcal{T})$ is the smallest positive eigenvalue of $L$, it follows that $a(\mathcal{T}) \leq 1/\rho(L[\bar{k}]^{-1})$. Therefore $a(\mathcal{T}) = 1/\rho(L[\bar{k}]^{-1})$. Since $k$ is adjacent to a vertex whose corresponding entry in $w$ is nonzero, it follows from Corollary 6.2.1 that $\mathcal{T}$ is of Type I with $k$ as its characteristic vertex.

Finally, we note that if $v$ is an eigenvector of $L$ corresponding to $a(\mathcal{T})$, then since $k$ is the unique characteristic vertex, it follows from Corollary 6.2.1 that $v_k = 0$. Hence the subvector of $v$ formed by deleting $v_k$ is an eigenvector for $L[\bar{k}]^{-1}$ corresponding to $\rho(L[\bar{k}]^{-1})$. Moreover, since $L[\bar{k}]^{-1}$ is (permutationally similar to) a block diagonal matrix with each block corresponding to a bottleneck matrix at a branch at $k$, it follows that $v$ can be permuted and partitioned so that the resulting nonzero subvectors are Perron vectors for the bottleneck matrices of Perron branches at $k$. □

**EXAMPLE 6.2.16** Consider the tree $\mathcal{T}$ below.

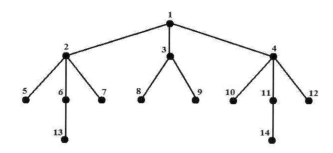

It can be calculated that $a(\mathcal{T}) = 0.151$ and that the corresponding Fiedler vector is

$$v = [0, -0.570, 0, 0.570, -0.672, -0.849, -0.672, 0, 0, 0.672, 0.849, 0.672, -1, 1]^T.$$

Since the characteristic valuation of vertex 1 is zero, and this vertex is the only vertex of valuation zero that is adjacent to vertices of nonzero valuation (namely vertices 2 and 4), it follows that $\mathcal{T}$ is a Type I tree with characteristic vertex 1. Observe that the bottleneck matrices, $M_1$ and $M_2$, for the branches at vertex 1 containing vertices 2 and 4, respectively, in each case are

$$M_1 = M_2 = \begin{bmatrix} 1 & 1 & 1 & 1 & 1 \\ 1 & 2 & 1 & 1 & 1 \\ 1 & 1 & 2 & 1 & 1 \\ 1 & 1 & 1 & 2 & 2 \\ 1 & 1 & 1 & 2 & 3 \end{bmatrix}$$

while the bottleneck matrix for the branch at vertex 1 containing vertex 3 is

$$M_3 = \begin{bmatrix} 1 & 1 & 1 \\ 1 & 2 & 1 \\ 1 & 1 & 2 \end{bmatrix}$$

Since $\rho(M_1) = \rho(M_2) > \rho(M_3)$, it follows that the branches containing vertices 2 and 4 are the Perron branches of $\mathcal{T}$ at vertex 1. Also observe that $\rho(M_1) = \rho(M_2) = 6.61 = 1/a(\mathcal{T})$.

We can combine Corollary 6.2.14 and Theorem 6.2.15 to obtain the following important theorem from [49]:

**THEOREM 6.2.17** *Let $\mathcal{T}$ be a weighted tree. $\mathcal{T}$ is a Type I tree if and only if there is exactly one vertex at which there are two or more Perron branches. $\mathcal{T}$ is a Type II tree if and only if at each vertex there is a unique Perron branch.*

We now present a theorem from [35] which gives us useful information about the number of Perron branches that a Type $I$ tree has. For notational simplicity, we will write $y(v)$ instead of $y_v$ to denote the characteristic valuation of vertex $v$.

**THEOREM 6.2.18** *Let $\mathcal{T}$ be a Type I tree with characteristic vertex $w$ and algebraic connectivity $a$. Let $L$ be a Laplacian matrix for $\mathcal{T}$. Suppose that as an eigenvalue of $L$, $a$ has multiplicity $k-1$. Then $w$ has exactly $k$ Perron branches.*

**Proof**: Let $y$ be an eigenvector of $L$ corresponding to $a$ such that $y(v) = 0$ if and only if $v = w$ or $v$ is a vertex on a non-Perron branch at $w$. Let $B_1, B_2, \ldots, B_k$ be the Perron branches at $w$ and $B_{k+1}, \ldots, B_d$ be the non-Perron branches. We will show that $a$ has a multiplicity of $k-1$ and an eigenvalue of $L$. Without loss of generality, we can order the vertices of $\mathcal{T}$ so that $L$ is of the form

$$\begin{bmatrix} L[B_1] & 0 & \cdots & 0 & z_1 \\ 0 & L[B_2] & \cdots & 0 & z_2 \\ \cdots & \cdots & \cdots & \cdots & \cdots \\ 0 & 0 & \cdots & L[B_d] & z_d \\ z_1^T & z_2^T & \cdots & z_d^T & d \end{bmatrix}$$

where for each $1 \leq i \leq d$, the vector $z_i$ is the column vector of appropriate length whose only nonzero entry is $-1$ in the last row. Further, for each $1 \leq i \leq d$, let $r_i$ be the vertex in $B_i$ that is adjacent to $w$ and let the vertices in $B_i$ be labeled so that $r_i$ is labeled the largest of all vertices in $B_i$. We see that for each $B_i$, $1 \leq i \leq k$, there exists a characteristic valuation $f_i \in \xi(\mathcal{T})$ such that $f_i(r_i) \neq 0$. Thus by Corollary 6.2.1, we may assume $f_i(r_i) = 1$ and $f_i(v) > 1$ for all vertices $v \in B_i$ where $v \neq r_i$. Again by Corollary 6.2.1, there must exist an eigenvector $f \in \xi(\mathcal{T})$ such that $f(v) = 0$ if and only if $v = w$ or $v$ lies on a non-Perron branch at $w$. Hence we may partition $f$ such that

$$f = [x^{(1)}, x^{(2)}, \ldots, x^{(k)}, 0, \ldots, 0]^T$$

where each $x^{(i)}$ corresponds to the branch $B_i$. By Corollary 6.2.1 each $x^{(i)}$ is such that either $x^{(i)}$ is strictly positive or $x^{(i)}$ is strictly negative. For each $1 \leq i \leq k$, create the vector $y^{(i)}$ by rescaling each $x^{(i)}$ so that the entry in the $r_i$ position is 1 (this is always possible because there are no zero entries in $x^{(i)}$). Observe for each $i$ that $y^{(i)}$ is a positive eigenvector of $L[B_i]$ corresponding to $a$. Partitioning in the same manner as $f$ is partitioned, define

$$g_i = [y^{(1)}, 0, 0, \ldots, 0, -y^{(i)}, 0, 0, \ldots, 0]^T$$

for $2 \leq i \leq k$. For each $i$, observe by matrix multiplication that $Lg_i = ag_i$. Thus $\{g_2, \ldots, g_k\}$ is a linearly independent set of $k-1$ eigenvectors of $L$ corresponding to $a$.

It remains to show that any other eigenvector of $L$ corresponding to $a$ is in the span of $\{g_2, \ldots, g_k\}$. Let $h \in \xi(\mathcal{T})$ be an arbitrary but fixed eigenvector corresponding to $a$. Then $h(v) = 0$ for all vertices such that $v = w$ or $v \in B_j$ for $k+1 \leq j \leq d$. Thus we can partition $h$ in accordance to the branches at $w$:

$$h = [u^{(1)}, u^{(2)}, \ldots, u^{(k)}, 0, \ldots, 0]^T.$$

It follows from Corollary 6.2.1 that for a fixed $1 \leq i \leq k$ that exactly one of the following holds: all entries in $u^{(i)}$ are zero, all entries in $u^{(i)}$ are positive, or all entries in $u^{(i)}$ are negative. Moreover, since $Lh = ah$ and since each $L[B_i]$ is an irreducible M-matrix, it follows by block multiplication that $u^{(i)}$ is an eigenvector of $L[B_i]$ corresponding to the simple eigenvalue $a$. Thus there exist constants $c_i$ such that $u^{(i)} = c_i y^{(i)}$, $1 \leq i \leq k$. Considering the product $Lh = ah$ and looking at the portion of the matrix multiplication when we take the dot product of the last row of $L$ with $h$, we see that since the last entry in each $y^{(i)}$ is 1, then the sum of the $c_i$'s must be zero. Therefore

$$-c_1 y^{(1)} = \sum_{i=2}^{k} c_i y^{(i)}.$$

If we consider the sum

$$c_2 g_2 + c_3 g_3 + \ldots c_k g_k$$

we see that

$$h = -c_2 g_2 - c_3 g_3 - \ldots - c_k g_k.$$

So $\{g_2, \ldots, g_k\}$ spans $\xi(\mathcal{T})$.

Since we have shown that $\{g_2, \ldots, g_k\}$ is a linearly independent set of eigenvectors that spans $\xi(\mathcal{T})$, it follows that the algebraic multiplicity of $a$ as an eigenvalue of $L$ is $k-1$. □

We have now completed our investigation of characteristic vertices of weighted trees. We now consider vertices of weighted trees that are not characteristic vertices. The following theorem from [49] helps us understand the structure behind the Perron branches at non-characteristic vertices.

**THEOREM 6.2.19** *Let $\mathcal{T}$ be a weighted tree on vertices $1, \ldots, n$ and suppose that $m$ is not a characteristic vertex of $\mathcal{T}$. Then the unique Perron branch of $\mathcal{T}$ at $m$ is the branch which contains all of the characteristic vertices of $\mathcal{T}$.*

**Proof:** Since $m$ is not a characteristic vertex of $\mathcal{T}$, then by Theorem 6.2.15 there is a unique Perron branch at $m$. Let $m_1$ be the vertex adjacent to $m$ on that Perron branch. For $i \geq 1$, consider the following inductively constructed path $P$: If there are two or more Perron branches at $m_i$, then stop the construction. If there is a unique Perron branch at $m_i$, then let $m_{i+1}$ be the vertex adjacent to $m_i$ on that Perron branch, provided that $m_{i+1}$ is distinct from $m_1, \ldots m_i$; otherwise, stop the construction.

Since $\mathcal{T}$ has only finitely many vertices and since our construction does not allow vertices to be repeated, the construction must terminate. Observe that the construction must terminate in one of two ways: (1) $m_i$ has at least two Perron branches or (2) $m_{i+1}$ has been encountered before. In Case (1), the vertex $m_i$ is the characteristic vertex of the Type I tree $\mathcal{T}$. In Case (2), since $\mathcal{T}$ has no cycles, it follows that $m_{i+1} = m_{i-1}$. Thus $m_{i-1}$ and $m_i$ are the characteristic vertices of the Type II tree $\mathcal{T}$. In both cases, $P$ is a path in the Perron branch at $m$ that begins at $m$ and ends at the characteristic vertex (vertices) of $\mathcal{T}$. Thus the Perron branch at $m$ contains the characteristic vertices of $\mathcal{T}$. □

Continuing with the theme of valuations and Perron branches at non-characteristic vertices, we saw in the proof of Theorem 6.1.10 that in a tree $\mathcal{T}$, if we traverse a path beginning at a characteristic vertex such that the path does not contain the other characteristic vertex of $\mathcal{T}$ (if one exists), then the valuations of the vertices either increase or decrease as we traverse such a path. We close this section with the following theorem from [49] where we gain more insight into the rate at which the valuations increase or decrease in the case where $\mathcal{T}$ is unweighted.

**THEOREM 6.2.20** *Let $\mathcal{T}$ be an unweighted tree with Laplacian matrix $L$ and algebraic connectivity $a := a(\mathcal{T})$.*

*(a) If $\mathcal{T}$ is a Type I tree with characteristic vertex $k$, and $y$ is an eigenvector of $L$ corresponding to $a$, then along any path in $\mathcal{T}$ beginning at $k$, the corresponding entries in $y$ are either increasing and concave down, decreasing and concave up, or identically zero.*

*(b) If $\mathcal{T}$ is a Type II tree with characteristic vertices $i$ and $j$, and $y$ is an eigenvector of $L$ corresponding to $a$, then one of two things occur: either (i) the corresponding entries in $y$ are increasing and concave down along any path starting at $i$ and not including $j$, while the corresponding entries in $y$ are decreasing and concave up along any path starting at $j$ and not including $i$, or (ii) the corresponding entries in $y$ are decreasing and concave up along any path starting at $i$ and not including $j$, while the corresponding entries in $y$ are increasing and concave down along any path starting at $j$ and not including $i$.*

**Proof**: Let $\mathcal{T}$ be a Type I tree with characteristic vertex $k$, and choose a path $Q$ starting at $k$ along which the entries in $y$ are all positive. Suppose that $i$, $i+1$, and $i+2$ are consecutive vertices on such a path, with vertex $i$ being closest to vertex $k$. Since we already saw in the proof of Theorem 6.2.12 that the values of such vertices in $y$ are increasing, we want to show that $y_{i+2} - y_{i+1} < y_{i+1} - y_i$ to establish the fact that the values are increasing in a concave down fashion. To that end, let $M$ be the bottleneck matrix for the branch $B$ at $k$ containing $Q$. Recall from Theorem 6.2.15 that the subvector $w$ of $y$ corresponding to the vertices of $B$ is a Perron vector for $M$. Thus $Mw = (1/a)w$. From Theorem 6.2.12, it follows that the entry in position $j$ of the row vector $(e_{i+1}^T - e_i^T)M$ is the number of edges (since $\mathcal{T}$ is unweighted) which are on the path from $i+1$ to $j$ but not on the path from $i$ to $j$. Since $i$ and $i+i$ are consecutive vertices on $Q$, this number is 1 if the path from $j$ to $k$ goes through $i+1$, and 0 if not. Letting $S_{i+1}$ be the set of vertices whose path to $k$ goes through $i+1$, we see that the $j^{th}$ entry of $(e_{i+1}^T - e_i^T)M$ is 1 if and only if $j \in S_{i+1}$, and 0 otherwise. Using the fact that $Mw = (1/a)w$ and that $w$ is a positive subvector of $y$, it follows that

$$y_{i+1} - y_i = \frac{1}{a} \sum_{j \in S_{i+1}} y_j.$$

Similarly,

$$y_{i+2} - y_{i+1} = \frac{1}{a} \sum_{j \in S_{i+2}} y_j.$$

Since $S_{i+2}$ is a proper subvector of $S_{i+1}$, it follows that $y_{i+2} - y_{i+1} < y_{i+1} - y_i$. Hence on a branch of $\mathcal{T}$ at $k$, the corresponding positive entries in $y$ are increasing and concave down. Since $-y$ is also an eigenvector of $L$ corresponding to $a$, we see that on a branch of $\mathcal{T}$ at $k$, the corresponding negative entries in $y$ are decreasing and concave up.

If $\mathcal{T}$ is a Type II tree, then by appealing to Theorem 6.2.12, we can apply a similar argument to the matrix $M - \gamma J$ for the appropriate $0 < \gamma < 1$ in order to establish a similar result. □

## Exercises:

**1.** Find the bottleneck matrix for the tree in Example 6.2.8 at vertex 7.

**2.** Find the bottleneck matrix for the weighted tree in Example 6.2.9 at vertex 7.

**3.** (See [49]) Suppose that $\mathcal{T}_1$ and $\mathcal{T}_2$ are weighted trees and let $\mathcal{T}_3$ be the weighted tree which results from identifying a vertex of $\mathcal{T}_1$ with a vertex of $\mathcal{T}_2$. Prove that any characteristic vertices lie on the path joining the characteristic vertices of $\mathcal{T}_1$ to the characteristic vertices of $\mathcal{T}_2$.

**4.** (See [49]) Take two copies of a weighted tree $\mathcal{T}$ and create a new tree $\mathcal{T}'$ by joining a vertex of one copy of $\mathcal{T}$ to the corresponding vertex of the other copy of $\mathcal{T}$ by a weighted edge. Prove that $\mathcal{T}'$ is of Type II. In the language of Theorem 6.2.12, find $\gamma$.

**5.** (See [49]) Consider the sequence of Fibonacci trees $\mathcal{T}_i$, $i \geq 1$, where $\mathcal{T}_1$ consists of a single vertex $v_1$, $\mathcal{T}_2$ consists of a single vertex $v_2$, $\mathcal{T}_3$ is formed by taking a new vertex $v_3$ and making it adjacent to both $v_1$ and $v_2$, and $\mathcal{T}_i$ ($i \geq 4$) is formed by taking a new vertex $v_i$ and making it adjacent to vertices $v_{i-1}$ of $\mathcal{T}_{i-1}$ and $v_{i-2}$ of $\mathcal{T}_{i-2}$. Show that for $i \geq 4$, $\mathcal{T}_i$ is a Type II tree with characteristic vertices $v_{i-1}$ and $v_i$.

**6.** (See [35]) Let $\mathcal{T}$ be a Type I tree with characteristic vertex $v$. Create a tree $\mathcal{T}'$ by adding a vertex $p$ adjacent to $v$. Show that $\mathcal{T}'$ is a Type I tree with characteristic vertex $v$.

**7.** (See [35]) Let $\mathcal{T}$ be a Type I tree with characteristic vertex $w$ and algebraic connectivity $a$. Let $B$ be a non-Perron branch at $w$. Prove that $\lambda_1(L[B]) > a$ where $\lambda_1$ denotes the smallest eigenvalue of a matrix.

## 6.3 Excursion: Nonisomorphic Branches in Type I Trees

In Example 6.2.16, we saw an unweighted Type I tree in which the Perron branches at the characteristic vertex were isomorphic. From Theorem 6.2.17, it is clear that the easiest way to construct a Type I tree $\mathcal{T}$ is to attach two isomorphic trees to a vertex $v$. Then $v$ has two Perron branches and hence $v$ is the unique characteristic vertex of $\mathcal{T}$, thus making $\mathcal{T}$ a Type I tree. In this section based largely on [42], we show how to construct unweighted Type I trees in which the Perron branches at the characteristic vertex are not isomorphic. To this end, we first recall a matrix-theoretic definition.

**DEFINITION 6.3.1** *For an $m \times n$ martix $A$ and any matrix $B$, the Kronecker product $A \otimes B$ is defined as*

$$A \otimes B = \begin{bmatrix} a_{11}B & \cdots & a_{1n}B \\ \cdots & & \cdots \\ a_{m1}B & \cdots & a_{mn}B \end{bmatrix}.$$

We now begin with a useful lemma from [42]. In this lemma, let $J(k)$ denote the $k \times k$ matrix of all ones, and let $e(k)$ denote the $k$-dimensional vector of all ones.

**LEMMA 6.3.2** *Let $\mathcal{G}$ be a connected weighted graph with Laplacian matrix $L$ and let $C$ be a proper subset of vertices of $\mathcal{G}$. Suppose that $C_1 \subseteq C$ and that the vertices of $\mathcal{G}$ are numbered so that those in $C_1$ come before those in $C \setminus C_1$. Partition $L[C]^{-1}$ as*

$$L[C]^{-1} = \begin{bmatrix} L_1 & L_2 \\ L_2^T & L_3 \end{bmatrix}$$

*where the $(1,1)$ block corresponds to the vertices in $C_1$ while the $(2,2)$ block corresponds to the vertices in $C \setminus C_1$. Now form a new graph as follows: for each vertex $v$ of $C_1$, add $j$ new vertices adjacent to $v$, giving each new edge a weight of 1. Let $A$ denote the set of new vertices, and let $\hat{L}$ be the Laplacian matrix of the new graph.*

Then $\hat{L}[A \cup C]^{-1}$ is permutationally similar to

$$\left[\begin{array}{c|c|c} I + L_1 \otimes J(j) & L_1 \otimes e(j) & L_2 \otimes e(j) \\ \hline L_1 \otimes e(j)^T & L_1 & L_2 \\ \hline L_2^T \otimes e(j)^T & L_2^T & L_3 \end{array}\right]$$

where the $(1,1)$, $(2,2)$, and $(3,3)$ blocks correspond to the vertices in $A$, $C_1$, and $C \setminus C_1$, respectively.

**Proof:** Partitioning $L[C]$ as we did $L[C]^{-1}$ in the statement of the lemma, we have

$$L[C] = \left[\begin{array}{c|c} U & V \\ \hline V^T & W \end{array}\right]$$

for appropriate matrices $U$, $V$, and $W$. Partitioning $\hat{L}[A \cup C]$ as we did $\hat{L}[A \cup C]^{-1}$ in the statement of lemma, we have

$$\hat{L}[A \cup C] = \left[\begin{array}{c|c|c} I & -I \otimes e(j) & 0 \\ \hline -I \otimes e(j)^T & U + jI & V \\ \hline 0 & V^T & W \end{array}\right].$$

Using the fact that

$$\left[\begin{array}{c|c} U & V \\ \hline V^T & W \end{array}\right] \left[\begin{array}{c|c} L_1 & L_2 \\ \hline L_2^T & L_3 \end{array}\right] = I,$$

it is a straightforward computation to verify that

$$\left[\begin{array}{c|c|c} I & -I \otimes e(j) & 0 \\ \hline -I \otimes e(j)^T & U + jI & V \\ \hline 0 & V^T & W \end{array}\right] \left[\begin{array}{c|c|c} I + L_1 \otimes J(j) & L_1 \otimes e(j) & L_2 \otimes e(j) \\ \hline L_1 \otimes e(j)^T & L_1 & L_2 \\ \hline L_2^T \otimes e(j)^T & L_2^T & L_3 \end{array}\right] = I.$$

□

To proceed with our creation of Type I trees with nonisomorphic Perron branches, we first inductively create the rooted trees that we will be needing. For positive integers $k_1, \ldots, k_m$, let $\mathcal{T}(k_1)$, be the star on $k_1 + 1$ vertices, and for $m \geq 2$, let $\mathcal{T}(k_1, \ldots, k_m)$ be the tree created from $\mathcal{T}(k_1, \ldots, k_{m-1})$ by adding $k_m$ pendant vertices adjacent to each of the pendant vertices of $\mathcal{T}(k_1, \ldots, k_{m-1})$. For example, the tree $\mathcal{T}(4, 2, 3)$ is:

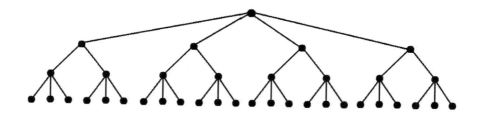

Clearly $\mathcal{T}(k_1,\ldots,k_m)$ is not necessarily isomorphic to $\mathcal{T}(k_m,\ldots,k_1)$. Our goal will be to show that if we create a tree consisting of a vertex $v$ with two branches: $\mathcal{T}(k_1,\ldots,k_m)$ and $\mathcal{T}(k_m,\ldots,k_1)$, then these branches will both be Perron branches. Hence by Theorem 6.2.17 we will have created a Type I tree with nonisomorphic Perron branches at characteristic vertex $v$. Thus we need to prove that the bottleneck matrices for $\mathcal{T}(k_1,\ldots,k_m)$ and $\mathcal{T}(k_m,\ldots,k_1)$ have the same spectral radius. To this end, we prove another lemma and corollary from [42].

**LEMMA 6.3.3** *Let $M$ be a bottleneck matrix of a branch $B$ at vertex $v$ in some weighted tree. Modify $B$ as follows: for each vertex $x \in B$, take a distinct copy of $\mathcal{T}(k_1,\ldots,k_m)$, and identify its root vertex with $x$. Then the bottleneck matrix of the modified branch at $v$ is permutationally similar to the $(m+1) \times (m+1)$ symmetric block matrix where the $(i,i)$ block is*

$$I + \{(\ldots((I + M \otimes J(k_1)) \otimes J(k_2) + I) \otimes J(k_3) \ldots + I)\} \otimes J(k_i),$$

*where $1 \leq i \leq m$, and the $(i,j)$ block is*

$$I + \{(\ldots((I + M \otimes J(k_1)) \otimes J(k_2) + I) \otimes J(k_3) \ldots + I)\} \otimes J(k_i) \otimes e(k_{i+1}) \otimes \ldots \otimes e(k_j),$$

*where $1 \leq i < j \leq m$, and where the $(i, m+1)$ block is $M \otimes e(k_1) \otimes e(k_2) \otimes \ldots \otimes e(k_i)$.*

**Proof**: We will prove this by induction on $m$. When $m = 1$, then by Lemma 6.3.2 it follows that the modified bottleneck matrix is permutationally similar to

$$\left[ \begin{array}{c|c} I + M \otimes J(k_1) & M \otimes e(k_1) \\ \hline M \otimes e(k_1)^T & M \end{array} \right]. \quad (6.3.1)$$

Thus the statement holds for $m = 1$. Now suppose the statement holds for some $m_0 \geq 1$. Observe that for each vertex $x \in B$, taking a distinct copy of $\mathcal{T}(k_1,\ldots,k_{m_0},k_{m_0+1})$ and identifying $x$ as its root vertex is the same as taking a distinct copy of $\mathcal{T}(k_1,\ldots,k_{m_0})$, identifying $x$ as its root vertex, and then adding $k_{m_0+1}$ pendant vertices adjacent to every pendant vertex of each of the new copies of $\mathcal{T}(k_1,\ldots,k_{m_0})$. Applying the inductive hypothesis and using (6.3.1) yields the desired result. $\square$

**COROLLARY 6.3.4** *Let $B$ be a branch at a vertex $v$ in a weighted tree, and let the Perron value of $B$ be $\rho$. Modify $B$ as described in Lemma 6.3.3. Let $f(k_1,\ldots,k_i) = \rho k_i \ldots k_1 + k_i \ldots k_2 + \ldots + k_i + 1$ and form the matrix $A$ of order $m+1$ whose entries are as follows:*

*(1) In the $(i,j)$ position where $1 \leq i \leq j \leq m$, the entries are $f(k_1,\ldots,k_{m+1-i})$. (Note these entries depend only on $i$.)*

*(2) In the $(i,j)$ position where $1 \leq j < i \leq m$, the entries are $k_{m+1-j} \ldots k_{m+2-i} f(k_1,\ldots,k_{m+1-i})$.*

(3) In the $(m+1, i)$ and $(i, m+1)$ positions, the entries are $\rho k_{m+1-i} \ldots k_1$ and $\rho$, respectively.

Then the Perron value of the modified branch at $v$ is the same as the Perron value of $A$.

**Proof**: Let $y$ be the Perron vector for the bottleneck matrix for $B$, and let $[a_1, \ldots, a_{m+1}]^T$ be a Perron vector for $A$. From the block formula given in Lemma 6.3.3, we find that the vector

$$\begin{bmatrix} a_1 y \otimes e(k_1 \ldots k_m) \\ \ldots \\ a_m y \otimes e(k_1) \\ a_{m+1} y \end{bmatrix}$$

is a Perron vector for the bottleneck matrix of the modified branch, and that its Perron value is the same as that of $A$. □

We are now ready to prove the main result of this section which is a theorem from [42]. This theorem tells us how to construct unweighted Type I trees with nonisomorphic Perron branches.

**THEOREM 6.3.5** *For positive integers $k_1, \ldots, k_m$, form an unweigted tree $\mathcal{T}$ by taking a vertex $x$ and making it adjacent to the root vertices of both $\mathcal{T}(k_1, \ldots, k_m)$ and $\mathcal{T}(k_m, \ldots, k_1)$. Then $\mathcal{T}$ is a Type I tree with characteristic vertex $x$.*

**Proof**: From Theorem 6.2.17, it suffices to prove that the Perron value for the rooted branch $\mathcal{T}(k_1, \ldots, k_m)$ is equal to the Perron value of the rooted branch $\mathcal{T}(k_m, \ldots, k_1)$. From Corollary 6.3.4, the Perron value of the rooted branch $\mathcal{T}(k_1, \ldots, k_m)$ is the same as that of the matrix $A$ whose entries are described in that corollary where $\rho = 1$. It now follows that $A$ can be factored as $XY_1$ where

$$X = \begin{bmatrix} 1 & 1 & \ldots & 1 \\ 0 & 1 & \ldots & 1 \\ \ldots & \ldots & \ldots \\ 0 & 0 & \ldots & 1 \end{bmatrix} \text{ and } Y_1 = \begin{bmatrix} 1 & 0 & \ldots & 0 & 0 \\ k_m & 1 & & & \\ k_m k_{m-1} & k_{m-1} & & \ldots & \ldots \\ \ldots & \ldots & & 1 & 0 \\ k_m \ldots k_1 & k_{m-1} \ldots k_1 & \ldots & k_1 & 1 \end{bmatrix}$$

Similarly, the Perron value of the rooted branch $\mathcal{T}(k_m, \ldots, k_1)$ is the same as that of $XY_2$, where

$$Y_2 = \begin{bmatrix} 1 & 0 & \ldots & 0 & 0 \\ k_1 & 1 & & & \\ k_1 k_2 & k_2 & & \ldots & \ldots \\ \ldots & \ldots & & 1 & 0 \\ k_1 \ldots k_m & k_2 \ldots k_m & \ldots & k_m & 1 \end{bmatrix}.$$

# The Fiedler Vector and Bottleneck Matrices for Trees

But note that each of $XY_1$, $Y_1^T X^T$, and $X^T Y_1^T$ has the same Perron value. Further, if $P$ is the permutation matrix with 1's on the back diagonal, then the common Perron value coincides with the Perron value of $PX^T P^T P Y_1^T P^T$. But $PX^T P^T = X$ and $PY_1^T P^T = Y_2$, so we see that $XY_1$ and $XY_2$ have the same Perron value. □

Note that the Perron branches at $x$ are nonisomorphic if and only if $k_i \neq k_{m-i+1}$ for some $1 \leq i \leq m$. Thus we have shown how we can create unweighted Type I trees in which the two Perron branches are nonisomorphic.

**EXAMPLE 6.3.6** Consider the tree $\mathcal{T}$ below. Observe that the branches at vertex $x$ are $\mathcal{T}(1,3)$ and $\mathcal{T}(3,1)$. These branches are clearly nonisomorphic. Yet by Theorem 6.3.5, these branches will have the same Perron value, thus making $\mathcal{T}$ a Type I tree.

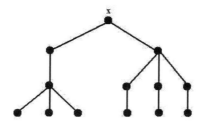

Note that the bottleneck matrices at $x$ are

$$\begin{bmatrix} 1 & 1 & 1 & 1 & 1 \\ 1 & 2 & 2 & 2 & 2 \\ 1 & 2 & 3 & 2 & 2 \\ 1 & 2 & 2 & 3 & 2 \\ 1 & 2 & 2 & 2 & 3 \end{bmatrix} \text{ and } \begin{bmatrix} 1 & 1 & 1 & 1 & 1 & 1 & 1 \\ 1 & 2 & 1 & 1 & 2 & 1 & 1 \\ 1 & 1 & 2 & 1 & 1 & 2 & 1 \\ 1 & 1 & 1 & 2 & 1 & 1 & 2 \\ 1 & 2 & 1 & 1 & 3 & 1 & 1 \\ 1 & 1 & 2 & 1 & 1 & 3 & 1 \\ 1 & 1 & 1 & 2 & 1 & 1 & 3 \end{bmatrix}$$

and that the spectral radius of each of these matrices is 9.26. Also note by Theorem 6.2.15 that $a(\mathcal{T}) = 1/9.26 = 0.108$.

## Exercise:

1. Explain why it would be trivial to describe *weighted* Type I trees with nonisomorphic Perron branches.

## 6.4 Perturbation Results Applied to Extremizing the Algebraic Connectivity of Trees

We saw in Section 6.2 that bottleneck matrices at characteristic vertices are closely related to the algebraic connectivity of a tree. In this section, we use bottleneck matrices to show which trees on a fixed number of vertices and fixed diameter have the largest and smallest algebraic connectivities. We then broaden these results by showing the effects that perturbing branches has on the algebraic connectivity of a tree. This will enable us to order trees by their algebraic connectivities. To achieve this goal, we begin with an important theorem from [47] which explores the effects of the algebraic connectivity of a tree and the location of its characteristic vertices by altering a branch at a vertex not containing all of the characteristic vertices of the tree. In this theorem, let $B_x(y)$ and $\hat{B}_x(y)$ denote the branches at $x$ in $\mathcal{T}$ and $\hat{\mathcal{T}}$, respectively, containing $y$ and we let $\rho(B_x(y))$ and $\rho(\hat{B}_x(y))$ denote the Perron values of the bottleneck matrices of such branches.

**THEOREM 6.4.1** *Suppose that $\mathcal{T}$ is a weighted tree with algebraic connectivity $a$. Let $x$ be a vertex of $\mathcal{T}$, and suppose $B$ is a branch of $\mathcal{T}$ at $x$ which does not contain all of the characteristic vertices of $\mathcal{T}$. Form a new tree $\hat{\mathcal{T}}$ from $\mathcal{T}$ by replacing the branch $B$ at $x$ by some other branch $\hat{B}$ at $x$ (and possibly changing the weight of the edge between $x$ and the vertex in $B$ which is adjacent to $x$). Let $\hat{a}$ be the algebraic connectivity of $\hat{\mathcal{T}}$. Let $M$ be the bottleneck matrix for $B$ at $x$ in $\mathcal{T}$, and let $\hat{M}$ be the bottleneck matrix for $\hat{B}$ at $x$ in $\hat{\mathcal{T}}$. If $M << \hat{M}$ then $\hat{a} \leq a$. Furthermore, the characteristic vertices of $\hat{\mathcal{T}}$ are either on the path joining the characteristic vertices of $\mathcal{T}$ to $x$, or they are located on the new branch $\hat{B}$.*

**Proof**: We first begin with the location of the characteristic vertices of $\hat{\mathcal{T}}$. Let $P$ be the path in $\hat{\mathcal{T}}$ joining the characteristic vertices in $\mathcal{T}$ to $x$. Our goal is to show that the characteristic vertices of $\hat{\mathcal{T}}$ lie in $P \cup \hat{B}$. Suppose that $y$ is a vertex in $\hat{\mathcal{T}}$ that is not in $P \cup \hat{B}$; we will show that $y$ cannot be a characteristic vertex of $\hat{\mathcal{T}}$ by deducing a contradiction. First note that by Theorem 6.2.19, the unique Perron branch at $y$ in $\mathcal{T}$ is the branch containing $P$. Since $M << \hat{M}$, it follows that the unique Perron branch at $y$ in $\hat{\mathcal{T}}$ is the branch containing $P \cup \hat{B}$.

Suppose that $y$ is a characteristic vertex of $\hat{\mathcal{T}}$. Since $y$ has only one Perron branch in $\hat{\mathcal{T}}$, it follows by Theorem 6.2.17 that $\hat{\mathcal{T}}$ is a Type II tree and that the other characteristic vertex, say $z$, is adjacent to $y$ (see Corollary 6.2.14). Suppose $z \notin P \cup \hat{B}$, then by the above argument, the unique Perron branch at $z$ in $\hat{\mathcal{T}}$ would be the branch containing $P \cup \hat{B}$. However, by Corollary 6.2.14 if $y$ and $z$ were both characteristic vertices of $\hat{\mathcal{T}}$, then the unique Perron branch at each vertex should contain the other, contradicting the earlier assertion that the unique Perron branch at $y$ in $\hat{\mathcal{T}}$ contains $P$. Therefore $z \in P \cup \hat{B}$; more specifically $z \in P$ (hence the path from $y$ to $x$ contains $z$). Furthermore, since $y$ is a characteristic vertex of $\hat{\mathcal{T}}$ it follows that the unique Perron branch at $z$ in $\hat{\mathcal{T}}$ must contain $y$. Moreover, since $M << \hat{M}$, it follows that the branch at $z$ containing $B$ cannot be a Perron branch at $z$ in $\mathcal{T}$. Hence the unique Perron branch at $z$ in $\mathcal{T}$ must contain $y$. But since the unique Perron branch at $z$ in $\mathcal{T}$ must contain all characteristic vertices of $\mathcal{T}$ (Theorem 6.2.19), it follows that $y$ must lie on $P$. Yet this contradicts our assumption that $y \notin P \cup \hat{B}$. Hence our assumption that $y$ is a characteristic vertex of $\hat{\mathcal{T}}$ is false. Therefore any characteristic vertex of $\hat{\mathcal{T}}$ is in $P \cup \hat{B}$.

We will now show that $\hat{a} \leq a$. We consider two cases: when $\mathcal{T}$ is of Type I and when $\mathcal{T}$ is of Type II.

*Case I:* Let $\mathcal{T}$ be of Type I with characteristic vertex $v$. Thus by Theorem 6.2.15, we have $\rho(B_v(x)) \leq 1/a$. We will consider the cases when $\rho(\hat{B}_v(x)) \leq 1/a$ and when $\rho(\hat{B}_v(x)) > 1/a$. First, suppose $\rho(\hat{B}_v(x)) \leq 1/a$. Since $M << \hat{M}$, we have $\rho(B_v(x)) < \rho(\hat{B}_v(x))$. Therefore since $\rho(B_v(x)) < 1/a$ in this case, there must be at least two Perron branches at $v$ in $\mathcal{T}$, none of which contains $B$. Thus, in $\hat{\mathcal{T}}$ there are at least two Perron branches at $v$ not containing $\hat{B}$ that each have Perron value $1/a$. Hence, $\hat{\mathcal{T}}$ is also Type I with characteristic vertex $v$ and thus $\hat{a} = a$.

Now we suppose that $\rho(\hat{B}_v(x)) > 1/a$, and let $z$ be the characteristic vertex of $\hat{\mathcal{T}}$ which is furthest from $v$. If $\hat{\mathcal{T}}$ is a Type I tree, say with Perron branch $B_1$ at $z$, then
$$\frac{1}{\hat{a}} = \rho(B_1) \geq \rho(\hat{B}_z(v)) > \rho(\overline{B}_v) = \frac{1}{a},$$
where $\overline{B}_v$ is some Perron branch at $v$ in $\mathcal{T}$ which does not contain $z$. In particular, $a > \hat{a}$. If $\hat{\mathcal{T}}$ is a Type II tree, with $w$ as the other characteristic vertex, and $\theta$ as the weight of the edge between $w$ and $z$, then by Theorem 6.2.12 we have
$$\frac{1}{\hat{a}} = \rho\left(\hat{B}_z(w) - \frac{\gamma}{\theta}J\right)$$
for some $0 < \gamma < 1$. But then we have
$$\frac{1}{\hat{a}} > \rho\left(\hat{B}_z(w) - \frac{1}{\theta}J\right) \geq \rho(\overline{B}_v) = \frac{1}{a},$$
and again $a > \hat{a}$.

*Case II:* Let $\mathcal{T}$ be a Type II tree with characteristic vertices $u$ and $v$. Let $\theta$ be the weight of the edge joining $u$ and $v$. Without loss of generality, we can suppose one of the following:

(i) $x$ is on a branch at $v$ which does not contain $u$.
(ii) $x = v$ and $u \notin B$, or
(iii) $x = u$ and $v \in B$

Before continuing, note that the first two cases cover the possibility that $B$ contains no characteristic vertices of $\mathcal{T}$, while the last is the case that $B$ contains one characteristic vertex of $\mathcal{T}$. In any of these three cases, since $M << \hat{M}$, it follows that the unique Perron branch at $u$ in $\hat{\mathcal{T}}$ is the branch containing $v$. Also note that in any of these cases we have $\hat{B}_v(u) = B_v(u)$. By Theorem 6.2.12, for some $0 < \gamma_0 < 1$, we have:
$$\frac{1}{a} = \rho\left(B_u(v) - \frac{\gamma_0}{\theta}J\right) = \rho\left(B_v(u) - \frac{1-\gamma_0}{\theta}J\right).$$

Suppose that the unique Perron branch at $v$ in $\hat{\mathcal{T}}$ is the branch containing $u$. Then for some $0 < \gamma_1 < 1$, we have
$$\frac{1}{\hat{a}} = \rho\left(\hat{B}_u(v) - \frac{\gamma_1}{\theta}J\right) = \rho\left(\hat{B}_v(u) - \frac{1-\gamma_1}{\theta}J\right). \qquad (6.4.1)$$

Since $M \ll \hat{M}$, we see that

$$\rho\left(\hat{B}_u(v) - \frac{\gamma}{\theta}J\right) > \rho\left(B_u(v) - \frac{\gamma}{\theta}J\right)$$

for all $0 < \gamma < 1$. Since

$$\rho\left(\hat{B}_v(u) - \frac{1-\gamma_0}{\theta}J\right) = \rho\left(B_v(u) - \frac{1-\gamma_0}{\theta}J\right) = \rho\left(B_u(v) - \frac{\gamma_0}{\theta}J\right) < \rho\left(\hat{B}_u(v) - \frac{\gamma_0}{\theta}J\right),$$

we see from (6.4.1) that $\gamma_1 > \gamma_0$. Hence

$$\frac{1}{\hat{a}} = \rho\left(\hat{B}_v(u) - \frac{1-\gamma_1}{\theta}J\right) = \rho\left(B_v(u) - \frac{1-\gamma_1}{\theta}J\right) > \rho\left(B_v(u) - \frac{1-\gamma_0}{\theta}J\right) = \frac{1}{a}.$$

Thus $a > \hat{a}$.

Now suppose that the branch at $v$ containing $u$ is not the unique Perron branch at $v$ in $\hat{\mathcal{T}}$. Then, as above, we let $z$ be the characteristic vertex of $\hat{\mathcal{T}}$ which is furthest from $v$. If $\hat{\mathcal{T}}$ is of Type I, then

$$\frac{1}{\hat{a}} \geq \rho(\hat{B}_z(u)) > \rho\left(\hat{B}_v(u) - \frac{\gamma}{\theta}J\right) = \frac{1}{a}.$$

If $\hat{\mathcal{T}}$ is of Type II, then

$$\frac{1}{\hat{a}} \geq \rho\left(\hat{B}_z(v) - \frac{1}{\theta}J\right) \geq \rho(\hat{B}_v(u)) > \frac{1}{a}.$$

In either case we find that $a > \hat{a}$. □

We should observe from the proof above that the only case where $a = \hat{a}$ is when $\mathcal{T}$ is of Type I and $\rho(\hat{B}_v(x)) \leq 1/a$. We now obtain several corollaries from Theorem 6.4.1, all of which are from [47].

**COROLLARY 6.4.2** *Suppose that $\mathcal{T}$ is a weighted tree with algebraic connectivity $a$. Form $\hat{\mathcal{T}}$ from $\mathcal{T}$ by adding a new pendant vertex and a weighted edge, and denote the algebraic connectivity of $\hat{\mathcal{T}}$ by $\hat{a}$. Then $\hat{a} \leq a$ and the characteristic vertices of $\hat{\mathcal{T}}$ lie on the path between the characteristic vertices of $\mathcal{T}$ and the new pendant vertex.*

Applying Corollary 6.4.2 we obtain the following:

**COROLLARY 6.4.3** *Suppose that a weighted tree $\mathcal{T}$ is a weighted subtree of $\hat{\mathcal{T}}$. Denoting the algebraic connectivities of $\mathcal{T}$ and $\hat{\mathcal{T}}$ by $a$ and $\hat{a}$, respectively, we have $\hat{a} \leq a$.*

We also can easily obtain a corollary discussing the effects of perturbing the weight of an existing edge on a tree:

**COROLLARY 6.4.4** *Let $\mathcal{T}$ be a weighted tree and let $e$ be an edge of $\mathcal{T}$. Form $\hat{\mathcal{T}}$ from $\mathcal{T}$ by decreasing the weight of $e$. Then the algebraic connectivity of $\hat{\mathcal{T}}$ is at most that of $\mathcal{T}$, and the characteristic vertices of $\hat{\mathcal{T}}$ lie on the path joining the characteristic vertices of $\mathcal{T}$ to the vertices incident with $e$.*

We now apply these results to show which unweighted trees of diameter $d$, for fixed $d$, have the maximum and minimum algebraic connectivity. Let $\mathcal{T}(k,l,d)$ be the unweighted tree on $n$ vertices constructed by taking the path on vertices $1,\ldots,d$, and adding $k$ pendant vertices adjacent to vertex 1 and $l$ vertices adjacent to vertex $d$:

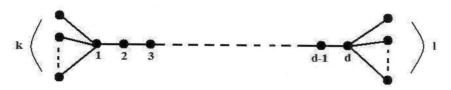

Let $\pi_{r,s}$ denote the unweighted tree formed by taking the path on $r$ vertices and adding $s$ pendant vertices, all of which adjacent to one end of the path:

By induction on $s$, we can easily prove the following claim from [47]:

**CLAIM 6.4.5** *Let $A$ be the bottleneck matrix for any unweighted branch with diameter $r+1$ having $r+s$ vertices. If $M$ is the bottleneck matrix for $\pi_{r,s}$ based at the vertex adjacent to the end point of the path which is not adjacent to the $s$ extra vertices, then $A << M$.*

Observe that the diameter of $\mathcal{T}(k,l,d)$ is $d+1$. We now prove a theorem from [47] that shows that trees of the form $\mathcal{T}(k,l,d)$ have the smallest algebraic connectivity over all unweighted trees on $n$ vertices having diameter $d+1$.

**THEOREM 6.4.6** *Consider the set $\tau_{n,d+1}$ of unweighted trees on $n$ vertices having diameter $d+1$. The algebraic connectivity is minimized over $\tau_{n,d+1}$ by $\mathcal{T}(k,l,d)$ for some $1 \leq k, l \leq n$.*

**Proof**: Suppose that $\mathcal{T} \in \tau_{n,d+1}$ but is not of the form $\mathcal{T}(k,l,d)$. Let the algebraic connectivity of $\mathcal{T}$ be $a$. We will show that some tree of the form $\mathcal{T}(k,l,d)$ has algebraic connectivity at most $a$. Suppose that at some characteristic vertex of $\mathcal{T}$, there is a branch $B$ with $r$ vertices and diameter $f$. If that branch is not of the form $\pi_{f-1,r-f+1}$, then by Claim 6.4.5 and Theorem 6.4.1 the algebraic connectivity will not increase if we replace $B$ by $\pi_{f-1,r-f+1}$. Therefore, we can assume that each branch at a characteristic vertex of $\mathcal{T}$ is of the form $\pi_{f-1,r-f+1}$. Hence if $\mathcal{T}$ is a Type II tree or a Type I tree with just two branches at its characteristic vertex, then $\mathcal{T}$ is of the form $\mathcal{T}(k,l,d)$.

Now suppose that $\mathcal{T}$ is of Type I with three or more branches at its characteristic vertex $v$. Form a new tree $\mathcal{T}_1$ from $\mathcal{T}$ by discarding all of the non-Perron branches at $v$, and all but two of the Perron branches at $v$. By Theorem 6.2.15, note that the algebraic connectivity of $\mathcal{T}_1$ is still $a$. Also note that the two branches at $v$ in $\mathcal{T}_1$ are of the form $\pi_{r_1,s_1}$ and $\pi_{r_2,s_2}$. From the fact that $\mathcal{T}$ has diameter $d+1$, we find that $r_1 + r_2 \leq d-1$; since $\mathcal{T}$ has $n$ vertices, it follows that $s_1 + s_2 \leq n-d$. Thus form a new tree $\mathcal{T}_2$ by replacing the branch $\pi_{r_2,s_2}$ at $v$ by $\pi_{d-1-r_1,n-d-s_1}$ and note by Theorem 6.4.1 that the algebraic connectivity of $\mathcal{T}_2$ is at most $a$. However, $\mathcal{T}_2$ is of the form $\mathcal{T}(k,l,d)$. Thus for any tree in $\mathcal{T}_{n,d+1}$ with algebraic connectivity $a$, there is a tree of the form $\mathcal{T}(k,l,d)$ whose algebraic connectivity is at most $a$. Hence, the algebraic connectivity is minimized over $\mathcal{T}_{n,d+1}$ by $\mathcal{T}(k,l,d)$ for some $1 \leq k,l \leq n$. $\square$

In the next theorem which is from [18], we determine the values of $k$ and $l$ in which the minimum algebraic connectivity over all trees of the form $\mathcal{T}(k,l,d)$ occurs.

**THEOREM 6.4.7** *Among all unweighted trees in $\mathcal{T}_{n,d+1}$, the minimum algebraic connectivity is attained by $\mathcal{T}(\lceil \frac{n-d}{2} \rceil, \lfloor \frac{n-d}{2} \rfloor, d)$.*

**Proof:** By Theorem 6.4.6, we can restrict our attention to trees of the form $\mathcal{T}(k,l,d)$. The Laplacian matrix for such a tree is

$$L = \begin{bmatrix} I_k & -e & 0 & 0 & \cdots & 0 & 0 \\ -e^T & k+1 & -1 & 0 & \cdots & 0 & 0 \\ 0 & -1 & 2 & -1 & \cdots & 0 & 0 \\ 0 & 0 & \cdots & \cdots & \cdots & 0 & 0 \\ 0 & 0 & 0 & -1 & 2 & -1 & 0 \\ \cdots & \cdots & 0 & 0 & -1 & l+1 & -e^T \\ 0 & \cdots & 0 & 0 & 0 & -e & I_l \end{bmatrix}.$$

Observe that the eigenvalues of $L$ are: 1 with algebraic multiplicity $k+l-2$, along with the eigenvalues of the $(d+2) \times (d+2)$ matrix

$$M_{k,l,d} = \begin{bmatrix} 1 & -1 & 0 & 0 & \cdots & 0 & 0 \\ -k & k+1 & -1 & 0 & \cdots & 0 & 0 \\ 0 & -1 & 2 & -1 & 0 & & \cdots \\ 0 & 0 & \cdots & \cdots & \cdots & 0 & 0 \\ 0 & 0 & 0 & -1 & 2 & -1 & 0 \\ \cdots & \cdots & 0 & 0 & -1 & l+1 & -l \\ 0 & \cdots & 0 & 0 & 0 & -1 & 1 \end{bmatrix}$$

To determine the eigenvalues of $M_{k,l,d}$, we need its characteristic polynomial which we will denote by $p_{k,l,d}$. Since we want to observe the effects of keeping the diameter fixed but changing the number of vertices appended to each end vertex of the path (i.e., vertices 1 and $d$), we will consider the expression $p_{k,l,d} - p_{k-1,l+1,d}$. Since the determinant is multilinear, it follows that

$$p_{k,l,d} - p_{k-1,l,d} = -\lambda \det \begin{bmatrix} \lambda - 2 & 1 & 0 & 0 & 0 & 0 \\ 1 & \lambda - 2 & 1 & 0 & 0 & 0 \\ 0 & \cdots & \cdots & \cdots & 0 & 0 \\ 0 & 0 & 1 & \lambda - 2 & 1 & 0 \\ 0 & 0 & 0 & 1 & \lambda - l - 1 & l \\ 0 & 0 & 0 & 0 & 1 & \lambda - 1 \end{bmatrix}$$

where the above matrix is $d \times d$. Similarly, we obtain

$$p_{k-1,l,d} - p_{k-1,l+1,d} = \lambda \det \begin{bmatrix} \lambda - 1 & 1 & 0 & 0 & 0 & 0 \\ k-1 & \lambda - k & 1 & 0 & 0 & 0 \\ 0 & 1 & \lambda - 2 & 1 & 0 & 0 \\ 0 & 0 & \cdots & \cdots & \cdots & 0 \\ 0 & 0 & 0 & 1 & \lambda - 2 & 1 \\ 0 & 0 & 0 & 0 & 1 & \lambda - 2 \end{bmatrix}$$

which, by permutational similarity, is equal to

$$\lambda \det \begin{bmatrix} \lambda - 2 & 1 & 0 & 0 & 0 & 0 \\ 1 & \lambda - 2 & 1 & 0 & 0 & 0 \\ 0 & \cdots & \cdots & \cdots & 0 & 0 \\ 0 & 0 & 1 & \lambda - 2 & 1 & 0 \\ 0 & 0 & 0 & 1 & \lambda - k & k - 1 \\ 0 & 0 & 0 & 0 & 1 & \lambda - 1 \end{bmatrix}$$

Thus

$$p_{k,l,d} - p_{k-1,l+1,d} = \lambda^2 (l + 1 - k) \det \begin{bmatrix} \lambda - 2 & 1 & 0 & 0 \\ 1 & \cdots & \cdots & 0 \\ 0 & \cdots & \cdots & 1 \\ 0 & 0 & 1 & \lambda - 2 \end{bmatrix} \quad (6.4.2)$$

where this matrix is $(d-2) \times (d-2)$.

Observe that the linear term in $\det(\lambda I - L)$ is the sum of all the $(n-1) \times (n-1)$ principal minors of $-L$. By the Matrix Tree Theorem (Theorem 3.2.2), this sum is equal to $(-1)^{n-1} n$ (recall $n = k + l + d$). Furthermore, since 1 is an eigenvalue of $L$ with algebraic multiplicity $k + l - 2$ and since the other eigenvalues of $L$ are the eigenvalues of $M_{k,l,d}$, it follows that $\det(\lambda I - L) = (\lambda - 1)^{k+l-2} p_{k,l,d}$. Since $p_{k,l,d}$ has a factor of $\lambda$ (since $L$ has a simple eigenvalue of 0), we find that the linear term of $p_{k,l,d}$ is $(-1)^{d+1} n$.

Let $a_{k,l,d}$ and $a_{k-1,l+1,d}$ be the algebraic connectivities of $\mathcal{T}(k,l,d)$ and $\mathcal{T}(k-1,l+1,d)$, respectively. Since we want to show that the algebraic connectivity is minimized when $k = \lceil \frac{n-d}{2} \rceil$ and $l = \lfloor \frac{n-d}{2} \rfloor$, the proof will be complete if we can show that $a_{k,l,d} < a_{k-1,l+1,d}$ when $k < l+1$, and $a_{k,l,d} > a_{k-1,l+1,d}$ when $k > l+1$. To this end, we recall that since the path on $d+2$ vertices is a subgraph of each of $\mathcal{T}(k,l,d)$

and $\mathcal{T}(k-1, l+1, d)$, it follows from Theorem 5.1.7 that the algebraic connectivity of each tree is at most that of such a path, i.e., $2(1-\cos(\frac{\pi}{d+2}))$. Applying Theorem 5.1.7 again, we can determine that the smallest eigenvalue of the $(d-2) \times (d-2)$ matrix

$$\begin{bmatrix} 2 & -1 & 0 & 0 \\ -1 & \ldots & \ldots & 0 \\ 0 & \ldots & \ldots & -1 \\ 0 & 0 & -1 & 2 \end{bmatrix}$$

is $2(1-\cos(\frac{\pi}{d-1}))$. Hence if $0 < \lambda < \min(a_{k,l,d}, a_{k-1,l+1,d})$, then we have

$$\operatorname{sgn}\left(\det \begin{bmatrix} \lambda - 2 & 1 & 0 & 0 \\ 1 & \ldots & \ldots & 0 \\ 0 & \ldots & \ldots & 1 \\ 0 & 0 & 1 & \lambda - 2 \end{bmatrix}\right) = (-1)^{d-2}. \qquad (6.4.3)$$

Suppose that $0 < \lambda < a_{k,l,d}$ and note that $p_{k,l,d}(\lambda)$ is increasing (decreasing) at $0$ when $d$ is odd (even), as is $p_{k-1,l+1,d}(\lambda)$. Thus from (6.4.2) and (6.4.3), we see that for such $\lambda$

$$\operatorname{sgn}(p_{k,l,d} - p_{k-1,l+1,d}) = \operatorname{sgn}((l+1-k)\lambda^2(-1)^{d-2}).$$

If $k < l+1$, then $\operatorname{sgn}(p_{k,l,d} - p_{k-1,l+1,d})$ is negative (positive) if $d$ is odd (even) and hence $a_{k,l,d} < a_{k-1,l+1,d}$. Likewise, If $k > l+1$, then $\operatorname{sgn}(p_{k,l,d} - p_{k-1,l+1,d})$ is positive (negative) if $d$ is odd (even) and hence $a_{k,l,d} > a_{k-1,l+1,d}$. Thus we conclude that for a tree on $n$ vertices with diameter $d+1$, the tree which attains minimum algebraic connectivity is $\mathcal{T}(\lceil \frac{n-d}{2} \rceil, \lfloor \frac{n-d}{2} \rfloor, d)$. □

Now that we know which trees on $n$ vertices of diameter $d+1$ yield the minimum algebraic connectivity, we now determine the value of this algebraic connectivity. Once this is done, we will have determined a lower bound on the algebraic connectivity of all trees on $n$ vertices with diameter $d+1$. From Theorem 6.4.7, we see that such a lower bound can be achieved by computing the algebraic connectivity of $\mathcal{T}(\lceil \frac{n-d}{2} \rceil, \lfloor \frac{n-d}{2} \rfloor, d)$, or at least obtain a lower bound for such a tree. We do this in a theorem based on results in [47]:

**THEOREM 6.4.8** *Let $\mathcal{T}$ be an unweighted tree on $n$ vertices with diameter $d+1$ and algebraic connectivity $a$. Then $a \geq 1/p$ where*

$$p = \frac{(\lceil \frac{d}{2} \rceil + 1)(\lceil \frac{d}{2} \rceil + 2)}{2} + \left(\left\lfloor \frac{n-d}{2} \right\rfloor - 1\right)\left\lceil \frac{d}{2} \right\rceil.$$

**Proof**: From Theorem 6.4.7, we see that we need to compute a lower bound on the algebraic connectivity of $\mathcal{T}(\lceil \frac{n-d}{2} \rceil, \lfloor \frac{n-d}{2} \rfloor, d)$. We will do so using bottleneck matrices. There are four cases to consider: (i) $n$ and $d$ are both odd, (ii) $n$ and $d$ are both even, (iii) $n$ is even and $d$ is odd, and (iv) $n$ is odd and $d$ is even. In all four cases, it can be seen that vertex $\lceil \frac{d}{2} \rceil$ is a characteristic vertex of $\mathcal{T}(k, l, d)$ and that a Perron branch at this vertex contains vertex $d$ and the $l$ pendant vertices adjacent

*The Fiedler Vector and Bottleneck Matrices for Trees* 247

to it. Letting $M$ be the bottleneck matrix for this branch at vertex $\lceil \frac{d}{2} \rceil$, we see that if $\mathcal{T}(k,l,d)$ is of Type I then $a(\mathcal{T}(k,l,d)) = 1/\rho(M)$ by Theorem 6.2.15, while if $\mathcal{T}(k,l,d)$ is of Type II, then for some $0 < \gamma < 1$ it follows that $a(\mathcal{T}(k,l,d)) = 1/\rho(M - \gamma J) > 1/\rho(M)$ by Theorem 6.2.12. In either case, $a(\mathcal{T}(k,l,d)) \geq 1/\rho(M)$. We now find an upper bound on $\rho(M)$. Note that $M$ is permutationally similar to

$$\begin{bmatrix} & & & & \lceil \frac{d}{2} \rceil & \lceil \frac{d}{2} \rceil - 1 & \ldots & 2 & 1 \\ & I + \lceil \frac{d}{2} \rceil J & & & \ldots & & & & \ldots \\ & & & & \lceil \frac{d}{2} \rceil & \lceil \frac{d}{2} \rceil - 1 & \ldots & 2 & 1 \\ \hline \lceil \frac{d}{2} \rceil & \ldots & \lceil \frac{d}{2} \rceil & \lceil \frac{d}{2} \rceil & \lceil \frac{d}{2} \rceil - 1 & \ldots & 2 & 1 \\ \lceil \frac{d}{2} \rceil - 1 & \ldots & \lceil \frac{d}{2} \rceil - 1 & \lceil \frac{d}{2} \rceil - 1 & \lceil \frac{d}{2} \rceil - 1 & \ldots & 2 & 1 \\ \ldots & & \ldots & & & & & \ldots \\ 2 & \ldots & 2 & 2 & 2 & \ldots & 2 & 1 \\ 1 & \ldots & 1 & 1 & 1 & \ldots & 1 & 1 \end{bmatrix}$$

where the $(1,1)$ block of this matrix is $\lfloor \frac{n-d}{2} \rfloor \times \lfloor \frac{n-d}{2} \rfloor$ and corresponds to the pendant vertices of $\pi_{\lceil \frac{d}{2} \rceil, \lfloor \frac{n-d}{2} \rfloor}$ which are adjacent to vertex $d$. Thus we find that the maximum row sum of $M$ (achieved from the first row) is $p$. Since $M$ is a positive matrix, by Theorem 1.1.13 this is an upper bound on $\rho(M)$. Hence by the arguments in the first paragraph of this proof, it follows that if $\mathcal{T}$ is a tree on $n$ vertices with diameter $d+1$, then $a(\mathcal{T}) \geq 1/p$. □

Now that we have shown which trees of fixed diameter have the smallest algebraic connectivity, our goal for the remainder of this section will be to show the effects that varying the diameter has on the algebraic connectivity of a tree. We will then be able to order trees of fixed diameter and number of vertices by their algebraic connectivities. To this end, we will continue to use the perturbation results that we have established. First, we recall from Theorem 6.2.17 that a Type I tree $\mathcal{T}$ has at least two Perron branches and consider the effects on the algebraic connectivity of $\mathcal{T}$ when we keep at least two of the Perron branches intact but perturb the rest of the tree. We do this in the following theorem from [36]:

**THEOREM 6.4.9** *Let $\mathcal{T}$ be a Type I tree whose characteristic vertex $w$ has degree $d$. Let $B_1, \ldots, B_d$ be the branches of $\mathcal{T}$ at $w$. Let $B_1$ and $B_2$ be Perron branches. Suppose $B_3, \ldots, B_d$ (each of which may be Perron or non-Perron) contain a total of $p$ vertices. Let $\Gamma$ be any graph on $p$ vertices. Let $\mathcal{G}$ be the graph obtained from $\mathcal{T}$ by discarding branches $B_3, \ldots, B_d$ and then adjoining $w$ to some or all of the vertices of $\Gamma$ by new edges. Then $a(\mathcal{G}) \leq a(\mathcal{T})$.*

**Proof**: We may assume that $\mathcal{T}$ and $\mathcal{G}$ share the same vertex set $V$. Thus let $\mathcal{T} = (V, E)$ and $\mathcal{G} = (V, F)$ where $E$ and $F$ are the edges of $\mathcal{T}$ and $\mathcal{G}$, respectively. Then by Theorem 1.2.7:

$$a(\mathcal{G}) = \min_{\substack{x^T e = 0 \\ x^T x = 1}} x^T L(\mathcal{G}) x,$$

where $x$ is a real-valued vector with $n$ entries. We saw in Theorem 6.2.18 that there exists a characteristic valuation $y$ of $L(\mathcal{T})$ which is positive on $B_1$, negative on

$B_2$, and zero elsewhere. Without loss of generality, assume $y$ is a unit vector. Let $i_3, \ldots, i_d$ be the vertices of $B_3, \ldots, B_d$, respectively, that are adjacent to $w$ in $\mathcal{T}$. Let $F$ be obtained from $E$ by deleting the edges $\{w, i_r\}$, $3 \leq r \leq d$, and adjoining edges $\{w, j_s\}$, $1 \leq s \leq q$, where $j_1, \ldots, j_q$ are the vertices of $\Gamma$ joined to vertex $w$ by these new edges to form $\mathcal{G}$. Then

$$y^T L(\mathcal{G})y = y^T L(\mathcal{T})y - \sum_{r=3}^{d}(y_w - x_{i_r})^2 + \sum_{r=1}^{q}(y_w - x_{j_r})^2$$

$$= y^T L(\mathcal{T})y = a(\mathcal{T}),$$

because $y_w = y_{i_3} = \ldots y_{i_d} = y_{j_1} = \ldots y_{j_q} = 0$. Therefore

$$a(\mathcal{T}) = y^T L(\mathcal{G})y \geq \min_{\substack{x^T e = 0 \\ x^T x = 1}} x^T L(\mathcal{G})x = a(\mathcal{G}).$$

□

**EXAMPLE 6.4.10** Consider the Type I tree below:

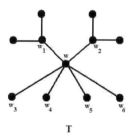

T

Note that $a(\mathcal{T}) = 0.268$. Observe that $w$ is the characteristic vertex, and that the branches containing vertices $w_1$ and $w_2$ are the Perron branches while the pendant vertices $w_3, \ldots, w_6$ are the non-Perron branches. Theorem 6.4.9 says that if we create a graph from vertices $w_3, \ldots, w_6$ in any fashion and let $w$ be adjacent to some or all of $w_3, \ldots, w_6$, then the algebraic connectivity of the resulting graph will not exceed 0.268. Looking at examples of such graphs below, observe that $a(\mathcal{G}_1) = 0.1495$ and $a(\mathcal{G}_2) = 0.268$.

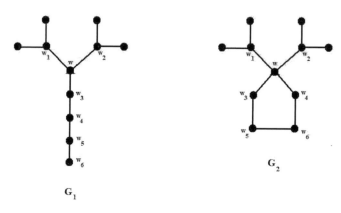

In the next theorem from [36], we illustrate the effect of the algebraic connectivity of a Type I tree by moving Perron branches further away from the characteristic vertex.

**THEOREM 6.4.11** *Let $\mathcal{T} = (V, E)$ be a Type I tree with characteristic vertex $w$. Let $B_1, \ldots, B_r$ be $r$ Perron branches at $w$. Suppose $i$ is the vertex of $B_i$ adjacent in $\mathcal{T}$ to $w$ for $1 \leq i \leq r$. Let $j_1, \ldots, j_r$ be, not necessarily distinct, vertices oo the non-Perron branches of $\mathcal{T}$. Form the tree $\mathcal{T}' = (V, E')$ by deleting edges $w$-$i$ from $\mathcal{T}$ and replacing them with new edges $j_i i$ for $1 \leq i \leq r$. Then $a(\mathcal{T}') \leq a(\mathcal{T})$, with strict inequality if the total number of Perron branches of $\mathcal{T}$ is at least $r+1$.*

**Proof**: Let $y$ be a unit vector of $L(\mathcal{T})$ corresponding to $a(\mathcal{T})$. Then

$$a(\mathcal{T}') = \min_{\substack{x^T e = 0 \\ x^T x = 1}} x^T L(\mathcal{T}') x \leq y^T L(\mathcal{T}') y = y^T L(\mathcal{T}) y - \sum_{i=1}^{r}(y_v - y_i)^2 + \sum_{i=1}^{r}(y_{j_i} - y_i)^2$$

$$= y^T L(\mathcal{T}) y = a(\mathcal{T}),$$

where the second-to-last equality follows from the fact that $y_w = 0$ (since $w$ is the characteristic vertex) and $y_{j_i} = 0$ for all $1 \leq i \leq r$ (since vertices $j_i$ lie on non-Perron branches). Thus $a(\mathcal{T}') \leq a(\mathcal{T})$.

To show strict inequality when there are at least $r+1$ Perron branches, let $B_{r+1}$ be the $(r+1)^{th}$ Perron branch with vertex $k \in B_{r+1}$ adjacent to $v$ in both $\mathcal{T}$ and $\mathcal{T}'$. As in the proof of Theorem 6.4.9 choose a unit eigenvector $y$ corresponding to $a(\mathcal{T})$ that is positive on $B_{r+1}$ and negative on $B_2$. If $a(\mathcal{T}) = a(\mathcal{T}')$, then $y$ must also be a characteristic valuation of $\mathcal{T}'$. Then $y_w = 0 = y_{j_2}$, but $y_k \neq 0 \neq y_2$, yet $wk$ and $j_2 2$ are edges of $\mathcal{T}'$, violating Corollary 6.2.1 which says there exists a unique vertex with zero valuation adjacent to a vertex with nonzero valuation. □

Theorem 6.4.11 states that if we delete edges in a Type I tree $\mathcal{T}$ joining the characteristic vertex $w$ to vertices in Perron branches and replace them with edges joining such vertices in the Perron branches to vertices in non-Perron branches, the algebraic connectivity of the resulting tree $\mathcal{T}'$ will never be more than the algebraic connectivity of the original tree $\mathcal{T}$. However, when making such a replacement, if we do not delete all of the edges joining $w$ to vertices in the Perron branches of $\mathcal{T}$, the the algebraic connectivity of $\mathcal{T}'$ will be strictly less than the algebraic connectivity of $\mathcal{T}$.

**EXAMPLE 6.4.12** Consider the tree $\mathcal{T}$ from Example 6.4.10 and recall $a(\mathcal{T}) = 0.268$. According to Theorem 6.4.11, if we eliminate edges $ww_1$ and $ww_2$, and join each of $w_1$ and $w_2$ to (not necessarily distinct) vertices of the non-Perron branches of $\mathcal{T}$, then the algebraic connectivity of the resulting tree will be less than or equal to $a(\mathcal{T})$. Observe the drawings below:

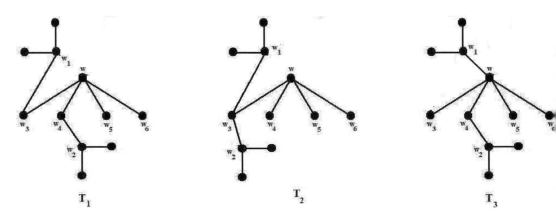

Note that $T_1$ was created by removing edges $ww_1$ and $ww_2$ from $T$ and replacing them with edges $w_3w_1$ and $w_4w_2$ (the branch at $w_2$ was moved below the pendant vertices as to make the drawing more clear). Observe that $a(T_1) = 0.1387 < a(T)$. The tree $T_2$ was created by removing edges $ww_1$ and $ww_2$ from $T$ and replacing them with edges $w_3w_1$ and $w_3w_2$. Observe that $a(T_2) = 0.268 = a(T)$. In creating both of these trees, we replaced *all* edges in $T$ joining the characteristic vertex of $T$ to a vertex in an Perron branch. However, in $T_3$, we did not replace all such edges as the edge $ww_1$ is kept. Observe that $a(T_3) = 0.1772 < a(T)$, i.e., we have strict inequality as expected.

Theorem 6.4.11 demonstrated the effects of moving Perron branches away from the characteristic vertex. Our next theorem from [36] demonstrates the effects of moving non-Perron branches away from the characteristic vertex.

**THEOREM 6.4.13** *Let $T$ be a Type I tree on $n$ vertices with characteristic vertex $w$. Let $B$ be a non-Perron branch and suppose $v$ is the vertex of $B$ adjacent to $w$. Let $j$ be a vertex of some Perron branch of $T$. Form a tree $T'$ by deleting the edge $wv$ and replacing it with a new edge $jv$. Then $a(T') < a(T)$.*

**Proof**: Let $z$ be a unit eigenvector of $L(T)$ corresponding to $a(T)$ such that $z_v = 0 = z_w < z_j$. Create $T'$ as stated in the statement of the theorem, with the labels of $z$ still intact. Then as in Theorem 6.4.11, we arrive at $z^T L(T')z = z^T L(T)z + z_j^2$. Suppose that $B$ has a total of $p$ vertices including vertex $v$. Form vector $y$ by changing each of the $p$ entries of $z$ corresponding to the vertices in $B$ from 0 to $z_j$. Summing $(y_s - y_t)^2$ over the edges $st$ of $T'$ produces $y^T L(T')y$. However, $y_s - y_t = 0$ for all pairs of vertices in $B$ since in $B$ we have $y_s = y_t = z_j$. Thus it follows that $y^T L(T')y = z^T L(T)z = a(T)$. Since $e^T z = 0$, it follows that $e^T y = pz_j$. Therefore, letting $x = y - (pz_j/n)e$, we have $x^T e = 0$. Thus

$$x^T L(T')x = y^T L(T')y - (pz_j/n)^2 e^T L(T')e = y^T L(T')y = a(T).$$

To complete the proof, we need only show that $x^T x > 1$ because then

$$a(T') = \min_{u \neq 0} \frac{u^T L(T')u}{u^T u} \leq \frac{x^T L(T')x}{x^T x} = \frac{a(T)}{x^T x} < a(T).$$

Observe that

$$x^T x = p\left(\frac{n-p}{n}z_j\right)^2 + \sum_{s \notin B}\left(z_s - \frac{p}{n}z_j\right)^2$$

$$= p\left(\frac{n-p}{n}z_j\right)^2 + \sum_{s=1}^{n}z_s^2 - \frac{2pz_j}{n}\sum_{s=1}^{n}z_s + (n-p)\frac{p^2}{n^2}z_j^2$$

$$= \sum_{s=1}^{n}z_s^2 - \frac{2pz_j}{n}\sum_{s=1}^{n}z_s + \frac{p(n-p)}{n}z_j^2$$

the second equality resulting from the fact that $z_s = 0$ for all $s \in B$. Since $\sum z_s^2 = z^T z = 1$ and $\sum z_s = z^T e = 0$, the right-hand side of this equation collapses to $1 + p(n-p)z_j^2/n$. But this expression is greater than one since $p < n$ and since $z_j \neq 0$. Thus $x^T x > 1$. □

**EXAMPLE 6.4.14** Again, consider the tree $\mathcal{T}$ from Example 6.4.10 and recall $a(\mathcal{T}) = 0.268$. Recall that $w_3, \ldots, w_6$ were the pendant vertices of the non-Perron branches of $\mathcal{T}$ adjacent to the characteristic vertex $w$. In the language of Theorem 6.4.13 let $v = w_3$. The tree $\mathcal{T}'$ drawn below was created from $\mathcal{T}$ by removing the edge $wv$ and replacing it with the edge $jv$. Observe $a(\mathcal{T}') = 0.208 < 0.268 = a(\mathcal{T})$.

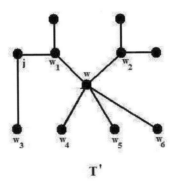

T'

Theorems 6.4.9, 6.4.11, and 6.4.13 lead us to an important theorem from [36] which will help us order trees using their algebraic connectivities. In this theorem, we investigate the effects on the algebraic connectivity of both Type I and Type II trees when we alter the Perron branches by deleting and replacing edges.

**THEOREM 6.4.15** (a) Let $\mathcal{T}$ be a Type I tree with characteristic vertex $w$ and let $B$ be a Perron branch at $w$. Let $i$ and $j$ be two vertices of $B$ such that $i$ is on the unique path from $w$ to $j$. Let $K$ be a branch at $i$ containing neither vertices $j$ nor $w$. Let $k$ be the vertex of $K$ adjacent, in $\mathcal{T}$, to $i$. If $\mathcal{T}'$ is the tree obtained from $\mathcal{T}$ by deleting the edge $i$-$k$ and adjoining a new edge $j$-$k$, then $a(\mathcal{T}') < a(\mathcal{T})$.

(b) Let $\mathcal{T}$ be a Type II tree with characteristic vertices $v$ and $w$. Let $i$ and $j$ be two vertices such that $v$ is on the unique path from $w$ to vertex $j$, and either $i = v$ or vertex $i$ is on the path from $v$ to $j$. Let $K$ be a branch at $i$ containing neither vertices $j$ nor $w$. Let $k$ be the vertex of $K$ adjacent, in $\mathcal{T}$, to $i$. If $\mathcal{T}'$ is the tree obtained from $\mathcal{T}$ by deleting the edge $i$-$k$ and adjoining a new edge $j$-$k$, then $a(\mathcal{T}') < a(\mathcal{T})$.

**Proof**: We prove parts (a) and (b) together. Corollary 6.2.1 guarantees the existence of a unit vector $z$ for $a(\mathcal{T})$ such that $0 < z_i < z_j$ and $z_i < z_k$. We proceed as in Theorem 6.4.11, obtaining

$$z^T L(\mathcal{T}')z = z^T L(\mathcal{T})z - (z_i - z_k)^2 + (z_j - z_k)^2.$$

Suppose that $z_k > (z_i + z_j)/2$. Then $a(\mathcal{T}') \leq z^T L(\mathcal{T}')z < z^T L(\mathcal{T})z = a(\mathcal{T})$ which would complete the proof. So for the remainder of the proof, we assume $z_k \leq (z_i + z_j)/2$.

Suppose $z_k = (z_i + z_j)/2$. Then $z^T L(\mathcal{T}')z = z^T L(\mathcal{T})z$. But then by Corollary 6.2.1, $z$ is not a characteristic valuation of $\mathcal{T}'$ because if it were, the path $v \to \ldots \to i \to \ldots \to j \to k$ would be increasing in $z$, meaning $z_i < z_k$ and $z_j < z_k$. This contradicts $z_k = (z_i + z_j)/2$. Thus $a(\mathcal{T}') \neq z^T L(\mathcal{T}')z$ and therefore

$$a(\mathcal{T}') = \min_{\substack{u^T u = 1 \\ u^T e = 0}} u^T L(\mathcal{T}')u < z^T L(\mathcal{T}')z = z^T L(\mathcal{T})z = a(\mathcal{T}).$$

Now suppose $z_k < (z_i + z_j)/2$. Let $\alpha := (z_j - z_k) - (z_k - z_i) > 0$. Suppose that branch $K$ has a total of $p$ vertices including vertex $k$. Form a new vector $y$ by adding $\alpha$ to each of the $p$ entries of $z$ corresponding to the $p$ vertices of $K$, and let $x = y - (p\alpha/n)e$. Then, as in the proof of Theorem 6.4.13, we obtain $x^T L(\mathcal{T}')x = a(\mathcal{T})$. But $x^T x > 1$ because $z_u > 0$ for every vertex $u$ of $K$. Hence by similar reasoning as in the proof of Theorem 6.4.13, it follows that $a(\mathcal{T}') < a(\mathcal{T})$. □

**EXAMPLE 6.4.16** To illustrate part (a) of Theorem 6.4.15, let us consider the Type I tree used in Example 6.4.10. We will redraw it below as to label the vertices in accordance with Theorem 6.4.15.

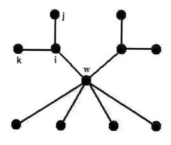

$\mathcal{T}$

Observe that $i$, $j$, and $k$ all lie on a Perron branch at $w$. Letting $K$ be the branch in $\mathcal{T}$ at $i$ containing $k$, then according to Theorem 6.4.15 removing edge $ij$ and replacing it with edge $jk$ will cause the algebraic connectivity to decrease.

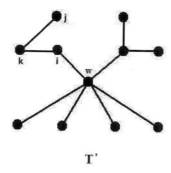

T'

Observe that $a(\mathcal{T}') = 0.2213 < a(\mathcal{T}) = 0.268$.

**EXAMPLE 6.4.17** To illustrate part (b) of Theorem 6.4.15, consider the following Type II tree with characteristic vertices $v$ and $w$.

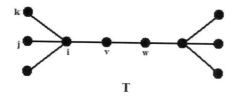

T

Note that $a(\mathcal{T}) = 0.144$ and observe that $i$, $j$, and $k$ are as stated in Theorem 6.4.15. Thus creating $\mathcal{T}'$ by removing edge $ik$ and replacing it with edge $jk$ will decrease the algebraic connectivity. Note that $a(\mathcal{T}') = 0.1353$.

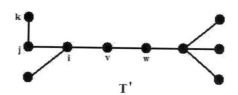

T'

Theorems 6.4.9 through 6.4.15 illustrate that for a fixed $n$, the algebraic connectivity of a tree tends to decrease as the diameter increases. Observe that this certainly holds true when considering all trees on $n = 7$ vertices [36]:

| T | diam(T) | a(T) |
|---|---|---|
| (star, 7 vertices) | 2 | 1 |
| | 3 | 0.466 |
| | 3 | 0.398 |
| | 4 | 0.382 |
| | 4 | 0.382 |
| | 4 | 0.322 |
| | 4 | 0.296 |
| | 4 | 0.268 |
| | 5 | 0.260 |
| | 5 | 0.225 |
| | 6 | 0.198 |

# Exercises:

1. Let $\mathcal{T}$ be a tree with diameter $d$. Prove that $a(\mathcal{T}) \leq 2(1 - \cos\frac{\pi}{d+1})$.

2. (See [47]) Let $\mathcal{T}_1$ be an unweighted tree with algebraic connectivity $a_1$. Let $B$ be a branch at vertex $v$ of $\mathcal{T}_1$ with $k$ vertices. Form $\mathcal{T}_2$ from $\mathcal{T}_1$ by replacing $B$ by a star on $k$ vertices, with the center of the star adjacent to $v$. Prove that $a_2 \geq a_1$ where $a_2$ denotes the algebraic connectivity of $\mathcal{T}_2$.

3. (See [47]) Let $\mathcal{T}_1$ be an unweighted tree with algebraic connectivity $a_1$. Let $B$ be a branch at vertex $v$ of $\mathcal{T}_1$ with $k$ vertices. Form $\mathcal{T}_2$ from $\mathcal{T}_1$ by replacing $B$ by a path on $k$ vertices, with a pendant vertex of the path adjacent to $v$. Prove that

$a_2 \leq a_1$ where $a_2$ denotes the algebraic connectivity of $\mathcal{T}_2$.

4. Using the theorems of this section, order all trees on $n = 6$ vertices by their algebraic connectivities.

5. (See [21]) Let $\mathcal{T}$ be a tree, $y$ be a Fiedler vector, $S$ be the set of characteristic vertices of $\mathcal{T}$, and $B$ a positively valuated component at some vertex of $\mathcal{T}$. Let $u, v \in B$ and suppose that $u$ is closer to $S$ than $v$. Let $B_u$ be a positively valuated component of $\mathcal{T}$ at $u$ such that $v \notin B_u$. Let $\mathcal{T}'$ be the tree obtained from $\mathcal{T}$ by moving $B_u$ to $v$, i.e., removing all edges between $u$ and the vertices in $B_u$ adjacent to $u$ and adding edges joining $v$ to such vertices in $B_u$. Prove the following:

(i) $a(\mathcal{T}') < a(\mathcal{T})$
(ii) The set $S'$ of characteristic vertices of $\mathcal{T}'$ lies on the path joining $S$ and $v$. If $S'$ consists only of a single vertex, then it is not $v$.

6. (See [21]) Suppose $\mathcal{T}$ is a tree, $y$ is a Fiedler vector and $S$ is the set of characteristic vertices of $\mathcal{T}$. Let $u$ and $v$ be two vertices of $\mathcal{T}$ such that $y_u > 0$ and $y_v < 0$, so that $S$ lies on the path $P$ joining $u$ and $v$. Let $B_1, \ldots, B_k$ be nonnegatively valuated components of $\mathcal{T} - P$ and $C_1, \ldots, C_r$ be the negatively valuated components of $\mathcal{T} - P$. Let $\mathcal{T}'$ be the tree obtained from $\mathcal{T}$ by moving all of the $B_i$'s to $u$ and all of the $C_j$'s to $v$. Assume that in $\mathcal{T}$, at least one $B_i$ is not a component at $u$, or at least one $C_j$ is not a component at $v$. Prove the following:

(i) $a(\mathcal{T}') < a(\mathcal{T})$ and the characteristic set $S'$ of $\mathcal{T}'$ lies on the path $P$.
(ii) If $S'$ consists of a single vertex, then it is neither $u$ nor $v$.

7. (See [21]) Let $\mathcal{T}$ be a tree such that exactly one vertex $v$ of $\mathcal{T}$ is of degree 3. Suppose that $P_1$, $P_2$, and $P_3$ are components of $\mathcal{T} - v$, where $P_i$ are paths such that $|P_3| \geq |P_2| \geq |P_1| \geq k$, for some positive integer $k$. Consider the tree $\mathcal{T}'$ described below:

(i) $|\mathcal{T}'| = |\mathcal{T}|$,
(ii) Exactly one vertex $u$ of $\mathcal{T}'$ is of degree 3,
(iii) There are 3 components $P_1', P_2', P_3'$ of $\mathcal{T}' - u$ such that each of them is a path and $|P_1'| = |P_2'| = k$.

Prove $a(\mathcal{T}') \leq a(\mathcal{T})$, and equality holds only when $|P_1| = |P_2| = k$.

8. (See [21]) Let $\mathcal{T}$ be a tree and let $v$ be a vertex of degree $k \geq 3$. Let $B_i$, $i = 1, 2, \ldots, k$ be the branches of $\mathcal{T}$ at the vertex $v$. Let $\mathcal{T}'$ be the tree created from $\mathcal{T}$ by replacing each branch $B_i$ with the path on $|B_i|$ vertices. Prove $a(\mathcal{T}') \leq a(\mathcal{T})$, and the inequality is strict if one of the Perron branches at $v$ in $\mathcal{T}$ is not a path.

## 6.5 Application: Joining Two Trees by an Edge of Infinite Weight

In this section based largely on [69], we let $\mathcal{T}_1$ and $\mathcal{T}_2$ be two trees, and we join the trees at vertices $x \in T_1$ and $y \in T_2$ by an edge $e$ of weight $w$ to obtain a new tree $\hat{\mathcal{T}}_w$. We apply the results obtained in this section on bottleneck matrices to investigate the algebraic connectivity of such a tree as the weight $w$ tends to infinity, i.e.,

$$\lim_{w \to \infty} a\left(\hat{\mathcal{T}}_w\right) \qquad (6.5.1)$$

We first need to define the concept of the type of tree at infinity.

**DEFINITION 6.5.1** Let $\mathcal{T}_1$ and $\mathcal{T}_2$ be trees and join them by an edge $e$ of weight $w$ to obtain a new tree $\hat{\mathcal{T}}_w$. Letting $w \to \infty$, we define the limit tree $\hat{\mathcal{T}}_\infty$ to be a *Type I tree at infinity* with characteristic vertex $v$ if there exists $w_0 > 0$ such that for all $w \in (w_0, \infty)$, $\hat{\mathcal{T}}_w$ is a Type I tree with characteristic vertex $v$. Similarly, we define $\hat{\mathcal{T}}_\infty$ to be a *Type II tree at infinity* with characteristic vertices $i$ and $j$ if there exists $w_0 > 0$ such that for all $w \in (w_0, \infty)$, $\hat{\mathcal{T}}_w$ is a Type II tree with characteristic vertices $i$ and $j$.

In the remainder of this section, we shall write $\hat{\mathcal{T}}$ for $\hat{\mathcal{T}}_\infty$. When using bottleneck matrices in our investigation, the notation $M_{(v-u)\mathcal{T}}$ will denote the bottleneck matrix for the branch at vertex $v$ containing vertex $u$ in tree $\mathcal{T}$. Likewise, $M_{(v-e)\mathcal{T}}$ will denote the bottleneck matrix for the branch at vertex $v$ containing edge $e$ in tree $\mathcal{T}$. $M_{v\mathcal{T}}$ will denote the bottleneck matrix for the Perron branch(es) at $v$ in $\mathcal{T}$. If $v$ has more than one Perron branch in $\mathcal{T}$, then $M_{v\mathcal{T}}$ will be (permutationally similar to) the block diagonal matrix where each block is the bottleneck matrix for a Perron branch at $v$ in $\mathcal{T}$.

By Definition 6.5.1, in order for $\hat{\mathcal{T}}$ to be of Type I or of Type II at infinity, there must exist an appropriate value $w_0 > 0$ such that in $(w_0, \infty)$, $\hat{\mathcal{T}}_w$ does not change its type nor its characteristic vertices. In the next theorem from [69] we show that such a $w_0$ always exists:

**THEOREM 6.5.2** *Let $\hat{\mathcal{T}}_w$ be a tree obtained from joining two trees by $\mathcal{T}_1$ and $\mathcal{T}_2$ by an edge $e$ of weight $w$. Then there exists a number $w_0 > 0$ such that in $(w_0, \infty)$, $\hat{\mathcal{T}}_w$ does not change its type nor its characteristic vertex (vertices).*

**Proof**: Let $x > 0$ and let $v$ be a (not necessarily unique) characteristic vertex of $\hat{\mathcal{T}}_x$. Suppose first that at $v$, $\hat{\mathcal{T}}_x$ has at least two Perron branches which do not contain $e$. Then if $y > x$, it follows from Exercise 1(ii) of Section 1.3 and Theorem 6.2.5 that

$$\rho\left(M_{(v-e)\hat{\mathcal{T}}_y}\right) < \rho\left(M_{(v-e)\hat{\mathcal{T}}_x}\right). \qquad (6.5.2)$$

Therefore, for all $y \geq x$, at least two of the Perron branches at $v$ in $\hat{T}_x$ are identical to at least two of the Perron branches at $v$ in $\hat{T}_y$. Thus $v$ has at least two Perron branches in $\hat{T}_y$ making $\hat{T}_y$ a Type I tree with characteristic vertex $v$ for all $y \geq x$.

Suppose now that at $v$, $\hat{T}_x$ has only one Perron branch which does not contain $e$. Then by (6.5.2), $v$ has only one Perron branch in $\hat{T}_y$ if $y > x$. Let $S_w$ be the set of vertices such that $s \in S_w$ if and only if any characteristic vertex of $\hat{T}_w$ is on the path from $s$ to $e$. As the weight $w$ of $e$ is increased, the spectral radii of the bottleneck matrices at all vertices $s \in S_w$ for the branch containing $e$ decrease. Since the spectral radii of the bottleneck matrices at all vertices $s \in S_w$ for branches not containing $e$ (and hence not containing a characteristic vertex of $\hat{T}_w$) remain constant as $w$ increases, and since in $\hat{T}_w$ the unique Perron branch at any non-characteristic vertex $s \in S_w$ contains a characteristic vertex of $\hat{T}_w$ by Theorem 6.2.19, it follows that if $b > a$, the characteristic vertex (vertices) of $\hat{T}_b$ lie in $S_a$. Thus if $b > a$ then $S_b \subseteq S_a$. Therefore as $w$ increases beyond $x$, the characteristic vertex (vertices) of $\hat{T}_w$ move from $v$ away from $e$. Since for each $w$ there are only finitely many vertices in $S_w$ and since $S_b \subseteq S_a$ if $b > a$, it follows that for some $w_0$ large enough, $S_w$ will equal $S_{w_0}$ for all $w > w_0$. Hence by definition of $S_w$, the characteristic vertex (vertices) will not change as $w$ is increased beyond $w_0$. □

**REMARK 6.5.3** In Theorem 6.5.2, $w_0$ cannot, in general, be taken to be 0. For example, let $\mathcal{T}_1$ be a path on 3 vertices and $\mathcal{T}_2$ be a path on 2 vertices and let $\hat{T}_w$, $w > 0$ be the tree created by joining a pendant vertex of $\mathcal{T}_1$ to a pendant vertex of $\mathcal{T}_2$ by an edge of weight $w$:

Observe $\hat{T}_w$ is a Type II tree with 3 and 4 as its characteristic vertices when $w \in (0,1)$. $\hat{T}_w$ is a Type I tree with 3 as its characteristic vertex when $w = 1$. Yet when $w \in (1, \infty)$, $\hat{T}_w$ is a Type II tree with 2 and 3 as its characteristic vertices. Thus $w_0 = 1$.

Since the main theorem of this section concerns the algebraic connectivity of the tree $\hat{\mathcal{T}} := \hat{T}_\infty$, it is important to elaborate on how $a(\hat{\mathcal{T}})$ is computed. Suppose first that $\hat{T}_w$ is a tree of Type I with characteristic vertex $v$ for all $w \in (w_0, \infty)$. Then for all $w \in (w_0, \infty)$, $v$ has at least two Perron branches in $\hat{T}_w$. Since $\rho(M_{(v-e)\hat{T}_w})$ decreases as $w$ increases, it follows that $e$ is not on a Perron branch at $v$ in $\hat{\mathcal{T}}$. Therefore by Theorem 6.2.15

$$a(\hat{\mathcal{T}}) = \frac{1}{\rho(M_{v_{\hat{T}_w}})}, \quad w \in (w_0, \infty).$$

Suppose next that $\hat{T}_w$ is of Type II with characteristic vertices $i$ and $j$ for all $w \in (w_0, \infty)$. If $e$ joins $i$ and $j$, then by Theorem 6.2.12

$$a\left(\hat{\mathcal{T}}\right) = \lim_{w \to \infty} a\left(\hat{\mathcal{T}}_w\right) = \lim_{w \to \infty} \frac{1}{\rho\left(M_{(i-j)_{\hat{\mathcal{T}}_w}} - \frac{\gamma_w}{w}J\right)} = \frac{1}{\rho\left(M_{(i-j)_{\hat{\mathcal{T}}}}\right)},$$

where $0 < \gamma_w < 1$.

**REMARK 6.5.4** If $\hat{\mathcal{T}}$ is of Type II and both its characteristic vertices $i$ and $j$ are incident to $e$, then it is possible for $i$ and $j$ to have other Perron branches, respectively, besides the branch that contains the other characteristic vertex. This is due to the fact that since the Perron branch at $i$ in $\hat{\mathcal{T}}_w$ contains $e$ when $w \in (w_0, \infty)$, $\rho(M_{i-j_{\hat{\mathcal{T}}}})$ decreases as $w$ increases. Hence at infinity, it is possible for $\rho(M_{(i-j)_{\hat{\mathcal{T}}}})$ to equal $\rho(M_{(i-k)_{\hat{\mathcal{T}}}})$, where $k$ is a vertex on another branch at $i$ on $\hat{\mathcal{T}}$. An example of this would be when $\mathcal{T}_1$ and $\mathcal{T}_2$ are each paths on 2 vertices. If we join vertex $i$ of $\mathcal{T}_1$ to vertex $j$ of $\mathcal{T}_2$ with an edge of weight $w$ to create $\hat{\mathcal{T}}_w$, we see that $\hat{\mathcal{T}}_w$ is a Type II tree with characteristic vertices $i$ and $j$ for all $w > 0$. Letting $w$ tend to infinity, we observe that $\hat{\mathcal{T}}$ is of Type II with both its characteristic vertices incident to $e$, yet each characteristic vertex has two Perron branches.

Suppose now $\hat{\mathcal{T}}$ is of Type II with characteristic vertices $i$ and $j$ for all $w \in (w_0, \infty)$ but $i$ is not joined to $j$ by $e$. Then either the path from $i$ to $e$ contains $j$ or the path from $j$ to $e$ contains $i$. Without loss of generality, let the path from $i$ to $e$ contain $j$. Then by Theorem 6.2.12

$$a\left(\hat{\mathcal{T}}\right) = \lim_{w \to \infty} \frac{1}{\rho\left(M_{(j-i)_{\hat{\mathcal{T}}_w}} - \frac{\gamma_w}{\theta}J\right)}$$

for some $0 < \gamma_w < 1$ ($\gamma_w$ depends on $w$) where $\theta$ is the weight of the edge joining $i$ and $j$.

As stated in the beginning of this section, our goal is to determine $a(\hat{\mathcal{T}})$. We will do this by comparing $a(\hat{\mathcal{T}})$ to $a(\mathcal{T}_1)$ and $a(\mathcal{T}_2)$. To this end, we begin with two claims from [69]:

**CLAIM 6.5.5** *Let $\mathcal{T}_1$ and $\mathcal{T}_2$ be trees with algebraic connectivities $a(\mathcal{T}_1)$ and $a(\mathcal{T}_2)$, respectively. Suppose that the tree $\hat{\mathcal{T}}_w$ is obtained by joining a vertex in $\mathcal{T}_1$ to a vertex in $\mathcal{T}_2$ by an edge of weight $w > 0$. Then*

$$a(\hat{\mathcal{T}}_w) \leq \min\{a(\mathcal{T}_1), a(\mathcal{T}_2)\}. \tag{6.5.3}$$

**Proof**: Let $L_1$ and $L_2$ be the Laplacian matrices for $\mathcal{T}_1$ and $\mathcal{T}_2$, respectively, and form the $2 \times 2$ block matrix $L$ with the $(1,1)$ and the $(2,2)$ blocks as $L_1$ and $L_2$, respectively, and with the $(1,2)$ and $(2,1)$ blocks being the zeros matrices of the appropriate sizes. Thus $\lambda_1(L) = \lambda_2(L) = 0$, while $\lambda_3(L) = \min\{a(\mathcal{T}_1), a(\mathcal{T}_2)\}$. Now let $\hat{L}_w$ be the Laplacian matrix of $\hat{\mathcal{T}}_w$. As $(\hat{L}_w)$ is a rank 1 perturbation of $L$ we have, by Corollary 1.2.9 that

$$a(\hat{\mathcal{T}}_w) = \lambda_2\left(L\left(\hat{\mathcal{T}}_w\right)\right) \leq \lambda_3(L) = \min\{a(\mathcal{T}_1), a(\mathcal{T}_2)\}.$$

As $w$ tends to infinity, our goal is to determine when equality holds in (6.5.3). The next claim from [69] will be useful in proving the three subsequent lemmas which will help us achieve this goal. In this claim and in the subsequent lemmas, let $M_{(a-b)\mathcal{T}}$ denote the bottleneck matrix of the branch at vertex $a \in \mathcal{T}$ containing vertex $b$.

**CLAIM 6.5.6** *Let $\mathcal{T}$ be a tree. Let $i$ and $j$ be adjacent vertices in $\mathcal{T}$ joined by an edge of weight $\alpha_1$; let $p$ and $q$ be adjacent vertices in $\mathcal{T}$ joined by an edge of weight $\alpha_2$. If the path from $i$ to $q$ contains $j$ and $p$, and if $0 < \gamma_1, \gamma_2 < 1$, then*

$$\rho\left(M_{(j-i)\mathcal{T}} - \frac{\gamma_1}{\alpha_1}J\right) < \rho\left(M_{(j-i)\mathcal{T}}\right) < \rho\left(M_{(q-p)\mathcal{T}} - \frac{\gamma_2}{\alpha_2}J\right) < \rho\left(M_{(q-p)\mathcal{T}}\right).$$

**Proof**: The first and third inequality follow from

$$M_{(j-i)\mathcal{T}} - \frac{\gamma_1}{\alpha_1}J < M_{(j-i)\mathcal{T}} \qquad (6.5.4)$$

and

$$M_{(q-p)\mathcal{T}} - \frac{\gamma_2}{\alpha_2}J < M_{(q-p)\mathcal{T}} \qquad (6.5.5)$$

and the fact that all matrices in (6.5.4) and (6.5.5) are nonnegative and irreducible.

To prove the second inequality, let $(j-i)_\mathcal{T}$ be the subgraph of $\mathcal{T}$ that consists of $j$ and the branch at $j$ containing $i$. Likewise, let $(q-p)_\mathcal{T}$ be the subgraph of $\mathcal{T}$ consisting of $q$ as well as its branch containing $p$. Let $\beta$ be the set of indices of $M_{(q-p)\mathcal{T}}$ that correspond to the vertices that determine $M_{(j-i)\mathcal{T}}$. Since $(j-i)_\mathcal{T}$ is a subgraph of $(q-p)_\mathcal{T}$ and since $p$ and $q$ are joined by an edge of weight $\alpha_2$, then

$$M_{(j-i)\mathcal{T}} \leq M_{(q-p)\mathcal{T}}[\beta] - \frac{1}{\alpha_2}J.$$

Since all the remaining entries of $M_{(q-p)\mathcal{T}} - (1/\alpha_2)J$ are positive, it follows that

$$\rho\left(M_{(j-i)\mathcal{T}}\right) \leq \rho\left(M_{(q-p)\mathcal{T}} - \frac{1}{\alpha_2}J\right).$$

But as

$$\rho\left(M_{(q-p)\mathcal{T}} - \frac{1}{\alpha_2}J\right) < \rho\left(M_{(q-p)\mathcal{T}} - \frac{\gamma_2}{\alpha_2}J\right),$$

the second inequality in this claim thus follows. □

We are now ready to begin examining three possible cases of joining two trees by an edge whose weight tends to infinity. We do this through three corresponding lemmas from [69]. The first lemma examines the case of joining two trees by an edge whose weight tends to infinity when one of the building trees, say $\mathcal{T}_1$, is of Type II. In this lemma, $\theta_1$ will denote the weight of the edge joining the characteristic vertices of $\mathcal{T}_1$, and $\gamma_1 \in (0,1)$ will denote the value of $\gamma$ referred to in Theorem 6.2.12(b).

**LEMMA 6.5.7** Let $\mathcal{T}_1$ be of Type II and let $\hat{\mathcal{T}}_w$ be obtained from $\mathcal{T}_1$ and $\mathcal{T}_2$ by joining a vertex $x$ of $\mathcal{T}_1$ to a vertex $y$ of $\mathcal{T}_2$ by an edge of weight $w$. Then $a(\hat{\mathcal{T}}) < a(\mathcal{T}_1)$.

**Proof**: Let $i$ and $j$ be the characteristic vertices of $\mathcal{T}_1$ such that the path from $i$ to $x$ contains $j$. Suppose $\hat{\mathcal{T}}$ is of Type I with $v$ as its characteristic vertex. Then, by Theorem 6.4.1, $v$ lies on the path from $i$ to any characteristic vertex of $\mathcal{T}_2$. Therefore,

$$a(\hat{\mathcal{T}}) = \lim_{w \to \infty} a(\hat{\mathcal{T}}_w) = \lim_{w \to \infty} \frac{1}{\rho(M_{(v-i)\hat{\mathcal{T}}_w})}$$

$$\leq \frac{1}{\rho(M_{(j-i)T_1})} < \frac{1}{\rho(M_{(j-i)T_1} - \frac{\gamma_1}{\theta_1}J)} = a(\mathcal{T}_1).$$

Suppose now that $\hat{\mathcal{T}}$ is of Type II. Let $p$ and $q$ be the characteristic vertices of $\hat{\mathcal{T}}$ and let $p$ be on the path from $q$ to $i$. Then

$$a(\hat{\mathcal{T}}) = \lim_{w \to \infty} a(\hat{\mathcal{T}}_w) = \lim_{w \to \infty} \frac{1}{\rho(M_{(q-p)\hat{\mathcal{T}}_w} - \frac{\gamma_w}{\theta}J)}$$

$$\leq \frac{1}{\rho(M_{(j-i)T_1} - \frac{\gamma_1}{\theta_1}J)} = a(\mathcal{T}_1), \tag{6.5.6}$$

where the inequality follows from the fact that $a(\hat{\mathcal{T}}) \leq a(\mathcal{T}_1)$. However, according to Claim 6.5.6 the inequality in (6.5.6) is strict. Thus $a(\hat{\mathcal{T}}) < a(\mathcal{T}_1)$ whenever $\mathcal{T}_1$ is a Type II tree. $\square$

The next two lemmas from [69] examine the relationship between $a(\hat{\mathcal{T}})$ and $a(\mathcal{T}_1)$ when $\mathcal{T}_1$ is of Type I. The first of these two lemmas concerns the case when $\mathcal{T}_1$ and $\mathcal{T}_2$ are joined by an edge $e$ of weight $w$ where $e$ is not incident to the characteristic vertex of $\mathcal{T}_1$. The second of these two lemmas will examine the case when $e$ is incident to the characteristic vertex of $\mathcal{T}_1$.

**LEMMA 6.5.8** Suppose $a(\mathcal{T}_1) \leq a(\mathcal{T}_2)$ and $\mathcal{T}_1$ is of Type I with characteristic vertex $i$. Let $\hat{\mathcal{T}}_w$ be obtained from $\mathcal{T}_1$ and $\mathcal{T}_2$ by joining a vertex $x$ of $\mathcal{T}_1$ to a vertex $y$ of $\mathcal{T}_2$ by an edge $e$ of weight $w$. If $e$ is not incident to $i$, then $a(\hat{\mathcal{T}}) = a(\mathcal{T}_1)$ if and only if $i$ is the only characteristic vertex of $\hat{\mathcal{T}}$.

**Proof**: Suppose $\hat{\mathcal{T}}$ is of Type I with characteristic vertex $v$. It follows from Theorem 6.4.1 that $v$ lies on the path from $i$ to any characteristic vertex of $\mathcal{T}_2$. Thus

$$a(\hat{\mathcal{T}}) = \lim_{w \to \infty} a(\hat{\mathcal{T}}_w) = \lim_{w \to \infty} \frac{1}{\rho(M_{v\hat{\mathcal{T}}_w})} = \frac{1}{\rho(M_{v\hat{\mathcal{T}}})} \leq \frac{1}{\rho(M_{iT_1})} = a(\mathcal{T}_1),$$

where the inequality follows from the fact that $a(\hat{\mathcal{T}}) \leq a(\mathcal{T}_1)$. Since $e$ is not incident to $i$, equality holds if and only if $v = i$.

Now suppose that $\hat{\mathcal{T}}$ is of Type II with characteristic vertices $p$ and $q$ with $p$ on the path from $q$ to $i$. Therefore,

$$a(\hat{\mathcal{T}}) = \lim_{w \to \infty} a(\hat{\mathcal{T}}_w) = \lim_{w \to \infty} \frac{1}{\rho(M_{(q-p)\hat{\mathcal{T}}_w} - \frac{\gamma_w}{\theta}J)}$$

$$= \frac{1}{\rho(M_{(q-p)\hat{\mathcal{T}}} - \frac{\gamma}{\theta}J)} \leq \frac{1}{\rho(M_{i_{T_1}})} = a(\mathcal{T}_1), \qquad (6.5.7)$$

where, again, the inequality follows from $a(\hat{\mathcal{T}}) \leq a(\mathcal{T}_1)$. However by Claim 6.5.6,

$$\rho\left(M_{i_{T_1}}\right) < \rho\left(M_{(q-p)\hat{\mathcal{T}}} - \frac{\gamma}{\theta}J\right).$$

Thus the inequality in (6.5.7) is strict. Hence $a(\hat{\mathcal{T}}) < a(\mathcal{T}_1)$ if $\hat{\mathcal{T}}$ has more than one characteristic vertex. □

We see from Lemma 6.5.8 that in order for $a(\hat{\mathcal{T}})$ to equal $a(\mathcal{T}_1)$ when $\mathcal{T}_1$ is a Type I tree and $e$ is not incident to the characteristic vertex of $\mathcal{T}_1$, the characteristic vertex of $\mathcal{T}_1$ must be the only characteristic vertex of $\hat{\mathcal{T}}$.

**REMARK 6.5.9** Since $y$ is the vertex in $\mathcal{T}_2$ that is incident to $e$, we see that $\rho(M_{yT_2})$ must be small if $a(\hat{\mathcal{T}}) = a(\mathcal{T}_1)$.

The following lemma examines the relationship between $a(\hat{\mathcal{T}})$ and $a(\mathcal{T}_1)$ when $e$ is incident to the characteristic vertex of $\mathcal{T}_1$.

**LEMMA 6.5.10** Suppose $a(\mathcal{T}_1) \leq a(\mathcal{T}_2)$ and $\mathcal{T}_1$ is of Type I with characteristic vertex $i$. Let $\hat{\mathcal{T}}_w$ be obtained from $\mathcal{T}_1$ and $\mathcal{T}_2$ by joining $i$ to a vertex $y$ of $\mathcal{T}_2$ by an edge $e$ of weight $w$. Then $a(\hat{\mathcal{T}}) = a(\mathcal{T}_1)$ if and only if $\rho(M_{yT_2}) \leq 1/a(\mathcal{T}_1)$.

**Proof**: We shall consider three cases:

*Case I*: If $\rho(M_{yT_2}) < 1/a(\mathcal{T}_1)$, then $i$ is the unique vertex in $\hat{\mathcal{T}}$ that has more than one Perron branch and so $\hat{\mathcal{T}}$ is of Type I and $a(\hat{\mathcal{T}}) = a(\mathcal{T}_1)$.

*Case II*: If $\rho(M_{yT_2}) = 1/a(\mathcal{T}_1)$, then as $w \to \infty$, it follows from Theorem 6.2.5 that

$$\rho(M_{(i-y)\hat{\mathcal{T}}}) = \rho(M_{(y-i)\hat{\mathcal{T}}}) = 1/a(\mathcal{T}_1).$$

Hence $\hat{\mathcal{T}}$ is of Type II with $i$ and $y$ as its characteristic vertices. Thus

$$a(\hat{\mathcal{T}}) = \lim_{w \to \infty} a(\hat{\mathcal{T}}_w) = \lim_{w \to \infty} \frac{1}{\rho(M_{(i-y)\hat{\mathcal{T}}_w} - \frac{\gamma_w}{w}J)}$$

$$= \frac{1}{\rho(M_{(i-u)\hat{\mathcal{T}}})} = a(\mathcal{T}_1).$$

*Case III*: If $\rho(M_{y_{T_2}}) > 1/a(\mathcal{T}_1)$, then by Theorem 6.4.1 the characteristic vertex (vertices) of $\hat{\mathcal{T}}$ lies in $\mathcal{T}_2$. If $\hat{\mathcal{T}}$ is of Type I with characteristic vertex $v$, then

$$a(\hat{\mathcal{T}}) = \frac{1}{\rho(M_{v_{\hat{\mathcal{T}}}})} < \frac{1}{\rho(M_{i_{T_1}})} = a(\mathcal{T}_1).$$

Now if $\hat{\mathcal{T}}$ is of Type II with characteristic vertices $p$ and $q$ where $p$ lies on the path from $q$ to $i$, then

$$a(\hat{\mathcal{T}}) = \lim_{w \to \infty} a(\hat{\mathcal{T}}_w) = \lim_{w \to \infty} \frac{1}{\rho(M_{(q-p)_{\hat{\mathcal{T}}_w}} - \frac{\gamma_w}{\theta}J)}$$

$$= \frac{1}{\rho(M_{(q-p)_{\hat{\mathcal{T}}}} - \frac{\gamma}{\theta}J)} < \frac{1}{\rho(M_{i_{T_1}})} = a(\mathcal{T}_1).$$

where the strict inequality is due to Claim 6.5.6. The conclusion of this lemma now follows. □

The results we have proved in this section permit us now to proceed to the proof of the main result of this section which is from [69]:

**THEOREM 6.5.11** *Let $\mathcal{T}_1$ and $\mathcal{T}_2$ be trees and suppose that $\hat{\mathcal{T}}_w$ is a tree formed from joining a vertex $v$ in $\mathcal{T}_1$ to a vertex $u$ in $\mathcal{T}_2$ by an edge $e$ of weight $w$. Then for the tree $\hat{\mathcal{T}}$ at infinity we have that $a(\hat{\mathcal{T}}) = \min\{a(\mathcal{T}_1), a(\mathcal{T}_2)\}$ if and only if the component tree, $\mathcal{T}_1$ or $\mathcal{T}_2$, with the lower algebraic connectivity, say $\mathcal{T}_1$, is of Type I and one of the following conditions holds:*

*(i) The characteristic vertex of $\mathcal{T}_1$ is the only characteristic vertex of $\hat{\mathcal{T}}$.*

*(ii) The characteristic vertex of $\mathcal{T}_1$ is incident to $e$ and $\rho(M_{u_{T_2}}) \leq 1/a(\mathcal{T}_1)$.*

**Proof**: Without loss of generality, let $a(\mathcal{T}_1) \leq a(\mathcal{T}_2)$. (Hence $\min\{a(\mathcal{T}_1), a(\mathcal{T}_2)\} = a(\mathcal{T}_1)$). The necessity of $\mathcal{T}_1$ being a Type I tree in order for $a(\hat{\mathcal{T}})$ to equal $a(\mathcal{T}_1)$ follows from Lemma 6.5.7. Let $i$ be the characteristic vertex of $\mathcal{T}_1$. If $e$ is not incident to $i$ then by Lemma 6.5.8 $a(\hat{\mathcal{T}}) = a(\mathcal{T}_1)$ if and only if $i$ is the characteristic vertex of $\hat{\mathcal{T}}$. If $e$ is incident to $i$, then by Lemma 6.5.10 $a(\hat{\mathcal{T}}) = a(\mathcal{T}_1)$ if and only if $\rho(M_{u_{T_2}}) \leq 1/a(\mathcal{T}_1)$. □

**REMARK 6.5.12** Suppose that equality holds in (6.5.3), namely, the algebraic connectivity of the tree $\hat{\mathcal{T}}$ at infinity is equal to the minimum of the algebraic connectivities of $\mathcal{T}_1$ and $\mathcal{T}_2$. Then it is possible to determine precisely what type of tree we obtain for $\hat{\mathcal{T}}$. Assume without loss of generality that $a(\hat{\mathcal{T}}) = a(\mathcal{T}_1) \leq a(\mathcal{T}_2)$. Then by Lemma 6.5.7, $\mathcal{T}_1$ is a Type I tree. Let $i$ be the characteristic vertex of $\mathcal{T}_1$. If $i$ is incident to $e$, then by Lemma 6.5.10, $\hat{\mathcal{T}}$ is of Type I if $\rho(M_{y_{T_2}}) < 1/a(\mathcal{T}_1)$, while $\hat{\mathcal{T}}$ is of Type II if $\rho(M_{y_{T_2}}) = 1/a(\mathcal{T}_1)$, where $y$ is the vertex in $\mathcal{T}_2$ which is incident to $e$. Finally, if $i$ is not incident to $e$, then by Lemma 6.5.8, $i$ is the only characteristic vertex of $\hat{\mathcal{T}}$ and thus $\hat{\mathcal{T}}$ is of Type I.

# Exercise:

1. For each edge in the tree below, let its weight tend to infinity while the weight of the remaining edges remains fixed. Use Theorem 6.5.11 to determine if the resulting tree at infinity is of Type I or Type II.

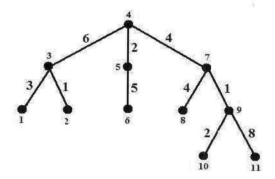

## 6.6 The Characteristic Elements of a Tree

In this section, we generalize the results from Section 6.2 to eigenvectors of the Laplacian matrix corresponding to eigenvalues other than the algebraic connectivity. Recall from Section 6.2 that if $y$ is an eigenvector corresponding to the algebraic connectivity of $\mathcal{G}$, then vertex $v$ is the unique characteristic vertex of $\mathcal{G}$ if $y_v = 0$ and there exists a vertex $w$ in $\mathcal{G}$ adjacent to $v$ such that $y_w \neq 0$. A graph $\mathcal{G}$ has two characteristic vertices if there exists adjacent vertices $v$ and $w$ such that $y_v y_w < 0$. The edge incident to both vertices $v$ and $w$ is often referred to as the *characteristic edge*. We now investigate the characteristic vertices and edges determined by eigenvectors corresponding to other eigenvalues of the Laplacian matrix. We restrict our attention to trees and begin with some definitions.

**DEFINITION 6.6.1** If $\lambda_k > \lambda_{k-1}$ ($k \geq 2$), a *k-vector* is an eigenvector corresponding to the $k^{th}$ smallest eigenvalue, $\lambda_k$, of $L(\mathcal{G})$.

**DEFINITION 6.6.2** The *characteristic set* of $\mathcal{T}$ with respect to $y$, denoted as $\mathcal{C}(\mathcal{T}, y)$, is the set of all characteristic vertices and edges of $\mathcal{T}$ with respect to the eigenvector $y$ of $L(\mathcal{T})$.

**DEFINITION 6.6.3** A tree is *k-simple* if $|\mathcal{C}(\mathcal{T}, y)| = 1$ for all k-vectors $y$.

We should note here that in a characteristic set, vertices incident to a characteristic edge are not considered part of the characteristic set (despite having been referred to as characteristic vertices of a Type II tree in the earlier sections). Only the characteristic edge incident to such vertices is considered to be part of the characteristic set. Therefore:

**REMARK 6.6.4** All trees are 2-simple.

We begin with a lemma and a claim from [70]. The lemma gives a lower bound on the nonzero eigenvalues of $L$. This lower bound is used to prove the subsequent claim which begins to generalize Theorem 6.2.4 regarding zero-valued vertices determined by various eigenvectors of the Laplacian matrix corresponding to the same eigenvalue.

**LEMMA 6.6.5** *Let $\mathcal{G}$ be a connected graph with Laplacian $L$ and let $y$ be an eigenvector of $L$ corresponding to a nonzero eigenvalue $\lambda$. Let $\mathcal{G}_1$ be a connected subgraph of $\mathcal{G}$ such that for each vertex $v \in \mathcal{G}_1$, $y_v > 0$ and for each vertex $w \notin \mathcal{G}_1$ adjacent to any vertex in $\mathcal{G}_1$, $y_w \leq 0$. Then $\lambda_1(L[\mathcal{G}_1]) \leq \lambda$. Further, the inequality is strict if for some such vertex $w$, $y_w < 0$.*

**Proof**: Let $\mathcal{G}_2 = \mathcal{G} \setminus \mathcal{G}_1$. Consider the eigenvalue-eigenvector equation

$$\begin{bmatrix} L[\mathcal{G}_1] & L_{12} \\ L_{12}^T & L[\mathcal{G}_2] \end{bmatrix} \begin{bmatrix} y[\mathcal{G}_1] \\ y[\mathcal{G}_2] \end{bmatrix} = \lambda \begin{bmatrix} y[\mathcal{G}_1] \\ y[\mathcal{G}_2] \end{bmatrix}.$$

From the fact that $L[\mathcal{G}_1]y[\mathcal{G}_1] + L_{12}y[\mathcal{G}_2] = \lambda y[\mathcal{G}_1]$ and the fact that $L_{12}y[\mathcal{G}_2] \geq 0$, we have

$$L[\mathcal{G}_1]y[\mathcal{G}_1] \leq \lambda y[\mathcal{G}_1] \qquad (6.6.1)$$

where stict inequality would hold if $L_{22}y[\mathcal{G}_2]$ is nonzero.

Observe that $L[\mathcal{G}_1]$ is a proper principal submatrix of the irreducible singular M-matrix $L$. Hence by Theorem 1.4.10 it follows that $L[\mathcal{G}_1]$ is a nonsingular M-matrix. Thus $\lambda_1(L[\mathcal{G}_1]) > 0$ and the corresponding eigenvector, say $z$, is positive. Thus multiplying (6.6.1) on the left by $z^T$ we obtain

$$\lambda z^T y[\mathcal{G}_1] \geq z^T L[\mathcal{G}_1]y[\mathcal{G}_1] = (L[\mathcal{G}_1]z)^T y[\mathcal{G}_1] = \lambda_1(L[\mathcal{G}_1])z^T y[\mathcal{G}_1].$$

Since $z^T y[\mathcal{G}_1] > 0$, it follows that $\lambda \geq \lambda_1(L[\mathcal{G}_1])$. If there is a vertex $w \notin \mathcal{G}_1$ that is adjacent to a vertex in $\mathcal{G}_1$ such that $y_w < 0$, then $L_{12}y[\mathcal{G}_2]$ is nonzero and hence strict inequality would follow. □

**CLAIM 6.6.6** *Let $\mathcal{G}$ be a connected graph with Laplacian $L$ and let $y$ be an eigenvector of $L$ corresponding to a nonzero eigenvalue $\lambda$. Let $v$ be a vertex in $\mathcal{C}(\mathcal{G}, y)$ such that $\mathcal{G} - v$ has a positive component with respect to $y$. If $z$ is any other eigenvector corresponding to $\lambda$, then $z_v = 0$.*

**Proof**: Since $v$ is a characteristic vertex with respect to the eigenvector $y$ and since $\mathcal{G} - v$ has a positive component, it follows that $v$ is a cut vertex, for otherwise $\mathcal{G} - v$ would only have one component and by a variation of Theorem 6.1.3 this component must contain negatively and positively $y$-valued vertices adjacent to $v$. Also, by the eigenvalue-eigenvector relationship and the fact that $y_v = 0$, it follows that $\lambda$ is an eigenvalue of $L[B]$ corresponding to a positive eigenvector. Since $L[B]$ is a nonsingular M-matrix, it follows from Observation 1.4.2 that $\lambda_1(L[B])) = \lambda$.

If $z_v \neq 0$, then without loss of generality let $z_v < 0$. Consider the vector $x = \epsilon z + y$

for some $\epsilon > 0$. First note that $x$ is an eigenvector of $L$ corresponding to $\lambda$ and that $x_v < 0$. Choose $\epsilon$ so small so that $x[B] > 0$. Applying Lemma 6.6.5, it follows that $\lambda_1(L[B]) < \lambda$ which is a contradiction. Thus $z_v = 0$. □

One of the goals of this section is to generalize Theorem 6.2.4 to all trees containing a unique characteristic vertex with respect to a $k$-vector. To this end, we prove the following lemmas from [25]. The first is a useful matrix theoretic result based on Observation 1.4.2 and the second gives bounds on $\lambda_k$ in terms of the eigenvalues of the submatrices of $L$ corresponding to components at a characteristic vertex determined by a $k$-vector.

**LEMMA 6.6.7** *Let $A$ be a symmetric, irreducible M-matrix. Then the smallest eigenvalue $\lambda_1$ of $A$ is simple and the corresponding eigenvector is positive, unique up to a scalar multiple. Moreover, if there exists a positive vector $x$ such that $Ax \leq \alpha x$, then $\lambda_1 \leq \alpha$ with equality if and only if $x$ is an eigenvector of $A$ corresponding to the eigenvalue $\alpha$.*

**Proof**: Write $A := sI - B$ where $B$ is nonnegative (and irreducible) and $s \geq \rho(B)$. Since $B$ is nonnegative and irreducible, it follows from Theorem 1.3.20 that $\rho(B)$ is a simple eigenvalue of $B$ and that the corresponding eigenvector $x$ is positive (up to scalar multiplication). Hence $\lambda_1 = s - \rho(B)$ is the smallest eigenvalue of $A$ with $x$ as the corresponding eigenvector. Moreover, if $Ax \leq \alpha x$, then $Bx \geq (s - \alpha)x$. The result now follows from Corollary 1.3.3. □

**LEMMA 6.6.8** *Suppose that $y$ is a $k$-vector ($k \geq 3$) of a tree $\mathcal{T}$ and $u$ is a vertex in $\mathcal{C}(\mathcal{T}, y)$. Suppose that $|\mathcal{C}(\mathcal{T}, y)| \leq 2$. Then there exists $m$ ($1 \leq m \leq k - 2$) components $T_0^1, \ldots, T_0^m$ of $\mathcal{T} - u$ for which*

$$\lambda_1(L[T_0^i]) \leq \lambda_{k-2}\left(\bigoplus_{i=1}^m L[T_0^i]\right) < \lambda_k \leq \lambda_{k-1}\left(\bigoplus_{i=1}^m L[T_0^i]\right)$$

*for $i = 1, \ldots, m$. Moreover, there is at least one component $T_1$ of $\mathcal{T} - u$ for which $\lambda_1(L[T_1]) = \lambda_k$. For the rest of the components $T_2, \ldots, T_s$, we have $\lambda_1(L[T_i]) \geq \lambda_k$. Further, if $\lambda_k$ is not an eigenvalue of $\oplus_{i=1}^m L[T_0^i]$, then there exists another component $T_2 \neq T_1$ such that $\lambda_1(L[T_2]) = \lambda_k$*

Before proceeding with the proof, we show, for visual clarity, the structure of the Laplacian matrix:

$$L = \begin{bmatrix} L[T_0^1] & 0 & \cdots & 0 & 0 & 0 & \cdots & 0 & x(0)^1 \\ 0 & L[T_0^2] & \cdots & 0 & 0 & 0 & \cdots & 0 & x(0)^2 \\ 0 & 0 & \cdots & 0 & 0 & 0 & \cdots & 0 & \cdots \\ 0 & 0 & \cdots & L[T_0^m] & 0 & 0 & \cdots & 0 & x(0)^m \\ 0 & 0 & \cdots & 0 & L[T_1] & 0 & \cdots & 0 & x(1) \\ 0 & 0 & \cdots & 0 & 0 & L[T_2] & \cdots & 0 & x(2) \\ 0 & 0 & \cdots & 0 & 0 & 0 & \cdots & 0 & \cdots \\ 0 & 0 & \cdots & 0 & 0 & 0 & \cdots & L[T_s] & x(s) \\ (x(0)^1)^T & (x(0)^2)^T & \cdots & (x(0)^m)^T & x(1)^T & x(2)^T & \cdots & x(s)^T & \deg(u) \end{bmatrix}$$

**Proof**: Let $y$ be a $k$-vector giving a valuation of the vertices of $\mathcal{T}$. Then $\mathcal{T} - u$ has at least two nonzero components $C_1, C_2$. As $|\mathcal{C}(\mathcal{T}, y)| \leq 2$, at least one of $C_1$ and $C_2$ is negative or positive, say $C_1$. Note that

$$L[C_i]y[C_i] = \lambda_k y[C_i] \qquad (6.6.2)$$

for $i = 1, 2$ which implies that $\lambda_k$ is an eigenvalue of $L[C_i]$. Moreover, by Lemma 6.6.7, we obtain $\lambda_1(L[C_1]) = \lambda_k$ as $y[C_1]$ is positive or negative.

Take $z$ to be a vector such that $z[C_1] = y[C_1]$ and $z_v = 0$ for $v \in V(\mathcal{T}) \setminus V(C_1)$. Let $w$ be the unique vertex of $C_1$ that is adjacent to the vertex $u$ in $\mathcal{T}$. Then

$$L[\mathcal{T} - u]z[\mathcal{T} - u] = \lambda_k z[\mathcal{T} - u]$$

and

$$z[\mathcal{T} - u]^T L[\mathcal{T} - u, u] = y[C_1]^T L[C_1, u] = -y_w \neq 0.$$

Considering the matrix $L - \lambda_k I$, its principal submatrix $L[T - u] - \lambda_k I$, and the vector $(L - \lambda_k I)[\mathcal{T} - u, u]$, and letting $\lambda_-(F)$ denote the number of negative eigenvalues (counting multiplicity) of a matrix $F$, it follows from Lemma 1.12 of [29] that

$$\lambda_-(L - \lambda_k I) = \lambda_-(L[\mathcal{T} - u] - \lambda_k I) + 1. \qquad (6.6.3)$$

Therefore, $\lambda_-(L[\mathcal{T} - u] - \lambda_k I) = \lambda_-(L - \lambda_k I) - 1 = k - 2$ which implies

$$\lambda_{k-2}(L[T - u]) < \lambda_k \qquad (6.6.4)$$

and

$$\lambda_{k-1}(L[\mathcal{T} - u]) \geq \lambda_k \qquad (6.6.5)$$

By (6.6.4) and the fact that $k \geq 3$, we may assume without loss of generality that $T_0^1, \ldots, T_0^m$ ($m \geq 1$) are those components of $\mathcal{T} - u$ such that $\lambda_1(L[T_0^i]) < \lambda_k$ for $i = 1, \ldots, m$. Then $m \leq k - 2$ by (6.6.5). For the other components $T_i$ of $\mathcal{T} - u$, we have $\lambda_1(L[T_i]) \geq \lambda_k$. So

$$\lambda_{k-2}(L[\mathcal{T} - u]) = \lambda_{k-2}\left(\bigoplus_{i=1}^m L[T_0^i]\right) < \lambda_k$$

and

$$\lambda_{k-1}\left(\bigoplus_{i=1}^m L[T_0^i]\right) \geq \lambda_k$$

and hence $\lambda_1(L[T_0^i]) \leq \lambda_{k-2}(\oplus_{i=1}^m L[T_0^i])$ since $\lambda_1(L[T_0^i]) < \lambda_k$.

To complete the proof, suppose that $\lambda_k$ is not an eigenvalue of $\oplus_{i=1}^m L[T_0^i]$. Then $C_2 \neq T_0^i$ for $i = 1, \ldots, m$, and hence $C_2$ is one of the components $T_i$ of $\mathcal{T} - u$ with $\lambda_1(L[T_i]) \geq \lambda_k$. Therefore, $\lambda_1(L[C_2]) \geq \lambda_k$, and hence $\lambda_1(L[C_2]) = \lambda_k$ since by (6.6.2) $\lambda_k$ is an eigenvalue of $L[C_2]$. □

**REMARK 6.6.9** *[25]* Suppose that $y$ is a $k$-vector of a tree $\mathcal{T}$ and $u$ is a vertex in $\mathcal{C}(\mathcal{T}, y)$. Suppose also that $|\mathcal{C}(\mathcal{T}, y)| \leq 2$. In the language of Lemma 6.6.8 consider the components of $\mathcal{T} - u$:

(i) $T_0^i$ for $i = 1, \ldots, m$ where $1 \leq m \leq k - 2$.
(ii) $T_1^j$ for $j = 1, \ldots, p$ where $p \geq 1$.
(iii) $T_2^\ell$ for $\ell = 1, \ldots, q$ where $q \geq 0$. (If $q = 0$, that means these components do not exist.)

By Lemma 6.6.8, these components can be ordered as follows:

$$\lambda_1(L[T_0^i]) \leq \lambda_{k-2}\left(\bigoplus_{i=1}^m L[T_0^i]\right) < \lambda_1(L[T_1^j]) = \begin{cases} \lambda_k \leq \lambda_{k-1}\left(\bigoplus_{i=1}^m L[T_0^i]\right) \\ \lambda_k < \lambda_1(L[T_2^\ell]) \end{cases}$$

**EXAMPLE 6.6.10** Consider the tree $\mathcal{T}$ below:

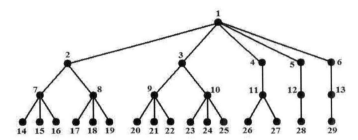

It can be calculated that $\lambda_5 = 0.1985 > \lambda_4 = 0.1649$. Thus let us investigate the eigenvector corresponding to $\lambda_5 = 0.1985$ (i.e., let $k = 5$). Upon further calculations, the corresponding eigenvector $y$ is such that $y_5 = 0.445$, $y_6 = -0.445$, $y_{12} = 0.802$, $y_{13} = -0.802$, $y_{28} = 1$, $y_{29} = -1$, and $y_i = 0$ for all other values $i = 1, \ldots, 29$. Thus vertex $u = 1$ is the unique characteristic element in $\mathcal{C}(\mathcal{T}, y)$. Considering the branches at vertex 1, upon further computations (the results of which we give in the next paragraph) we see that we can label the branch containing vertex 2 as $T_0^1$, the branch containing vertex 3 as $T_0^2$, the branch containing vertex 4 as $T_0^3$, the branch containing vertex 5 as $T_1^1$, and the branch containing vertex 6 as $T_1^2$.

We labeled the branches in the previous paragraph as such because the eigenvalues of $L[T_0^1] = L[T_0^2]$ are $0.075, 0.208, 1, 1, 1, 1, 2.447, 4.792, 5.478$. The eigenvalues of $L[T_0^3]$ are $0.1387, 1, 1.746, 4.115$. Finally, the eigenvalues of $L[T_1^1] = L[T_1^2]$ are $0.1985, 1.556, 3.247$. Observe that $\lambda_1(L[T_0^1]) = \lambda_1(L[T_0^2]) = 0.075$ while $\lambda_1(L[T_0^3]) = 0.1387$. Also observe that $\lambda_3\left(\bigoplus_{i=1}^3 L[T_0^i]\right) = 0.1387$. Thus $\lambda_1(L[T_0^i]) \leq \lambda_{k-2}\left(\bigoplus_{i=1}^3 L[T_0^i]\right)$ for $i = 1, 2, 3$ in accordance with Remark 6.6.9. Also observe that $\lambda_1(L[T_1^j]) = 0.1985 = \lambda_5$ for $j = 1, 2$. Therefore, $\lambda_{k-2}\left(\bigoplus_{i=1}^3 L[T_0^i]\right) < \lambda_1(L[T_1^j]) = \lambda_k$ for $i = 1, \ldots, m$ in accordance with Remark 6.6.9. Finally, $\lambda_4\left(\bigoplus_{i=1}^3 L[T_0^i]\right) = 0.208$. Hence $\lambda_k \leq \lambda_{k-1}\left(\bigoplus_{i=1}^m L[T_0^i]\right)$ as stated in Remark 6.6.9. It should be noted that there are no branches of the form $T_2^\ell$ in this example.

The following lemma from [25] gives us information as to the multiplicity $m_L(\lambda_k)$ of $\lambda_k$ as an eigenvalue of $L$. Note that if $k = 2$ then $p = 0$. Therefore this lemma is a generalization of Theorem 6.2.18.

**LEMMA 6.6.11** *Let $\mathcal{T}$ be a tree and $L$ be its Laplacian matrix. Suppose $y$ is a $k$-vector of a tree $\mathcal{T}$ and $u$ is a vertex in $\mathcal{C}(\mathcal{T}, y)$. Suppose also that $|\mathcal{C}(\mathcal{T}, y)| \leq 2$. Let $T_0^i$, $T_1^j$, $T_2^\ell$ be the components of $\mathcal{T} - u$ as described in Remark 6.6.9. Then*

$$m_L(\lambda_k) = p - 1 + m_{\oplus_{i=1}^m L[T_0^i]}(\lambda_k).$$

**Proof**: Let $y$ be a vector giving a valuation of the vertices of $\mathcal{T}$. Since $|\mathcal{C}(\mathcal{T}, y)| \leq 2$, it follows that $\mathcal{T} - u$ necessarily has a positive or a negative component. For any $k$-vector $x$, $x_u = 0$ by Claim 6.6.6, and for any component $C$ of $\mathcal{T} - u$ we have

$$L[C]x[C] = \lambda_k x[C]. \tag{6.6.6}$$

Let $w_0^i$ and $w_1^j$ be the unique vertices of $T_0^i$ and $T_1^j$, respectively, which are adjacent to $u$ in $\mathcal{T}$. We have the following two cases:

*Case I*: Suppose that $\lambda_k$ is not an eigenvalue of $\oplus_{i=1}^m L[T_0^i]$. Then for any $k$-vector $x$, by (6.6.6) we have $x\left[\cup_{i=1}^m T_0^i\right] = 0$. Additionally, $x[T_2^\ell] = 0$ as $\lambda_k$ is also not an eigenvalue for $\ell = 1, \ldots, q$.

By Lemma 6.6.8, $\mathcal{T} - u$ has at least two components $T_1$ and $T_2$ with $\lambda_1(L[T_i]) = \lambda_k$ for $i = 1, 2$. So using the order of components of $\mathcal{T} - u$ given in Remark 6.6.9, we have $p \geq 2$. For $i = 1, \ldots, p-1$, let $z_{i,p}$ be the vector such that $z_{i,p}[T_1^i]$ is a positive eigenvector of $L[T_1^i]$, $z_{i,p}[T_1^p]$ is a negative eigenvector of $L[T_1^p]$, $z_{i,p} = 0$ elsewhere, and $z_{i,p}$ is normalized so that $z_{i,p}[w_1^i] = -z_{i,p}[w_1^p]$. Clearly $x$ is in the span of the linear independent set of $k$-vectors $\{z_{i,p}, i = 1, \ldots, p-1\}$. The result follows.

*Case II*: Suppose that $\lambda_k$ is an eigenvalue of $\oplus_{i=1}^m L[T_0^i]$. Suppose that $w_1, \ldots, w_r$ are the linearly independent eigenvectors of $\oplus_{i=1}^m L[T_0^i]$ corresponding to $\lambda_k$. We need to show that $m_L(\lambda_k) = p - 1 + r$. To this end, for $j = 1, \ldots, r$, define $z_{j,p}$ to be the vector such that $z_{j,p}\left[\cup_{i=1}^m T_0^i\right] = w_j$, $z_{j,p}[T_1^p]$ is $\alpha$ times a positive eigenvector $v$ of $L[T_1^p]$, and $z_{j,p} = 0$ elsewhere, where $\alpha = -(\sum_{i=1}^m w_j[w_0^i])/v[w_1^p]$. Combining the vectors $z_{j,p}$ with the vectors $z_{i,p}$ defined in Case I, we get the set of vectors $\{z_{j,p}, z_{i,p} : j = 1, \ldots, r; i = 1, \ldots, p-1\}$. Clearly this set is a linearly independent set of $k$-vectors and any $k$-vector $x$ is a linear combination of the vectors in this set. This proves Case II. □

For the first main theorem of this section which is from [25], we now refine the above result to $k$-simple trees, i.e., where $|\mathcal{C}(\mathcal{T}, y)| = 1$. Observe that part (ii) is a generalization of Theorem 6.2.4.

**THEOREM 6.6.12** *Let $\mathcal{T}$ be a $k$-simple tree with Laplacian matrix $L$.*

*(i) For a given $k$-vector $y$ of $\mathcal{T}$ with $\mathcal{C}(\mathcal{T}, y) = \{u\}$, let $T_0^i$, $T_1^j$, and $T_2^\ell$ be components of $\mathcal{T} - u$ as described in Remark 6.6.9. Then $p \geq 2$ in the order of*

the components of Remark 6.6.9, and $\lambda_k$ is not an eigenvalue of $\oplus_{i=1}^m L(\mathcal{T}_0^i)$ so that $m_L(\lambda_k) = p - 1$ and $y\left[\cup_{i=1}^m T_0^i\right] = 0$, $y\left[\cup_{\ell=1}^q T_2^\ell\right] = 0$, $y[T_1^j]$ is either positive or negative or zero for $j = 1, \ldots, p$. Moreover,

$$\lambda_{k-2}\left(\bigoplus_{i=1}^m L[T_0^i]\right) < \lambda_k < \lambda_{k-1}\left(\bigoplus_{i=1}^m L[T_0^i]\right)$$

(ii) $\mathcal{C}(\mathcal{T}, y) = \{u\}$ for any $k$-vector $y$. In other words, $\mathcal{C}(\mathcal{T}, y)$ is invariant regardless of the choice of $k$-vectors $y$.

**Proof**: Let $y$ be a vector valuating the vertices of $\mathcal{T}$. As $\mathcal{C}(\mathcal{T}, y) = \{u\}$, $\mathcal{T} - u$ contains a positive component $T_1$ and a negative component $T_2$. Also, $L[T_i]y[T_i] = \lambda_k y[T_i]$ for $i = 1, 2$ so that $\lambda_k = \lambda_1(L[T_1]) = \lambda_1(L[T_2])$ by Theorem 1.3.20. Then in the order of the components of $\mathcal{T} - u$ in Remark 6.6.9 we have $p \geq 2$.

We now claim that $\lambda_k$ is not an eigenvalue of $\oplus_{i=1}^m L[T_0^i]$, and hence $m_L(\lambda_k) = p - 1$ by Lemma 6.6.11. If $\lambda_k$ were an eigenvalue of $\oplus_{i=1}^m L[T_0^i]$, then by Case II of the proof of Theorem 6.6.11, the eigenvectors $w_j$ ($1 \leq j \leq r$) of $\oplus_{i=1}^m L[T_0^i]$ corresponding to the eigenvalue $\lambda_k$ contain both positive and negative entries by Lemma 6.6.7 as $\lambda_k > \lambda_1(L[T_0^i])$ for $i = 1, \ldots, m$ by Lemma 6.6.8. Therefore $|\mathcal{C}(\mathcal{T}, z)| \geq 2$ for the $k$-vectors $z = z_{j,p}$ as defined in the proof of Theorem 6.6.11, which contradicts the definition of $k$-simple trees. Thus $\lambda_k$ is not an eigenvalue of $\oplus_{i=1}^m L[T_0^i]$, and hence $m_L(\lambda_k) = p - 1$.

For $i = 1, \ldots, m$, by the eigenvalue-eigenvector relationship $L[T_0^i]y[T_0^i] = \lambda_k y[T_0^i]$, we obtain $y[T_0^i] = 0$ as $\lambda_k$ is not an eigenvalue of $L[T_0^i]$ by the previous paragraph. Similarly, $y[T_2^\ell] = 0$ for $\ell = 1, \ldots, q$. Since $L[T_1^j]y[T_1^j] = \lambda_k y[T_1^j] = \lambda_1(L[T_1^j])y[T_1^j]$ for $j = 1, \ldots, p$, it follows by Lemma 6.6.7 that $y[T_1^j]$ is either positive, negative, or zero.

From the order of the component of $\mathcal{T} - u$ as in Remark 6.6.9, we obtain $\lambda_{k+1}(\oplus_{i=1}^m L[T_0^i]) \geq \lambda_k$. But since $\lambda_k$ is not an eigenvalue of $\oplus_{i=1}^m L[T_0^i]$, it follows that

$$\lambda_{k-1}\left(\bigoplus_{i=1}^m L[T_0^i]\right) > \lambda_k > \lambda_{k-2}\left(\bigoplus_{i=1}^m L[T_0^i]\right).$$

For any $k$-vector $z$, $z_u = 0$ by Claim 6.6.6. By a similar discussion as above, we obtain $z[T_0^i] = 0$ for $i = 1, \ldots, m$, and $z[T_2^\ell] = 0$ for $\ell = 1, \ldots, q$. Also by Lemma 6.6.7 we find that $z[T_1^j]$ is either positive of negative or zero for $j = 1, \ldots, p$. Thus with respect to $z$, $u$ is the only zero-valuated vertex adjacent to nonzero valuated vertices. Moreover, there are no pairs of adjacent vertices $a$ and $b$ in $\mathcal{T}$ such that $z_a z_b < 0$. Hence $\mathcal{C}(\mathcal{T}, z) = \{u\}$ and the result follows. □

**REMARK 6.6.13** *[25]* Since $\mathcal{C}(\mathcal{T}, y)$ is fixed for any $k$-vector $y$ by Theorem 6.6.12, we can let $\mathcal{T}$ be a $k$-simple tree and let $\mathcal{C}(\mathcal{T}, y) = \{u\}$ for any $k$-vector $y$. Then by Theorem 6.6.12, the components $T_0^i$, $T_1^j$, $T_2^\ell$, of $\mathcal{T} - u$ can be further ordered as:

$$\lambda_1(L[T_0^i]) \leq \lambda_{k-2}\left(\bigoplus_{i=1}^m L[T_0^i]\right) < \lambda_1(L[T_1^j])$$

$$= \lambda_k < \min\left\{\lambda_{k-1}\left(\bigoplus_{i=1}^m L[T_0^i]\right), \lambda_1(L[T_2^\ell])\right\}.$$

where $i = 1\ldots m$ with $1 \leq m \leq k-2$, $j = 1\ldots p$ with $p \geq 2$, and $\ell = 1\ldots q$ with $q \geq 0$.

At this point, we can characterize all $k$-simple trees. We do so here in the second main theorem of this section which is from [25]:

**THEOREM 6.6.14** *A tree $\mathcal{T}$ is $k$-simple if and only if there exists a cut vertex $u$ in $\mathcal{T}$ such that the components of $\mathcal{T} - u$ can be ordered as*

$$\lambda_1(L[T_0^i]) \leq \lambda_{k-2}\left(\bigoplus_{i=1}^m L[T_0^i]\right) < \lambda_1(L[T_1^1]) = \ldots = \lambda_1(L[T_1^p])$$

$$< \min\left\{\lambda_{k-1}\left(\bigoplus_{i=1}^m L[T_0^i]\right), \lambda_1(L[T_2^\ell])\right\}.$$

where $i = 1\ldots m$ with $1 \leq m \leq k-2$, $j = 1\ldots p$ with $p \geq 2$, and $\ell = 1\ldots q$ with $q \geq 0$.

**Proof**: The necessity follows readily from Corollary 6.6.12 and Remark 6.6.13. As for the sufficiency, let $\lambda_1(L(\mathcal{T}_1^j)) := \alpha$ for $j = 1, \ldots, p$. We need to show two claims: (i) that $\mathcal{T}$ actually has $k$-vectors, i.e., that $\lambda_k(\mathcal{T}) = \alpha > \lambda_{k-1}(\mathcal{T})$, and (ii) that $\mathcal{T}$ is $k$-simple, i.e., that $|\mathcal{C}(\mathcal{T}, y)| = 1$ for any $k$-vector $y$.

*Claim (i)*: To show that $\lambda_k(\mathcal{T}) = \alpha > \lambda_{k-1}(\mathcal{T})$, we first note that since $p \geq 2$, we have that $\lambda_{k-1}(L[T-u]) = \lambda_k(L[T-u])$. By Theorem 1.2.8, it follows that $\lambda_k(\mathcal{T}) = \alpha$. By Lemma 6.6.7, there exists a positive eigenvector $z_1$ such that $L[T_1^1]z_1 = \alpha z_1$. Let $z$ be a vector such that $z[T_1^1] = z_1$ and $z$ is zero elsewhere. Then by the discussion preceeding (6.6.3) it follows that

$$\lambda_-(L(\mathcal{T}) - \alpha I) = \lambda_-(L[T-u] - \alpha I) + 1.$$

By the ordering of the components of $\mathcal{T} - u$ in the statement of this theorem, it follows that $\lambda_-(L(\mathcal{T}) - \alpha I) = k - 2$ and therefore $\lambda_-(L[T-u] - \alpha I) = k - 1$. This implies that $\lambda_k(\mathcal{T}) \geq \alpha > \lambda_{k-1}(\mathcal{T})$. Since we already saw that $\lambda_k(\mathcal{T}) = \alpha$, we obtain $\lambda_k(\mathcal{T}) = \alpha > \lambda_{k-1}(\mathcal{T})$ as desired.

*Claim (ii)*: To show that $|\mathcal{C}(\mathcal{T}, y)| = 1$ for any $k$-vector $y$, we see from Case I of the proof of Theorem 6.6.11 that $z_{1,p}$ is a $k$-vector of $\mathcal{T}$ with $z_{1,p}[u] = 0$. Additionally, $z_{1,p}$ is positive on the component $\mathcal{T}_1^1$ of $\mathcal{T} - u$. By Claim 6.6.6, for any $k$-vector $y$ of $\mathcal{T}$, we have $y_u = 0$. As $\lambda_k$ is not an eigenvalue of $L[T_0^i]$ or $L[T_2^\ell]$, we get $y[T_0^i] = y[T_2^\ell] = 0$ for $i = 1, \ldots, m$ and $\ell = 1, \ldots, q$. By Lemma 6.6.7, $y[T_1^j]$ is either positive, negative, or zero. Hence $u$ is the unique vertex of $\mathcal{T}$ adjacent to a vertex with nonzero valuation via $y$. Moreover, there do not exist adjacent vertices $v$ and $w$ such that $y_v y_w < 0$. Hence $\mathcal{C}(\mathcal{T}, y) = \{u\}$, and thus $\mathcal{T}$ is $k$-simple. □

Recall from Corollary 5.1.13 that for a tree, $\lambda_2 \leq 1$ where equality holds if and only if the tree is a star. The following theorem from [25] generalizes this result to $\lambda_k$ for $k$-simple trees.

**THEOREM 6.6.15** *Let $\mathcal{T}$ be a $k$-simple tree with the unique characteristic element (vertex) $u$. Then $\lambda_k \leq 1$ with equality only if $\mathcal{T} - u$ consists of $p \geq 2$ isolated vertices and $m$ components each containing at least two vertices, where $1 \leq m \leq k - 2$.*

*Moreover, a tree $\mathcal{T}$ is $k$-simple with characteristic vertex $u$ and $\lambda_k = 1$ if and only if $\mathcal{T}$ is obtained from the vertex $u$ by appending to $u$, $p \geq 2$ pendant vertices and $m$ ($1 \leq m \leq k - 2$) trees $T_1, T_2, \ldots, T_m$ each containing at least two vertices such that for $i = 1, \ldots, m$*

$$\lambda_1(L[T_0^i]) \leq \lambda_{k-2}\left(\bigoplus_{i=1}^m L[T_0^i]\right) < 1 < \lambda_{k-1}\left(\bigoplus_{i=1}^m L[T_0^i]\right). \quad (6.6.7)$$

**Proof**: Let $C$ be a component of $\mathcal{T} - u$ and let $w$ be the unique vertex in $C$ that is adjacent to $u$ in $\mathcal{T}$. Letting $y := L[C]e$, then $y_w = 1$ and $y$ is zero on the remaining vertices of $C$. Thus $L[C]e \leq e$. So by Lemma 6.6.7, we have $\lambda_1(L[C]) \leq 1$ with equality if and only if $e$ is an eigenvector of $L[C]$ corresponding to the eigenvalue 1. However, this happens if and only if $C$ contains exactly one vertex.

Since for any component $C$ of $\mathcal{T} - u$ we have $\lambda_1(L[C]) \leq 1$, it follows that in the order of the components of $\mathcal{T} - u$ given in Remark 6.6.13, we have:

(i) For $\ell = 1, \ldots, q$, the trees $T_2^\ell$ do not exist, and hence $q = 0$.

(ii) For $j = 1, \ldots, p$ where $p \geq 2$, it holds that $\lambda_1(L[T_1^j]) = 1$, so that $T_1^j$ consists of an isolated vertex for each $j$.

(iii) For $1 \leq i \leq m$ where $1 \leq m \leq k - 2$, the trees $T_0^i$ each contain at least two vertices since $\lambda_1(L[T_0^i]) < \lambda_k = 1$.

Items (i) through (iii) prove the first result. The second result follows from above and from Theorem 6.6.14. □

**EXAMPLE 6.6.16** Consider the tree $\mathcal{T}$ below:

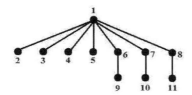

It can be calulated that $\lambda_5 = 1 > \lambda_4 = 0.581$. Thus let us investigate the eigenvector corresponding to $\lambda_5 = 1$ (i.e. let $k = 5$). Upon further calculations, vertex $u = 1$ is the unique characteristic element in $\mathcal{C}(\mathcal{T}, y)$. Considering the branches at vertex 1, clearly we can label the branches containing vertices 6, 7, and 8 as $T_0^1$, $T_0^2$, and $T_0^3$, respectively. The eigenvalues of $L[T_0^i]$ are 0.382 and 2.618 for each $i = 1, 2, 3$. Hence $\lambda_1(L[T_0^i]) = 0.382$ for each $i$, $\lambda_3\left(\bigoplus_{i=1}^3 L[T_0^i]\right) = 0.382$, and $\lambda_4\left(\bigoplus_{i=1}^m L[T_0^i]\right) = 2.618$. Thus (6.6.7) is satisfied.

We close this section with results from [70] and [25] which are generalizations of Theorem 6.2.20 involving the entries of the eigenvectors corresponding to $\lambda_k$.

**THEOREM 6.6.17** *Let $\mathcal{T}$ be a tree and let $y$ be an eigenvector of $L(\mathcal{T})$ corresponding to a nonzero eigenvalue $\lambda$. Let $i$-$j$ be an edge of $\mathcal{T}$. Suppose that the valuations of the vertices (with respect to $y$) in the branch $T_j$ at $i$ containing $j$ are positive. Then along any path that starts at $j$ and does not contain $i$, the entries of $y$ increase and are concave down.*

**Proof**: Consider the vector $u$ defined as follows: $u_s = 1$ if $s \in T_j$ and $u_s = 0$ if $s \notin T_j$. Computing $u^T L y$, we see on the one hand that

$$u^T L y = \lambda u^T y = \lambda \sum_{s \in T_j} y_s.$$

On the other hand, $u^T L y = z^T y$ where $z_i = -1$, $z_j = 1$, and $z_v = 0$ for all other vertices $v$. Thus $u^T L y = z^T y = y_j - y_i$ and hence

$$y_j - y_i = \lambda \sum_{s \in T_j} y_s.$$

Since $y$ is positive on $T_j$, it follows that $y_j > y_i$. Further, if $jk$ is an edge in $T_j$ then a similar argument shows that

$$y_k - y_j = \lambda \sum_{s \in T_k} y_s$$

where $T_k$ is the branch at $j$ containing $k$. Since $T_k$ is a proper subgraph of $T_j$, it follows that $y_k - y_j < y_j - y_i$. Hence the entries in $y$ are increasing in a concave down fashion. □

This along with Theorem 6.6.12 gives the following result from [25]:

**COROLLARY 6.6.18** *Let $\mathcal{T}$ be a $k$-simple tree with the unique characteristic element (vertex) $u$ and let $y$ be any $k$ vector. Then each component of $\mathcal{T} - u$ is either zero, positive, or negative, and along any path in $\mathcal{T}$ which starts at $u$, the entries of $y$ either increase and are concave down, decrease and are concave up, or identically zero.*

## Exercises:

**1.** (See [25]) Let $\mathcal{T}$ be a $k$-simple tree on $n$ vertices and let $w$ be the number of eigenvalues greater than 1 (counting multiplicity). Show that $k \leq n - w - m_L(\lambda_k) + 1$.

**2.** (See [25]) The matching number $\mu(\mathcal{T})$ for a tree is the cardinality of the largest set $S$ of edges of $\mathcal{T}$ in which no two edges in $S$ share a common vertex. Show that if $\mathcal{T}$ is a $k$-simple tree on $n$ vertices then $k \leq n - \mu(\mathcal{T})$.

## 6.7 The Spectral Radius of Submatrices of Laplacian Matrices for Trees

In Section 6.2, we investigated bottleneck matrices for weighted trees on $n$ vertices which were created by taking the inverse of the $(n-1) \times (n-1)$ submatrix of the Laplacian matrix formed by deleting a row and column corresponding to a vertex of the tree. In this section, based mainly on [66], we study the spectral radius of such matrices prior to the inverse being taken. We will find many interesting properties of the spectral radius that compare with the properties of bottleneck matrices.

Let $v$ be a vertex of a weighted tree $\mathcal{T}$ on $n$ vertices and let $L$ be its Laplacian. Define the function $r(v)$ to be the spectral radius of $L[\bar{v}]$. It is known from Corollary 1.2.10 that

$$\lambda_{n-1} \leq r(v) \leq \lambda_n \tag{6.7.1}$$

for each vertex $v \in \mathcal{T}$. This inequality has two components: $\lambda_{n-1} \leq r(v)$ and $r(v) \leq \lambda_n$. We begin our investigation by focusing on the latter component. To do this, we need the following matrix theoretic result based off Proposition 1 of [29] which we state without proof.

**PROPOSITION 6.7.1** *Let $L$ be the Laplacian matrix for a connected weighted tree $\mathcal{T}$ on $n$ vertices and let $\lambda$ be an eigenvalue of $L$. If $\lambda$ is simple and if all entries of the corresponding eigenvector are different from zero, then $\lambda$ cannot be an eigenvalue of any submatrix of $L$ of order $n-1$.*

Since $\lambda_n$ is simple by Theorem 4.3.12, it follows from Proposition 6.7.1 that $r(v)$ can equal $\lambda_n$ only if an eigenvector corresponding to $\lambda_n$ has a zero entry. To see that this can never happen in the Laplacian matrix for a tree, we recall that all nontrivial trees are bipartite and obtain the following definition:

**DEFINITION 6.7.2** *Let $\mathcal{G}$ be a bipartite graph and $L$ be its Laplacian matrix. Let $B$ be the matrix created from $L$ by taking the absolute value of each entry. Then $B$ is the bipartite complement of $L$.*

We proved in Theorem 4.3.12 that if $L$ is the Laplacian matrix for a bipartite graph, then $L$ and $B$ are similar. Recalling that all trees are bipartite, it follows that for any tree $\mathcal{T}$ that $L$ and $B$ are similar via the matrix $U$ in Theorem 4.3.12. Since similar matrices have the same eigenvalues, we obtain the following result from [66]:

**LEMMA 6.7.3** *Let $x$ be an eigenvector for $L$ corresponding to the eigenvalue $\lambda$. Then $Ux$ is an eigenvector of $B$ corresponding to $\lambda$. Moreover, for each entry $x_i$ of $x$, it follows that $x_i = 0$ if and only if $(Ux)_i = 0$.*

**Proof**: Since $L$ and $B$ are unitarily similar via the matrix $U$ from Theorem 4.3.12, we can write $UL = BU$. Hence $ULx = BUx$. Since $x$ is an eigenvector of $L$ corresponding to $\lambda$, it follows that $\lambda Ux = BUx$. Thus $Ux$ is an eigenvector of $B$ corresponding to $\lambda$. By the construction of $U$, it is clear that $x_i = 0$ if and only if $(Ux)_i = 0$. $\square$

Lemma 6.7.3 leads us to our main result concerning the relationship between $r(v)$ and $\lambda_n$, showing that the inequality is actually strict (see [66]):

**THEOREM 6.7.4** *Let $\mathcal{T}$ be a weighted tree and $L$ be its Laplacian matrix. Then $r(v) < \lambda_n$ for all vertices $v \in \mathcal{T}$.*

**Proof**: Let $x$ be an eigenvector of $L$ corresponding to $\lambda_n$. We see that since $\lambda_n$ is the largest eigenvalue of $L$, then by Theorem 4.3.12 it must also be the largest eigenvalue for the bipartite complement $B$ with $Ux$ as its corresponding eigenvector. Observe that $B$ is a nonnegative, irreducible matrix. Thus by Theorem 1.3.8(ii) it follows that $Ux$ does not have a zero entry. Hence, by Lemma 6.7.3, $x$ does not have a zero entry. The conclusion now follows from Proposition 6.7.1. $\square$

We now focus our attention on the vertices $v$ of $\mathcal{T}$ which yield a minimum value for $r(v)$. Recall from (6.7.1) that $r(v) \geq \lambda_{n-1}$. Unlike $\lambda_n$, we will see that there do exist trees which contain vertices in which this inequality is sharp. Consider the following example.

**EXAMPLE 6.7.5** *[66]* Each vertex $v$ in the following unweighted tree is labeled with its value $r(v)$:

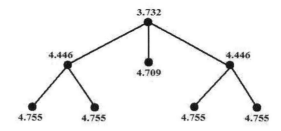

We see that $\lambda_{n-1} = 3.732$ and there is a vertex $v$ in this tree with $r(v) = 3.732$. This leads us to the following definition:

**DEFINITION 6.7.6** *A weighted tree is of Type A if there exists a vertex $v$ such that $r(v) = \lambda_{n-1}$. A weighted tree is of Type B if no such vertex exists.*

This definition is strikingly similar to Definition 6.2.2 which is Merris' classification of trees (see [58]) into Type I and Type II based on the entries in the Fiedler vector. As with the theorems concerning Merris' classification of trees and bottleneck matrices found in Section 6.2, we will make use of the block diagonal structure of the submatrices of $L$ (and $B$). As we saw in Section 6.2, $L[\overline{v}]$ is permutationally similar to a block diagonal matrix with each block corresponding to a branch at $v$. Thus for a vertex $v$, $r(v)$ attains its value from the block(s) of $L[\overline{v}]$ with the maximum spectral radius. This leads us to the following definition:

**DEFINITION 6.7.7** The block(s) of $L[\overline{v}]$ having the maximum spectral radius is the *spectral block(s)* at $v$ and the corresponding branch(es) of $\mathcal{T}$ is the *spectral branch(es)* at $v$.

We now investigate Type A trees and the vertex $v$ such that $r(v) = \lambda_{n-1}$. Before continuing, we need the following definition:

**DEFINITION 6.7.8** In a Type A tree, the vertex $v$ such that $r(v) = \lambda_{n-1}$ is the *spectral vertex* of the tree.

The following theorem from [66] gives useful properties of a spectral vertex:

**THEOREM 6.7.9** *Let $v$ be a vertex in a weighted tree $\mathcal{T}$. Then $v$ has more than one spectral branch if and only if $r(v) = \lambda_{n-1}$. Moreover, if such a vertex $v$ exists, then it is unique.*

**Proof**: First suppose that $r(v) = \lambda_{n-1}$ and that, to the contrary, $v$ has just one spectral branch. Without loss of generality, let $v = 1$. Per Theorem 4.3.12, we will prove this proposition using the bipartite complement, $B$. Permute $B$ such that $v = 1$ and recall $B[\overline{1}]$ is (permutationally similar to) a block diagonal with each block $B_i$ corresponding to the branch $b_i$ at $v = 1$. By Proposition 6.7.1, there exists an eigenvector $x$ of $B$ corresponding to $\lambda_{n-1}$ such that the first entry is zero. Partition $x$ in accordance with the blocks of $B$ as follows:

$$x = [\,0\,|\, y_1 \,|\, y_2 \,|\, \ldots \,|\, y_k\,]^T$$

where for each $i = 1, \ldots, k$, the vector $y_i$ is a Perron vector of $B_i$. Suppose that $v = 1$ has only one spectral branch, say $b_1$. Since each block $B_i$ is nonnegative and irreducible and since $B_i y_i = \lambda_{n-1} y_i$ for each $i$, it follows from Theorem 1.3.8(ii) that $y_1$ has strictly positive entries while $y_i = 0$ for $2 \leq i \leq k$. Hence

$$x = [\,0\,|\, y_1 \,|\, 0 \,|\, \ldots \,|\, 0\,]^T$$

Suppose $y_1$ has, say, $p$ entries. Since the $(1,1)$ entry of $B$ is positive and since there exists exactly one entry in the first $p+1$ entries of the first row of $B$ other than the $(1,1)$ entry that is positive, it follows by matrix multiplication that $(Bx)_1 > 0$ which contradicts the eigenvalue-eigenvector equation $Bx = \lambda_{n-1}x$. Hence $v$ must have at least two spectral branches.

Now suppose $v$ has $m \geq 2$ spectral branches for some $2 \leq m \leq k$, and show that $r(v) = \lambda_{n-1}$. Without loss of generality, let $v = 1$. Using the bipartite complement $B$, let $B_1, \ldots, B_m$ be the spectral blocks of $B$ at $v = 1$. Consider the vector

$$x = [\,0\mid y_1 \mid y_2 \mid \ldots \mid y_m \mid 0 \mid \ldots \mid 0\,]^T$$

where 0 is the first entry, $y_i$ are the Perron vectors for the spectral blocks of $B$, and the remaining entries of $x$ are all zero. Observe $B_i y_i = r(1) y_i$ for each $i$. Since $m \geq 2$, and since every entry in each subvector $y_i$ is nonzero, it follows by matrix-vector multiplication that we can normalize $y_1, \ldots, y_m$ to create a vector $\hat{x}$ so that $B\hat{x} = r(1)\hat{x}$. Thus $r(1)$ is an eigenvalue of $B$. So by (6.7.1), Theorem 4.3.12, and Theorem 6.7.4, it follows that $r(1) = \lambda_{n-1}$.

To show uniqueness, let $v, w \in T$ be such that $v \neq w$ and $r(v) = r(w) = \lambda_{n-1}$. Then $v$ and $w$ must each have at least two spectral branches. Thus there exists a spectral branch at $w$ that does not contain $v$. Let $C$ be the block of $B[\overline{w}]$ that corresponds to such a branch. If we let $C_w$ be the block in $B[\overline{v}]$ that contains the row/column corresponding to vertex $w$, then upon observing the fact that $C$ is a proper submatrix of $C_w$ and both matrices are nonnegative and irreducible, we obtain via Exercise 1(ii) of Section 1.3 that

$$r(v) \geq \rho(C_w) > \rho(C) = r(w) = \lambda_{n-1}$$

which contradicts the fact that $r(v) = \lambda_{n-1}$. Therefore, if there exists vertex such that $r(v) = \lambda_{n-1}$, then such a vertex is unique. □

Theorem 6.7.9 shows if a tree is of Type A, then the spectral vertex is the unique vertex that has more than one spectral branch.

**EXAMPLE 6.7.10** *[66]* Consider the tree $\mathcal{T}$ in Example 6.7.5. Since there exists a vertex $v$ such that $r(v) = \lambda_{n-1}$, we see that $\mathcal{T}$ is of Type A. Labeling the vertices of $\mathcal{T}$ as follows

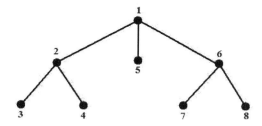

Considering the bipartite complement of $L[\overline{1}]$:

$$B[\overline{1}] = \begin{bmatrix} 3 & 1 & 1 & 0 & 0 & 0 & 0 \\ 1 & 1 & 0 & 0 & 0 & 0 & 0 \\ 1 & 0 & 1 & 0 & 0 & 0 & 0 \\ 0 & 0 & 0 & 1 & 0 & 0 & 0 \\ 0 & 0 & 0 & 0 & 3 & 1 & 1 \\ 0 & 0 & 0 & 0 & 1 & 1 & 0 \\ 0 & 0 & 0 & 0 & 1 & 0 & 1 \end{bmatrix}$$

Observe that the branches containing vertices 2 and 6 are the spectral branches at vertex 1 since the blocks of $B$ corresponding to these branches have the largest spectral radius, namely $\lambda_{n-1} = 3.732$. Note that of all vertices $v = 1, \ldots, 8$, $B[\overline{1}]$ is the only such matrix $B[\overline{v}]$ that has more than one spectral block, hence $v = 1$ is the unique vertex of $\mathcal{T}$ having more than one spectral branch per Theorem 6.7.9.

Observe that Theorem 6.7.9 is very similar in flavor to Theorem 6.2.15 concerning the characteristic vertex and Perron branches of a Type I tree. Type I trees have a unique characteristic vertex and that vertex has at least two Perron branches; Type A trees have a unique spectral vertex and that vertex has at least two spectral branches. Therefore, it is natural to investigate if there are analogous results for the spectral vertices for trees of Type B. In the following theorem from [66], we show such a result which parallels Theorem 6.2.17.

**THEOREM 6.7.11** *Let $\mathcal{T}$ be a weighted tree. Then there exists two adjacent vertices $v$ and $w$ such that a spectral branch at $v$ contains $w$ and a spectral branch at $w$ contains $v$. Moreover, if $\mathcal{T}$ is of Type B, then this pair of vertices is unique.*

**Proof**: Consider the graph $\mathcal{G}$ with vertices $1, 2, \ldots, n$ created as follows: For each vertex $i$, place an edge $i, i'$ where $i'$ is a vertex adjacent to $i$ that lies on a spectral branch at $i$. Since $\mathcal{T}$ has $n - 1$ edges and $\mathcal{G}$ has $n$ edges, it follows that at least one of these edges is duplicated. The vertices incident to a duplicated edge are such vertices $v$ and $w$, hence showing existence.

To show uniqueness in the Type B case, let $x$ and $y$ be another pair of adjacent vertices such that the spectral branch at each of these vertices contains the other vertex. Without loss of generality, let the path from $w$ to $x$ contain no other spectral vertices of $\mathcal{T}$. Let $B$ be the bipartite complement of $L$. For distinct vertices $i$ and $j$, let $B_i^j$ be the block of $B[\overline{i}]$ that contains the row/column corresponding to vertex $j$. Thus by the definition of spectral vertices and by the fact that $B_i^j$ is irreducible for all $i, j \in \mathcal{T}$, we see that

$$\rho(B_w^v) > \rho(B_w^x) > \rho(B_x^y) > \rho(B_x^w) > \rho(B_w^v) \qquad (6.7.2)$$

where the second and fourth inequalities follow from Exercise 1(ii) of Section 1.3. But (6.7.2) reduces to $\rho(B_w^v) > \rho(B_w^v)$, a contradiction. Hence in a Type B tree, such vertices $v$ and $w$ are unique. $\square$

Theorem 6.7.11 lends itself to the following definition which parallels Definition 6.2.3 concerning characteristic vertices.

**DEFINITION 6.7.12** *In a Type B tree, the spectral vertices are the unique pair of adjacent vertices $v$ and $w$ such that the spectral branch at $v$ contains $w$ and the spectral branch at $w$ contains $v$.*

The inequalities in (6.7.2) of Theorem 6.7.11 imply the following corollary from [66] which concerns vertices that are not spectral vertices. Observe that this corollary parallels Theorem 6.2.19 for characteristic vertices.

**COROLLARY 6.7.13** *Let $\mathcal{T}$ be a weighted tree and suppose that $m$ is not a spectral vertex of $\mathcal{T}$, then the unique spectral branch at $m$ is the branch which contains all of the spectral vertices of $\mathcal{T}$.*

At this point we summarize the results for Type A and Type B trees (see [66]) that parallel Theorem 6.2.17:

**THEOREM 6.7.14** *Let $\mathcal{T}$ be a weighted tree with $L$ as its Laplacian matrix. Then:*

*(a) If $k$ is the unique spectral vertex of a Type A tree, then $k$ is the unique vertex that has at least two spectral branches.*

*(b) If $i$ and $j$ are the spectral vertices of a Type B tree, then the unique spectral branch at $i$ contains $j$ while the unique spectral branch at $j$ contains $i$.*

*(c) If $m$ is not a spectral vertex in $\mathcal{T}$, then the unique spectral branch at $m$ contains all of the spectral vertices of $\mathcal{T}$.*

We can also summarize our results thus far (see [66]) which parallel Fiedler's famous result which is stated in Corollary 6.2.1.

**THEOREM 6.7.15** *Let $\mathcal{T}$ be a weighted tree on $n$ vertices labeled $1,\ldots,n$ with Laplacian matrix $L$. Then exactly one of the following occurs:*

*(a) For the function $r(v)$, we have $r(k) = \lambda_{n-1}$ for a unique vertex $k \in \mathcal{T}$. In addition, the value of the function $r(v)$ increases along any path in $\mathcal{T}$ which starts at $k$.*

*(b) The function $r(v)$ never attains $\lambda_{n-1}$ for any vertex $v \in \mathcal{T}$. In this case, there exist a unique pair of adjacent vertices $i$ and $j$ such that the value of $r(v)$ increases along any path in $\mathcal{T}$ which starts at $i$ and does not contain $j$, while the value of $r(v)$ increases along any path which starts at $j$ and does not contain $i$.*

Using the results from Theorem 6.7.14, observe that if condition (a) holds in Theorem 6.7.15, then $\mathcal{T}$ is a Type A tree with $k$ as its spectral vertex, while if condition (b) holds, then $\mathcal{T}$ is a Type B tree with $i$ and $j$ as its spectral vertices.

**REMARK 6.7.16** Theorem 6.7.15 implies that in any tree $\mathcal{T}$, the function $r(v)$ is minimized at a spectral vertex and maximized at a pendant vertex. However, it is important to note that the two spectral vertices of a Type B tree are not necessarily the two vertices in which the function $r(v)$ obtains its two smallest values. Observe the following tree from [66]:

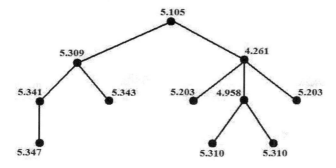

The vertices $v$ with values for $r(v)$ of 4.261 and 4.958 are the two vertices with the smallest values for $r(v)$, yet the vertices with values 4.261 and 5.105 are the spectral vertices.

Recall from Theorem 6.2.20 that as one travels away from the characteristic vertex / vertices of an unweighted tree, the characteristic valuations either increase in a concave down fashion, decrease in a concave up fashion, or are identically zero. Focusing on the concavity aspect, if such vertices are not valuated identically zero, then the absolute values of the valuations increase in a concave down fashion. Thus it is natural to ask if this same idea holds true the function $r(v)$ as we travel away from the spectral vertex / vertices of an unweighted tree. While it is often the case that the same idea holds true, it is not always the case as seen below in an example from [66]:

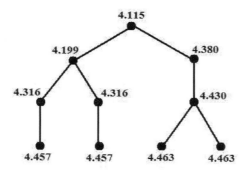

Observe that this is a Type A tree where the vertex $v$ with $r(v) = 4.115$ is the unique spectral vertex. However, the values of $r(v)$ for the vertices do not increase in a concave downward fashion as one travels away from the spectral vertex.

In all of the examples shown thus far, it has been that case that a tree is both Type I and Type A with one vertex being both the characteristic vertex and spectral vertex, or that a tree is both Type II and Type B with a pair of adjacent vertices being both the characteristic vertices and spectral vertices. Therefore, it is natural to ask the following questions:

(1) If a tree is both of Type I and Type A, is the characteristic vertex necessarily

the same as the spectral vertex?

(2) If a tree is both of Type II and Type B, is the pair of characteristic vertices necessarily the same as the pair of spectral vertices?

(3) Are all Type A trees necessarily Type I trees, and are all Type B trees necessarily Type II trees?

Questions (1) and (2) are similar, thus one would expect them to both have the same answer. The answer to both questions is "no." In fact, in regards to Question (1), we can create a tree that is of both Type I and Type A in which the distance from the characteristic vertex to the spectral vertex is arbitrarily large. We do so by considering the following unweighted tree from [66]:

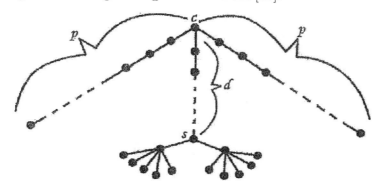

Observe that vertex $s$ has two branches each consisting of one vertex adjacent to five vertices (other than $s$). Letting $L$ be the Laplacian matrix for this tree, note that the spectral radius of the submatrix of $L$ corresponding to each of these branches is 6.854. Observe that in the remaining branch at $s$, the spectral radius of the submatrix of $L$ corresponding to this branch cannot be greater than 6 according to the Geršgorin Disc Theorem (Theorem 1.2.1). Hence $s$ has two spectral branches and thus is the unique spectral vertex in accordance with Theorem 6.7.14. However, by the construction of bottleneck matrices, as $p$ increases the spectral radius of each bottleneck matrix at vertex $c$ containing each branch not containing $s$ increases without bound. Hence, for each fixed $d$, we can make $p$ large enough such that each of these branches are the characteristic branches at $c$. By Theorem 6.2.17, $c$ would be the characteristic vertex. Thus we have created a tree that is of Type I and Type A where the distance between the spectral vertex and characteristic vertex is $d$. Below, we see the minimum value that $p$ must be in order for the distance between the characteristic vertex and spectral vertex to be $d$.

| $d$ | 1 | 2 | 3 | 4 | 5 |
|---|---|---|---|---|---|
| $p$ | 7 | 9 | 11 | 12 | 14 |

Similarly, if a tree is of Type II and Type B, we can create a tree in which the minimum distance between a characteristic vertex and a spectral vertex is as large as we would like. Consider the following unweighted tree from [66]:

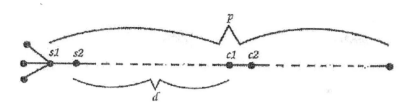

Observe that the spectral radius of the submatrix of the Laplacian matrix corresponding to the branch at $s_1$ containing the three pendant vertices is 1, while the spectral radius of the submatrix corresponding to the other branch is clearly larger than 1. Similarly, the spectral radius of the submatrix corresponding to the branch at $s_2$ containing the three pendant vertices is 4.792, while the submatrix corresponding to the other branch at $s_2$ is less than 4 by the Geršgorin Disc Theorem (Theorem 1.2.1). Thus the unique spectral branch at $s_1$ contains $s_2$, and vice versa. Therefore, $s_1$ and $s_2$ are the spectral vertices of the tree in accordance with Theorem 6.7.11. Letting $c_1$ and $c_2$ be the characteristic vertices, it is clear by the construction of bottleneck matrices that as $p$ increases without bound, the distance $d$ between $c_1$ and $s_2$ increases without bound. Below, we see appropriate values of $p$ for each given distance $d$ between $s_2$ and $c_1$:

| $d$ | 1 | 2 | 3 | 4 | 5 |
|---|---|---|---|---|---|
| $p$ | 7 | 9 | 11 | 13 | 15 |

Finally, the answer to Question (3) is also "no." Consider the trees $\mathcal{T}_1$ and $\mathcal{T}_2$ from [66]:

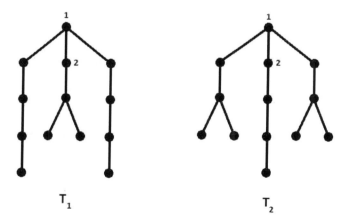

In $\mathcal{T}_1$, vertex 1 is the unique characteristic vertex, yet vertices 1 and 2 are the spectral vertices. Therefore $\mathcal{T}_1$ is of Type I and Type B. In $\mathcal{T}_2$, vertices 1 and 2 are the characteristic vertices yet vertex 1 is the unique spectral vertex. Hence $\mathcal{T}_2$ is of Type II and of Type A. As with the Type I-A trees and the Type II-B trees, we can create trees in which the distance between the characteristic vertices and spectral vertices is as large as we desire. Since the process is similar to the above processes, we omit the details.

# Exercises:

**1.** Classify each of the following trees as Type A or Type B and label the spectral vertex (vertices).

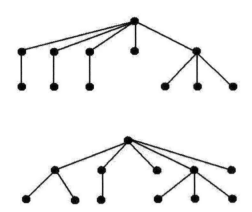

**2.** Let $\mathcal{T}$ be a weighted tree.

(i) Suppose $\mathcal{T}$ is symmetric about a vertex $v$. Show that $\mathcal{T}$ is of Type I and of Type A with $v$ being both the characteristic and spectral vertex.

(ii) Suppose $\mathcal{T}$ is symmetric about an edge $e$ incident to vertices $v$ and $w$. Show that $\mathcal{T}$ is of Type II and of Type B with $v$ and $w$ being both the characteristic and spectral vertices.

**3.** (See [66]) Suppose we create an unweighted tree $\mathcal{T}$ by taking a vertex $v$ and making it adjacent to the root vertices of $\mathcal{T}(k_1, \ldots, k_m)$ and $\mathcal{T}(k_m, \ldots, k_1)$ as in Section 6.3. Show that $\mathcal{T}$ is a Type A tree with $v$ as its spectral vertex. Thus we have shown it is possible for a Type A tree to have nonisomorphic spectral branches.

**4.** Show that the distance between the characteristic vertex and the spectral vertices of an unweighted Type I-B tree can be arbitrarily large. Show the same for an unweighted Type II-A tree.

# Chapter 7

# Bottleneck Matrices for Graphs

In the previous chapter we investigated the bottleneck matrices for trees. In this chapter we generalize this concept to graphs which are not trees. We provide a brief review of the group inverse of a matrix in Section 7.1 and discuss how bottleneck matrices for graphs are constructed. We then apply these ideas in Section 7.2 to show how the Perron components at cut vertices of a graph can be used to compute its algebraic connectivity. In this section, we generalize much of Section 6.2. In Sections 7.3 and 7.4 we use bottleneck matrices for graphs to show which graphs of fixed girth have the minimum and maximum algebraic connectivities. We see in Section 7.3 that the graphs which have the minimum algebraic connectivity are certain unicyclic graphs. Therefore in Section 7.4 we focus on determining the unicyclic graphs of fixed girth that yield the maximum algebraic connectivity. Since unicyclic graphs closely resemble trees, we will combine ideas from both this chapter and Chapter 6 in order to prove our results. In Section 7.5 we apply the idea of bottleneck matrices to find an upper bound on the algebraic connectivity of a graph in terms of the number of cut vertices. Recall from Section 5.1 that the maximum algebraic connectivity of a graph with a cut vertex is one. In this Section 7.5 we will refine these results. Finally in Section 7.6 we generalize the results of Section 6.7 by investigating the spectral radius of submatrices of the Laplacian matrices of graphs created by deleting a row and column corresponding to a vertex of the graph. As in Chapter 6, taking the inverse of these matrices yields bottleneck matrices. Hence we compare the results concerning these submatrices to those of bottleneck matrices.

## 7.1 Constructing Bottleneck Matrices for Graphs

In order to construct bottleneck matrices at vertices of graphs which are not trees, we will need to recall from Section 1.6 the concept of the group inverse of a square matrix. We will discuss the group inverse of the Laplacian matrix in greater detail in Chapter 8, but for now we will give the information necessary to construct bottleneck matrices for graphs. Recall from Section 1.6 that for an $n \times n$ matrix $A$, the group inverse of $A$, when it exists, is the unique $n \times n$ matrix $X$ that satisfies all of the following:

$$\begin{aligned}(i) & \quad AXA = A \\ (ii) & \quad XAX = X \\ (iii) & \quad AX = XA\end{aligned} \quad (7.1.1)$$

If the group inverse of $A$ exists, we denote it by $A^{\#}$. Clearly, if $A$ is nonsingular, then $A^{-1} = A^{\#}$. However, it is possible for $A^{\#}$ to exist even if $A$ is singular, as is the case for Laplacian matrices of connected graphs. While we can find the group inverse of a matrix using (1.6.3), if $A$ is an irreducible M-matrix we can apply the formula for the group inverse of an irreducible singular M-matrix found in [50]:

**THEOREM 7.1.1** *Let $A$ be a singular irreducible M-matrix with right null vector $x = [x_1, x_2, \ldots, x_n]^T$ and left null vector $y = [y_1, y_2, \ldots, y_n]^T$. Then*

$$A^{\#} = \frac{\hat{y}^T M \hat{x}}{(y^T x)^2} xy^T + \left[ \begin{array}{c|c} M - \frac{1}{y^T x} M \hat{x} \hat{y}^T - \frac{1}{y^T x} \hat{x} \hat{y}^T M & \frac{-y_n}{y^T x} M \hat{x} \\ \hline \frac{-x_n}{y^T x} \hat{y}^T M & 0 \end{array} \right] \quad (7.1.2)$$

*where $\hat{x} = [x_1, x_2, \ldots, x_{n-1}]^T$, $\hat{y} = [y_1, y_2, \ldots, y_n]^T$, and $M = A[\bar{n}]^{-1}$.*

Since Laplacian matrices of connected graphs are singular irreducible M-matrices, Theorem 7.1.1 is relevant. Recall from Chapter 6 that $M = L[\bar{i}]^{-1}$ is known as the bottleneck matrix of the Laplacian matrix $L$ at vertex $i$. Since $e$ is the left and right null vector of any Laplacian matrix, we have an immediate corollary from [50]:

**COROLLARY 7.1.2** *Let $L$ be the Laplacian matrix of a connected weighted graph with $n$ vertices. Then*

$$L^{\#} = \frac{e^T M e}{n^2} J + \left[ \begin{array}{c|c} M - \frac{1}{n} MJ - \frac{1}{n} JM & -\frac{1}{n} Me \\ \hline -\frac{1}{n} e^T M & 0 \end{array} \right] \quad (7.1.3)$$

*where $M$ is the bottleneck matrix at vertex $n$.*

Since $M = L[\bar{n}]^{-1} = L[\overline{\{n\}}, \overline{\{n\}}]^{-1}$, we can use the cofactor formula for the inverse to see that the $(i, j)$ entry of the bottleneck matrix $M$ of $L$ based at $n$ is

$$m_{i,j} = \frac{(-1)^{i+j} \det L[\overline{\{j,n\}}, \overline{\{i,n\}}]}{\det L[\overline{\{n\}}, \overline{\{n\}}]}. \quad (7.1.4)$$

Before exploring the combinatorial aspects of (7.1.4), we need to establish some notation. If $E$ is a set of edges of $\mathcal{G}$, the *weight* of $E$ is denoted by $w(E)$ and is the product of the weights of the edges in $E$. If $E$ is empty, then we define $w(E) := 1$. The weight of a subgraph $\mathcal{H}$ of $\mathcal{G}$ is the weight of the set of edges in $\mathcal{H}$. The set of all spanning trees of $\mathcal{G}$ is denoted by $S$. Letting $i$, $j$, and $k$ be (not necessarily distinct) vertices of $\mathcal{G}$, an $(\{i, j\}, k)$-*spanning forest* of $\mathcal{G}$ is a spanning forest of $\mathcal{G}$ which has exactly two connected components, one of which contains vertex $k$ and the other of which contains the vertices $i$ and $j$. The set of all $(\{i, j\}, k)$-spanning forests of $\mathcal{G}$ is denoted by $S_k^{\{i,j\}}$.

By the Matrix Tree Theorem (Theorem 3.2.2), we see that

$$\det L[\overline{\{n\}}, \overline{\{n\}}] = \sum_{T \in S} w(T)$$

and

$$\det L[\overline{\{j,n\}}, \overline{\{i,n\}}] = \sum_{F \in S_n^{\{i,j\}}} w(F)$$

where $1 \leq 1, j, \leq n - 1$. In light of this, we can rewrite (7.1.4) for entries of the bottleneck matrix at vertex $v$ as (see [50]):

$$m_{i,j} = \frac{\sum_{F \in S_v^{\{i,j\}}} w(F)}{\sum_{T \in S} w(T)}. \qquad (7.1.5)$$

If $\mathcal{G}$ is an unweighted graph, then (7.1.5) implies that the entries for the bottleneck matrix at vertex $v$ are:

$$m_{i,j} = \frac{|S_v^{\{i,j\}}|}{|S|}. \qquad (7.1.6)$$

**EXAMPLE 7.1.3** Consider the unweighted graph $\mathcal{G}$ below:

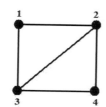

Observe that the Laplacian matrix is

$$L = \begin{bmatrix} 2 & -1 & -1 & 0 \\ -1 & 3 & -1 & -1 \\ -1 & -1 & 3 & -1 \\ 0 & -1 & -1 & 2 \end{bmatrix}$$

Suppose we want to find the bottleneck matrix at vertex 4. First observe that $\mathcal{G}$ contains 8 spanning trees:

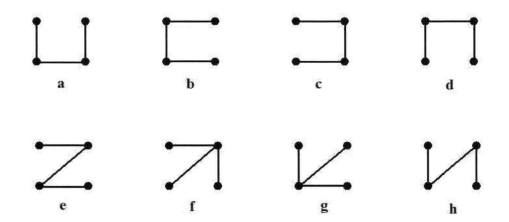

Thus $|S| = 8$. Now note all of the spanning forests of $\mathcal{G}$ that contain exactly two components:

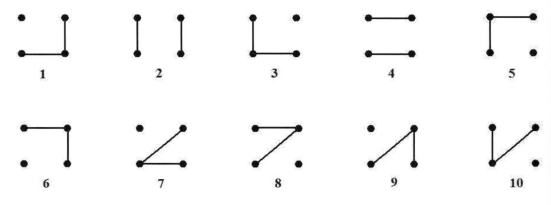

To compute the entries of the bottleneck matrix $L[\overline{4}]^{-1}$, we construct the following chart. Since $S_n^{\{i,j\}} = S_n^{\{j,i\}}$ for all $i, j$, we need only compute one of $S_n^{\{i,j\}}$ or $S_n^{\{j,i\}}$.

| set | cardinality | spanning forests |
|---|---|---|
| $S_4^{\{1,1\}}$ | 8 | 1, 2, 4, 5, 7, 8, 9, 10 |
| $S_4^{\{1,2\}}$ | 4 | 4, 5, 8, 10 |
| $S_4^{\{1,3\}}$ | 4 | 2, 5, 8, 10 |
| $S_4^{\{2,2\}}$ | 5 | 3, 4, 5, 8, 10 |
| $S_4^{\{2,3\}}$ | 3 | 5, 8, 10 |
| $S_4^{\{3,3\}}$ | 5 | 2, 5, 6, 8, 10 |

(7.1.7)

Thus

$$L[\overline{4}]^{-1} = \frac{1}{8} \begin{bmatrix} 8 & 4 & 4 \\ 4 & 5 & 3 \\ 4 & 3 & 5 \end{bmatrix}.$$

# Bottleneck Matrices for Graphs

**EXAMPLE 7.1.4** Consider the weighted graph $\mathcal{G}$ below:

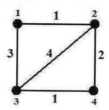

Observe that the Laplacian matrix is

$$L = \begin{bmatrix} 4 & -1 & -3 & 0 \\ -1 & 7 & -4 & -2 \\ -3 & -4 & 8 & -1 \\ 0 & -2 & -1 & 3 \end{bmatrix}$$

Suppose we want to find the bottleneck matrix at vertex 4. Recalling the 8 spanning trees from the previous example, we see the weights of these spanning trees are

| tree | a | b | c | d | e | f | g | h |
|---|---|---|---|---|---|---|---|---|
| weight | 6 | 3 | 2 | 6 | 4 | 8 | 12 | 24 |

Thus $\sum_{T \in S} w(T) = 65$. Recalling from the previous example the 10 spanning forests containing exactly two components, observe that the weights of each of these forests are

| forest | 1 | 2 | 3 | 4 | 5 | 6 | 7 | 8 | 9 | 10 |
|---|---|---|---|---|---|---|---|---|---|---|
| weight | 2 | 6 | 3 | 1 | 3 | 2 | 4 | 4 | 8 | 12 |

With the assistance of (7.1.7) and the fact that the numerator of the $(i,j)$ entry of the bottleneck matrix $L[4]^{-1}$ is $\sum_{F \in S_4^{\{i,j\}}} w(F)$, we see that

$$L[4]^{-1} = \frac{1}{65} \begin{bmatrix} 40 & 20 & 25 \\ 20 & 23 & 19 \\ 25 & 19 & 27 \end{bmatrix}.$$

**REMARK 7.1.5** Note that (7.1.5) can be used to simplify our approach to Theorem 6.2.5. Let $\mathcal{T}$ be a weighted tree on vertices $1, 2, \ldots, n$. Fix a vertex $v$ and let $i$ and $j$ be (not necessarily distinct) vertices of $\mathcal{T}$ other than $v$. Since $\mathcal{T}$ is a tree, every spanning forest $F$ of $\mathcal{T}$ with exactly two components can be obtained by removing exactly one edge $e$ from $\mathcal{T}$. Therefore, $F$ is an $(\{i,j\}, v)$-spanning forest of $\mathcal{T}$ if and only if $F$ is obtained from $\mathcal{T}$ by removing an edge $e$ that lies concurrently on the path from $i$ to $v$ and on the path from $j$ to $v$. For such an edge $e$, we recall that since $\mathcal{T} = F - e$, it follows that $w(\mathcal{T}) = w(e)w(F)$. Thus for such an edge $e$, we obtain $\frac{w(F)}{w(\mathcal{T})} = \frac{1}{w(e)}$. Thus (7.1.5) implies Theorem 6.2.5.

Clearly, if two vertices belong to different components at a cut vertex $v$, then the corresponding entry of the bottleneck matrix at $v$ is zero. The following proposition from [18] gives us insight into the construction of the bottleneck matrix at a vertex in a specific component of a graph containing a cut vertex.

**PROPOSITION 7.1.6** *Let $\mathcal{G}$ be a connected graph, and suppose that $v$ is a cut vertex of $\mathcal{G}$ with connected components $C_1, \ldots, C_{k+1}$ at $v$. Let $D$ be a proper subset of the vertices of $C_{k+1}$. Without loss of generality, we can write $L[C_1 \cup C_2 \cup \ldots \cup C_k \cup D \cup \{v\}]$ as*

$$\begin{bmatrix} L[C_1] & 0 & 0 & 0 & -L[C_1]e \\ 0 & \ldots & 0 & \ldots & \ldots \\ 0 & 0 & L[C_k] & 0 & -L[C_k]e \\ 0 & \ldots & 0 & L[D] & -\theta \\ \hline -e^T L[C_1] & \ldots & -e^T L[C_k] & -\theta^T & d \end{bmatrix}$$

*where $d \geq \sum_{i=1}^{k} e^T L[C_i]e$. Then $(L[C_1 \cup C_2 \cup \ldots \cup C_k \cup D \cup \{v\}])^{-1}$ can be written as*

$$\begin{bmatrix} M + \alpha J & \alpha e \theta^T (L[D])^{-1} & \alpha e \\ \hline \alpha (L[D])^{-1} \theta e^T & (L[D])^{-1} + \alpha (L[D])^{-1} \theta \theta^T (L[D])^{-1} & \alpha (L[D])^{-1} \theta \\ \hline \alpha e^T & \alpha \theta^T (L[D])^{-1} & \alpha \end{bmatrix}$$

*where $\alpha = 1/(d - \sum_{i=1}^{k} e^T L[C_i]e - \theta^T (L[D])^{-1}\theta)$, and*

$$M = \begin{bmatrix} (L[C_1])^{-1} & 0 & 0 \\ 0 & \ldots & 0 \\ 0 & 0 & (L[C_k])^{-1} \end{bmatrix}.$$

**Proof**: This follows by direct computation provided that $\alpha > 0$. To show $\alpha > 0$, it suffices to show that $d - \sum_{i=1}^{k} e^T L[C_i]e - \theta^T (L[D])^{-1}\theta$ is positive. Note that the entire Laplacian matrix can be written as

$$\begin{bmatrix} L[C_1] & 0 & 0 & 0 & -L[C_1]e & 0 \\ 0 & \ldots & 0 & \ldots & \ldots & \ldots \\ 0 & 0 & L[C_k] & 0 & -L[C_k]e & 0 \\ 0 & \ldots & 0 & L[D] & -\theta & -X \\ \hline -e^T L[C_1] & \ldots & -e^T L[C_k] & -\theta^T & d & -y^T \\ 0 & \ldots & 0 & -X^T & -y & L[C_{k+1} - D] \end{bmatrix}.$$

Since the row sums of $L$ are all zero, we see that $L[D]e = \theta + Xe$. Multiplying through on the left by $\theta^T (L[D])^{-1}$ and rearranging we obtain

$$\theta^T (L[D])^{-1}\theta = \theta^T e - \theta^T (L[D])^{-1} Xe. \tag{7.1.8}$$

Again using the fact that the row sums of $L$ are all zero, we see that $d = \sum_{i=1}^{k} e^T L[C_i]e + \theta^T e + y^T e$. Using this together with (7.1.8) we see that

$$d - \sum_{i=1}^{k} e^T L[C_i]e - \theta^T (L[D])^{-1}\theta = y^T e + \theta^T (L[D])^{-1} Xe.$$

At this point, the proof is complete if we can show that $y^T e + \theta^T (L[D])^{-1} X e$ is positive. Observe that $y^T e \geq 0$ and the inequality is strict if and only if $v$ is adjacent to some vertex in $C_{k+1} - D$. Looking at $\theta^T (L[D])^{-1} X e$, since $L[D]$ is a nonsingular M-matrix, it follows from Theorem (1.4.7) that $(L[D])^{-1} > 0$. Thus $\theta^T (L[D]^{-1}) X e \geq 0$ where the inequality is strict if and only if there is a walk from $v$ to a vertex in $C_{k+1} - D$ going through a vertex in $D$ (since $\theta$ or $X$ would be zero otherwise). But since $\mathcal{G}$ is connected, it follows that at least one of these two quantities is positive, thus $y^T e + \theta^T (L[D])^{-1} X e$ is positive. □

We end this section with a theorem from [43] concerning the diagonal entries of bottleneck matrices. This theorem will be useful to us later in the chapter.

**THEOREM 7.1.7** *Let $\mathcal{G}$ be an unweighted graph and let $C$ be a component at a cut vertex $v$ of $\mathcal{G}$ where $C$ has $q$ vertices. If $M$ is the bottleneck matrix for $C$, then each diagonal entry of $M$ is at least $2/(q+1)$. Moreover, if $m_{ii} = 2/(q+1)$, then vertex $i$ is adjacent to $v$ and to every vertex of $C$.*

**Proof**: We will first show that adding an edge into $C$ or between $v$ and a vertex of $C$ cannot increase any diagonal entry of $M$. To verify this claim, suppose that $M'$ is the bottleneck matrix arising from the addition of the edge between vertices $i$ and $j$ of $C$. Then $(M')^{-1} = M^{-1} + (e_i - e_j)(e_i^T - e_j^T)$, and therefore

$$M' = M - (Me_i - Me_j)(e_i^T M - e_j^T M)/(1 + m_{ii} + m_{jj} - 2m_{ij}).$$

Thus for any vertex $p$, we have

$$m'_{pp} = m_{pp} - \frac{(m_{pi} - m_{pj})^2}{1 + m_{ii} + m_{jj} - 2m_{ij}} \leq m_{pp}$$

where the inequality follows from the fact that since $M$ is positive definite, the expression $m_{ii} + m_{jj} - 2m_{ij}$ is nonnegative. This proves the claim in the case of adding an edge joining two nonadjacent vertices of $C$. The proof of the case of adding an edge joining $v$ to a vertex in $C$ is similar.

From the preceeding paragraph, since adding edges cannot increase any diagonal entry of $M$, it follows that the diagonal entries of $M$ are bounded below by the diagonal entries of the bottleneck matrix for a component at $v$ on $q$ vertices where such a component is the complete graph, $K_q$. But the bottleneck matrix for such a component is clearly $\frac{1}{q+1}(I + J)$. Hence the diagonal entries are bounded below by $2/(q+1)$.

To complete the proof, suppose that vertex $i$ of $C$ is not adjacent to a vertex $j$ of $C$. We claim then that $m_{ii} > 2/(q+1)$. To verify this, we see from the preceeding paragraph that the diagonal of $M$ is bounded below by that of the bottleneck matrix, call it $N$, of the component in which the edge $ij$ is missing, but all other possible edges (including those involving $v$) are present. But

$$N = \frac{1}{q+1}(I + J) + \frac{1}{q^2 - 1}(e_i - e_j)(e_i - e_j)^T.$$

In particular,
$$m_{ii} = \frac{2}{q+1} + \frac{1}{q^2-1} > \frac{2}{q+1},$$
thus proving the theorem if $i$ is not adjacent to $j$. A similar argument holds for $i$ not adjacent to $v$. □

## Exercises:

**1.** Find the bottleneck matrix at vertex $v$ in the graph below. Then verify Theorem 7.1.7.

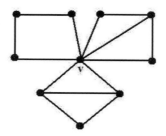

**2.** Find the bottleneck matrix at any vertex in the graph above that is not a cut vertex.

**3.** (See [45]) Let $v$ be a cut vertex of a weighted graph $\mathcal{G}$ and let $C$ be a component at $v$ with bottleneck matrix $M$. Fix two vertices $x$ and $y$ of $C$. Show that as the weight of the edge joining $x$ and $y$ increases (assuming the weight is zero if $x$ and $y$ are not adjacent) the Perron value of $M$ is nonincreasing.

## 7.2 Perron Components of Graphs

In this section, we generalize many of the results of Section 6.2 to graphs which are not trees. We use the bottleneck matrices described in the previous section to compute the algebraic connectivity of a graph as we did with trees in Section 6.2. We then use these results to generalize the perturbation results for trees found in Section 6.4 to graphs. This will enable us to find upper and lower bounds on the valuations of vertices that are not cut vertices. Since we are generalizing many of the results found in Chapter 6, we need several definitions, many of which mimic those found in that chapter. In these definitions we let $y$ be a characteristic valuation (Fiedler vector) of $\mathcal{G}$ and we consider only graphs which contain a cut vertex. We refer the reader to Theorem 6.1.10 to show that these definitions are well defined.

**DEFINITION 7.2.1** A graph $\mathcal{G}$ is of *Type I* with *characteristic vertex* $z$ if $z$ is the unique cut vertex such that $y_z = 0$ and $z$ is adjacent to vertices with nonzero valuation.

**DEFINITION 7.2.2** A graph $\mathcal{G}$ is of Type II with *characteristic block B* if $B$ is the unique block of $\mathcal{G}$ that contains both positively and negatively valuated vertices.

Since we are considering graphs with cut vertices, the idea of a Perron component becomes relevant. We generalize the definition of Perron branch from Chapter 6 (Definition 6.2.10) as follows:

**DEFINITION 7.2.3** Let $v$ be a cut vertex of a graph $\mathcal{G}$. A *Perron component at* $v$ is a component of $\mathcal{G} - v$ whose bottleneck matrix has the largest Perron value.

The following theorem from [45] generalizes some of the theorems from Section 6.2. Part (i) of the following theorem generalizes Theorem 6.2.15 saying that a graph $\mathcal{G}$ with a cut vertex is of Type I if and only if there are at least two Perron components at the cut vertex. Moreover, the algebraic connectivity of such graphs are computed in a similar fashion as to Type I trees. Part (ii) of the theorem generalizes Theorems 6.2.4 and 6.2.18 which give the multiplicity of the algebraic connectivity in relation to the number of Perron components. This part of the theorem also shows that the characteristic vertex of a Type I tree is invariant of the Feidler vector used. Part (iii) of the theorem generalizes Theorem 6.2.19 which says that if $v$ is not a characteristic vertex of a Type I graph $\mathcal{G}$, then the unique Perron component at $v$ contains the characteristic vertex of $\mathcal{G}$.

**THEOREM 7.2.4** *Let $\mathcal{G}$ be a weighted graph with algebraic connectivity a.*

*(i) $\mathcal{G}$ is of Type I with characteristic vertex $z$ if and only if there are two or more Perron components at $z$. Further, $a = 1/\rho(L[C]^{-1})$ for any Perron component $C$ at $z$.*

*(ii) If the Perron components at $z$ are $C_1, \ldots, C_m$, let $x_i$ be the Perron vector of $L[C_i]^{-1}$ for $1 \leq i \leq m$ normalized so that the entries sum to 1. For each $2 \leq i \leq m$, let $b_{i-1}$ be the Fiedler vector which valuates the vertices of $C_1$ by $x_1$, valuates the vertices of $C_i$ by $-x_i$, and valuates all other vertices 0. Then $b_1, \ldots, b_{m-1}$ is a basis for the eigenspace corresponding to a. In particular, if there are $m$ Perron components at $z$, then the multiplicity of a is $m - 1$, and every Fiedler vector has zeros in the positions corresponding to $z$ and to the vertices of the non-Perron components at $z$.*

*(iii) For any cut vertex $v \neq z$, the unique Perron component at $v$ is the component containing $z$.*

**Proof**: To prove (i), first suppose that $\mathcal{G}$ is of Type I. Let $C_1, \ldots, C_k$ be the connected components of $\mathcal{G} - z$ (note $k \geq m$), and let $y$ be the relevant Fiedler vector. Without loss of generality, we write the Laplacian matrix as

$$L = \begin{bmatrix} L[C_1] & 0 & 0 & 0 & f_1 \\ 0 & L[C_2] & 0 & 0 & f_2 \\ 0 & 0 & \ldots & 0 & \ldots \\ 0 & 0 & \ldots & L[C_k] & f_k \\ f_1^T & f_2^T & \ldots & f_k^T & d \end{bmatrix}$$

and we partition $y$ conformally as $[y(1)^T, \ldots, y(k)^T, 0]^T$. By matrix multiplication, we then have for each $1 \leq i \leq k$, $L[C_i]y(i) = ay_i$. Since $L[C_i]$ is an M-matrix and $a \leq \lambda_1(L[C_i])$, it follows from Observation 1.4.2 that $y(i)$ is either all positive, all negative, or all zero. In the first two instances, we see that $y(i)$ must be a Perron vector for $L[C_i]^{-1}$ which necessarily has Perron value $1/a$. Further, since $e^T y = 0$, we find there are at least two components, without loss of generality, $C_1$ and $C_2$, whose Perron values are $1/a$.

Suppose that $C_1$ and $C_2$ are not Perron components at $z$. Then there is another component, without loss of generality $C_3$, such that the Perron value of $C_3$ is greater than $1/a$. Let $x$ be a Perron vector for $L[C_3]^{-1}$, normalized so that $e^T x = 1$, and let $u = y(1)/e^T y(1)$. Consider the vector $w = [u, 0, -x, 0, 0, \ldots, 0]^T$. Observe that $e^T u = e^T x = 1$ since $w$ is orthogonal to $e$. Now

$$w^T L w = a u^T u + \frac{1}{\rho(L[C_3]^{-1})} x^T x < a u^T u + a x^T x = a w^T w,$$

which contradicts Theorem 1.2.7. Thus we find that $C_1$ and $C_2$ must be Perron components at $z$ which proves one direction of (i).

To prove the converse of (i), suppose that there are at least two Perron components at $z$, say $C_1, \ldots, C_m$. Let $x_1$ and $x_2$ be Perron vectors for $L[C_1]^{-1}$ and $L[C_2]^{-1}$, respectively, and normalize them so $f_1^T x_1 + f_2^T x_2 = 0$ (this is possible since $f_1$ and $f_2$ are nonpositive and neither vector is the zero vector). Then by matrix multiplication, the vector $y = [x_1, x_2, 0, 0, \ldots, 0]^T$ is an eigenvector for $L$ corresponding to the eigenvalue $1/\rho(L[C_1]^{-1})$. It then follows from Theorem 1.2.7 and Corollary 1.2.10 that $1/\rho(L[C_1]^{-1})$ is the smallest positive eigenvalue of $L$, i.e., it is the algebraic connectivity of $\mathcal{G}$. Hence $y$ is a Fiedler vector of the type described in Definition 7.2.1 for Type I graphs. This proves the converse of (i).

To prove (ii), let $z$ be the characteristic vertex of $\mathcal{G}$ and let $y$ be any Fiedler vector. Suppose that $C_1$ is a Perron component at $z$. Since

$$L[C_1]y(1) + y_z f_1 = \frac{1}{\rho(L[C_1]^{-1})} y(1),$$

we multiply through on the left by $\rho(L[C_1]^{-1})L[C_1]^{-1}$ and rearrange to obtain

$$L[C_1]^{-1}y(1) - \rho(L[C_1]^{-1})y_z L[C_1]^{-1} f_1 = \rho(L[C_1]^{-1})y(1).$$

Multiplying on the left by a left Perron vector for $L[C_1]^{-1}$ now yields that $y_z$ must be zero. Let $C_i$ be a component at $z$. Then $L[C_i]y(i) = 1/\rho(L[C_i]^{-1})y(i)$, which is equivalent to $L[C_i]^{-1}y(i) = \rho(L[C_i]^{-1})y(i)$. Thus if $C_i$ is a Perron component, then $y(i)$ is a nonzero scalar multiple of the Perron vector for $L[C_i]^{-1}$, while if $C_i$ is not a Perron component, then necessarily $y(i)$ must be the zero vector. It now follows that for any Perron component $C_i$ at $z$ (with $i \geq 2$), if $x_1$ and $x_i$ are Perron vectors for $L[C_1]^{-1}$ and $L[C_i]^{-1}$, respectively, and both are normalized so that $e^T x_1 = e^T x_i = 1$, then $[x_1, 0, 0, \ldots, 0, x_i, 0, \ldots, 0]^T$ is a Fiedler vector. Moreover, the collection of all such vectors ranging over all Perron components distinct from $C_1$ is a basis for the eigenspace corresponding to the algebraic connectivity. The statements on multiplicity and the zero entries now follow.

To prove (iii), suppose $v$ is a vertex of $\mathcal{G}$ with $v \neq z$, and that there are at least two Perron components at $z$. Let $C_v$ be the component of $\mathcal{G} - z$ which contains $v$. Then for any component $A$ of $\mathcal{G} - v$ which does not contain $z$, we have $A \subset C_v$. Let $B$ be the component at $v$ which contains $z$. Since there are at least two Perron components at $z$, one of them, say $C_0$, is strictly contained in $B$. Since $A \subset C_v$, we find that $L[A]^{-1}$ is strictly dominated entrywise by the principle submatrix of $L[C_v]^{-1}$. Thus by Exercise 1(ii) of Section 1.3, it follows that $\rho(L[A]^{-1}) < \rho(L[C_v]^{-1})$. A similar assertion follows from the fact that $C_0 \subset B$. Thus

$$\rho(L[A]^{-1}) < \rho(L[C_v]^{-1}) \leq \rho(L[C_0]^{-1}) < \rho(L[B]^{-1}).$$

Hence the unique Perron component at $v$ is $B$, the component containing $z$. This proves (iii). □

We now have an immediate corollary from [45] which generalizes Theorem 6.2.17:

**COROLLARY 7.2.5** *Let $\mathcal{G}$ be a weighted graph. Then $\mathcal{G}$ is of Type II if and only if there is a unique Perron component at every vertex of $\mathcal{G}$. Moreover, $\mathcal{G}$ is of Type I if and only if there is a unique vertex at which there are two or more Perron components.*

**Proof**: From Theorem 7.2.4(i), we see that if there are two or more Perron components at some vertex, then Case (a) of Theorem 6.1.10 cannot hold for any Fiedler vector. Conversely, if there is a unique Perron component at every vertex, then Case (b) of Theorem 6.1.10 cannot hold for any Fiedler vector. Thus Case (a) of Theorem 6.1.10 must hold for every Fiedler vector. The result follows. □

We see from Theorem 7.2.4 that the algebraic connectivity of a Type I graph is the reciprocal of the spectral radius of the bottleneck matrix of a Perron branch at the characteristic vertex. This generalizes Theorem 6.2.15 regarding the algebraic connectivity of Type I trees. Our goal now is to generalize such a result for Type II graphs. In Theorem 6.2.12, we saw that if $\mathcal{T}$ is a Type II tree with characteristic vertices $i$ and $j$ (hence the edge $ij$ is the characteristic block), then the reciprocal of the algebraic connectivity is the spectral radius of the matrix $M - (\gamma/\theta)J$ where $M$ is the bottleneck matrix at the branch at $i$ containing $j$, $\theta$ is the weight of the edge joining $i$ and $j$, and $\gamma$ is an appropriate value between 0 and 1. To generalize our results to Type II graphs, we focus on an arbitrary cut vertex $v$ (not necessarily incident to the characteristic block as we considered in trees) and investigate the component at $v$ not containing the characteristic block. To this end, we will prove a lemma and theorem from [45]. The observation from [45] below will be useful in proving the lemma that follows and can be verified by direct calculation:

**OBSERVATION 7.2.6** Suppose that $A$ is an invertible matrix and that

$$B = \begin{bmatrix} A & -c \\ -c^T & d \end{bmatrix}$$

Then

$$B^{-1} = \begin{bmatrix} A^{-1} & 0 \\ 0^T & 0 \end{bmatrix} + \frac{1}{(d - c^T A^{-1} c)} \begin{bmatrix} A^{-1} c \\ 1 \end{bmatrix} [c^T A^{-1} \mid 1].$$

**LEMMA 7.2.7** Let $\mathcal{G}$ be a connected weighted graph and let $C$ be a subset of vertices which induces a proper connected subgraph of $\mathcal{G}$. Let $v \in C$ be a cut vertex of $\mathcal{G}$ such that every path from a vertex in $C$ to a vertex in $\mathcal{G} - C$ passes through $v$. Let $w$ be the sum of the weights of all edges between $v$ and the vertices in $\mathcal{G} - C$ adjacent to $v$. If the vertices are labeled so that $v$ is the last among the vertices in $C$, then

$$L[C]^{-1} = \begin{bmatrix} L[C-v]^{-1} & 0 \\ 0^T & 0 \end{bmatrix} + \frac{1}{w} J$$

**Proof**: Suppose that the connected components of $C - v$ are $A_1, \ldots, A_m$. Then

$$L[C] = \begin{bmatrix} L[A_1] & & & -L[A_1]e \\ \hline & \ldots & & \ldots \\ \hline & & L[A_m] & -L[A_m]e \\ \hline -e^T L[A_1] & \ldots & -e^T L[A_m] & d \end{bmatrix}$$

where

$$d = w + e^T L[A_1] e + \ldots + e^T L[A_m] e = w + e^T L[C - v] e.$$

The result now follows by applying Observation 7.2.6 with $A = L[C - v]$ and $c = L[C - v]e$. □

**EXAMPLE 7.2.8** Consider the graph $\mathcal{G}$ below.

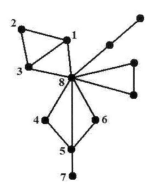

# Bottleneck Matrices for Graphs

Let $v = 8$ and let $C$ be the set of vertices that are numbered. Observe that $C$ and $v$ satisfy the conditions of Lemma 7.2.19. Note

$$L[C] = \begin{bmatrix} 3 & -1 & -1 & 0 & 0 & 0 & 0 & -1 \\ -1 & 2 & -1 & 0 & 0 & 0 & 0 & 0 \\ -1 & -1 & 3 & 0 & 0 & 0 & 0 & -1 \\ 0 & 0 & 0 & 2 & -1 & 0 & 0 & -1 \\ 0 & 0 & 0 & -1 & 4 & -1 & -1 & -1 \\ 0 & 0 & 0 & 0 & -1 & 2 & 0 & -1 \\ 0 & 0 & 0 & 0 & -1 & 0 & 1 & 0 \\ -1 & 0 & -1 & -1 & -1 & -1 & 0 & 8 \end{bmatrix}$$

Observe that since $\mathcal{G}$ is unweighted and since $v$ is adjacent to three vertices not in $C$, we have $w = 3$. Observing the matrices $L[C - v]^{-1}$ and $L[C]^{-1}$ below, we see that we obtain $L[C]^{-1}$ from $L[C - v]^{-1}$ by placing $1/w = 1/3$ as the entries in row 8 and column 8, and then adding $1/3$ to the entries of $L[C - v]^{-1}$.

$$L[C - v]^{-1} = \frac{1}{8} \begin{bmatrix} 5 & 4 & 3 & 0 & 0 & 0 & 0 \\ 4 & 8 & 4 & 0 & 0 & 0 & 0 \\ 3 & 4 & 5 & 0 & 0 & 0 & 0 \\ 0 & 0 & 0 & 5 & 2 & 1 & 2 \\ 0 & 0 & 0 & 2 & 4 & 2 & 4 \\ 0 & 0 & 0 & 1 & 2 & 5 & 2 \\ 0 & 0 & 0 & 2 & 4 & 2 & 12 \end{bmatrix}$$

$$L[C]^{-1} = \frac{1}{24} \begin{bmatrix} 23 & 20 & 17 & 8 & 8 & 8 & 8 & 8 \\ 20 & 32 & 20 & 8 & 8 & 8 & 8 & 8 \\ 17 & 20 & 23 & 8 & 8 & 8 & 8 & 8 \\ 8 & 8 & 8 & 23 & 14 & 11 & 14 & 8 \\ 8 & 8 & 8 & 14 & 20 & 14 & 20 & 8 \\ 8 & 8 & 8 & 11 & 14 & 23 & 14 & 8 \\ 8 & 8 & 8 & 14 & 20 & 14 & 44 & 8 \\ 8 & 8 & 8 & 8 & 8 & 8 & 8 & 8 \end{bmatrix}$$

**THEOREM 7.2.9** *Suppose that $\mathcal{G}$ is a Type II weighted graph (thus there is a unique Perron component at every vertex of $\mathcal{G}$) with algebraic connectivity $a$. Let $y$ be a Fiedler vector and let $B_0$ be the unique block (the characteristic block) of $\mathcal{G}$ containing both positively and negatively valuated vertices in $y$. Let $v$ be a cut vertex of $\mathcal{G}$, let $C_0$ denote the set of vertices in the connected component of $\mathcal{G} - v$ which contains vertices in $B_0$, and let $C_1$ denote the vertices in $\mathcal{G} - C_0$. Permute* and partition the Laplacian matrix as

$$
\begin{bmatrix}
 & & & 0 & \cdots & 0 \\
 & L[C_1] & & \cdots & & \cdots \\
 & & & 0 & \cdots & 0 \\
 & & & -\theta^T & & \\
\hline
0 & \cdots & 0 & & & \\
\cdots & & \cdots & & & \\
\cdots & & \cdots & -\theta & L[C_0] & \\
0 & \cdots & 0 & & &
\end{bmatrix}
\qquad (7.2.1)
$$

where vertex $v$ corresponds to the last row of $L[C_1]$, and partition $y$ as $y = [y(1)^T, y(0)^T]^T$. Then

(i)
$$L[C_1]^{-1} + \frac{\theta^T y(0)}{\theta^T e (\theta^T e y_v - \theta^T y(0))} J$$

is a positive matrix whose Perron value is $1/a$ and whose Perron vector is a scalar multiple of $y(1)$.

(ii) At every vertex $x$ of $\mathcal{G}$, the unique Perron component at $x$ is the component containing vertices in $B_0$.

**Proof:** To prove (i), first note that both $C_1$ and $C_0$ induce subgraphs of $\mathcal{G}$. Since $\mathcal{G}$ is of Type II, every block of $\mathcal{G}$ distinct from $B_0$ is valuated all positively, all negatively, or all zero by $y$. Without loss of generality, suppose $y_v > 0$. Then every block in $C_1$ to which $v$ belongs is valuated positively. Therefore, every vertex in $C_1$ is valuated positively by $y$. Thus $y(1) > 0$. Suppose there are $m$ vertices in $C_1$. Then by the eigenvalue-eigenvector relationship, we have

$$L[C_1]y(1) - \theta^T y(0) e_m = a y(1). \qquad (7.2.2)$$

Since the column sums of $L$ are each zero, we have $e^T L[C_1] = \theta^T e e_m^T$. Dividing through by $\theta^T e$, taking the transpose of both sides, and then multiplying on the left by $L[C_1]^{-1}$, we obtain

$$\frac{1}{\theta^T e} e = L[C_1]^{-1} e_m. \qquad (7.2.3)$$

We also see from the eigenvalue-eigenvector relationship, focusing on row $m$, that $\theta^T e y_v - \theta^T y(0) = a e^T y(1)$. Dividing through by $e^T y(1)$ and taking the reciprocal we obtain

$$\frac{1}{a} = \frac{e^T y(1)}{\theta^T e y_v - \theta^T y(0)}, \qquad (7.2.4)$$

Further, dividing (7.2.2) by $a$ and multiplying through on the left by $L[C_1]^{-1}$, we have

$$\frac{1}{a} y(1) - \frac{\theta^T y(0)}{a} L[C_1]^{-1} e_m = L[C_1]^{-1} y(1). \qquad (7.2.5)$$

Rearranging (7.2.5) and using (7.2.3) and (7.2.4) we find that

$$\left(L[C_1]^{-1} + \frac{\theta^T y(0)}{\theta^T e(\theta^T e y_v - \theta^T y(0))} J\right) y(1) = \frac{1}{a} y(1). \qquad (7.2.6)$$

By Lemma 7.2.7, we know $L[C_1]^{-1} - 1/(\theta^T e)J$ is positive. Further, since $y_v$, $\theta^T e$, and $\theta^T e y_v - \theta^T y(0)$ are positive (recall from (7.2.4) that the latter is equal to $ae^T y(1)$), it follows that

$$\frac{\theta^T y(0)}{\theta^T e(\theta^T e y_v - \theta^T y(0))} > -\frac{1}{\theta^T e}$$

and so by Lemma 7.2.7

$$L[C_1]^{-1} + \frac{\theta^T y(0)}{\theta^T e(\theta^T e y_v - \theta^T y(0))} J$$

is a positive matrix. Consequently from (7.2.6), $a$ is the reciprocal of the Perron value of that matrix and $y(1)$ is a corresponding Perron vector.

To prove (ii), we begin by showing that for any cut vertex of $\mathcal{G}$ that lies in $B_0$, the unique Perron component is the one containing the vertices in $B_0$. Let $v \in B_0$ be a cut vertex of $\mathcal{G}$ where $y_v > 0$. From the above, we see that $1/a$ is larger than the maximum eigenvalue of $L[C_1]^{-1} - 1/(\theta^T e)J$; by Lemma 7.2.7 this last quantity is the maximum of the Perron values of the components at $v$ which do not have vertices in $B_0$. From Theorem 1.2.8, we know that $a \geq 1/\rho(L[\mathcal{G}-v]^{-1})$, so that we must have

$$\rho(L[\mathcal{G}-v]^{-1}) \geq \frac{1}{a} > \rho\left(L[C_1]^{-1} - \frac{1}{\theta^T e}J\right).$$

Since $\rho(L[\mathcal{G}-v]^{-1})$ is the Perron value of the Perron component at $v$, we find that the Perron component at $v$ must be the one containing vertices in $B_0$. A similar argument follows if $v \in B_0$ is a cut vertex of $\mathcal{G}$ such that $y_v < 0$.

Now suppose that $v$ is a cut vertex of $B_0$ such that $y_v = 0$. Since $\mathcal{G}$ is of Type II, we find that $y(1) = 0$ and so $a$ is an eigenvalue of $L[C_0]$. In particular, $a \geq 1/\rho(L[C_0]^{-1})$. Suppose that some component $C_2$ at $v$ has a larger Perron value than that of $C_0$. Let $c$ be the Perron vector $L[C_0]^{-1}$ normalized so that $c^T e = 1$, and let $b$ be the Perron vector for $L[C_2]^{-1}$ normalized so that $b^T e = 1$. Form the vector $w$ by placing the entries in $c$ in the positions corresponding to the vertices in $C_0$, placing the entries of $-b$ in the positions correspondiing to the vertices in $C_2$, and placing zeros elsewhere. Then $w^T e = 0$, and

$$w^T L w = \frac{1}{\rho(L[C_0]^{-1})} c^T c + \frac{1}{\rho(L[C_2]^{-1})} b^T b < \frac{1}{\rho(L[C_0]^{-1})} w^T w \leq a w^T w,$$

contradicting Theorem 1.2.7. Hence it must be the case that the Perron component at $v$ is the one containing the vertices in $B_0$.

Finally, let $x$ be any vertex of $\mathcal{G}$. If $x$ is not a cut vertex of $\mathcal{G}$, then there is just one connected component at $x$, it is the Perron component, and it contains the vertices in $B_0$. If $x$ is a cut vertex of $\mathcal{G}$ and $x \in B_0$, the result follows from the arguments above. Finally, if $x$ is a cut vertex of $\mathcal{G}$ that is not in $B_0$, let $A$ be a component at $x$ not containing vertices of $B_0$, and let $B$ be the component which does contain the vertices of $B_0$. There is a cut vertex $v$ of $\mathcal{G}$ lying in $B_0$ such that (using the earlier notation) $A \subset C_1$ and $C_0 \subset B$. But then we have

$$\rho(L[A]^{-1}) < \rho(L[C_1]^{-1}) < \rho(L[C_0]^{-1}) < \rho(L[B]^{-1}).$$

Hence the unique Perron component at $x$ is $B$. □

**REMARK 7.2.10** *[45]* Suppose that $\mathcal{G}$ is of Type I and that $v$ is any cut vertex such that $y_v \neq 0$. Let $C_0$ be the vertex set of the connected component at $v$ containing the characteristic vertex $z$, and let $C_1$ be the vertex set of $\mathcal{G} - C_0$. A minor modification of the proof of Theorem 7.2.9 shows that

$$L[C_1]^{-1} + \frac{\theta^T y(0)}{\theta^T e(\theta^T e y_v - \theta^T y(0))} J$$

is a positive matrix with Perron value $1/a$ and Perron vector $y(1)$.

**REMARK 7.2.11** Since the property of being a Perron component at a vertex $v$ relies solely on the spectral radius of the block(s) of $L[\mathcal{G} - v]^{-1}$ and not on the particular Fiedler vector, we see from Corollary 7.2.5 and Theorem 7.2.9 that for a given graph $\mathcal{G}$, any Fiedler vector will identify the same characteristic vertex if $\mathcal{G}$ is of Type I, and the same characteristic block if $\mathcal{G}$ is of Type II. Thus we have broadened Theorem 7.2.4 to Type II graphs.

At this point, we broaden the previous theorems to other eigenvalues of the Laplacian matrix. We do this in a lemma from [18]:

**LEMMA 7.2.12** *Suppose $\mathcal{G}$ is a Type II weighted graph on $n$ vertices with Laplacian $L$. Let $B_0$ be the characteristic block of $\mathcal{G}$. Suppose $v$ is a cut vertex of $\mathcal{G}$. Let $C_0$ denote the set of vertices in the connected component of $\mathcal{G} - v$ which contains the vertices in $B_0$, and let $C_1$ denote the vertices in $G - C_0$. Assume, without loss of generality, that $L$ is partitioned as in (7.2.1) with $v$ corresponding to the last row of $L[C_1]$. If there exists an $\gamma > 0$ such that*

$$\frac{1}{\alpha} = \rho\left(L[C_1]^{-1} - \frac{1-\gamma}{\theta^T e}J\right) = \rho\left(L[C_0]^{-1} - \frac{\gamma}{\theta^T e}J\right).$$

*then $\alpha$ is an eigenvalue of $L$.*

**Proof**: Let $y(1)$ and $y(0)$ be eigenvectors corresponding to $\frac{1}{\alpha}$ for the matrices $L[C_1]^{-1} - \frac{1-\gamma}{\theta^T e}J$ and $L[C_0]^{-1} - \frac{\gamma}{\theta^T e}J$, respectively. First suppose $e^T y(0) = 0$. Then $cJy(0) = 0$ for any constant $c$. Hence

$$\frac{1}{\alpha}y(0) = \left(L[C_0]^{-1} - \frac{\gamma}{\theta^T e}J\right)y(0) = L[C_0]^{-1}y(0). \qquad (7.2.7)$$

Multiplying through on the left by $\theta^T$ we obtain
$$\frac{1}{\alpha}\theta^T y(0) = \theta^T L[C_0]^{-1} y(0) = e^T y(0) = 0.$$

where the second-to-last equality follows from the fact that $\theta^T L[C_0]^{-1} = e^T$ (since $e^T L[C_0] = \theta^T$). Hence $\theta^T y(0) = 0$ and $L[C_0]y(0) = \alpha y(0)$. Thus by matrix multiplication, we see that

$$L\begin{bmatrix} 0 \\ y(0) \end{bmatrix} = \begin{bmatrix} L[C_1] & \begin{array}{c} 0 \\ -\theta^T \end{array} \\ \hline 0 \quad -\theta & L[C_0] \end{bmatrix} \begin{bmatrix} 0 \\ y(0) \end{bmatrix} = \alpha \begin{bmatrix} 0 \\ y(0) \end{bmatrix}.$$

Thus $[0^T, y(0)^T]^T$ is an eigenvector of $L$ corresponding to $\alpha$.

Now suppose $e^T y(0) \neq 0$, and normalize $y(1)$ and $y(0)$ so that $e^T y(0) = e^T y(1)$. We have
$$L[C_1]^{-1} y(1) - \frac{1-\gamma}{\theta^T e} J y(1) = \frac{1}{\alpha} y(1). \tag{7.2.8}$$

Recalling that $L[C_1]e = (\theta^T e)e_v$, we see that by multiplying (7.2.8) through on the left by $\alpha L[C_1]$ and rearranging, we obtain
$$\alpha y(1) = L[C_1]y(1) - (\gamma - 1)\alpha e_v e^T y(1). \tag{7.2.9}$$

Also recall from (7.2.7) that
$$\left( L[C_0]^{-1} - \frac{\gamma}{\theta^T e} J \right) y(0) = \frac{1}{\alpha} y(0) \tag{7.2.10}$$

which implies
$$\alpha y(0) = L[C_0]y(0) + \frac{\alpha \gamma e^T y(0)}{\theta^T e}\theta \tag{7.2.11}$$

when multiplying through on the left by $\alpha L[C_0]$ and rearranging. Multiplying (7.2.10) through on the left by $\theta^T$ we see that
$$\theta^T L[C_0]^{-1} y(0) - \gamma e^T y(0) = \frac{1}{\alpha}\theta^T y(0).$$

Since $\theta^T L[C_0]^{-1} = e^T$, we obtain
$$(1-\gamma)e^T y(0) = \frac{1}{\alpha}\theta^T y(0).$$

Recalling that $e^T y(0) = -e^T y(1)$, we now have from (7.2.9) that
$$\alpha y(1) = L[C_1]y(1) - (\gamma-1)\alpha e_v e^T y(1) = L[C_1]y(1) - (\theta^T y(0))e_v. \tag{7.2.12}$$

Also, multiplying (7.2.8) on the left by $e_v^T$ we obtain
$$\frac{\gamma}{\theta^T e} e^T y(1) = \frac{1}{\alpha} y_v.$$

Plugging this into (7.2.11) and recalling that $e^T y(1) = -e^T y(0)$ we obtain

$$\alpha y(0) = L[C_0]y(0) - y_v\theta. \tag{7.2.13}$$

Putting together (7.2.12) and (7.2.13), we see that $L[y(1)^T, y(0)^T]^T = \alpha[y(1)^T, y(0)^T]^T$. Hence $[y(1)^T, y(0)^T]^T$ is an eigenvector of $L$ corresponding to the eigenvalue $\alpha$. □

Lemma 7.2.12 can now be specialized to the algebraic connectivity of $\mathcal{G}$. This is done in the following lemma from [18]:

**LEMMA 7.2.13** *Suppose $\mathcal{G}$ is a Type II weighted graph on $n$ vertices with Laplacian $L$ and algebraic connectivity $a$. Let $B_0$ be the characteristic block of $\mathcal{G}$. Suppose $v$ is a cut vertex of $\mathcal{G}$. Let $C_0$ denote the set of vertices in the connected component of $\mathcal{G} - v$ which contains the vertices in $B_0$, and let $C_1$ denote the vertices in $G - C_0$. Assume, without loss of generality, that $L$ is partitioned as in (7.2.1) where $v$ corresponds to the last row of $L[C_1]$. Then there exists an $\gamma > 0$ such that*

$$\frac{1}{a} = \rho\left(L[C_1]^{-1} - \frac{1-\gamma}{\theta^T e}J\right) = \rho\left(L[C_0]^{-1} - \frac{\gamma}{\theta^T e}J\right).$$

**Proof**: Let $y = [y(1)^T, y(0)^T]^T$ be a Fiedler vector and suppose $y_v \neq 0$. By Theorem 7.2.9,

$$\rho\left(L[C_1]^{-1} + \frac{\theta^T y(0)}{\theta^T e(\theta^T e y_v - \theta^T y(0))}J\right) = \frac{1}{a}. \tag{7.2.14}$$

From the eigenvalue-eigenvector relationship,

$$L[C_0]y(0) = ay(0) + y_v\theta. \tag{7.2.15}$$

Since $e^T L[C_0] = \theta^T$, it follows that by multiplying (7.2.15) on the left through by $\frac{1}{a}L[C_0]^{-1}$ and rearranging we obtain

$$\frac{1}{a}y(0) = L[C_0]^{-1}y(0) + \frac{y_v}{a}e. \tag{7.2.16}$$

Also, multiplying (7.2.15) on the left through by $e^T$ we obtain

$$e^T L[C_0]y(0) = \theta^T y(0) = ae^T y(0) + e^T \theta y_v. \tag{7.2.17}$$

Note that $e^T y(0) \neq 0$ (for otherwise $\mathcal{G}$ would be disconnected) and therefore rearranging (7.2.17) we obtain

$$\frac{1}{a} = \frac{e^T y(0)}{\theta^T y(0) - \theta^T e y_v}. \tag{7.2.18}$$

Therefore, substituting (7.2.18) into (7.2.16) and normalizing $y$ so that $e^T y(0) = 1$, we obtain

$$\left(L[C_0]^{-1} + \frac{y_v}{\theta^T y(0) - \theta^T e y_v}J\right)y(0) = \frac{1}{a}y(0). \tag{7.2.19}$$

Next observe that

$$\frac{y_v}{\theta^T y(0) - \theta^T e y_v} + \frac{\theta^T y(0)}{\theta^T e(\theta^T e y_v - \theta^T y(0))} = \frac{\theta^T e y_v - \theta^T y(0)}{\theta^T e(\theta^T y(0) - \theta^T e y_v)} = -\frac{1}{\theta^T e}.$$

So setting

$$\gamma := -\theta^T e \left( \frac{y_v}{\theta^T y(0) - \theta^T e y_v} \right),$$

we have from (7.2.14) and (7.2.19)

$$\frac{1}{a} = \rho \left( L[C_1]^{-1} + \frac{\theta^T y(0)}{\theta^T e(\theta^T e y_v - \theta^T y(0))} J \right)$$

$$= \rho \left( L[C_1]^{-1} - \frac{1-\gamma}{\theta^T e} J \right) \quad (7.2.20)$$

$$= \lambda_i \left( L[C_0]^{-1} - \frac{\gamma}{\theta^T e} J \right),$$

for some integer $1 \leq i \leq n$ (here $\lambda_n$ is the largest eigenvalue). Now if $\lambda_i \left( L[C_0]^{-1} - \frac{\gamma}{\theta^T e} J \right) < \lambda_n \left( L[C_0]^{-1} - \frac{\gamma}{\theta^T e} J \right)$, then there exists $\hat{\gamma} > \gamma$ such that

$$\rho \left( L[C_1]^{-1} - \frac{1-\hat{\gamma}}{\theta^T e} J \right) = \lambda_n \left( L[C_0]^{-1} - \frac{\hat{\gamma}}{\theta^T e} J \right) = \frac{1}{\alpha},$$

since the left-hand side of (7.2.20) is increasing in $\gamma$ and the right-hand side of (7.2.20) is nonincreasing in $\hat{\gamma}$, and so $\alpha$ is an eigenvalue of $L$ by Lemma 7.2.12. But then $0 < \alpha < a$ which contradicts the fact that $\lambda_1(L) = 0$ and $\lambda_2(L) = a$. Hence $i = n$ as desired.

Finally, suppose that $y_v = 0$. Then by Theorem 6.1.10, $y(1) = 0$ and thus $0 = e^T y = e^T y(0)$. We then have that $L[C_0]y(0) = ay(0)$, but more importantly $L[C_0]^{-1}y(0) = \frac{1}{a}y(0)$. Moreover, for all $\gamma > 0$,

$$\left( L[C_0]^{-1} - \frac{\gamma}{\theta^T e} J \right) y(0) = \frac{1}{a} y(0)$$

so that

$$\lambda_n \left( L[C_0]^{-1} - \frac{\gamma}{\theta^T e} J \right) \geq \frac{1}{a} \quad (7.2.21)$$

for all $\gamma > 0$. Now

$$\rho \left( L[C_1]^{-1} - \frac{1-\gamma}{\theta^T e} J \right) < \lambda_n \left( L[C_0]^{-1} - \frac{\gamma}{\theta^T e} J \right)$$

at $\gamma = 0$ since $C_0$ is the unique Perron component at $v$. Observe the left-hand side tends to $\infty$ as $\gamma \to \infty$, while the right-hand side is nonincreasing in $\gamma$. Thus for some $\gamma$

$$\rho \left( L[C_1]^{-1} - \frac{1-\gamma}{\theta^T e} J \right) = \lambda_n \left( L[C_0]^{-1} - \frac{\gamma}{\theta^T e} J \right) = \frac{1}{\alpha}$$

where $\alpha \neq 0$ is an eigenvalue of $L$ (by Lemma 7.2.12). Hence from (7.2.21) we see that $\alpha \leq a$. However, since $\lambda_2(L) = a$ and $\alpha \neq 0 = \lambda_1(L)$, it follows that $\alpha \geq a$. Thus we conclude $\alpha = a$. □

**OBSERVATION 7.2.14** Lemma 7.2.13 is a generalization of Theorem 6.2.12 concerning the algebraic connectivity of Type II trees since $\theta^T e$ in Theorem 6.2.12 is merely the weight of the edge joining the characteristic vertices of the tree.

We now use the previous two lemmas to generalize Theorem 6.4.1 to graphs. The following two theorems from [18] investigate the effects on the algebraic connectivity of graphs when we perturb components at certain cut vertices. The first theorem deals with Type I graphs, while the second deals with Type II graphs.

**THEOREM 7.2.15** *Let $\mathcal{G}$ be a Type I weighted graph with Laplacian $L$ and $z$ as the characteristic vertex. Let $v$ be any cut vertex with connected components $C_1, C_2, \ldots, C_k$ at $v$. Let $C_{i_1}, C_{i_2}, \ldots C_{i_j}$ be any collection of connected components at $v$ such that the vertex set $C = \cup_{\ell=1}^{j} C_{i_\ell}$ does not contain the vertex set of every Perron component at $v$. Form a new graph $\hat{\mathcal{G}}$ by replacing $C$ with a single connected component $\hat{C}$ at $v$. Let $M := L[C]^{-1}$ and let $\hat{M}$ denote the bottleneck matrix of $\hat{C}$. Denote the algebraic connectivities of $\mathcal{G}$ and $\hat{\mathcal{G}}$ by $a$ and $\hat{a}$, respectively. If $M << \hat{M}$, then $\hat{a} \leq a$.*

**Proof**: First, suppose $v = z$. If $\rho(\hat{M}) \leq 1/a$, then there are still two or more Perron components at $z$. Thus $z$ is still the characteristic vertex of $\hat{\mathcal{G}}$, and hence $a = \hat{a}$. Suppose now that $\rho(\hat{M}) > 1/a$ (and still $v = z$). Then in $\hat{\mathcal{G}}$, the unique Perron component at $z$ is $\hat{C}$. If $\hat{\mathcal{G}}$ is of Type I, then there exists a cut vertex $w \in \hat{C}$ such that $\rho(D) = 1/\hat{a}$ for some Perron component at $w$ with bottleneck matrix $D$. But if $C'$ is the component at $w$ containing $z$ with bottleneck matrix $D'$, then

$$\frac{1}{\hat{a}} = \rho(D) \geq \rho(D') > \rho(D_0) = \frac{1}{a},$$

where $D_0$ is the bottleneck matrix for some Perron component at $z$ in $\mathcal{G}$ not containing $w$ (the last inequality follows from the fact that $D_0 << D'$). In this case $\hat{a} < a$. Finally, if $\hat{\mathcal{G}}$ is of Type II, then at $z$, $\hat{C}$ is the characteristic block. Letting $\overline{C} = \hat{\mathcal{G}} - \hat{C}$, then by Lemma 7.2.13 there exists an $\gamma > 0$ such that

$$\frac{1}{\hat{a}} = \rho\left(L[\overline{C}]^{-1} - \frac{1-\gamma}{\theta^T e}J\right) > \rho\left(L[\overline{C}]^{-1} - \frac{1}{\theta^T e}J\right) \geq \rho(D_0) = \frac{1}{a}$$

where $D_0$ is the bottleneck matrix for some Perron component at $z$ in $\mathcal{G}$ not containing the vertices in $\hat{C}$. Hence, again, $\hat{a} < a$.

Finally, suppose $v \neq z$ and assume without loss of generality that $z \in C_k$. By the definition of $C$ it follows that $z \notin C$. Let $C'$ be the component at $z$ containing $v$, and let $D = C' \cap C_k$.

# Bottleneck Matrices for Graphs

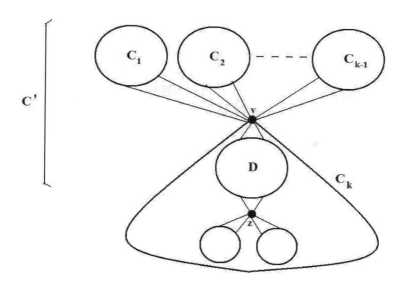

Observe that

$$L[C]^{-1} = M = \begin{bmatrix} L[C_{i_1}]^{-1} & 0 & 0 \\ 0 & \cdots & 0 \\ 0 & 0 & L[C_{i_j}]^{-1} \end{bmatrix}$$

Consider the matrix $N = (L[C_1 \cup \ldots C_k \cup D \cup \{v\}])^{-1}$ which is the bottleneck matrix for the component at $z$ containing $v$ in $\mathcal{G}$. Let $K$ be the bottleneck matrix for the component at $z$ containing $v$ in $\hat{\mathcal{G}}$. Using Proposition 7.1.6 and the assumption that $M << \hat{M}$, we see that $K >> N$. The result follows by implementing the same analysis as used above. □

**EXAMPLE 7.2.16** Consider the graph $\mathcal{G}$ below with $z$ and $v$ labeled in accordance with Theorem 7.2.15.

Suppose we replace $C$ with $\hat{C}$ to create the graph $\hat{\mathcal{G}}$:

Observe that

$$M = \frac{1}{5}\begin{bmatrix} 2 & 1 & 1 & 1 \\ 1 & 2 & 1 & 1 \\ 1 & 1 & 2 & 1 \\ 1 & 1 & 1 & 2 \end{bmatrix} \quad \text{and} \quad \hat{M} = \frac{1}{5}\begin{bmatrix} 4 & 3 & 2 & 1 \\ 3 & 6 & 4 & 2 \\ 2 & 4 & 6 & 3 \\ 1 & 2 & 3 & 4 \end{bmatrix}$$

and that $M << \hat{M}$. Hence Theorem 7.2.15 says that $\hat{a} \leq a$. Note that $\hat{a} = 0.1438$ and $a = 0.1590$.

**REMARK 7.2.17** *[21]* Suppose that in Theorem 7.2.15 that $\mathcal{G}$ has the single characteristic vertex $z$, that one of the components $C_{i_\ell}$ is a Perron component at $z$, but $C$ does not include all of the Perron components at $z$. If $M << \hat{M}$, then in fact $a(\hat{\mathcal{G}}) < a(\mathcal{G})$. Similarly, suppose that $\mathcal{G}$ has a single characteristic edge which is not on any cycle, and that $z$ is a vertex incident to that edge. If $C$ is the Perron component at $z$ and $M << \hat{M}$, then in fact $a(\hat{\mathcal{G}}) < a(\mathcal{G})$.

We now prove a similar result to that of Theorem 7.2.15 for Type II graphs (see [18]):

**THEOREM 7.2.18** *Let $\mathcal{G}$ be a Type II weighted graph with Laplacian $L$ and $B_0$ as the characteristic block. Let $v$ be any cut vertex with connected components $C_0, A_1, A_2, \ldots, A_k$ at $v$, where $C_0$ is the unique Perron component at $v$ (containing the vertices in $B_0$). Let $C = \cup_{\ell=1}^{j} A_{i_\ell}$ where $i_1, i_2, \ldots, i_j \in \{1, 2, \ldots, k\}$. Form a new graph $\hat{\mathcal{G}}$ by replacing $C$ with a single connected component $\hat{C}$ at $v$. Let $M := L[C]^{-1}$ and let $\hat{M}$ denote the bottleneck matrix of $\hat{C}$. Denote the algebraic connectivities of $\mathcal{G}$ and $\hat{\mathcal{G}}$ by $a$ and $\hat{a}$, respectively. If $M << \hat{M}$, then $\hat{a} \leq a$.*

**Proof**: We will prove this by cases: (i) $\rho(\hat{M}) = \rho(L[C_0]^{-1})$, (ii) $\rho(\hat{M}) < \rho(L[C_0]^{-1})$, and (iii) $\rho(\hat{M}) > \rho(L[C_0]^{-1})$. Let $C_1 = \mathcal{G} - C_0$. Then by Lemma 7.2.13, there exists $\gamma > 0$ such that

$$\frac{1}{a} = \rho\left(L[C_1]^{-1} - \frac{1-\gamma}{\theta^T e} J\right) = \lambda_n\left(L[C_0]^{-1} - \frac{\gamma}{\theta^T e} J\right).$$

Suppose $\rho(\hat{M}) = \rho(L[C_0]^{-1})$. Then $\hat{\mathcal{G}}$ is a Type I graph with characteristic vertex $v$, and $C_0$ and $\hat{C}$ as Perron components. Thus

$$\frac{1}{\hat{a}} = \rho(L[C_0]^{-1}) > \lambda_n\left(L[C_0]^{-1} - \frac{\gamma}{\theta^T e} J\right) = \frac{1}{a}.$$

Hence $\hat{a} < a$.

Now suppose $\rho(\hat{M}) < \rho(L[C_0]^{-1})$. Then the Perron component at $v$ is still $C_0$ and the Perron component at every other vertex contains vertices in $B_0$. Letting $\hat{C}_1 = \hat{\mathcal{G}} - C_0$, we see from Lemma 7.2.13 that there exists $\hat{\gamma} > 0$ such that

$$\frac{1}{\hat{a}} = \rho\left(L[\hat{C}_1]^{-1} - \frac{1-\hat{\gamma}}{\theta^T e} J\right) = \lambda_n\left(L[C_0]^{-1} - \frac{\hat{\gamma}}{\theta^T e} J\right). \qquad (7.2.22)$$

Noting that

$$\rho\left(L[\hat{C}_1]^{-1} - \frac{1-\hat{\gamma}}{\theta^T e} J\right) > \rho\left(L[C_1]^{-1} - \frac{1-\hat{\gamma}}{\theta^T e} J\right), \qquad (7.2.23)$$

we see that in order for the right-hand side of (7.2.22) to equal the right-hand side of (7.2.23), it must be that $\hat{\gamma} < \gamma$ since the left-hand side of (7.2.22) is increasing in $\hat{\gamma}$ and the right-hand side of (7.2.22) is nonincreasing in $\hat{\gamma}$. Consequently,

$$\frac{1}{\hat{a}} = \lambda_n\left(L[C_0]^{-1} - \frac{\hat{\gamma}}{\theta^T e} J\right) \geq \lambda_n\left(L[C_0]^{-1} - \frac{\gamma}{\theta^T e} J\right) = \frac{1}{a}.$$

Hence $\hat{a} \leq a$.

Finally, suppose $\rho(\hat{M}) > \rho(L[C_0]^{-1})$. Then if $\hat{\mathcal{G}}$ is a Type I graph where the characteristic vertex $z \in \hat{C}$, then

$$\frac{1}{\hat{a}} = \rho(D) \geq \rho(L[V]^{-1}),$$

where $D$ is the bottleneck matrix for some Perron component at $z$ in $\hat{\mathcal{G}}$ and $V$ is the component at $z$ containing $v$. But $C_0 \subset V$, so

$$\frac{1}{\hat{a}} \geq \rho(L[V]^{-1}) > \rho(L[C_0]^{-1}) \geq \lambda_n\left(L[C_0]^{-1} - \frac{\gamma}{\theta^T e}J\right) = \frac{1}{a}, \quad (7.2.24)$$

hence $\hat{a} < a$. If $\hat{\mathcal{G}}$ is of Type II, then at $v$, $\hat{C}$ is the characteristic block. Thus by Lemma 7.2.13 there exists $t > 0$ such that

$$\frac{1}{\hat{a}} = \lambda_n\left(L[\hat{C}]^{-1} - \frac{t}{\theta^T e}J\right) = \rho\left(L[\hat{\mathcal{G}} - \hat{C}]^{-1} - \frac{1-t}{\theta^T e}J\right) > \rho(L[C_0]^{-1}),$$

where the last inequality follows from the fact that $C_0 \cup \{v\} \subset \hat{\mathcal{G}} - \hat{C}$, so that by Proposition 7.1.6 $(L[\hat{\mathcal{G}} - \hat{C}]^{-1} - \frac{1-t}{\theta^T e}J) \gg L[C_0]^{-1}$. Continuing with (7.2.24), note that

$$\rho(L[C_0]^{-1}) \geq \lambda_n\left(L[C_0]^{-1} - \frac{\gamma}{\theta^T e}J\right) = \frac{1}{a}.$$

Hence $\hat{a} < a$. □

We close this section by sharpening the results of Theorem 6.1.10 which shows how the valuations of the cut vertices of a graph $\mathcal{G}$ increase / decrease as we travel throughout the graph. In the lemma and theorem from [45] below, we investigate the valuations of the vertices of $\mathcal{G}$ that are not cut vertices and their relationship to the valuations of the cut vertices.

**LEMMA 7.2.19** *Let $v$ be a cut vertex of a connected weighted graph $\mathcal{G}$, and let $C_1, \ldots, C_{k+1}$ be the connected components of $\mathcal{G} - v$. Let $C$ be the set of vertices of $\mathcal{G} - C_{k+1}$. Then for each vertex of $C_i$, $1 \leq i \leq k$, the corresponding row of $L[C]^{-1}$ dominates entrywise the row corresponding to the vertex $v$.*

**Proof**: Let $w$ be the sum of the weights of the edges which join $v$ to a vertex in $C_{k+1}$. From Lemma 7.2.7, we find that without loss of generality, $L[C]^{-1}$ can be written as

$$L[C]^{-1} = \left[\begin{array}{c|c} L[C-v]^{-1} & 0 \\ \hline 0^T & 0 \end{array}\right] + \frac{1}{w}J$$

where the last row and column correspond to vertex $v$. The result now follows from the fact that $L[C - v]^{-1}$ is nonnegative. □

**THEOREM 7.2.20** *Let $\mathcal{G}$ be a weighted graph and suppose $B$ is a block of $\mathcal{G}$ such that every vertex in $B$ has positive valuation. Let $v$ be the cut vertex in $B$ such that every path from a vertex in $B$ to the characteristic vertex/block passes through $v$. Then $v$ is the unique vertex in $B$ with minimum valuation.*

**Proof**: From Theorems 7.2.4 and 7.2.9, and Remark 7.2.10, the vector of valuations of $B$ is obtained from the Perron vector of a positive matrix of the form $L[C]^{-1} + xJ$ where $C$ is the set of vertices corresponding to the complement of the connected component at $v$ containing vertices in $B_0$ if $\mathcal{G}$ is of Type II, or containing the vertex $z$ if $\mathcal{G}$ is of Type I. From Lemma 7.2.19 we have that for each vertex $u \neq v$ of $C$, the row of $L[C]^{-1}$ corresponding to $u$ entrywise dominates the row of $L[C]^{-1}$ which corresponds to vertex $v$. It now follows that the entry in the Perron vector for $L[C]^{-1} + xJ$ corresponding to the vertex $u$ is larger than the entry corresponding to $v$. □

**REMARK 7.2.21** We saw in Theorem 6.1.10 that as one travels away from the characteristic vertex / block of a graph, the valuations of the vertices either increase, decrease, or are identically zero. Theorem 7.2.20 expands upon this by showing that the valuations of the vertices within a block $B$ of a graph are bounded above and below by the cut vertices of $\mathcal{G}$ in $B$ that have maximum and minimum valuations, respectively.

**EXAMPLE 7.2.22** Consider the graph below where each vertex is labeled with its characteristic valuation. Observe that the valuations of each non-cut vertex $x$ are bounded above and below by the valuations of the cut vertices of the block in which $x$ lies whose valuations are maximum and minimum.

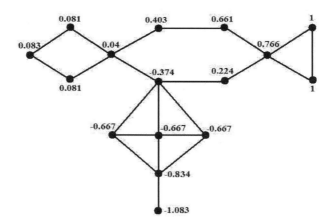

# Exercises:

1. (See [44]) (a) Let $\mathcal{G}$ be a connected graph having cut vertex $v$. Suppose that the components at $v$ are $C_1, \ldots, C_m$, with bottleneck matrices $B_1, \ldots, B_m$, respectively. If $C_m$ is a Perron component at $v$, show that there exists a unique $\gamma \geq 0$ such

that

$$\rho\left(\begin{bmatrix} B_1 & 0 & \cdots & 0 & 0 \\ 0 & B_2 & \cdots & 0 & 0 \\ \cdots & \cdots & \cdots & \cdots & \cdots \\ 0 & \cdots & 0 & B_{m-1} & 0 \\ 0 & 0 & \cdots & 0 & 0 \end{bmatrix} + \gamma J\right) = \lambda_n(B_m - \gamma J) = \frac{1}{a(\mathcal{G})},$$

and that $\gamma = 0$ if and only if there are two or more Perron components at $v$.

(b) Show that $y$ is a Fiedler vector for $\mathcal{G}$ if and only if $y$ can be written $[y(1)|y(2)]^T$ where $y(1)$ is an eigenvector of

$$B := \begin{bmatrix} B_1 & 0 & \cdots & 0 & 0 \\ 0 & B_2 & \cdots & 0 & 0 \\ \cdots & \cdots & \cdots & \cdots & \cdots \\ 0 & \cdots & 0 & B_{m-1} & 0 \\ 0 & 0 & \cdots & 0 & 0 \end{bmatrix} + \gamma J$$

corresponding to $\rho(B)$, $y(2)$ is an eigenvector of $B_m - \gamma J$ corresponding to $\lambda_n$, and where $e^T y(1) + e^T y(2) = 0$.

**2.** (See [44]) Let $\mathcal{G}$ be a connected graph with cut vertex $v$. Suppose that we have two components $C_1, C_2$ at $v$ with corresponding Perron values $\rho_1$ and $\rho_2$, respectively. Show the following:

(a) If $\rho_1 \leq \rho_2$, then $a(\mathcal{G}) \leq 1/\rho_1$.

(b) If $a(\mathcal{G}) = 1/\rho_1$, then $\rho_1 = \rho_2$ and both $C_1$ and $C_2$ are Perron components at $v$.

**3.** (See [43]) Let $\mathcal{G}$ be a connected graph with algebraic connectivity $a$. Let $v$ be a cut vertex of $\mathcal{G}$ with components $C_1, \ldots, C_m$ at $v$, and suppose that $C_1$ is a Perron component at $v$. Then there is a unique $\gamma \geq 0$ such that

$$\lambda_n(L[C_1]^{-1} - \gamma J) = \rho(L[C_2]^{-1} \oplus \ldots \oplus L[C_m]^{-1} \oplus [0] + \gamma J) = \frac{1}{a}$$

**4.** (See [43]) Let $\mathcal{G}$ be a connected graph with cut vertex $v$. Suppose that the components at $v$ are $C_1, \ldots, C_m$, with bottleneck matrices $B_1, \ldots, B_m$, respectively.

(a) Suppose $\rho(B_i) \geq \beta$ for at least two distinct $i$'s between 1 and $m$. Show that $a(\mathcal{G}) \leq 1/\beta$.

(b) Suppose $\rho(B_i) > \beta$ for at least one $i$. Show that $a(\mathcal{G}) < 1/\beta$.

## 7.3 Minimizing the Algebraic Connectivity of Graphs with Fixed Girth

Now that we are able to use bottleneck matrices to compute the algebraic connectivity of graphs, we use these results along with the perturbation results from the previous section to determine which graphs on $n$ vertices and girth $g$, with $n \geq 3g - 1$, will have the smallest algebraic connectivity. To this end we require two important definitions:

**DEFINITION 7.3.1** The *girth* of a graph is the length of its smallest cycle.

**DEFINITION 7.3.2** A graph is *unicyclic* if it has exactly one cycle.

To achieve our goal of minimizing the algebraic connectivity of graphs with $n$ vertices and girth $g$, we will make use of the following observation from [18] which is based off Theorems 7.2.15 and 7.2.18:

**OBSERVATION 7.3.3** *Let $\mathcal{G}$ be a connected unicyclic graph with Laplacian matrix $L$ and let $C$ be a connected component at some vertex $u$ which contains the vertices on the cycle. For each vertex $i$ on the cycle, suppose there are $m_i$ components at vertex $i$ not including the vertices on the cycle. Let $B_i$ be the collection of vertices of the $m_i$ components at vertex $i$. Modify $C$ to form $\hat{C}$ by replacing those $m_i$ components by a single path on $|B_i|$ vertices at vertex $i$. Let the new graph be denoted by $\hat{\mathcal{G}}$, and let $\hat{L}$ denote its Laplacian. Then $L[C]^{-1} << \hat{L}[\hat{C}]^{-1}$.*

We now present a general proposition from [18] which shows which graphs of fixed girth $g$ have the minimum algebraic connectivity:

**PROPOSITION 7.3.4** *Among all connected graphs on $n$ vertices with fixed girth $g$, the algebraic connectivity is minimized by a unicyclic graph with girth $g$ having the following property: There are at most two connected components at every vertex on the cycle, and the component not including the vertices on the cycle (if such a component exists) is a path.*

Below we present an illustration of what such a graph would look like:

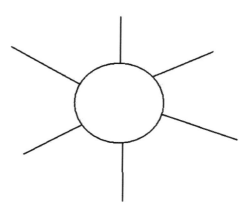

(7.3.1)

**Proof**: Recall from Theorem 5.1.2 that removing an edge from a graph cannot increase the algebraic connectivity. Thus when considering graphs of girth $g$, we may restrict ourselves to unicyclic graphs whose only cycle contains $g$ vertices. Therefore, let $\mathcal{G}$ be any connected unicyclic graph on $n$ vertices with girth $g$. Suppose the vertices on the cycle are labeled $1, 2, \ldots, g$. Since $\mathcal{G}$ is unicyclic, for each vertex $j$, $(1 \leq j \leq g)$, the connected components at $j$ not containing the vertices on the cycle are trees (possible empty). Thus the union of these components is a forest. Let $F_j$ be the union of the connected components at $j$ not containing the other vertices on the cycle. Fix $j$ and suppose that $F_j$ is not a path. If $\mathcal{G}$ is a Type I graph and the characteristic vertex $z$ is such that $z \notin F_j \cup \{j\}$, then replace $F_j$ by a path on $|F_j|$ vertices. Similarly, if $\mathcal{G}$ is a Type II graph and the characteristic block $B_0$ is such that $B_0 \not\subseteq F_j$, then replace $F_j$ by a path on $|F_j|$ vertices. In either case, apply Theorems 7.2.15 and 7.2.18 to produce a unicyclic graph whose algebraic connectivity is at most that of $\mathcal{G}$.

Suppose now that $\mathcal{G}$ is of Type I and $z \in F_j \cup \{j\}$. Then one of the following holds:

(i) some component $C$ at $z$ not containing the vertices on the cycle is not a path,

(ii) $z \neq j$ and in the component $C$ at $z$ containing the cycle, the tree at $j$ containing $z$ is not a path, or

(iii) all of the components at $z$ are either of form (i) or (ii), and $\deg z \geq 3$ (otherwise $F_j$ is a path).

In case (i), we apply Theorem 7.2.15 and replace such a component at $z$ with a path on $|C|$ vertices. In case (ii), we apply Theorem 7.2.18 and and replace $C$ by a path in accordance with Observation 7.3.3. In case (iii), since $\mathcal{G}$ is of Type I, $z$ has at least two Perron components, thus at least one of which must be a path. So replacing the remaining components at $z$ by a single unicyclic graph in which each of the remaining paths are adjoined to the end of the path connected to the cycle, we form a unicyclic graph whose algebraic connectivity by Theorem 7.2.15 is at most that of $\mathcal{G}$.

Now suppose $\mathcal{G}$ is of Type II and $B_0 \subseteq F_j$. In this case, $B_0$ is an edge $uv$ in $F_j$ (without loss of generality, let $d(u,j) < d(v,j)$). Since $F_j$ is not a path, either the component at $u$ containing $v$ is not a path, or in the component at $v$ containing $u$, the tree at $j$ not containing the vertices on the cycle is not a path. In either case, we can replace the corresponding components with a path and apply Theorem 7.2.18 to produce a unicyclic graph whose algebraic connectivity is at most that of $\mathcal{G}$. □

At this point, we know so far that of all connected graphs on $n$ vertices with girth $g$, the graph with the minimum algebraic connectivity will be in the form of (7.3.1). The goal of this section is to refine Proposition 7.3.4 by showing which subset of graphs of the form (7.3.1) have minimum algebraic connectivity. We will show that of all graphs of girth $g$ on $n$ vertices, the graph with minimum algebraic connectivity is the graph $C_{n,g}$ below which is the $g$-cycle with a path of length $n - g$

joined at exactly one vertex on the cycle:

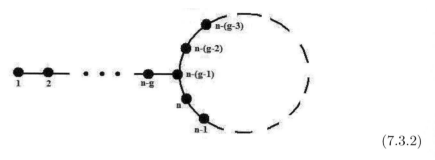

(7.3.2)

We prove this theorem for several cases. We begin with the case $g = 3$. If $g = 3$, then by Proposition 7.3.4, the graph with the minimum algebraic connectivity will be of the form $\mathcal{G}_{k,\ell}$ pictured below:

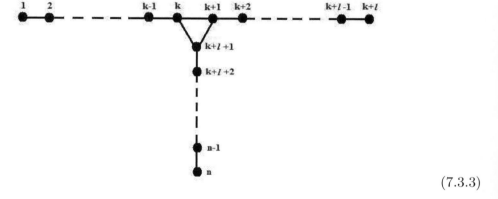

(7.3.3)

Thus for the case $g = 3$, our aim is to show that the graph $\mathcal{G}_{n-2,1}$ is the graph on $n$ vertices with minimum algebraic connectivity:

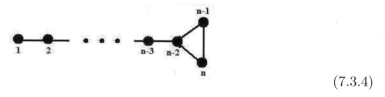

(7.3.4)

We now prove a lemma and corollary from [18] concerning the Perron components of $\mathcal{G}_{n-2,1}$.

**LEMMA 7.3.5** *For the graph $\mathcal{G}_{n-2,1}$, if $n$ is even, then vertices $\frac{n}{2}$ and $\frac{n+2}{2}$ have mutual Perron components. If $n$ is odd, then vertices $\frac{n-1}{2}$ and $\frac{n+1}{2}$ have mutual Perron components.*

**Proof**: First we note that the Perron value of the bottleneck matrix for the component $B$ at vertex $k$ ($k = 1, 2, \ldots, n-2$) containing vertices $n-1$ and $n$ is the same as that for the component $B'$ at $k$ containing vertices $n-1$ and $n$ of the graph $\mathcal{G}$ with the edge $(n-1), n$ deleted. This is because in $B'$, vertices $n-1$ and $n$ are isomorphic, hence the entries in the Perron vector are equal, from which it follows that the Perron vector for $B'$ also serves as the Perron vector for $B$. For even $n$, the components at vertex $\frac{n}{2}$ are a path on $\frac{n-2}{2}$ vertices and the following graph:

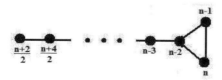

The Perron value of the second component remains the same even if the edge $(n-1), n$ is deleted. But this component (with the edge deleted) is a tree which contains a path on $\frac{n-2}{2}$ vertices. Thus it follows that the Perron component at vertex $\frac{n}{2}$ contains this component (pictured above) and this contains vertex $\frac{n+2}{2}$. At vertex $\frac{n+2}{2}$, the components are a path on $\frac{n}{2}$ vertices, and the graph on $\frac{n-2}{2}$ vertices below:

Applying similar arguments as in the case above for vertex $\frac{n}{2}$, it follows that the path on $\frac{n}{2}$ vertices is a Perron component. So the Perron component at vertex $\frac{n+2}{2}$ contains vertex $\frac{n}{2}$. The argument for the case where $n$ is odd is similar. □

In the following corollary from [18], let $P_k$ be the bottleneck matrix for a component which is a path on $k$ vertices. This corollary follows directly from Theorem 6.2.12.

**COROLLARY 7.3.6** *Let $a$ be the algebraic connectivity of $\mathcal{G}_{n-2,1}$. If $n$ is even, then $a = 1/\rho(P_{\frac{n}{2}} - \gamma J)$ for some $0 < \gamma < 1$, while if $n$ is odd then $a = 1/\rho(P_{\frac{n-1}{2}} - \gamma J)$ for some $0 < \gamma < 1$.*

In our attempt to show that $\mathcal{G}_{n-2,1}$ is the graph on $n$ vertices with girth $g = 3$ having minimum algebraic connectivity, we prove the following lemma from [18] which gives us important insight as to the algebraic connectivity of $\mathcal{G}_{k,k}$.

**LEMMA 7.3.7** *Suppose that some Fiedler vector of the Laplacian for $\mathcal{G}_{k,\ell}$ gives vertices $1, 2, \ldots, k$ a positive valuation, vertices $k+1, k+2, \ldots, k+\ell$ a negative valuation, and vertices $k+\ell+1, k+\ell+2, \ldots, n$ a zero valuation. Then $k = \ell$ and the algebraic connectivity of $\mathcal{G}_{k,k}$ is given by $a = 1/\rho(P_k - \frac{2}{3}J)$.*

**Proof**: The Laplacian $L$ for $\mathcal{G}_{k,\ell}$ can be written as

$$L = \begin{bmatrix} P_k^{-1} + e_k e_k^T & -e_k e_\ell^T & -e_k e_{n-\ell-k}^T \\ -e_\ell e_k^T & P_\ell^{-1} + e_\ell e_\ell^T & -e_\ell e_{n-\ell-k}^T \\ -e_{n-\ell-k} e_k^T & -e_{n-\ell-k} e_\ell^T & P_{n-\ell-k}^{-1} + e_{n-\ell-k} e_{n-\ell-k}^T \end{bmatrix}$$

and the Fiedler vector as

$$\begin{bmatrix} u \\ -v \\ 0 \end{bmatrix}$$

where $u$ and $v$ are entry-wise positive. Then by the eigenvalue-eigenvector relationship, we have

$$P_k^{-1} u + (u_k + v_\ell) e_k = au. \tag{7.3.5}$$

Mulitiplying both sides by $\frac{1}{a} P_k$ and recalling that $P_k e_k = e$, we obtain

$$P_k u - \left(\frac{u_k + v_\ell}{a}\right) e = \frac{1}{a} u. \tag{7.3.6}$$

Similarly,

$$P_\ell v - \left(\frac{u_k + v_\ell}{a}\right) e = \frac{1}{a} v. \tag{7.3.7}$$

Further, multiplying both sides of (7.3.5) by $e^T$ and recalling that $e^T e_k = 1$, we have

$$e^T P_k^{-1} u + (u_k + v_\ell) = ae^T u.$$

Since $e^T P_k^{-1} u = u_k$, it follows that $2u_k + v_\ell = ae^T u$, and similarly $u_k + 2v_\ell = ae^T v$. Thus we have

$$\begin{bmatrix} 2 & 1 \\ 1 & 2 \end{bmatrix} \begin{bmatrix} u_k \\ v_\ell \end{bmatrix} = a \begin{bmatrix} e^T u \\ e^T v \end{bmatrix}.$$

Since $e^T u = e^T v$, it follows that $u_k = v_\ell = \frac{a}{3} e^T u$. Consequently, $\frac{u_k + v_\ell}{a} = \frac{2}{3} e^T u$. From (7.3.6) and (7.3.7) we obtain

$$\left(P_k - \frac{2}{3} J\right) u = \frac{1}{a} u \quad \text{and} \quad \left(P_\ell - \frac{2}{3} J\right) v = \frac{1}{a} v.$$

Since $u$ and $v$ are Perron vectors and the corresponding Perron values are both $\frac{1}{a}$, it follows that $k = \ell$ and $a = 1/\rho(P_k - \frac{2}{3} J)$. □

We now use the above lemma to prove the following proposition from [18] which shows that if $\mathcal{G}_{k,\ell}$ is of Type II with the 3-cycle as the characteristic block, then $\mathcal{G}_{n-2,1}$ is the unique graph of the form $\mathcal{G}_{k,\ell}$ having minimum algebraic connectivity.

**PROPOSITION 7.3.8** *Suppose that $\mathcal{G}_{k,\ell}$ is of Type II and that some Fiedler vector has all nonzero entries, with the 3-cycle being the characteristic block. Let $a$ be the algebraic connectivity of $\mathcal{G}_{k,\ell}$ and let $a'$ be the algebraic connectivity of $\mathcal{G}_{n-2,1}$. Then $a' < a$.*

**Proof**: The Laplacian $L$ for $\mathcal{G}_{k,\ell}$ can be written as

$$L = \begin{bmatrix} P_k^{-1} + e_k e_k^T & -e_k e_\ell^T & -e_k e_{n-\ell-k}^T \\ -e_\ell e_k^T & P_\ell^{-1} + e_\ell e_\ell^T & -e_\ell e_{n-\ell-k}^T \\ -e_{n-\ell-k} e_k^T & -e_{n-\ell-k} e_\ell^T & P_{n-\ell-k}^{-1} + e_{n-\ell-k} e_{n-\ell-k}^T \end{bmatrix}$$

and the Fiedler vector as

$$\begin{bmatrix} u \\ -v \end{bmatrix}$$

where $u$ is positive and of order $k + \ell$, while $v$ is positive and of order $n - k - \ell$. Then by the eigenvalue-eigenvector relationship, we have

$$\left( \begin{bmatrix} P_k^{-1} & 0 \\ 0 & P_\ell^{-1} \end{bmatrix} + (e_k - e_{k+1})(e_k - e_{k+1})^T \right) u + v_1(e_k + e_{k+1}) = au \quad (7.3.8)$$

and

$$P_{n-k-\ell}^{-1} v + (v_1 + u_k + u_{k+1})e_1 = av. \quad (7.3.9)$$

Mulitplying (7.3.9) through on the left by $\frac{1}{a} P_{n-k-\ell}$ and rearranging, we find that

$$P_{n-k-\ell} v - \frac{(v_1 + u_k + u_{k+1})}{a} e = \frac{1}{a} v. \quad (7.3.10)$$

Multiplying (7.3.9) through on the left by $e^T$ we also find that $e^T P_{n-k-\ell}^{-1} v + (v_1 + u_k + u_{k+1}) = ae^T v$. Recalling that $e^T P_{n-k-\ell}^{-1} v = v_1$, it follows that $2v_1 + u_k + u_{k+1} = ae^T v$. Substituting this into (7.3.10) we obtain

$$\left( P_{n-k-\ell} - \left( \frac{v_1 + u_k + u_{k+1}}{2v_1 + u_k + u_{k+1}} \right) J \right) v = \frac{1}{a} v. \quad (7.3.11)$$

Since $v_1$, $u_k$, and $u_{k+1}$ are all positive, we have the coefficient of $J$ is between $\frac{1}{2}$ and 1. Therefore $a = 1/\rho(P_{n-k-\ell} - tJ)$ for some $\frac{1}{2} < t < 1$. Observe that for fixed $t$, the value of $a$ increases as $n - k - \ell$ decreases. In particular, for all $\frac{1}{2} < t < 1$, if

$$n - k - \ell \leq \begin{cases} \frac{n-2}{2}, & \text{if } n \text{ is even} \\ \frac{n-3}{2}, & \text{if } n \text{ is odd} \end{cases} \quad (7.3.12)$$

then $a' < a$ by Corollary 7.3.6. Now consider (7.3.8) and observe that

$$\left( \begin{bmatrix} P_k^{-1} & 0 \\ 0 & P_\ell^{-1} \end{bmatrix} + (e_k - e_{k+1})(e_k - e_{k+1})^T \right)^{-1} = \begin{bmatrix} P_k & 0 \\ 0 & P_\ell \end{bmatrix} - \frac{1}{3} \begin{bmatrix} J & -J \\ -J & J \end{bmatrix}$$

From (7.3.8), we obtain

$$\left( \begin{bmatrix} P_k & 0 \\ 0 & P_\ell \end{bmatrix} - \frac{1}{3} \begin{bmatrix} J & -J \\ -J & J \end{bmatrix} \right) u - \frac{v_1}{a} e = \frac{1}{a} u.$$

Using similar reasoning that we used to obtain (7.3.11), we see that

$$\left(\begin{bmatrix} P_k & 0 \\ 0 & P_\ell \end{bmatrix} - \frac{1}{3}\begin{bmatrix} J & -J \\ -J & J \end{bmatrix} - \frac{v_1}{2v_1 + u_k + u_{k+1}}J\right)u = \frac{1}{a}u.$$

Since $v_1$, $u_k$, and $u_{k+1}$ are all positive, clearly the coefficient for $J$ is between 0 and $\frac{1}{2}$. Therefore

$$a = 1/\rho\left(\begin{bmatrix} P_k & 0 \\ 0 & P_\ell \end{bmatrix} - \frac{1}{3}\begin{bmatrix} J & -J \\ -J & J \end{bmatrix} - tJ\right)$$

for some $0 < t < \frac{1}{2}$. Note that

$$a = 1/\rho\left(\begin{bmatrix} P_k & 0 \\ 0 & P_\ell \end{bmatrix} - \frac{1}{3}\begin{bmatrix} J & -J \\ -J & J \end{bmatrix} - tJ\right)$$

$$< \rho\left(\begin{bmatrix} P_k & 0 \\ 0 & P_\ell \end{bmatrix} - \frac{1}{3}\begin{bmatrix} J & -J \\ -J & J \end{bmatrix}\right) \le \rho\left(\begin{bmatrix} P_k & 0 \\ 0 & P_\ell \end{bmatrix}\right).$$

Observing that the last expression is an increasing function of $k$ and $\ell$, it follows that if

$$\max(k, \ell) \le \begin{cases} \frac{n-2}{2}, & \text{if } n \text{ is even} \\ \frac{n-3}{2}, & \text{if } n \text{ is odd} \end{cases} \quad (7.3.13)$$

then $a' < a$ by Corollary 7.3.6. On the other hand, if

$$\max(k, \ell) \ge \begin{cases} \frac{n}{2}, & \text{if } n \text{ is even} \\ \frac{n-1}{2}, & \text{if } n \text{ is odd} \end{cases}$$

then we have

$$n - k - \ell \le n - 1 - \max(k, \ell) \le \begin{cases} \frac{n-2}{2}, & \text{if } n \text{ is even} \\ \frac{n-1}{2}, & \text{if } n \text{ is odd} \end{cases}. \quad (7.3.14)$$

By (7.3.12) and (7.3.14), we see that the only case we have not considered is when $n$ is odd, $\min(k, \ell) = 1$, and $\max(k, \ell) = \frac{n-1}{2}$. In this case, we have the graph (without loss of generality, $k = 1$ and $\ell = \frac{n-1}{2}$):

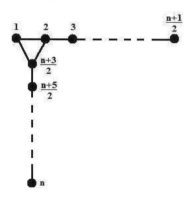

It is not difficult to see that every Fiedler vector yields a valuation of zero for the vertex on the triangle which is not a cut vertex, contrary to our hypothesis. Hence $a' < a$ as desired. □

At this point, to achieve our goal of proving that $\mathcal{G}_{n-2,1}$ is the graph on $n$ vertices and girth $g = 3$ with minimum algebraic connectivity, we still must prove this for the cases where (i) $\mathcal{G}_{k,\ell}$ is of Type I, and (ii) $\mathcal{G}_{k,\ell}$ is of Type II but the 3-cycle is not the characteristic block. To this end, we need the following lemma from [18]:

**LEMMA 7.3.9** *Consider the graph $\mathcal{H}$ given below with Laplacian matrix $L_1$, and let $C = \{1, 2, \ldots, k + \ell\}$. Consider also the graph given in (7.3.4) but relabeled via $i \to n - i + 1$ for $1 \leq i \leq n$, with Laplacian matrix $L_2$. Then $L_1[C]^{-1} << L_2[C]^{-1}$.*

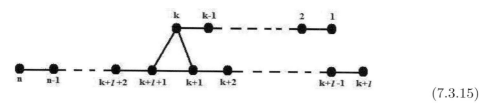

(7.3.15)

**Proof**: We have

$$L_1[C] = \begin{bmatrix} P_k^{-1} & 0 \\ 0 & P_\ell^{-1} \end{bmatrix} + (e_k - e_{k+1})(e_k - e_{k+1})^T,$$

so that

$$L_1[C]^{-1} = \begin{bmatrix} P_k - \frac{1}{3}J & \frac{1}{3}J \\ \frac{1}{3}J & P_\ell - \frac{1}{3}J \end{bmatrix}.$$

Also

$$L_2[C]^{-1} = F - \frac{1}{3}F(e_1 - e_2)(e_1 - e_2)^T F,$$

where

$$F = \begin{bmatrix} \begin{array}{cc} k+\ell-1 & k+\ell-2 \\ k+\ell-2 & k+\ell-1 \end{array} & ee_1^T P_{k+\ell+2} \\ \hline P_{k+\ell+2} e_1 e^T & P_{k+\ell+2} \end{bmatrix}.$$

It is now straightforward to verify that $(L_1[C])^{-1} << (L_2[C])^{-1}$. □

We now have enough information to prove that $\mathcal{G}_{n-2,1}$ is the graph on $n$ vertices and girth $g = 3$ with minimum algebraic connectivity. We do so in the following theorem from [18]:

**THEOREM 7.3.10** *The graph on $n$ vertices with girth 3 of minimum algebraic connectivity is $\mathcal{G}_{n-2,1}$.*

**Proof**: From Proposition 7.3.4, we have that the graph of girth 3 with minimum algebraic connectivity has the form of $\mathcal{G}_{k,\ell}$ as in (7.3.3). If some Fiedler vector valuates the 3-cycle with both positive and negative valuations, then the algebraic connectivity of $\mathcal{G}_{k,\ell}$ exceeds that for $\mathcal{G}_{n-2,1}$ by Lemma 7.3.7 and Proposition 7.3.8. It follows that the minimizer is either of Type I, or that there is an edge not on the 3-cycle in which the vertices incident to that edge have mutual Perron components. In either case, there is a vertex $x$ on the 3-cycle with the property that we may replace the component containing the other vertices on the 3-cycle at $x$ by the corresponding component at $x$ of $\mathcal{G}_{n-2,1}$ which, by Lemma 7.3.9 and Theorem 7.2.15, will lower the algebraic connectivity. Hence $\mathcal{G}_{n-2,1}$ has minimum algebraic connectivity. □

We devote the remainder of this section to generalizing Theorem 7.3.10 to graphs whose girth is $g \geq 4$ (restricting ourselves to $n \geq 3g - 1$.) To this end, we recall the graph $C_{n,g}$ in (7.3.2) and we begin with a lemma from [21] regarding the algebraic connectivity of such graphs.

**LEMMA 7.3.11** *Fix $n$ and suppose that $n \geq (3g-1)/2$ with $g \geq 4$. Then $a(C_{n,g}) > a(C_{n,g-1})$.*

**Proof**: Consider the graph $C_{n,g}$ with $n \geq (3g-1)/2$, and let $v$ be the only vertex of degree 3. At $v$ there are exactly two components, say $C_1$ and $C_2$, and suppose $C_2$ is the component that is a path. Since $|C_2| = n - g$, it follows that

$$\rho(L[C_1]^{-1}) = \frac{1}{2[1 - \cos(\frac{\pi}{g})]},$$

and

$$\rho(L[C_2]^{-1}) = \frac{1}{2[1 - \cos(\frac{\pi}{2n-2g+1})]}.$$

(See the proof of Theorem 5.1.7 for ideas relating to these quantities.) Since $n \geq (3g-1)/2$, it follows that $C_2$ is the Perron component at $v$ and so by Theorem 7.2.4 it follows that the characteristic vertex/block of $C_{n,g}$ lies on the path. Replace $C_1$ in $C_{n,g}$ with $C'$ pictured below, thus creating $C_{n,g-1}$.

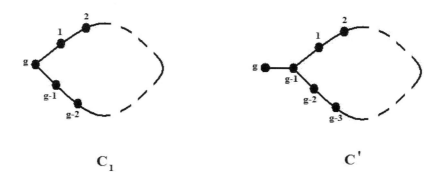

Let $M$ and $M'$ be the bottleneck matrices for $C_1$ and $C'$, respectively.

Letting $Q_{g-1}$ be the bottleneck matrix for any vertex on a cycle of $g-1$ vertices, we see that

$$M' = \left[\begin{array}{c|c} J + Q_{g-1} & e \\ \hline e^T & 1 \end{array}\right]$$

It is easy to check that $M'$ dominates $M$ entrywise and the domination is strict in the last row. Thus by Remark 7.2.17, it follows that $a(C_{n,g}) > a(C_{n,g-1})$. □

Our next lemma from [21] gives us insight as to how the algebraic connectivity of a tree changes when paths within a tree are altered. Since unicylic graphs closely resemble trees, this lemma will be useful to us in proving subsequent theorems. For notational simplicity, we will let $y(i)$ denote the $i^{th}$ entry of the vector $y$ (instead of $y_i$ as we normally do).

**LEMMA 7.3.12** *Let $\mathcal{T}$ be a tree with vertex $v$ such that there are least two components at $v$ which are paths. Assume that there is a Fiedler vector $y$ such that $y(v) > 0$ and that both of these paths are positive components at $v$. Let the path lengths be $\ell_1$ and $\ell_2$ with $\ell_1 \geq \ell_2$. Then modifying $\mathcal{T}$ so that those paths at $v$ have lengths $\ell_1 + 1$ and $\ell_2 - 1$, respectively, strictly decreases the algebraic connectivity.*

**Proof**: Suppose $v$ is a vertex of $\mathcal{T}$ such that two of the components at $v$ are paths, say $C_1 = P_{\ell_1}$ and $C_2 = P_{\ell_2}$. Label the vertices in $C_1$ as $v_1, v_2, \ldots, v_{\ell_1}$ (in increasing order according to the distance from $v$), and the vertices in $C_2$ as $w_1, w_2, \ldots, w_{\ell_2}$ (in the same manner). Suppose $y$ is a Fiedler vector of $\mathcal{T}$ that valuates the vertices of $C_1$ and $C_2$ positively (the existence of such a vector is guaranteed by the hypothesis). We claim that for each $i = \ell_1 - \ell_2 + 1, \ldots, \ell_1$ that

$$y(v_i) \geq y(w_j), \tag{7.3.16}$$

where $j = 1, \ldots, \ell_2 - \ell_1 + i$ if $\ell_1 > \ell_2$, or $j = 1, \ldots, \ell_2 - \ell_1 + i - 1$ if $\ell_1 = \ell_2$. Since the vertices in $C_1$ and $C_2$ are valuated positively, we have the following list of equations: $y(v_{\ell_1-1}) = c_1 y(v_{\ell_1})$, $y(v_{\ell_1-2}) = c_2 y(v_{\ell_1-1}), \ldots, y(v) = c_{\ell_1} y(v_1)$, where $c_1, c_2, \ldots, c_\ell$ depend only on $a$ and the entries of the Laplacian matrix $L$. It follows from Theorem 6.2.20 that $0 \leq c_j \leq 1$ for $1 \leq j \leq \ell_1$. Similarly we have $y(w_{\ell_2-1}) = c_1 y(w_{\ell_2})$, $y(w_{\ell_2-2}) = c_2 y(w_{\ell_2-1}), \ldots, y(v) = c_{\ell_2} y(w_1)$. We can now do back-substitution into the previous equations in order to verify the claim.

Note that by the hypothesis, the characteristic vertex (vertices) of $\mathcal{T}$ does not contain any vertices on $C_1$ or $C_2$. If $\mathcal{T}$ is of Type I let $u$ be the characteristic vertex while if $\mathcal{T}$ is of Type II let $u$ be the characteristic vertex which is farther from $v$. Let $B$ be the bottleneck matrix at $u$ containing $v$ (and $C_1$ and $C_2$). Let $\hat{B}$ be the bottleneck matrix for the component obtained by deleting the edge $w_{\ell_2}, w_{\ell_2-1}$ and adding the edge $w_{\ell_2}, v_{\ell_1}$. We claim that $\rho(\hat{B} - \gamma J) > \rho(B - \gamma J)$ for all $0 < \gamma < 1$. Let $z$ be an eigenvector associated with $\rho(B - \gamma J) := r$. Since $z$ is a subvector of the Fiedler vector corresponding to $\mathcal{T}$ in the positions given by the component being perturbed, it follows that for a vertex $x$ in that component we have $z(x) = y(x)$. Observe that we may write

$$B = \left[\begin{array}{c|c} 1 + M_{w_{\ell_2-1}, w_{\ell_2-1}} & e_{w_{\ell_2-1}}^T M \\ \hline M e_{w_{\ell_2-1}}^T & M \end{array}\right] \text{ and } \hat{B} = \left[\begin{array}{c|c} 1 + M_{v_{\ell_1}, v_{\ell_1}} & e_{w_{\ell_1}}^T M \\ \hline M e_{w_{\ell_1}}^T & M \end{array}\right]$$

where $M$ is a fixed entry-wise nonnegative and positive definite matrix that corresponds to the vertices which are not incident to the edge deleted or added (here the first row of $B$ and $\hat{B}$ corresponds to the vertex $w_{\ell_2}$). Hence

$$z^T(\hat{B} - \gamma J)z - z^T(B - \gamma J)z = y(w_{\ell_2})^2 (M_{v_{\ell_1}, v_{\ell_1}} - M_{w_{\ell_2-1}, \ell_2-1})$$
$$+ 2y(w_{\ell_2})(e_{v_{\ell_1}}^T M \hat{y} - e_{v_{\ell_1}}^T M \hat{y}), \quad (7.3.17)$$

where $z = [y(w_{\ell_2}) | \hat{y}]^T$ is partitioned conformally with $B$ and $\hat{B}$. Moreover, from the eigenvalue-eigenvector equation for $r$ and $B - \gamma J$, it follows that

$$y(w_{\ell_2})(M e_{w_{\ell_2-1}} - \gamma e) + (M - \gamma J)\hat{y} = r\hat{y}.$$

Rearranging and looking at the rows corresponding to $v_{\ell_1}$ and $w_{\ell_2-1}$, respectively, we obtain

$$e_{v_{\ell_1}}^T M \hat{y} = r\hat{y}(v_{\ell_1}) - y(w_{\ell_2}) M_{v_{\ell_1}, w_{\ell_2-1}} + r + r e^T \hat{y}$$

and

$$e_{v_{\ell_2-1}}^T M \hat{y} = r\hat{y}(v_{\ell_2-1}) - y(w_{\ell_2}) M_{v_{\ell_2-1}, w_{\ell_2-1}} + r + r e^T \hat{y}.$$

Substituting these into (7.3.17) yields

$$z^T(\hat{B} - \gamma J)z - z^T(B - \gamma J)z =$$

$$y(w_{\ell_2})^2 (M_{v_{\ell_1}, v_{\ell_1}} + M_{w_{\ell_2-1}, \ell_2-1} - 2M_{v_{\ell_1}, w_{\ell_2-1}}) + 2y(w_{\ell_2})r(\hat{y}(v_{\ell_1}) - \hat{y}(w_{\ell_2-1})) > 0$$

since $\hat{y}(v_{\ell_1}) \geq \hat{y}(w_{\ell_2-1})$ by (7.3.16), and $M_{v_{\ell_1}, v_{\ell_1}} + M_{w_{\ell_2-1}, \ell_2-1} > 2M_{v_{\ell_1}, w_{\ell_2-1}}$ since $M$ is positive definite. Thus $\rho(\hat{B} - \gamma J) > \rho(B - \gamma J)$. Therefore by Lemma 7.2.13, the algebraic connectivity decreases. $\square$

We are now ready to prove two important claims from [21]. These claims consider unicyclic graphs like those in Figure (7.3.1) and investigate the paths beginning at specific vertices of the cycle. Since unicyclic graphs closely resemble trees, the proofs of these claims make use of Exercises 6 to 8 from Section 6.4.

**CLAIM 7.3.13** *Let $\mathcal{G}$ be a unicyclic graph on $n$ vertices with cycle $[1, 2, \ldots, g, 1]$ and assume that $\mathcal{G}$ is not isomorphic to $C_{n,g}$. Suppose that for $i = 1, \ldots, g$, the component at vertex $i$ not containing any vertex of the cycle is a path, say $P_i$. Let $|P_1| = \max\{|P_i| : i = 1, \ldots, g\}$, and suppose that $|P_1| \geq (g-1)/2$.*

*(i) If $g$ is odd, then $a(\mathcal{G}) \geq a(C_{n,g})$. The inequality is strict if either*
  *(a) $|P_1| > (g-1)/2$ or*
  *(b) $|P_1| = (g-1)/2$ and $|P_k| > 0$, for some $k \in \{2, \ldots, g\}$, $k \neq (g+1)/2, (g+3)/2$.*
*(ii) If $g$ is even, assume that at least one of $|P_i|$, $i = 1, \ldots, g$, $i \neq (g+2)/2$, is positive. Then $a(\mathcal{G}) > a(C_{n,g})$.*

**Proof**: (i) Let $\mathcal{T}$ be the tree obtained by deleting the edge $(g+1)/2, (g+3)/2$ from $\mathcal{G}$; note that $a(\mathcal{G}) \geq a(\mathcal{T})$. Let $\mathcal{T}^*$ be obtained from $\mathcal{T}$ by replacing the components of $\mathcal{T}$ at vertex 1 by paths on the same number of vertices; applying Exercise 8 of Section 6.4 to $\mathcal{T}$ at vertex 1 we get $a(\mathcal{T}^*) \leq a(\mathcal{T})$. Let $\mathcal{T}_1$ be a tree on $n$ vertices such that $\mathcal{T}_1$ has exactly one vertex, call it 1, of degree 3 and the components of $\mathcal{T}_1$ at vertex 1 are paths $P_1'$, $P_2'$, and $P_3'$, where $|P_3'| > |P_2'| = |P_1'| = (g-1)/2$. Applying Exercise 7 of Section 6.4 we have $a(\mathcal{T}_1) \leq a(\mathcal{T}^*)$. Thus we have $a(\mathcal{T}_1) \leq a(\mathcal{T}^*) \leq a(\mathcal{T})$. Note that if (a) holds then necessarily there are at least two components in $\mathcal{T}$ (and hence in $\mathcal{T}^*$) at vertex 1 with more than $(g-1)/2$ vertices, so that in fact $a(\mathcal{T}_1) < a(\mathcal{T}^*)$ by Exercise 7 of Section 6.4. If (b) holds and there are at least two components in $\mathcal{T}$ at vertex 1 with more than $(g-1)/2$ vertices, then again $a(\mathcal{T}_1) < a(\mathcal{T}^*)$, while if (b) holds and there is just one component in $\mathcal{T}$ at vertex 1 with more than $(g-1)/2$ vertices, then by Exercise 8 of Section 6.4 it follows that $a(\mathcal{T}^*) < a(\mathcal{T})$ since the Perron component at vertex 1 in $\mathcal{T}$ is not a path. Thus we see that if either (a) or (b) holds, then $a(\mathcal{T}_1) < a(\mathcal{T})$. Case (i) is proven if we can show $a(C_{n,g}) = a(\mathcal{T}_1)$. To this end, observe that the characteristic vertex (vertices) of $\mathcal{T}_1$ lies on $P_3'$ since $P_3'$ is the longest path ($|P_3'| > (g-1)/2$) among the three. Thus any Fiedler vector valuates the end vertices of $P_1'$ and $P_2'$ equally. Therefore adding an edge between those two end vertices yields a graph with the same algebraic connectivity since by the eigenvalue-eigenvector relationship the Fiedler vector would be the same. But the graph created is precisely $C_{n,g}$, hence $a(C_{n,g}) = a(\mathcal{T}_1)$, yielding the result in Case (i).

(ii) Let $|P_k| > 0$ for some $2 \leq k \leq g/2$. Let $\mathcal{T}$ be the tree obtained by deleting the edge $g/2, (g+2)/2$ from $\mathcal{G}$ and note that $a(\mathcal{G}) \geq a(\mathcal{T})$. Now we proceed as in (i) above to obtain $a(\mathcal{T}) \geq a(C_{n,g+1})$. Applying Lemma 7.3.11 yields $a(C_{n,g}) < a(C_{n,g+1})$. □

**CLAIM 7.3.14** *Let $\mathcal{G}$ be a unicyclic graph of girth $g$ on $n$ vertices with cycle $[1, 2, \ldots, g, 1]$. Suppose that $n \geq 3g - 1$ and that for $i = 1, \ldots, g$, the component at vertex $i$ not containing any vertex of the cycle is a path, say $P_i$. Suppose that $\max\{|P_i| : i = 1, \ldots, g\} < g/2$. Then $a(\mathcal{G}) \geq a(C_{n,g})$ where equality holds if and only if $\mathcal{G}$ is isomorphic to $C_{n,g}$.*

**Proof**: Suppose that $\mathcal{G}$ is not isomorphic to $C_{n,g}$. Let $\mathcal{T}$ be the tree obtained from $\mathcal{G}$ by deleting the edge $1, g$ from the cycle, and note that $a(\mathcal{T}) \leq a(\mathcal{G})$. Observe that $\mathcal{T}$ is simply the path $[1, \ldots, g]$ with different paths $P_i$ attached to each vertex $i$, $i = 1, \ldots, g$. Since $|P_i| < g/2$ for each $i$, it follows that the Perron branch at vertex $i$ is not $P_i$. Thus the characteristic vertex (vertices) of $\mathcal{T}$ is located on the path $[1, \ldots, g]$. Applying Exercise 6 of Section 6.4 reveals that there exists a vertex $k$ such that if $\mathcal{T}'$ is the tree obtained by taking the path $[1, \ldots, g]$ and then attaching the paths $P_2, \ldots, P_k$ to vertex 1 and attaching paths $P_{k+1}, \ldots, P_{g-1}$ to vertex $g$, then $a(\mathcal{T}') \leq a(\mathcal{T})$. Since $n \geq 3g - 1$, it follows that at least one of $P_2, \ldots P_{g-1}$ is nonempty. So by Remark 7.2.17 we actually have strict inequality, i.e. $a(\mathcal{T}') < a(\mathcal{T})$. Exercise 6 of Section 6.4 also shows that the characteristic vertex (vertices) of $\mathcal{T}'$ lies on the path $[1, \ldots, g]$ and if it is a single vertex, it is neither 1 or $g$. In particular, for a Fiedler vector $y$, we have $y_1 y_g < 0$. Moreover, the vertices on the (non-Perron) paths attached to vertex 1 all have positive characteristic valuation if and only if $y_1 > 0$. Similarly, the vertices on the (non-Perron) paths attached to vertex $g$ all have negative characteristic valuation if and only if $y_g < 0$.

Let $n_1$ and $n_2$ be the number of vertices in $\{P_1, \ldots, P_k\}$ and $\{P_{k+1}, \ldots, P_g\}$ respectively. Since $n \geq 3g - 1$, it follows that either $n_1 \geq g$ or $n_2 \geq g$. Without loss of generality, suppose $n_1 \geq g$ and that $y_1 > 0$ so that the vertices on the paths $\{P_1, \ldots, P_k\}$ are all valuated positively. Again using the fact that $n \geq 3g - 1$ but also recalling $\max\{|P_i| : i = 1, \ldots, g\} \leq (g-1)/2$, we have $k \geq 3$. Without loss of generality, suppose that $|P_1| \geq |P_2| \geq \ldots \geq |P_k|$. Since the vertices on these paths all have the same sign in their characteristic valuations, it follows from Lemma 7.3.12 that extending the length (by one) of a longer path ($P_1$ or $P_2$) while decreasing the length (by one) of a shorter path (any of $P_3, \ldots, P_k$) strictly decreases the algebraic connectivity. Iterate this process until we have constructed a tree $\mathcal{T}''$ whose non-Perron components at vertex 1 consist of two paths $P_1''$ and $P_2''$ each of length $\lfloor g/2 \rfloor$, and possibly some other paths $P_3''', \ldots, P_j'''$ each of length less than $g/2$, and such that the vertex sets $P_1 \cup \ldots \cup P_k$ and $P_1'' \cup \ldots \cup P_j''$ have the same number of vertices. Note that $a(\mathcal{T}'') < a(\mathcal{T}')$. Denote the Perron component at vertex 1 in $\mathcal{T}''$ by $C$, and form $\mathcal{T}^*$ by replacing $C$ at vertex 1 by a component consisting of a path on $|C|$ vertices; note that this path is the unique Perron component at vertex 1 in $\mathcal{T}^*$. Applying Exercise 8 of Section 6.4 at vertex 1 of $\mathcal{T}''$ we find that $a(\mathcal{T}^*) \leq a(\mathcal{T}'')$. Finally, construct $\hat{\mathcal{T}}$ from $\mathcal{T}^*$ by replacing the components $P_1'' \cup \ldots \cup P_j'''$ by a single component $D$ on the same number of vertices such that $D$ has a single vertex $v$ of degree 3, while all others have smaller degree, and such that the two branches at $v$ not containing vertex 1 are $P_1''$ and $P_2''$. By Theorem 7.2.15, we conclude that $a(\hat{\mathcal{T}}) \leq a(\mathcal{T}^*)$. Further, at vertex $v$ of $\hat{\mathcal{T}}$, the two paths $P_1''$ and $P_2''$ of length $\lfloor g/2 \rfloor$ are isomorphic and the vertices of which each have positive valuation. Hence each pair of isomorphic vertices have the same valuation. Therefore adding an edge joining their respective pendant vertices yields a graph with the same algebraic connectivity since by the eigenvalue-eigenvector relationship the Fiedler vector would be the same. But the graph created is precisely $C_{n,g}$. Consequently, if $g$ is odd we have $a(C_{n,g}) = a(\hat{\mathcal{T}}) < a(\mathcal{G})$, while if $g$ is even we have $a(C_{n,g}) < a(C_{n,g+1}) = a(\hat{\mathcal{T}}) < a(\mathcal{G})$, the first inequality following from Lemma 7.3.11. □

Before we can prove the theorem that these claims are leading up to, we must define the following graph: The graph $\mathcal{G}_{\ell_1, \ell_2, g}$ refers to the graph with a cycle on $g$ vertices ($g \geq 4$ and even) labelled $1, \ldots, g$, with a path on $\ell_1$ vertices attached at vertex 1, and a path on $\ell_2$ vertices attached at vertex $\frac{g+2}{2}$.

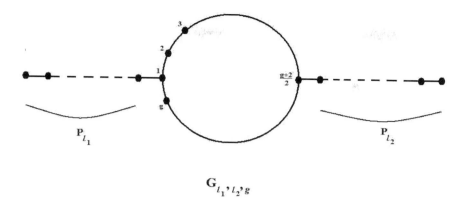

$G_{\ell_1, \ell_2, g}$

We may now state the next lemma which concerns unicyclic graphs of the form above. Since the proof of the following lemma is tedious and adds little to our discussion, we omit it. The proof can be found in [21].

**LEMMA 7.3.15** *Suppose that $k > \left(\frac{2}{\sqrt{2}} + \frac{1}{2}\right) g$ with $g \geq 4$, and let $\ell_1 + \ell_2 = k$, where $1 \leq \ell_1, \ell_2 \leq k - 1$. Then $a(\mathcal{G}_{\ell_1, \ell_2, g}) > a(C_{g+k, g})$.*

We are now ready to prove the second main theorem of this section (from [21]) which concerns minimizing the algebraic connectivity of graphs with girth $g \geq 4$ where $n \geq 3g - 1$.

**THEOREM 7.3.16** *Let $\mathcal{G}$ be a connected graph on $n$ vertices with girth $g \geq 4$ such that $n \geq 3g - 1$. Then $a(\mathcal{G}) \geq a(C_{n,g})$. Equality holds if and only if $\mathcal{G}$ is isomorphic to $C_{n,g}$.*

**Proof**: Let $\mathcal{G}_{n,g}$ denote the class of all connected unicyclic graphs on $n$ vertices with fixed girth $g$ that are of the from (7.3.1): there are at most two connected components at every vertex on the cycle and the component not including the vertices on the cycle (if such a component exists) is a path. From Proposition 7.3.4 we know that there exists a graph $H \in \mathcal{G}_{n,g}$ such that $a(H) \leq a(\mathcal{G})$. Thus it is sufficient to prove the desired result for any graph in $\mathcal{G}_{n,g}$.

Let $\mathcal{G} \in \mathcal{G}_{n,g}$, $[1, 2, \ldots, g, 1]$ be the cycle in $\mathcal{G}$ and $P_1, P_2, \ldots, P_g$ be the paths attached to the vertices $1, 2, \ldots, g$, respectively. Suppose that $\mathcal{G}$ is not isomorphic to $C_{n,g}$. We will prove this theorem by cases on the order of $P_i$.

Case I: $|P_i| \geq g/2$ for at least one $i \in \{1, 2, \ldots, g\}$: Without loss of generality, let $|P_1| \geq g/2$. If $g$ is odd, then necessarily $|P_1| > (g-1)/2$. The result now follows immediately from Claim 7.3.13(i-a). Now suppose $g$ is even. First suppose the only nonempty paths are $P_1$ and $P_{(g+2)/2}$. In the language of Lemma 7.3.15, observe that $k \geq 2g - 1$. Thus if $2g - 1 > \left(\frac{2}{\sqrt{2}} + \frac{1}{2}\right) g$, i.e., $g \geq 12$, then the result follows from Lemma 7.3.15. If $g \leq 10$ then since extending either path only decreases the algebraic connectivity of the graph, it follows that the only other cases are: $g \in \{4, 6, 8, 10\}$ with $n = 3g - 1$ for each such $g$. These four cases can be verified by direct computation (Exercise 1 below). Now supposing that there is a nonempty path $P_k$ with $k \neq (g+2)/2$, the result follows immediately from Claim 7.3.13(ii).

*Case II*: $|P_i| < g/2$ for each $i = 1, \ldots, g$.: This case is precisely Claim 7.3.14. □

## Exercise:

1. In each of the following pairs below, verify Theorem 7.3.16.
    (a) $g = 4$, $n = 11$
    (b) $g = 6$, $n = 17$
    (c) $g = 8$, $n = 23$
    (d) $g = 10$, $n = 29$

## 7.4 Maximizing the Algebraic Connectivity of Unicyclic Graphs with Fixed Girth

We saw in the previous section that given a graph on $n$ vertices and girth $g$, the graph that minimizes the algebraic connectivity is a unicyclic graph. To this end, it is natural to ask which unicyclic graphs on $n$ vertices with fixed girth $g$ maximize the algebraic connectivity. Let $U_{n,g}$ denote the class of connected unicyclic graphs on $n$ vertices with girth $g$. For a graph $\mathcal{G} \in U_{n,g}$, let $\mathcal{G}_{g,n_1,n_2,\ldots,n_g}$ be the graph with a g-cycle whose vertices are labeled $1, 2, \ldots, g$ and with $n_j$ pendant vertices adjacent to vertex $j$ (for $1 \leq j \leq g$) where $\sum n_j = n - g$. For notational simplicity, $\mathcal{G}_{g,n-g} = \mathcal{G}_{g,n-g,0,0,\ldots,0}$.

We first present a lemma from [18] which will assist us in determining which unicyclic graphs of girth $g$ have maximum algebraic connectivity.

**LEMMA 7.4.1** *Among all graphs in $U_{n,g}$, the algebraic connectivity is maximized by a graph of the form $\mathcal{G}_{g,n_1,n_2,\ldots,n_g}$.*

**Proof**: Suppose $\mathcal{G} \in U_{n,g}$ and let $a$ be the algebraic connectivity of $\mathcal{G}$. Label the vertices on the cycle $1, 2, \ldots, g$. Let $F_j$ denote the union of the connected components at vertex $j$ (for $1 \leq j \leq g$) except for the unique component at $j$ which contains the vertices on the cycle. Fix $j$. Suppose $F_j$ is not the union of $|F_j|$ pendant vertices. Form $\hat{\mathcal{G}}$ by replacing $F_j$ with $|F_j|$ pendant vertices each adjacent to vertex $j$; let $\hat{a}$ denote the algebraic connectivity of $\hat{\mathcal{G}}$. To complete the proof, we need to show that $a \leq \hat{a}$. We will do so by cases:

*Case I*: Suppose in $\hat{\mathcal{G}}$ that at least one of the $|F_j|$ pendant vertices adjacent to $j$ is a Perron component at $j$. If $\hat{\mathcal{G}}$ is of Type I then $\hat{a} = 1$ by Theorem 7.2.4(i). If $\hat{\mathcal{G}}$ is of Type II, then by Lemma 7.2.13 it follows that $\hat{a} > 1$. However, since $\mathcal{G}$ has a cut vertex, it follows from Theorem 5.1.10 that $a \leq 1$. Therefore $a \leq \hat{a}$.

*Case II*: Suppose in $\hat{\mathcal{G}}$ that none of the $|F_j|$ pendant vertices adjacent to $j$ is a Perron component at $j$. In this case, we can use entry-wise domination of the bottleneck matrices for $F_j$ for the $|F_j|$ pendant vertices and apply Theorems 7.2.15 and 7.2.18 to obtain $a \leq \hat{a}$. □

Lemma 7.4.1 shows that when given a graph on $n$ vertices of girth $g$, the candidates for graphs that would yield the maximum algebraic connectivity are those graphs with $n - g$ pendant vertices. In the case of $g = 3$, we show in the following theorem from [18] that the maximum occurs when all $n - 3$ pendant vertices are adjacent to the same vertex on the 3-cycle.

**THEOREM 7.4.2** *Let $\mathcal{G} = \mathcal{G}_{3,n-3}$. Then $a(\mathcal{G}) = 1$. Moreover, if $\mathcal{H} \in U_{n,3}$ is not isomorphic to $\mathcal{G}$, the $a(\mathcal{H}) < 1$.*

Note that this theorem says that $\mathcal{G}$ is the unique unicyclic graph on $n$ vertices with girth $g = 3$ of maximum algebraic connectivity, namely 1.

**Proof**: Let $x$ denote the vertex of degree $n - 1$ in $\mathcal{G}$. Clearly there are $n - 2$ Perron components at $x$, each with Perron value 1. Hence $\mathcal{G}$ is of Type I and by Theorem 7.2.15 we have $a(\mathcal{G}) = 1$. Now let $\mathcal{H}$ be a unicyclic graph on $n$ vertices with girth $g = 3$ that is not isomorphic to $\mathcal{G}$. We may assume $n \geq 4$. Since $\mathcal{H}$ has a cut vertex, we know by Theorem 5.1.10 that $a(\mathcal{H}) \leq 1$. By Lemma 7.4.1, we need only consider the case in which each vertex not on the 3-cycle is pendant, i.e., graphs of the form $\mathcal{G}_{g,n_1,n_2,n_3}$. If there are two or more vertices on the 3-cycle which are adjacent to pendant vertices, then by considering the Perron components we see that the 3-cycle is the characteristic block of $\mathcal{H}$, hence $\mathcal{H}$ being of Type II. By Lemma 7.2.13, we see that $a(\mathcal{H}) < 1$. □

We now direct our attention to graphs in $U_{n,4}$. It is not surprising that $\mathcal{G}_{4,n-4}$ will be the graph in $U_{n,4}$ whose algebraic connectivity is maximum. What is surprising though is how different the proof is from that of Theorem 7.4.2. We begin with a lemma from [20]:

**LEMMA 7.4.3** *For $n \geq 5$, $a(\mathcal{G}_{4,n-4}) > 2 - \sqrt{2}$.*

**Proof**: Let vertex 1 be the vertex on the cycle that is adjacent to the $n - 4$ pendant vertices. Applying Exercise 3 from Section 7.2 at vertex 1, there exists $\gamma \geq 0$ such that

$$\frac{1}{a(\mathcal{G}_{4,n-4})} = \rho\left(\left[\begin{array}{c|c} I_{n-4} + xJ & \gamma e \\ \hline \gamma e^T & \gamma \end{array}\right]\right) = \lambda_n \left(\begin{bmatrix} 1 & \frac{1}{2} & \frac{1}{2} \\ \frac{1}{2} & \frac{3}{4} & \frac{1}{4} \\ \frac{1}{2} & \frac{1}{4} & \frac{3}{4} \end{bmatrix} - \gamma J\right). \quad (7.4.1)$$

Let $\rho$ denote the number in the middle of (7.4.1), and let $\lambda_n$ be the number on the right-hand side of (7.4.1). Observing the characteristic polynomials of these matrices, note that $\rho$ solves the equation $z^2 - [1 + (n-3)\gamma]z + \gamma = 0$ and $\lambda_n$ solves the equation $z^2 - (2 - 3\gamma)z + 1/2 - \gamma = 0$. Hence setting these equations equal to each other and simplifying, we see that $(n\gamma - 1)z = 2\gamma - 1/2$.

Observe that $\gamma \neq 0$, for otherwise the spectral radius of the middle matrix is

1, while the Perron value of the matrix on the right is approximately 1.707. Also observe that if $\gamma \geq 1/4$, then $\rho > 1$ but $\lambda_n \leq 1$, thus making equality in (7.4.1) impossible. Therefore $0 < \gamma < 1/4$. This implies $(n\gamma - 1)z < 0$, but recalling that $z$ is positive (since it equals $\rho$ and $\lambda_n$), it follows that $(n\gamma - 1) < 0$, or equivalently, $\gamma < 1/n$. Therefore

$$\rho < \rho\left(\left[\begin{array}{c|c} I_{n-4} + \gamma J & (1/n)e \\ \hline (1/n)e^T & (1/n) \end{array}\right]\right) = \frac{1 + (n-3)/n + \sqrt{(1 + (n-3)/n)^2 - 4/n}}{2}$$

$$< \frac{2 + \sqrt{2}}{2}$$

Using (7.4.1) we conclude that $a(\mathcal{G}_{4,n-4}) > 2 - \sqrt{2}$. □

Before proving our main result concerning graphs in $U_{n,4}$, we have the following observations from [20]. The first observation can be checked by direct calculations; the second is a direct consequence of Corollary 1.2.10.

**OBSERVATION 7.4.4** $a(\mathcal{G}_{4,1,0,1,0}) = 2 - \sqrt{2}$.

**OBSERVATION 7.4.5** *Given a graph $\mathcal{G}$, form a new graph $\hat{\mathcal{G}}$ by adding a pendant vertex adjacent to some vertex of $\mathcal{G}$. Then $a(\hat{\mathcal{G}}) \leq a(\mathcal{G})$.*

Now for the main result from [20] regarding graphs of girth $g = 4$:

**THEOREM 7.4.6** *The unique maximizer of the algebraic connectivity over the graphs in $U_{n,4}$ is $\mathcal{G}_{4,n-4}$.*

**Proof**: From Proposition 7.4.1, it suffices to consider graphs of the form $\mathcal{G}_{4,n_1,n_2,n_3,n_4}$, where, without loss of generality, $n_1 \geq 1$. Applying Observations 7.4.4 and 7.4.5, and Lemma 7.4.3, we see that if $n_3 \geq 1$, then $\mathcal{G}_{4,n_1,n_2,n_3,n_4}$ cannot maximize the algebraic connectivity. A similar argument applies if both $n_2 \geq 1$ and $n_4 \geq 1$. Thus we deduce that up to a relabeling of the vertices, the only candidates for maximizing the algebraic connectivity over graphs in $U_{n,4}$ are of the form $\mathcal{G}_{4,n_1,n_2,0,0}$.

Suppose that $n_2 \geq 1$. Note that $L\mathcal{G}_{4,n_1,n_2,0,0})$ has $n - 6$ eigenvalues equal to 1, while the rest coincide with those of the matrix

$$A = \begin{bmatrix} n_1 + 2 & -1 & -1 & 0 & -n_1 & 0 \\ -1 & n_2 + 2 & 0 & -1 & 0 & -n_2 \\ -1 & 0 & 2 & -1 & 0 & 0 \\ 0 & -1 & -1 & 2 & 0 & 0 \\ -1 & 0 & 0 & 0 & 1 & 0 \\ 0 & -1 & 0 & 0 & 0 & 1 \end{bmatrix}.$$

Let $f_{n_1,n_2}(\lambda)$ be the characteristic polynomial of $A$. Then

$$f_{n_1,n_2} - f_{n_1+1,n_2-1} = \lambda^2(\lambda - 1)(\lambda - 3)(n_1 + 1 - n_2).$$

Being that we are interested in the eigenvalue associated with the algebraic connectivity for these graphs which by Theorem 5.1.10 is between 0 and 1 (since these graphs have a cut vertex), we assume $0 < \lambda < 1$. In such a case, if $n_1 \geq n_2$, then $f_{n_1,n_2}(\lambda) > f_{n_1+1,n_2-1}(\lambda)$. Since $f_{n_1,n_2}$ is a sixth order polynomial which has 0 as its smallest root, and since 0 is a simple root, we see that $f_{n_1,n_2}(\lambda) < 0$ for $\lambda$ between 0 and the smallest positive root of $f_{n_1,n_2}$. Similarly, we see that $f_{n_1+1,n_2-1}(\lambda) < 0$ for $\lambda$ between 0 and the smallest positive root of $f_{n_1+1,n_2-1}$. Using the fact that $f_{n_1,n_2}(\lambda) > f_{n_1+1,n_2-1}(\lambda)$, it follows that the smallest positive root of $f_{n_1+1,n_2-1}$ exceeds that of $f_{n_1,n_2}$. Consequently $a(\mathcal{G}_{4,n_1+1,n_2-1,0,0}) > a(\mathcal{G}_{4,n_1,n_2,0,0})$, which implies that $\mathcal{G}_{4,n-4}$ is the unique maximizer of algebraic connectivity over $U_{n,4}$. $\square$

At this point, we have seen that $\mathcal{G}_{g,n-g}$ maximizes the algebraic connectivity over $U_{n,g}$ when $g = 3$ or when $g = 4$. So it is natural to ask whether $\mathcal{G}_{g,n-g}$ is the unique maximizer of algebraic connectivity when $g \geq 5$. This turns out to be false. For example, take $n = 7$ and $g = 5$. Observe that $a(\mathcal{G}_{5,2}) = 0.6086$, yet $a(\mathcal{G}_{5,1,1}) = 0.6228$, thus $a(\mathcal{G}_{5,2}) < a(\mathcal{G}_{5,1,1})$ (see [20]). However, we will see that the trend of $\mathcal{G}_{g,n-g}$ being the maximizer of algebraic connectivity over $U_{n,g}$ continues for fixed $g \geq 5$ when $n$ is sufficiently large. We begin with a useful lemma from [20]:

**LEMMA 7.4.7** *Let $\mathcal{G}$ be a graph with vertex $v$ that is not a cut vertex. Let $B$ be the bottleneck matrix of $\mathcal{G}$ at $v$, i.e., $B = L[\overline{v}]^{-1}$. Construct $\mathcal{G}_k$ from $\mathcal{G}$ by adding $k$ pendant vertices adjacent to $v$. Then $a(\mathcal{G}_k)$ is a decreasing sequence with limit $1/\rho(B)$.*

**Proof**: By Observation 7.4.5, it is clear that $a(\mathcal{G}_k)$ is nonincreasing in $k$. In particular, by Exercise 3 of Section 7.2, for all natural numbers $k$, there exists $\gamma_k \geq 0$ such that

$$\frac{1}{a(\mathcal{G}_k)} = \rho\left(\left[\begin{array}{c|c} I_k + \gamma_k J & \gamma_k e \\ \hline x_k e^T & x_k \end{array}\right]\right) = \lambda_n(B - \gamma_k J)$$

Since

$$\rho\left(\left[\begin{array}{c|c} I_{k+1} + \gamma_k J & \gamma_k e \\ \hline \gamma_k e^T & \gamma_k \end{array}\right]\right) \geq \rho\left(\left[\begin{array}{c|c} I_k + \gamma_k J & \gamma_k e \\ \hline \gamma_k e^T & \gamma_k \end{array}\right]\right) = \lambda_n(B - \gamma_k J)$$

we see that necessarily $\gamma_{k+1} \leq \gamma_k$.

To finish the proof, we need to show that $\gamma_k \to 0$. Suppose not, then there exists $z > 0$ such that $\gamma_k > z$ for all natural numbers $k$. In this case, for all $k$ we have

$$\lambda_n(B - zJ) \geq \lambda_n(B - \gamma_k J) = \rho\left(\left[\begin{array}{c|c} I_k + \gamma_k J & \gamma_k e \\ \hline \gamma_k e^T & \gamma_k \end{array}\right]\right)$$

$$= \rho\left(\left[\begin{array}{c|c} 1 + k\gamma_k & \gamma_k \\ \hline k\gamma_k & \gamma_k \end{array}\right]\right) \geq \rho\left(\left[\begin{array}{c|c} 1 + kz & z \\ \hline kz & z \end{array}\right]\right)$$

This is a contradiction since the rightmost number increases without bound. Hence $\gamma_k \to 0$, so that $a(\mathcal{G}_k) = 1/\lambda_n(B - \gamma_k J) \to 1/\rho(B)$. □

We now prove a result from [20] which shows that when $n$ is large enough, the graph $\mathcal{G}_{g,n-g}$ is the graph in $U_{n,g}$ with maximum algebraic connectivity when $g \geq 5$.

**THEOREM 7.4.8** *Fix $g \geq 5$. Then there exists $N$ such that $n \geq N$ implies $\mathcal{G}_{g,n-g}$ is the unique maximizer of algebraic connectivity over graphs in $U_{n,g}$.*

**Proof**: Let $M$ be the bottleneck matrix for the component at vertex 1 which contains the vertices of the $g$-cycle. Thus by Lemma 7.4.7, $a(\mathcal{G}_{g,n-g})$ decreases to $1/\rho(M)$ as $n \to \infty$. Fix $2 \leq \ell \leq g$ and consider the graph $\mathcal{H}_{k,\ell}$ which is obtained from $\mathcal{G}_{g,k}$ by adding one pendant vertex to the vertex $\ell$. By Lemma 7.4.7,

$$\lim_{k \to \infty} a(\mathcal{H}_{k,\ell}) = \frac{1}{\rho\left(\begin{bmatrix} 1 + m_{\ell,\ell} & e_\ell^T M \\ M e_\ell & M \end{bmatrix}\right)} < \frac{1}{\rho(M)} \qquad (7.4.2)$$

In particular, there exists $p$ such that if $k \geq p$, then for each $2 \leq \ell \leq g$ we have $a(\mathcal{H}_{k,\ell}) \leq 1/\rho(M)$.

Let $n_i$, $i = 1 \ldots, g$, be nonnegative integers such that $\sum_1^g n_i = n - g$ and assume $n_1 = \max\{n_1, n_2, \ldots, n_g\}$, thus necessarily $n_1 \geq (n-g)/g$. By Proposition 7.4.1, we need only consider the unicyclic graphs $\mathcal{G}_{g,n_1,n_2,\ldots,n_g}$. If $n_\ell > 0$ for some $2 \leq \ell \leq g$, then $a(\mathcal{G}_{g,n_1,\ldots,n_g}) \leq a(\mathcal{H}_{n_1,\ell})$ by Observation 7.4.5. Observe that if $n \geq g(p+1)$, then we have $n_1 \geq (n-g)/g \geq p$. Since $n_1 \geq p$, we have from (7.4.2) and the line following that $a(\mathcal{H}_{n_1,\ell}) < 1/\rho(M)$. However, by Lemma 7.4.7, $1/\rho(M) \leq a(\mathcal{G}_{g,n-g})$. Putting these inequalities together, we see that if $n \geq N := g(p+1)$ we have

$$a(\mathcal{G}_{g,n_1,\ldots,n_g}) \leq a(\mathcal{H}_{n_1,\ell}) < \frac{1}{\rho(M)} \leq a(\mathcal{G}_{g,n-g}).$$

Thus $\mathcal{G}_{g,n-g}$ maximizes the algebraic connectivity over all graphs in $U_{n,g}$ when $n \geq g(p+1)$. □

The final goal if this section is to find the explicit value for $g(p+1)$. This can be done once we find the value for $p$. To this end, we introduce the following variable:

$$\hat{\alpha} = \frac{1}{2g + g^2/2(1 - \cos \pi/g) + (g^2/2)\left(\sqrt{(1 + (g-1)/g)^2 + (2/3g)(g-1)(2g-1)} - (1 + (g-1)/g)\right)}$$

The derivation of this variable can be found in [20]. In an effort to find $p$ explicitly, we need the following lemma from [20].

**LEMMA 7.4.9** *Consider the graph $\mathcal{H}_{k,i}$, $i \geq 2$, and let its algebraic connectivity be $a$. Set $c = 1/2(1 - \cos \pi/g) = \rho(M)$, where $M$ is the bottleneck matrix for the $g$-cycle. If $k \geq (c^2 - c - \hat{\alpha}/c\alpha) - 1$, then $a < 1/\rho(M)$.*

**Proof**: Note that the Perron value of

$$\left[\begin{array}{c|c} I_k + \hat{\alpha}J & \hat{\alpha}e \\ \hline \hat{\alpha}e^T & \hat{\alpha} \end{array}\right]$$

satisfies $\lambda^2 - (1 + (k+1)\hat{\alpha})\lambda + \hat{\alpha} = 0$, hence the Perron value is

$$\frac{1 + (k+1)\hat{\alpha} + \sqrt{(1 + (k+1)\hat{\alpha})^2 - 4\hat{\alpha}}}{2}.$$

Since our hypothesis states that $k \geq (c^2 - c - \hat{\alpha}/c\alpha) - 1$, we deduce that the Perron value of the above matrix is at least $c$. Applying Exercise 3 of Section 7.2 at vertex 1 we see there exists $\beta \geq 0$ such that

$$\frac{1}{a} = \lambda_n(A_i - \beta J) = \rho\left(\left[\begin{array}{c|c} I_k + \beta J & \beta e \\ \hline \beta e^T & \beta \end{array}\right]\right)$$

where

$$A_i = \left[\begin{array}{c|c} 1 + m_{ii} & e_i^T M \\ \hline M e_i & M \end{array}\right].$$

If $\beta \leq \hat{\alpha}$, we have

$$\frac{1}{a} = \lambda_n(A_i - \beta J) \geq \lambda_n(A_i - \hat{\alpha} J) > c,$$

while if $\beta > \hat{\alpha}$ we have

$$\frac{1}{a} = \rho\left(\left[\begin{array}{c|c} I_k + \beta J & \beta e \\ \hline \beta e^T & \beta \end{array}\right]\right) > \rho\left(\left[\begin{array}{c|c} I_k + \hat{\alpha}J & \hat{\alpha}e \\ \hline \beta e^T & \hat{\alpha} \end{array}\right]\right) \geq c.$$

In either case, $a < 1/c = 1/\rho(M)$. □

We are now ready to close this section by giving a precise value for $p$ such that $n \geq g(p+1)$ implies $\mathcal{G}_{g,n-g}$ is the graph which uniquely maximizes the algebraic connectivity over all unicyclic graphs on $n$ vertices with girth $g$. We do this in the following theorem from [20]:

**THEOREM 7.4.10** *Let $p = (c^2 - c - \hat{\alpha}/c\hat{\alpha}) - 1$. If $n \geq g(p+1)$, then $\mathcal{G}_{g,n-g}$ uniquely maximizes the algebraic connectivity over all graphs in $U_{n,g}$.*

**Proof**: Let $M$ be the bottleneck matrix for the $g$-cycle in $\mathcal{G}_{g,n-g}$. Since $n \geq g(p+1)$, there exists a vertex on the $g$-cycle, say vertex 1, with degree at least $p$. By Proposition 7.4.1, we need only consider graphs of the form $\mathcal{G}_{g;n_1,\ldots,n_g}$ in $U_{n,g}$. Consider such a graph and suppose $n_i \geq 1$ for some $2 \leq i \leq g$. Then we have

$$a(\mathcal{G}_{g;n_1,\ldots,n_g}) \leq a(\mathcal{H}_{k,i}) < 1/\rho(M)$$

where the first inequality follows from Observation 7.4.5 and the second inequality follows from Lemma 7.4.9. However, by Lemma 7.4.7, $a(\mathcal{G}_{g,n-g}) \geq 1/\rho(M)$. The result now follows. □

Setting $N = g(p+1)$ as in the language of Theorem 7.4.8, we see in the following chart the various values for $N$, i.e., the minimum number of vertices that are required in order to guarantee that $\mathcal{G}_{g,n-g}$ maximizes the algebraic connectivity over graphs in $U_{n,g}$ for fixed $g$.

| $g$ | 5 | 6 | 7 | 8 | 9 | 10 | 20 | 50 | 100 | 1000 |
|---|---|---|---|---|---|---|---|---|---|---|
| $N$ | 9 | 11 | 13 | 14 | 16 | 18 | 35 | 88 | 175 | 1750 |

## Exercises:

**1.** (See [20]) Prove that if $g \geq 4$, then $a(\mathcal{G}_{g,1}) < 2(1 - \cos(2\pi/g))$.

**2.** (See [20]) Let $L$ be the Laplacian matrix for $\mathcal{G}_{g,k}$ where $g \geq 4$ and $k > 1$, and let $y$ be a Fiedler vector. Show that $y_1 \neq 0$ where vertex 1 is the vertex on the cycle adjacent to the $k$ pendant vertices. Conclude that $a(\mathcal{G}_{g,1})$ is a simple eigenvalue of $L$.

## 7.5 Application: The Algebraic Connectivity and the Number of Cut Vertices

In Theorem 5.1.10, we learned that if $\mathcal{G}$ is a connected unweighted graph with a cut vertex, then $a(\mathcal{G}) \leq 1$. In this section, we use the concepts of Perron components and bottleneck matrices to generalize this theorem to unweighted graphs with many cut vertices. We begin the process of generalizing Theorem 5.1.10 in the following theorem and immediate corollary from [43]:

**THEOREM 7.5.1** *Let $\mathcal{G}$ be a connected unweighted graph with Laplacian matrix $L$ and cut vertex $v$. Let $C$ be a connected component at $v$ and let $B$ be the bottleneck matrix for $C$. Then $\rho(B) \geq 1$ with equality holding if and only if $v$ is adjacent to every vertex of $C$.*

**Proof**: Suppose $C$ has $k$ vertices and that $v$ is adjacent to $j$ of them. Let $L[C]$ be the Laplacian matrix for $C$. Observe that $L[C]$ can be written as $A + D$ where $A$ is the Laplacian matrix for the subgraph of $\mathcal{G}$ induced by the vertices of $C$, and $D$ is the diagonal matrix with 1's in the $j$ diagonal entries corresponding to the vertices of $C$ to which $v$ is adjacent, and 0's elsewhere. Let $w$ be the Perron vector for $B$. Then we have

$$\frac{e^T w}{\rho(B)} = e^T L[C] w = e^T A w + e^T D w = e^T D w$$

where the last equality follows from the fact that $e^T A = 0$. Therefore $\rho(B) = e^T w / e^T D w \geq 1$. Since $w$ is positive, we see that equality holds if and only if $D = I$, i.e., if $v$ is adjacent to every vertex in $C$. □

**COROLLARY 7.5.2** *Let $\mathcal{G}$ be a connected graph with cut vertex $v$. Then $a(\mathcal{G}) \leq 1$ with equality holding if and only if $v$ is adjacent to very other vertex of $\mathcal{G}$.*

In this section, we consider connected unweighted graphs that have more than one cut vertex. The goal of this section is to show that if $\mathcal{G}$ is a connected graph on $n$ vertices with $k \geq 2$ cut vertices, then

$$a(\mathcal{G}) \leq \frac{2(n-k)}{n-k+2+\sqrt{(n-k)^2+4}}.$$

We will also show the conditions when equality holds. In order to derive this upper bound, we need to divide this problem into two cases. The first case will be when $2 \leq k \leq n/2$, while the second case will be when $k > n/2$. We begin with the first case, i.e., where $2 \leq k \leq n/2$. To this end, we require two definitions:

**DEFINITION 7.5.3** *Let $\mathcal{G}$ be a connected graph with cut vertex $v$. Then $C$ is a pendant component at $v$ if $C$ contains no other cut vertices of $\mathcal{G}$ other than $v$.*

**DEFINITION 7.5.4** *Let $\mathcal{G}$ be a connected graph with cut vertex $v$. Then $C$ is a complete component at $v$ if the vertices in $\mathcal{G}$ induced by $C$ form a complete graph and if $v$ is adjacent to all vertices in $C$.*

Before proceeding, we recall from Section 6.2 that the bottleneck matrix for a path on $q$ vertices is

$$P_q = \begin{bmatrix} q & q-1 & q-2 & \cdots & 2 & 1 \\ q-1 & q-1 & q-2 & \cdots & 2 & 1 \\ \cdots & \cdots & \cdots & & \cdots & \cdots \\ 2 & 2 & & \cdots & 2 & 1 \\ 1 & 1 & & \cdots & 1 & 1 \end{bmatrix}$$

From this, define $\tilde{P}_s$ to be the $(s+1) \times (s+1)$ matrix

$$\tilde{P}_s := \begin{bmatrix} P_s & O \\ \hline O & 0 \end{bmatrix}.$$

We now begin with a lemma from [43]:

**LEMMA 7.5.5** *Let $\mathcal{G}$ be a connected graph with cut vertex $v$. Let $C$ be a component at $v$ constructed as follows: start with a complete component on $p+q$ vertices, and for each vertex $i \in C$ with $1 \leq i \leq q$, we add a complete component on $m_i$ vertices (see drawing below). Then the bottleneck matrix for $C$ is permutationally similar to*

$$\begin{bmatrix} A_1 + \frac{2}{p+q+1}J & & \frac{1}{p+q+1}J & & \frac{1}{p+q+1}ee_1^T(I+J) \\ \cdots & \cdots & & \cdots & \cdots \\ \frac{1}{p+q+1}J & & A_q + \frac{2}{p+q+1}ee_q^T(I+J) & & \frac{1}{p+q+1}ee_q^T(I+J) \\ \frac{1}{p+q+1}(I+J)e_1e^T & \cdots & \frac{1}{p+q+1}(I+J)e_qe^T & & \frac{1}{p+q+1}(I+J) \end{bmatrix} \quad (7.5.1)$$

*where for each $1 \leq i \leq q$, $A_i = \frac{1}{m_i+1}(I+J)$ and $A_i$ is $m_i \times m_i$.*

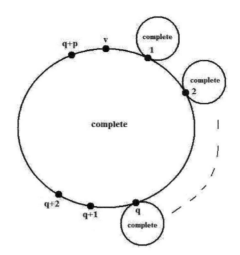

**Proof**: The submatrix of the Laplacian matrix for $\mathcal{G}$ corresponding to the vertices of $C$ can be written as

$$\left[\begin{array}{cccc|c} (m_1+1)I - J & 0 & \ldots & 0 & -ee_1^T \\ \ldots & \ldots & \ldots & & \ldots \\ 0 & \ldots & 0 & (m_q+1)I-J & -ee_q^T \\ \hline -e_1 e^T & \ldots & & -e_q e^T & (p+q+1)I - J + D \end{array}\right]$$

where $D = \text{diag}\{m_1, \ldots, m_q, 0 \ldots, 0\}$. The result now follows by direct computation. □

For the next two lemmas (see [43]), we need the following quantity

$$a_m := 1/\rho\left(\tilde{P}_1 + \frac{1}{m}J\right) = \frac{2m}{m+2+\sqrt{m^2+4}}$$

Our goal is to show that if $\mathcal{G}$ is a graph on $n$ vertices with $k$ cut vertices where $2 \leq k \leq n/2$, then $a(\mathcal{G}) \leq a_{n-k}$.

**LEMMA 7.5.6** *Suppose that $\mathcal{G}$ is a graph on $n$ vertices with $k$ cut vertices and Laplacian $L$. Let $v$ be a cut vertex of $\mathcal{G}$. If $C$ is a connected component at $v$ which contains a cut vertex, and $M$ is the bottleneck matrix at $v$ for $C$, then $\rho(M) > 1/a_{n-k}$.*

**Proof**: First suppose that there exists a cut vertex $w$ of $\mathcal{G}$ in $C$ such that a component $C'$ at $w$ not containing $v$ is nonpendant. In such a case, we can find a cut vertex $x$ of $\mathcal{G}$ in $C'$ such that all components at $x$ not containing $v$ are pendant. Observe then that

$$\rho(M) = \rho(L[C]^{-1}) \geq \rho(L[C']^{-1})$$

where the inequality follows from the fact that $L[C']$ is a principal submatrix of $L[C]$ which is an irreducible M-matrix. Therefore, for the remainder of the proof we can assume that at each cut vertex of $\mathcal{G}$ in $C$, the components not containing $v$ are pendant.

Suppose that $C$ contains $q$ cut vertices of $\mathcal{G}$. Since, by Exercise 3 of Section 7.1, adding edges into $C$ cannot increase its Perron value, we can assume that at each cut vertex $w$ in $C$, there is a single component not containing $v$ (thus making this a pendant component at $w$) and that each such component is complete. Suppose also that there are $p$ vertices in $C$ which are neither cut vertices nor in any pendant component. Again, since adding edges into $C$ cannot increase its Perron value, we can assume that every pair of vertices in this collection of $p$ vertices is adjacent and that every cut vertex in $C$, as well as $v$ itself, is adjacent to all $p$ of these vertices.

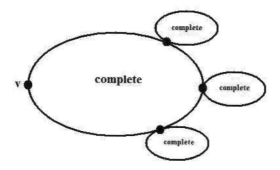

Since $C$ is of the form described in Lemma 7.5.5, we see that the bottleneck matrix $M$ for $C$ is of the form in (7.5.1) where each $A_i$ is of the form $(1/(m_i+1))(I+J)_{m_i}$ where $m_i$ is the number of vertices in the complete component at the $i^{th}$ cut vertex. Observe that $M$ has the following matrix $S$ has a principal submatrix:

$$ S = \begin{bmatrix} A_1 + \frac{2}{p+q+1}J & \frac{2}{p+q+1}e \\ \frac{2}{p+q+1}e^T & \frac{2}{p+q+1} \end{bmatrix}. $$

Clearly each block of $S$ contains constant row sums. Therefore the Perron value of $S$ is the Perron value of the $2 \times 2$ matrix $R$ where each entry of $R$ is the row sum of the corresponding block of $S$, i.e.,

$$ R = \begin{bmatrix} 1 + \frac{2m_1}{p+q+1} & \frac{2}{p+q+1} \\ \frac{2m_1}{p+q+1} & \frac{2}{p+q+1} \end{bmatrix}. $$

Therefore $\rho(M) \geq \rho(R)$.

Since the Perron value of a positive matrix increases as each entry is increased, the proof will be complete when we show that $2/(p+q+1) > 1/(n-k)$, or equivalently $2n - 2k > p + q + 1$. To see this, recall that $n - k$ is the number of vertices of $\mathcal{G}$ that are not cut vertices. In $C$, there is a set of $p$ vertices that are not cut vertices of $\mathcal{G}$. Moreover, for each of the $q$ cut vertices in $C$, there is at least one corresponding vertex which is not a cut vertex, i.e., a vertex on a pendant component at $q$. Therefore, $C$ contains at least $p + q$ vertices which are not cut vertices of $\mathcal{G}$. Additionally, every other component at $v$ must contain at least one vertex that is not a cut vertex of $\mathcal{G}$. Therefore, there are at least $p+q+1$ vertices in $\mathcal{G}$ which are not cut vertices. Hence

$$2n - 2k > n - k \geq p + q + 1.$$

□

Before we can proceed to the next lemma, we need to introduce another class of graphs. Fix $k \geq 2$ and fix natural numbers $m_1, \ldots, m_k$, and suppose that $n \geq m_1 + \ldots + m_k + k$. Define $H_{m_1,\ldots,m_k,n}$ to be the graph constructed by (1) starting with a complete graph on $n - m_1 - \ldots - m_k$ vertices, and (2) at each vertex $i$, $1 \leq i \leq k$ add a complete connected component on $m_i$ vertices. For example, the graph $H_{2,3,4,13}$ is as follows (note $k = 3$):

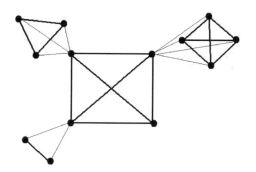

Observe that the condition $n \geq m_1 + \ldots + m_k + k$ necessarily means that $k \leq n/2$. To continue our investigation of the case $k \leq n/2$, we need the following lemma from [43]:

**LEMMA 7.5.7** $a(H_{m_1,\ldots,m_k,n}) \leq a_{n-k}$. Moreover, the inequality is strict if there exists $i$ $(1 \leq i \leq k)$ such that $m_i \geq 2$.

**Proof**: Label $H_{m_1,\ldots,m_k,n}$ in accordance with the labeling used in Lemma 7.5.5 and without loss of generality let $m_i$ be minimized at $i = 1$. Hence at vertex 1, the non-Perron component has bottleneck matrix $A_1 = \frac{1}{m_1+1}(I + J)$ (note $A_1$ is $m_1 \times m_1$) while by Lemma 7.5.5, the bottleneck matrix $B$ for the Perron component is given by

$$B = \begin{bmatrix} A_2 + \frac{2}{b}J & \cdots & \frac{1}{b}J & \frac{1}{b}ee_2^T(I+J) \\ \cdots & \cdots & \cdots & \cdots \\ \frac{1}{b}J & \cdots & A_k + \frac{2}{b}J & \frac{1}{b}ee_k^T(I+J) \\ \frac{1}{b}(I+J)e_2e^T & \cdots & \frac{1}{b}(I+J)e_ke^T & \frac{1}{b}(I+J) \end{bmatrix}$$

where $b = n - m_1 - \ldots - m_k$. Observe that

$$\lambda_n\left(B - \frac{1}{b}J\right) \geq \rho\left(\begin{bmatrix} A_2 + \frac{1}{b}J & \frac{1}{b}e \\ \frac{1}{b}e^T & \frac{1}{b} \end{bmatrix}\right)$$

since the latter matrix is a submatrix of the former. Observing that the blocks of
$$\begin{bmatrix} A_1 + \frac{1}{b}J & \frac{1}{b}e \\ \frac{1}{b}e^T & \frac{1}{b} \end{bmatrix} \text{ and } \begin{bmatrix} A_2 + \frac{1}{b}J & \frac{1}{b}e \\ \frac{1}{b}e^T & \frac{1}{b} \end{bmatrix}$$
have constant row sums, we see as in Lemma 7.5.6 that Perron value of these matrices is the Perron value of the $2 \times 2$ respective matrices whose entries correspond to row sums of the individual blocks, i.e.,
$$\rho\left(\begin{bmatrix} A_1 + \frac{1}{b}J & \frac{1}{b}e \\ \frac{1}{b}e^T & \frac{1}{b} \end{bmatrix}\right) = \rho\left(\begin{bmatrix} 1 + \frac{m_1}{b} & \frac{1}{b} \\ \frac{m_1}{b} & \frac{1}{b} \end{bmatrix}\right)$$
and
$$\rho\left(\begin{bmatrix} A_2 + \frac{1}{b}J & \frac{1}{b}e \\ \frac{1}{b}e^T & \frac{1}{b} \end{bmatrix}\right) = \rho\left(\begin{bmatrix} 1 + \frac{m_2}{b} & \frac{1}{b} \\ \frac{m_2}{b} & \frac{1}{b} \end{bmatrix}\right)$$
Since $\rho(A_2) \geq \rho(A_1)$, it follows that
$$\lambda_n\left(B - \frac{1}{b}J\right) \geq \rho\left(\begin{bmatrix} A_1 + \frac{1}{b}J & \frac{1}{b}e \\ \frac{1}{b}e^T & \frac{1}{b} \end{bmatrix}\right).$$
By Exercise 1 of Section 7.2 it follows that
$$a(H_{m_1,\ldots,m_k,n}) \leq 1/\rho\left(\begin{bmatrix} A_1 + \frac{1}{b}J & \frac{1}{b}e \\ \frac{1}{b}e^T & \frac{1}{b} \end{bmatrix}\right).$$
However, since
$$\rho\left(\begin{bmatrix} A_1 + \frac{1}{b}J & \frac{1}{b}e \\ \frac{1}{b}e^T & \frac{1}{b} \end{bmatrix}\right) \geq \frac{1}{a_b},$$
it follows that $a(H_{m_1,\ldots,m_k,n}) \leq a_b$. Yet since $b \leq n - k$, we have $a_b \leq a_{n-k}$ which is the desired result. Moreover, if $m_i \geq 2$ for some $i$, then $b < n - k$ which yields $a(H_{m_1,\ldots,m_k,n}) < a_b$. □

We are now ready to prove the first main result of this section (see [43]) which gives an upper bound on the algebraic connectivity of a graph with $k \leq n/2$ cut vertices.

**THEOREM 7.5.8** *Suppose $\mathcal{G}$ is a graph on $n$ vertices with $k$ cut vertices where $2 \leq k \leq n/2$. Then $a(\mathcal{G}) \leq a_{n-k}$.*

**Proof:** Let $v$ be a cut vertex of $\mathcal{G}$ and suppose that there are at least two components at $v$ which contain cut vertices. Then by Lemma 7.5.6, the Perron values of these components are bounded below by $1/a_{n-k}$. The result then follows by Exercise 4 of Section 7.2.

Now label the cut vertices of $\mathcal{G}$ as $1, \ldots, k$. In light of our arguments from the previous paragraph, we can now assume that at each cut vertex $i$, there is exactly one component which contains the other $k-1$ cut vertices of $\mathcal{G}$, and that the remaining components at $i$, comprising a total of, say, $m_i$ vertices are pendant components. Observe that we can form $H_{m_1,\ldots,m_k,n}$ from $\mathcal{G}$ by adding edges to $\mathcal{G}$. Since by Theorem 4.1.2, adding edges to $\mathcal{G}$ cannot decrease the algebraic connectivity, we see that

$$a(\mathcal{G}) \leq a(H_{m_1,\ldots,m_k,n}) \leq a_{n-k}$$

where the last inequality follows from Lemma 7.5.7. □

At this point, it is natural to ask for which graphs on $n$ vertices and $k \leq n/2$ cut vertices is equality achieved in Theorem 7.5.8. We answer this question in the following theorem from [43]:

**THEOREM 7.5.9** *Suppose that $\mathcal{G}$ is a graph on $n$ vertices with $k$ cut vertices where $2 \leq k \leq n/2$. Then $a(\mathcal{G}) = a_{n-k}$ if and only if both of the following hold:*
*(i) Each cut vertex is adjacent to every nonpendant vertex of $\mathcal{G}$.*
*(ii) There are exactly two components at each cut vertex of $\mathcal{G}$, one component of which is a pendant vertex.*

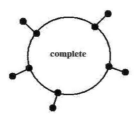

**Proof**: Suppose first that $a(\mathcal{G}) = a_{n-k}$. We first claim that there are exactly two components at each cut vertex, one of which is a pendant vertex. Suppose this were not the case. Then recalling the proof of Theorem 7.5.8 we have $a(\mathcal{G}) \leq a(H_{m_1,\ldots,m_k,n})$ for some $m_1,\ldots,m_k$ where for at least one $i$ we have $m_i \geq 2$. Lemma 7.5.7 then yields a contradiction. Hence our claim is established.

From here on, we now assume that there are exactly two components at each cut vertex, one component of which is a pendant vertex. Let $v$ be a cut vertex of $\mathcal{G}$. Then the bottleneck matrix for the non-Perron component (i.e., the pendant vertex) is the $1 \times 1$ matrix $[1]$, while the bottleneck matrix $M$ of the Perron component $C$ at $v$ has the form

$$M = \begin{bmatrix} I+B & B & U \\ B & B & U \\ U^T & U^T & X \end{bmatrix}$$

where

$$S := \begin{bmatrix} B & U \\ U^T & X \end{bmatrix}$$

is the bottleneck matrix of the component at $v$ obtained by deleting the pendant vertices of $C$, and where the rows and columns of $B$ correspond to the $k-1$ quasi-pendant vertices of $C$. From Theorem 7.1.7 we know that each diagonal entry of $B$

is at least $2/(n-k)$. Therefore, observing the $(1,1)$, $(1,2)$, $(2,1)$, and $(2,2)$ blocks of $M$, we see that $M - (1/(n-k))J$ has $k-1$ principal submatrices which entrywise dominate

$$T := \begin{bmatrix} 1 + \frac{1}{n-k} & \frac{1}{n-k} \\ \frac{1}{n-k} & \frac{1}{n-k} \end{bmatrix}.$$

Therefore

$$\lambda_n\left(M - \frac{1}{n-k}J\right) \geq \rho(T) = \rho\left(\tilde{P}_1 + \frac{1}{n-k}J\right),$$

with strict inequality if some diagonal entry of $M$ is larger than $2/(n-k)$. However, since we are assuming that $a(\mathcal{G}) = a_{n-k}$, Exercise 1 of Section 7.2 implies that each diagonal entry of $M$ must equal $2/(n-k)$. Hence by Theorem 7.1.7, it follows that each cut vertex must be adjacent to every nonpendant vertex of $\mathcal{G}$. Since $v$ was an arbitrary cut vertex, the structure of $\mathcal{G}$ now follows.

Conversely, suppose that $\mathcal{G}$ has the structure described in (i) and (ii) of the statement of this theorem. Let $v$ be a cut vertex of of $\mathcal{G}$. Then the bottleneck matrix for the Perron component $C$ at $v$ is $M$ as above, where $S$ and $B$ are as above. Let $A$ be the Laplacian matrix of the subgraph of $\mathcal{G}$ that is induced by the $n-2k$ vertices of $C$ which are neither pendant nor quasipendant. Observe from the construction of $\mathcal{G}$ that

$$S^{-1} = \begin{bmatrix} (n-k)I - J & -J \\ -J & (A+kI) \end{bmatrix},$$

i.e., the principal submatrix of $L$ corresponding to the vertices of $C$. By direct computation, we find that

$$S = \begin{bmatrix} \frac{1}{n-k}(I+J) & \frac{1}{n-k}J \\ \frac{1}{n-k}J & (A+kI)^{-1} + \frac{k-1}{k(n-k)}J \end{bmatrix}.$$

Therefore, $M - (1/(n-k))J$ is the direct sum of

$$N := \begin{bmatrix} \left(1 + \frac{1}{n-k}\right)I & \frac{1}{n-k}I \\ \frac{1}{n-k}I & \frac{1}{n-k}I \end{bmatrix}$$

and $(A+kI)^{-1} - 1/(k(n-k))J$. Since $A$ is the Laplacian matrix for a graph on $n-2k$ vertices, it follows that all eigenvalues of $A$ fall in the interval $[0, n-2k]$ (by Corollary 4.1.4). Hence the eigenvalues of $A+kI$ fall in the interval $[k, n-k]$, which means that the eigenvalues of $(A+kI)^{-1}$ fall in the interval $[\frac{1}{n-k}, \frac{1}{k}]$. Hence all eigenvalues of $(A+kI)^{-1} - 1/(k(n-k))J$ fall in the same interval (the smallest eigenvalue being $1/(n-k)$). Notably, the eigenvalues of $(A+kI)^{-1} - 1/(k(n-k))J$ are bounded above by $1/k$. Therefore

$$\lambda_n\left(M - \frac{1}{n-k}J\right) = \rho(T) = \rho\left(\tilde{P}_1 + \frac{1}{n-k}J\right).$$

Recalling that the non-Perron component at $v$ has bottleneck matrix $[1]$, it follows from Exercise 1 of Section 7.2 that $a(\mathcal{G}) = a_{n-k}$. $\square$

This settles the case of $n \leq k/2$. We now turn our attention to graphs on $n$ vertices and $k$ cut vertices where $n > k/2$. We begin with a useful observation from [18]:

**OBSERVATION 7.5.10** *Let $\mathcal{G}$ be a connected graph and suppose that $C$ is a connected component at some vertex $v$ of $\mathcal{G}$. Suppose further that $C$ has $p$ vertices, and let $M = [m_{i,j}]_{1 \leq i,j \leq p}$ be the bottleneck matrix for $C$. Construct a new component $\hat{C}$ at $v$ as follows: fix some integer $1 \leq k \leq p$ and select vertices $i = 1, \ldots, k$ of $C$ and for each $1 \leq i \leq k$, add a component $C_i$ with bottleneck matrix $B_i$ at vertex $i$ and suppose that in $C_i$, vertices $v_{j_1(i)}, v_{j_2(i)}, \ldots, v_{j_{m_i}(i)}$ are adjacent to vertex $i$. Then the resulting component at $v$ has bottleneck matrix $L[\hat{C}]^{-1}$ given by*

$$\begin{bmatrix} B_1 + m_{1,1}J & m_{1,2}J & \cdots & m_{1,k}J & ee_1^T M \\ m_{2,1}J & B_2 + m_{2,2}J & \cdots & m_{2,k}J & ee_2^T M \\ \cdots & & \cdots & & \cdots \\ m_{k,1}J \cdots & m_{k,k-1}J & \cdots & B_k + M_{k,k}J & ee_k^T M \\ \hline Me^1 e^T & Me_2 e^T & \cdots & Me_k e^T & M \end{bmatrix}$$

**Proof**: Note that $L[\hat{C}]$ can be written as

$$\begin{bmatrix} L[C_1] & 0 & \cdots & 0 & -\left(\sum_{q=1}^{m_1} e_{j_q(1)}\right) e_1^T \\ 0 & L[C_2] & \cdots & 0 & -\left(\sum_{q=1}^{m_2} e_{j_q(2)}\right) e_1^T \\ \cdots & \cdots & \cdots & 0 & \cdots \\ 0 & 0 & \cdots & L[C_k] & -\left(\sum_{q=1}^{m_k} e_{j_q(k)}\right) e_1^T \\ \hline -e_1 \left(\sum_{q=1}^{m_1} e_{j_p(1)}^T\right) & -e_2 \left(\sum_{q=1}^{m_2} e_{j_p(2)}^T\right) & \cdots & -e_k \left(\sum_{q=1}^{m_k} e_{j_p(k)}^T\right) & M^{-1} + \sum_{q=1}^{k} m_q e_q e_q^T \end{bmatrix}$$

The result now follows from direct computation recalling that $L[C_i]e = \sum_{q=1}^{m_i} e_{j_q(i)}$. □

In order to find an upper bound on the algebraic connectivity in terms of the number of cut vertices, we need the following classes of graphs found in [44]. In each class, $q$ and $m$ are natural numbers with $m \geq 2$.

(i) $E_0(q, m)$ is the graph formed by attaching a path on $q$ vertices to each vertex of the complete graph on $m$ vertices. When convenient, we will refer to $E_0(q, m)$ as a class of graphs.

**EXAMPLE 7.5.11** Below is the graph $E_0(3, 5)$.

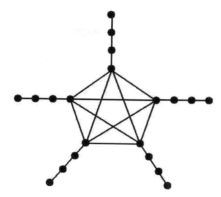

(ii) $E_1(q,m)$ is the class of graphs created by taking a graph $H$ on $m+1$ vertices having a special cut vertex $v_0$ which is adjacent to all other vertices of $H$ and attaching a path on $q$ vertices to each vertex of $H$ other than $v_0$.

**EXAMPLE 7.5.12** Consider the following graph $H$ with the vertex $v_0$ labeled.

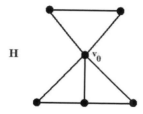

Then the graph $E_1(3,5)$ is

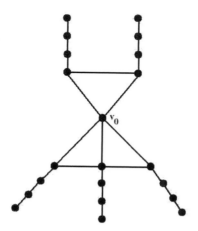

(iii) For a fixed $\ell$ where $m \geq \ell \geq 2$, the class $E_\ell(q,m)$ is the class of graphs formed as follows: Begin with a graph $H$ on $m$ vertices which has at least $r$ vertices of degree $m-1$ for some fixed $r$ where $m \geq r \geq \ell$. Then select $r$ such vertices of degree $m-1$, and at each attach a path on $q+1$ vertices. Finally, at each remaining vertex $i$ of $H$ with $1 \leq i \leq m - r$, attach a path on $j_i$ vertices where $0 \leq j_i \leq q$ so that

$$r + \sum_{i=1}^{m-r}(j_i - q) = \ell \qquad (7.5.2)$$

Since the classes of graphs descibed in (iii) are more difficult to visualize, we provide several examples. For each of these examples, regard $H$ as the graph below on $m = 5$ vertices. Vertices labeled $r = 1, 2, 3$ are the $r = 3$ vertices of degree $m-1$, and $i = 1, 2$ are the vertices labeled in accordance with the above explanation:

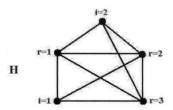

**EXAMPLE 7.5.13** Let's first construct graphs in $E_2(3, 5)$, i.e., $\ell = 2$ and $q = 3$. In this case, (7.5.2) becomes $3 + \sum_{i=1}^{2}(j_i - 3) = 2$. Hence we require $j_1 = 2$ and $j_2 = 3$, or $j_1 = 3$ and $j_2 = 2$. Below is each such graph.

$$(7.5.3)$$

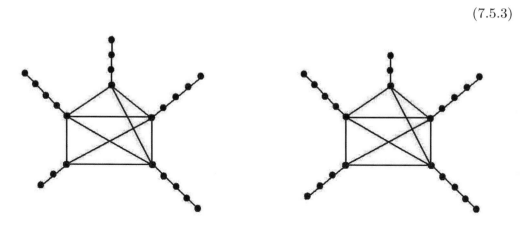

**EXAMPLE 7.5.14** Let's now construct a graph in $E_3(3, 5)$, i.e., $\ell = 3$ now. In this case, (7.5.2) becomes $3 + \sum_{i=1}^{2}(j_i - 3) = 3$. This forces $j_1 = j_2 = 3$ hence giving us the following graph:

# Bottleneck Matrices for Graphs

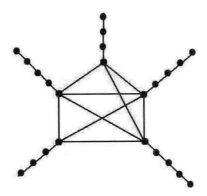

Before continuing, it is worth noting that we are not required to label all vertices whose degree is $m-1$ as an "$r$" vertex. For example, we can relabel the vertices in $H$ as follows:

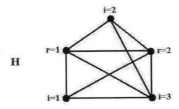

Thus we are letting $r = 2$. Hence if we wanted to create a graph in $E_2(3,5)$, then (7.5.2) forces $j_1 = j_2 = j_3 = 3$ giving us the following graph:

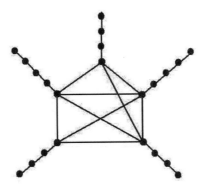

**OBSERVATION 7.5.15** [44] Consider any graph $\mathcal{G} \in E_\ell(q, m)$ where $\ell \geq 0$. Let $n$ denote the number of vertices of $\mathcal{G}$ and let $k$ denote the number of cut vertices. Then $\mathcal{G}$ necessarily has exactly $k = (qn + \ell)/(q+1)$ cut vertices.

In order to establish the desired upper bounds on the algebraic connectivity of graphs with $k$ cut vertices, we need the following quantities which can be found in [44].

$$a_{0,q,m} = 1/\rho\left(\tilde{P}_q + \frac{1}{m}J\right)$$

$$a_{1,q,m} = 1/\rho(P_{q+1})$$

and for $2 \leq \ell \leq m$:

$$a_{\ell,q,m} = 1/\rho\left(\tilde{P}_{q+1} + \frac{1}{m}J\right).$$

We are now ready to compute the algebraic connectivities of the graphs in classes $E_\ell(q,m)$ for $\ell \geq 0$. We do so in the following proposition from [44]:

**PROPOSITION 7.5.16** *(i) $a(E_0(q,m)) = a_{0,q,m}$. Further, if any edge is deleted from $E_0(q,m)$, then the resulting graph has algebraic connectivity strictly less than $a_{0,q,m}$.*
*(ii) For any graph $\mathcal{G} \in E_1(q,m)$, we have $a(\mathcal{G}) = a_{1,q,m}$. Further, if any edge incident with the special cut vertex $v_0$ is deleted from $\mathcal{G}$, then the resulting graph has algebraic connectivity strictly less than $a_{1,q,m}$.*
*(iii) If $\ell \geq 2$, then for any graph $\mathcal{G} \in E_\ell(q,m)$ we have $a(\mathcal{G}) = a_{\ell,q,m}$. Further, if any edge incident with a vertex of degree $m$ is deleted from $\mathcal{G}$, then the resulting graph has algebraic connectivity strictly less than $a_{\ell,q,m}$.*

**Proof**: (i) We first show that $a(E_0(q,m)) = a_{0,q,m}$. To do this, let $u$ be a vertex of $E_0(q,m)$ which has degree $m$. Then the non-Perron component at $u$ is the path on $q$ vertices, which has bottleneck matrix $P_q$. Thus it follows from Observation 7.5.10 that the bottleneck matrix for the Perron component at $u$ is

$$B = \left[\begin{array}{cccc|c} P_q + \frac{2}{m}J & \frac{1}{m}J & \cdots & \frac{1}{m}J & \frac{1}{m}ee_1^T(I+J) \\ \frac{1}{m}J & P_q + \frac{2}{m}J & \cdots & \frac{1}{m}J & \frac{1}{m}ee_2^T(I+J) \\ \cdots & \cdots & \cdots & \cdots & \cdots \\ \frac{1}{m}J & \cdots & \frac{1}{m}J & P_q + \frac{2}{m}J & \frac{1}{m}ee_{m-1}^T(I+J) \\ \hline \frac{1}{m}(I+J)e_1e^T & \frac{1}{m}(I+J)e_2e^T & \cdots & \frac{1}{m}(I+J)e_{m-1}e^T & \frac{1}{m}(I+J) \end{array}\right].$$

Observe that $B - \frac{1}{m}J$ is permutationally similar to a direct sum of $m-1$ copies of $\tilde{P}_q + \frac{1}{m}J$. Thus by Exercise 1(a) of Section 7.2, it follows that $a(E_0(q,m)) = a_{0,q,m}$.

We now show that the deletion of any edge from $E_0(q,m)$ results in a graph with algebraic connectivity strictly less than $a_{0,q,m}$. Removing any edges in one of the paths on $q$ vertices results in a disconnected graph, thus yielding a resulting algebraic connectivity of zero. Hence we will only consider removing an edge from the portion of $E_0(q,m)$ that is part of the complete graph. Let $w \neq u$ be another vertex of $E_0(q,m)$ with degree $m$ and note in particular that $w$ and $u$ are adjacent. We will consider the graph $E_0(q,m) - wu$. From Exercise 1(b) of Section 7.2 we see that the following construction yields a Fiedler vector $y$ of $E_0(q,m)$: Let $z$ be a positive Perron vector of $\tilde{P}_q + \frac{1}{m}J$. Now let the subvector of $y$ corresponding to the vertices in the Perron component at $u$, along with $u$ itself, be given by $z$; let the subvector of $y$ corresponding to the direct summand of $B - \frac{1}{m}J$ that includes vertex $w$ be given by $-z$; and let the remaining entries of $y$ be zero. Note that $y_u > 0 > y_w$. Therefore, if $L$ is the Laplacian matrix of the graph $E_0(q,m) - wu$, then

$$y^T L y = a_{0,q,m} y^T y - (y_u - y_w)^2 < a_{0,q,m} y^T y.$$

Hence
$$a(E_0(q,m) - wu) = \frac{y^T L y}{y^T y} < a_{0,q,m}.$$

(ii) Consider the graph $D_1$ formed by attaching $m$ paths on $q+1$ vertices to a $v_0$. Observe that $D_1 \in E_1(q,m)$ (where $H$ is the star on $m+1$ vertices whose center is $v_0$). From Exercise 1(a) of Section 7.2, it follows that $a(D_1) = a_{1,q,m}$. Moreover, since any graph $\mathcal{G} \in E_1(q,m)$ can be formed by adding edges to $D_1$, it follows that $a(\mathcal{G}) \geq a_{1,q,m}$. Next, let $C$ be a connected component at $v_0$ in $\mathcal{G}$. We claim that the Perron value of $C$ is at least $\rho(P_{q+1})$. Once this claim is proven, it will follow by Exercise 2(a) of Section 7.2 that $a(\mathcal{G}) \leq a_{1,q,m}$, thus yielding $a(\mathcal{G}) = a_{1,q,m}$ which is our desired result.

To prove the claim, begin by recalling from Exercise 1 of Section 7.1 that adding edges to $C$ can only decrease its Perron value. Therefore, we need only establish the claim for the case that the vertices in $C$ adjacent to $v_0$ induce a complete subgraph, say on $c-1$ vertices. In this case, Observation 7.5.10 dictates that the bottleneck matrix for $C$ is

$$B = \left[\begin{array}{cccc|c} P_q + \frac{2}{c}J & \frac{1}{c}J & \cdots & \frac{1}{c}J & \frac{1}{c}ee_1^T(I+J) \\ \frac{1}{c}J & P_q + \frac{2}{c}J & \cdots & \frac{1}{c}J & \frac{1}{c}ee_2^T(I+J) \\ \cdots & \cdots & \cdots & \cdots & \cdots \\ \frac{1}{c}J & \cdots & \frac{1}{c}J & P_q + \frac{2}{c}J & \frac{1}{c}ee_c^T(I+J) \\ \hline \frac{1}{c}(I+J)e_1 e^T & \frac{1}{c}(I+J)e_2 e^T & \cdots & \frac{1}{c}(I+J)e_c e^T & \frac{1}{c}(I+J) \end{array}\right].$$

which is permutationally similar to

$$\left[\begin{array}{ccccc} qI + \frac{1}{c}(I+J) & (q-1)I + \frac{1}{c}(I+J) & \cdots & I + \frac{1}{c}(I+J) & \frac{1}{c}(I+J) \\ (q-1)I + \frac{1}{c}(I+J) & (q-1)I + \frac{1}{c}(I+J) & \cdots & I + \frac{1}{c}(I+J) & \frac{1}{c}(I+J) \\ \cdots & \cdots & \cdots & \cdots & \cdots \\ \frac{1}{c}(I+J) & \frac{1}{c}(I+J) & \cdots & \frac{1}{c}(I+J) & \frac{1}{c}(I+J) \end{array}\right]$$

where each block is $(c-1) \times (c-1)$. Since the rows of each block of this last matrix sum to the corresponding entry of $P_{q+1}$, it follows that the Perron value of $C$ is $\rho(P_{q+1})$. By the arguments in the previous paragraph, we can now conclude that $a(\mathcal{G}) = a_{1,q,m}$.

We now show that the deletion of any edge from a graph $\mathcal{G} \in E_1(q,m)$ results in a graph with algebraic connectivity strictly less than $a_{1,q,m}$. Removing any edges in one of the paths on $q$ vertices results in a disconnected graph, thus yielding a resulting algebraic connectivity of zero. Hence we will only consider removing an edge from the portion of $E_1(q,m)$ that is part of the graph $H$. Let $w$ be a vertex adjacent to $v_0$. We will consider the graph $\mathcal{G} - wv_0$. From Exercise 1(b) of Section 7.2, we see that the following construction yields a Fiedler vector $y$ for $\mathcal{G}$: Let $z_1$ be a positive Perron vector for the bottleneck matrix of the component at $v_0$ containing $w$, and let $z_2$ be a negative Perron vector for the bottleneck matrix of some other component at $v_0$ normalized so that $e^T z_1 + e^T z_2 = 0$. Now let the subvectors of $y$ corresponding to those components at $v_0$ be $z_1$ and $z_2$, respectively, and let the remaining entries of $y$ be zero. Note that $y_w > 0 = y_{v_0}$. Therefore, if $L$ is the Laplacian matrix of the graph $E_1(q,m) - wv_0$, then

$$y^T L y = a_{1,q,m} y^T y - (y_{v_0} - y_w)^2 < a_{1,q,m} y^T y.$$

Hence

$$a(E_1(q,m) - wv_0) = \frac{y^T L y}{y^T y} < a_{1,q,m}.$$

(iii) Let $\ell \geq 2$ and let $\mathcal{G} \in E_\ell(q,m)$. Then $\mathcal{G}$ can be constructed by starting with a graph $H$ on $m$ vertices in which vertices $1, \ldots, r$ have degree $m-1$ and attaching paths of length $q+1$ to vertices $1, \ldots, r$. Next attach paths of length $0 \leq j_i \leq q$ to vertex $i$ for each $i = r+1, \ldots, m$ (the values $j_i$ are determined from (7.5.2)). Let $H_1$ be the complete graph on $m$ vertices and construct $\mathcal{G}_1 \in E_\ell(q,m)$ from $H_1$ via the procedure parallel to the construction of $\mathcal{G}$. Let $H_2$ be the graph on $m$ vertices in which vertices $1, \ldots, r$ have degree $m-1$ and vertices $r+1, \ldots, m$ have degree $r$, and construct $\mathcal{G}_2 \in E_\ell(q,m)$ from $H_2$ via the procedure parallel to the construction of $\mathcal{G}$. Observing that $\mathcal{G}$ can be formed by adding edges to $\mathcal{G}_2$ or by deleting edges from $\mathcal{G}_1$, we see that $a(\mathcal{G}_1) \geq a(\mathcal{G}) \geq a(\mathcal{G}_2)$. To prove our desired result, we will prove $a(\mathcal{G}_1) = a(\mathcal{G}_2) = a_{\ell,q,m}$.

[*Aside*: For example, let $m = 5$, $q = 3$, $r = 3$, and $\ell = 2$ as in (7.5.3). Observe each of the graphs in (7.5.3) can be created by removing edges from $\mathcal{G}_1$ or adding edges to $\mathcal{G}_2$.]

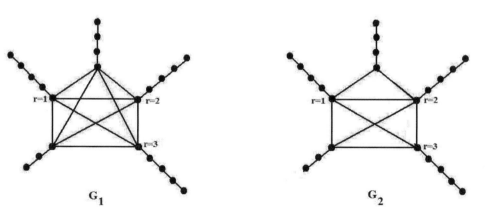

To prove $a(\mathcal{G}_1) = a_{\ell,q,m}$, let $u$ be a vertex of $\mathcal{G}_1$ of degree $m$. Then the non-Perron component at $u$ is the path on $q+1$ vertices, which has bottleneck matrix $P_{q+1}$. Moreover, it follows from Observation 7.5.10 that the bottleneck matrix $B_1$ for the

Perron component at $u$ is of the form

$$\left[\begin{array}{ccc|ccc|c} & & & & & & \frac{1}{m}ee_1^T(I+J) \\ & A & & & \frac{1}{m}J & & \frac{1}{m}ee_2^T(I+J) \\ & & & & & & \cdots \\ \hline & & & & & & \frac{1}{m}ee_{r-1}^T(I+J) \\ & & & & & & \frac{1}{m}ee_r^T(I+J) \\ & \frac{1}{m}J & & & B & & \frac{1}{m}ee_{r+1}^T(I+J) \\ & & & & & & \cdots \\ & & & & & & \frac{1}{m}ee_{m-1}^T(I+J) \\ \hline \frac{1}{m}(I+J)e_1e^T & \frac{1}{m}(I+J)e_2e^T & \cdots & \cdots & & \frac{1}{m}(I+J)e_{m-1}e^T & \frac{1}{m}(I+J) \end{array}\right]$$

where

$$A = \begin{bmatrix} P_{q+1}+\frac{2}{m}J & \frac{1}{m}J & \cdots & \frac{1}{m}J \\ \frac{1}{m}J & P_{q+1}+\frac{2}{m}J & \cdots & \frac{1}{m}J \\ \cdots & \cdots & \cdots & \cdots \\ \frac{1}{m}J & \cdots & \frac{1}{m}J & P_{q+1}+\frac{2}{m}J \end{bmatrix}$$

$$B = \begin{bmatrix} P_{j_1}+\frac{2}{m}J & \frac{1}{m}J & \cdots & \frac{1}{m}J \\ \frac{1}{m}J & P_{j_2}+\frac{2}{m}J & \cdots & \frac{1}{m}J \\ \cdots & \cdots & \cdots & \cdots \\ \frac{1}{m}J & \cdots & \frac{1}{m}J & P_{j_{m-r}}+\frac{2}{m}J \end{bmatrix}$$

Note that $B_1 - \frac{1}{m}J$ is permutationally similar to a direct sum of $r-1$ copies of $\tilde{P}_{q+1} + \frac{1}{m}J$ along with the matrices $\tilde{P}_{j_i} + \frac{1}{m}J$ for $1 \leq i \leq m-r$. It now follows from Exercise 1(a) of Section 7.2 that $a(\mathcal{G}_1) = a_{\ell,q,m}$.

To show that deleting an edge from $\mathcal{G}_1$ yields a graph whose algebraic connectivity is strictly less than $a_{\ell,q,m}$, we see from Exercise 1(b) of Section 7.2 that we can create a Fiedler vector $y$ for $\mathcal{G}_1$ as follows: Let $z_1$ be a positive Perron vector for $\tilde{P}_{q+1} + \frac{1}{m}J$, and let $z_2$ be the eigenvector of $B_1 - \frac{1}{m}J$ corresponding to $\lambda_n$ with all nonpositive entries normalized so that $e^T z_1 + e^T z_2 = 0$. Let the subvector of $y$ corresponding to the vertices in the Perron component at $u$, along with $u$ itself, be $z_2$ and let the remaining subvector of $y$ be $z_1$. Thus for each vertex $w$ in the Perron component at $u$ in $\mathcal{G}_1$ we have $y_u > 0 \geq y_w$, it now follows from similar arguments from earlier in this proof that $a(\mathcal{G}_1 - uw) < a_{\ell,q,m}$.

Next we consider the graph $\mathcal{G}_2$. Let $u$ be a vertex of $\mathcal{G}_2$ of degree $m$. Like before, the non-Perron component at $u$ is a path on $q+1$ vertices. Letting

$$M = \left[\begin{array}{c|c} \frac{1}{m}(I_{r-1}+J) & \frac{1}{m}J \\ \hline \frac{1}{m}J & \frac{1}{r}I_{m-r} + \frac{r-1}{mr}J \end{array}\right]$$

we see from Observation 7.5.10 that the bottleneck matrix $B_2$ for the Perron component at $u$ can be written as

$$\begin{bmatrix} A & B & \vdots \\ & & ee_{r-1}^T M \\ \hline C & D & \vdots \\ & & ee_{m-1}^T M \\ \hline Me_1 e^T \; Me_2 e^T \; \cdots & \cdots \; Me_{m-1} e^T & M \end{bmatrix}$$

where

$$A = \begin{bmatrix} P_{q+1} + M_{1,1}J & M_{1,2}J \cdots & M_{1,r-1}J & M_{1,r}J \\ M_{2,1}J & P_{q+1} + M_{2,2}J & \cdots & M_{2,r-1}J \\ \cdots & & & M_{r-2,r-1}J \\ M_{r-1,1}J & \cdots & M_{r-1,r-2}J & P_{q+1} + M_{r-1,r-1}J \end{bmatrix}$$

$$B = \begin{bmatrix} M_{1,r+1}J & \cdots & M_{1,m-1}J & ee_1^T M \\ M_{2,r}J & M_{2,r+1}J & \cdots & M_{2,m-1}J \\ \cdots & & \cdots & \\ M_{r-1,r}J & \cdots & & M_{r-1,m-1}J \end{bmatrix}$$

$$C = \begin{bmatrix} M_{1,r}J & M_{2,r}J & \cdots & M_{r-1,r}J \\ M_{1,r+1}J & M_{2,r+1}J & \cdots & M_{r-1,r+1}J \\ \cdots & \cdots & \cdots & \\ M_{1,m-1}J & \cdots & & M_{r-1,m-1}J \end{bmatrix}$$

$$D = \begin{bmatrix} P_{j_1} + M_{r,r}J & M_{r,r+1}J \cdots & M_{r,m-1}J & ee_r^T M \\ M_{r+1,r}J & P_{j_2} + M_{r+1,r+1}J & \cdots & M_{r+1,m-1}J \\ & & M_{m-2,m-1}J & \\ M_{m-1,r}J & \cdots & M_{m-1,m-2}J & P_{j_{m-r}} + M_{m-1,m-1}J \end{bmatrix}$$

and $M_{i,j}$ denotes the entry in row $i$ and column $j$ of $M$. Observe that $B_2 - \frac{1}{m}J$ is permutationally similar to a direct sum of $r-1$ copies of $\tilde{P}_{q+1} + \frac{1}{m}J$ along with the matrix

$$R = \begin{bmatrix} P_{j_1} + \frac{1}{r}J & O & \cdots & O & \frac{1}{r}ee_1^T \\ O & \cdots & & & \cdots \\ \cdots & \cdots & P_{j_{m-r}} + \frac{1}{r}I & \frac{1}{r}ee_{m-r}^T \\ \hline \frac{1}{r}e_1 e^T & \cdots & \cdots & \frac{1}{r}e_{m-r}e^T & \frac{1}{r}I \end{bmatrix} - \frac{1}{mr}J$$

Observe that $R + \frac{1}{mr}J$ is permutationally similar to a direct sum of the matrices $\tilde{P}_{j_i} + \frac{1}{r}J$ for $1 \leq i \leq m - r$. Therefore

$$\lambda_n(R) \leq \lambda_n\left(R + \frac{1}{mr}J\right) < \rho\left(\tilde{P}_{q+1} + \frac{1}{m}J\right).$$

Therefore
$$\lambda_n\left(B_2 - \frac{1}{m}J\right) = \rho\left(\tilde{P}_{q+1} + \frac{1}{m}J\right).$$

By considering the bottleneck matrices for the components at $u$, we apply Exercise 1(a) of Section 7.2 to see that $a(\mathcal{G}_2) = a_{\ell,q,m}$. Since $a(\mathcal{G}_1) \geq a(\mathcal{G}) \geq a(\mathcal{G}_2)$, it follows that $a(\mathcal{G}) = a_{\ell,q,m}$. Moreover, by a similar discussion as in the discussion of $\mathcal{G}_1$, removing an edge from $\mathcal{G}_2$ incident with a vertex of degree $m$ will result in a graph with strictly lower algebraic connectivity. Again, the fact that $a(\mathcal{G}_1) \geq a(\mathcal{G}) \geq a(\mathcal{G}_2)$ shows that the same holds true for $\mathcal{G}$. □

From the proof of Proposition 7.5.16, we are beginning to see that the graphs $E_\ell(q,m)$ are likely candidates for yielding an attainable upperbound on the algebraic connectivity of graphs on $n$ vertices with $k > n/2$ cut vertices. The following theorem from [44] confirms this.

**THEOREM 7.5.17** *Let $\mathcal{G}$ be a connected graph on $n$ vertices which has $k$ cut vertices. Suppose that $k > n/2$, say with $k = (qn + \ell)/(q+1)$ for some positive integer $q$ and nonnegative integer $\ell$. Then $a(\mathcal{G}) \leq a_{\ell,q,n-k}$ with equality holding if and only if $\mathcal{G} \in E_\ell(q, n-k)$.*

Since the proof of this theorem is rather lengthy, we omit it but refer the reader to [44] for the details. To assist the reader, we provide an outline of the proof here. The proof is done by induction on $n$. Since $(n+1)/2 \leq k \leq n-2$, the smallest admissible case is $n = 5$. This forces $k = 3$ which consequently yields $q = \ell = 1$. Therefore $\mathcal{G}$ is the path on 5 vertices. Observe $a(\mathcal{G}) = 1/\rho(P_2) = a_{1,1,2} = a_{\ell,q,n-k}$. Also observe, $\mathcal{G} \in E_1(1,2) = E_\ell(q, n-k)$.

We then continue by assuming $n \geq 6$ and that the result holds for all graphs on at most $n-1$ vertices. We investigate two major cases:

*Case I*: There exists a cut vertex $v$ of $\mathcal{G}$ at which there is a component $C$ containing no cut vertices of $\mathcal{G}$ and where $C$ has at least 2 vertices.

*Case II*: For any component $C$ at a cut vertex $v$ which contains no cut vertex of $\mathcal{G}$, $C$ is necessarily a single vertex.

Within Case II, we have two cases: Case IIA is where $\ell \geq 3$, and Case IIB is where $0 \leq \ell \leq 2$. Subsequently, within Case IIB there are two subcases: Case IIB1 is the case where there exists a cut vertex of $\mathcal{G}$ having at least three components, while Case IIB2 is the case where each cut vertex has exactly two components. Finally, Case IIB2 is divided into two further subcases. Both subcases deal with the case where there exists a cut vertex $u$ of $\mathcal{G}$ in which there exists a component $K$ at $u$ which is not the unique Perron component at $u$. Case IIB2a is the case where $K$ is not a path, while in Case IIB2b we assume $K$ is a path.

Summarizing the results of this section, suppose that we have a graph on $n$

vertices with $k$ cut vertices. If $k = 1$, then $a(\mathcal{G}) \leq 1$ where equality holds if and only if the cut vertex is adjacent to all other vertices of $\mathcal{G}$ (Corollary 7.5.2). Hence for equality to hold, $\mathcal{G}$ must be of the form $E_1(0, n-1)$. Thus in the language of Theorem 7.5.17, $a(\mathcal{G}) \leq 1/\rho(P_1) = 1$ as expected. Suppose now that $\mathcal{G}$ has $2 \leq k < n/2$ cut vertices. Observe that the upper bound on the algebraic connectivity of such graphs occurs when $\mathcal{G}$ is of the form $E_k(0, n-k)$ and note that this upper bound is $1/\rho(\tilde{P}_1 + \frac{1}{n-k}J)$ which equals $a_{n-k}$. If $\mathcal{G}$ has $k = \frac{n}{2}$ cut vertices, then the maximum algebraic connectivity of $a_{\frac{n}{2}}$ is achieved for graphs of the form $E_{\frac{n}{2}}(0, \frac{n}{2}) = E_0(1, \frac{n}{2})$. Finally, if $\mathcal{G}$ has $k = (qn + \ell)/(q+1)$ cut vertices for integers $q$ and $\ell$ such that $k > n/2$, then the upper bound is achieved when we take a complete graph on $n-k$ vertices and attach paths of $q$ vertices to each vertex of this complete graph. This yields an algebraic connectivity of $a_{\ell, q, n-k} = 1/\rho(\tilde{P}_q + \frac{1}{n-k}J)$. However, since this quantity decreases as $q$ increases, we see that it is maximized at $q = 1$ which yields the value $a_{n-k}$. Therefore:

**THEOREM 7.5.18** *For all $1 \leq k \leq n$, if $\mathcal{G}$ is a graph on $n$ vertices with $k$ cut vertices, then $a(\mathcal{G}) \leq a_{n-k}$.*

## 7.6 The Spectral Radius of Submatrices of Laplacian Matrices for Graphs

This chapter has dealt primarily with generalizing the results of Chapter 6 to graphs. We close this chapter with this section, based largely on [68], which generalizes the results of Section 6.7. Recall that in Section 6.7 we investigated the spectral radius of submatrices of Laplacian matrices at various vertices of a tree. For a vertex $v$ in a tree, we defined the function $r(v)$ to be the spectral radius of the submatrix of the Laplacian matrix $L$ obtained by deleting the row and column of $L$ corresponding to $v$. In this section, we consider the values of the function $r(v)$ at various vertices of graphs which are not trees. Since these submatrices are inverses of the bottleneck matrices that we have been studying, we investigate the values of $r(v)$ in the context of comparing such results with Fiedler's famous theorem, Theorem 6.1.10. Before we begin proving our results, let us generalize Definition 6.7.6:

**DEFINITION 7.6.1** *Let $\mathcal{G}$ be a weighted graph containing at least one cut vertex and where $\lambda_n \neq \lambda_{n-1}$. Then $\mathcal{G}$ is of Type A if there exists a cut vertex $v$ such that $r(v) = \lambda_{n-1}$. A graph $\mathcal{G}$ is of Type B is no such cut vertex exits.*

Note that when determining if a graph is of Type A or Type B, we consider only the cut vertices. Consider the graph from [68]:

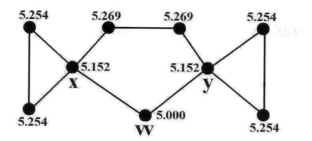

which is labeled with its values for $r(v)$. In this graph, we have $\lambda_{n-1} = 5$. Yet this graph is of Type B despite the fact that $r(w) = \lambda_{n-1}$ because $w$ is not a cut vertex.

The goal of this section is to prove results for graphs that are analagous to the results in Section 6.7 for trees. However, it is important to note that since not all graphs are bipartite, we are unable to use the bipartite complement of the Laplacian matrix as freely and as often as we did in Section 6.7. Thus the proofs of the theorems in this section will be quite different. We begin with a theorem from [68] which gives us an overview of the values of $r(v)$ at cut vertices.

**THEOREM 7.6.2** *Let $\mathcal{G}$ be a connected weighted graph with Laplacian matrix $L$. Let $P$ be a path that begins at a cut vertex $a$ in $\mathcal{G}$ in which the next cut vertex $b$ on $P$ is such that $r(b) > r(a)$. Then $P$ has the property that the values of $r(v)$ at the cut vertices contained in $P$ form an increasing sequence.*

<u>Proof</u>: Consider the representation of a graph $\mathcal{G}$ below:

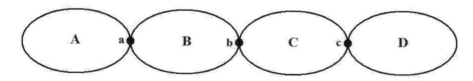

and consider its corresponding Laplacian matrix:

$$L = \begin{bmatrix} A & x & 0 & 0 & 0 & 0 & 0 \\ x^T & a & y^T & \alpha & 0 & 0 & 0 \\ 0 & y & B & z & 0 & 0 & 0 \\ 0 & \alpha & z^T & b & w^T & \beta & 0 \\ 0 & 0 & 0 & w & C & v & 0 \\ 0 & 0 & 0 & \beta & v^T & c & u^T \\ 0 & 0 & 0 & 0 & 0 & u & D \end{bmatrix}.$$

Suppose that $r(b) > r(a)$. Then it suffices to show that $r(c) > r(b)$. For visual clarity, observe the matrices $L[\bar{a}]$, $L[\bar{b}]$, and $L[\bar{c}]$ as follows:

$$L[\bar{a}] = \begin{bmatrix} A & 0 & 0 & 0 & 0 & 0 \\ 0 & B & z & 0 & 0 & 0 \\ 0 & z^T & b & w^T & \beta & 0 \\ 0 & 0 & w & C & v & 0 \\ 0 & 0 & \beta & v^T & c & u^T \\ 0 & 0 & 0 & 0 & u & D \end{bmatrix}.$$

$$L[\bar{b}] = \begin{bmatrix} A & x & 0 & 0 & 0 & 0 \\ x^T & a & y^T & 0 & 0 & 0 \\ 0 & y & B & 0 & 0 & 0 \\ 0 & 0 & 0 & C & v & 0 \\ 0 & 0 & 0 & v^T & c & u^T \\ 0 & 0 & 0 & 0 & u & D \end{bmatrix}.$$

$$L[\bar{c}] = \begin{bmatrix} A & x & 0 & 0 & 0 & 0 \\ x^T & a & y^T & \alpha & 0 & 0 \\ 0 & y & B & z & 0 & 0 \\ 0 & \alpha & z^T & b & w^T & 0 \\ 0 & 0 & 0 & w & C & 0 \\ 0 & 0 & 0 & 0 & 0 & D \end{bmatrix}.$$

Each of these matrices can be partitioned into a $2 \times 2$ block matrix in which the off-diagonal blocks are zero. Let $A_1$ and $A_2$ be the $(1,1)$ and $(2,2)$ blocks of $L[\bar{a}]$. Define $B_1$, $B_2$, $C_1$, and $C_2$ in a similar fashion for $L[\bar{b}]$ and $L[\bar{c}]$, respectively. Observe that $r(b) > r(a)$ implies that

$$r(b) = \max\{\rho(B_1), \rho(B_2)\} > \max\{\rho(A_1), \rho(A_2)\} = r(a).$$

At this point, we will show that $\rho(A_2) > \rho(B_2)$ which necessarily implies $r(b) = \rho(B_1)$. Let $\hat{A}$ be the matrix created from $A_2$ by taking all entries which are also entries in $B_2$ and replacing them with zero. Let $\hat{B}$ be the matrix that is the same size as $A$ that created from $B_2$ by bordering the matrix with zeros that correspond to the non-$B_2$ entries of $\hat{A}$. In other words:

$$\hat{A} = \begin{bmatrix} B & z & 0 & 0 & 0 \\ z^T & b & w^T & \beta & 0 \\ 0 & w & 0 & 0 & 0 \\ 0 & \beta & 0 & 0 & 0 \\ 0 & 0 & 0 & 0 & 0 \end{bmatrix}.$$

and

$$\hat{B} = \begin{bmatrix} 0 & 0 & 0 & 0 & 0 \\ 0 & 0 & 0 & 0 & 0 \\ 0 & 0 & C & v & 0 \\ 0 & 0 & v^T & c & u^T \\ 0 & 0 & 0 & u & D \end{bmatrix}.$$

Observe that $\lambda_1(\hat{A}) = 0$ since the set of vectors of the form $e_i$, where each $i$ corresponds to a row/column corresponding to the matrix $D$ of $\hat{B}$, forms a basis for the eigenspace of $\hat{A}$ corresponding to the eigenvalue zero. By Theorem 1.2.4 we see that

$$\begin{aligned}
\rho(A_2) &= \rho(\hat{A} + \hat{B}) &&= \max_{x^T x = 1} [x^T(\hat{A} + \hat{B})x] \\
&&&= \max_{x^T x = 1} [x^T \hat{A} x + x^T \hat{B} x] \\
&&&\geq \max_{x^T x = 1} [\lambda_1(\hat{A}) + x^T \hat{B} x] \\
&&&= \max_{x^T x = 1} [x^T \hat{B} x] \\
&&&= \rho(\hat{B}) \\
&&&= \rho(B_2).
\end{aligned}$$

However, observe that the inequality in this case will be strict since no unit vector which is in the eigenspace of $\hat{A}$ corresponding to $\lambda_1(\hat{A}) = 0$ is an eigenvector of $\hat{B}$. Thus $\rho(\hat{A} + \hat{B}) > \rho(\hat{B})$ and hence $\rho(A_2) > \rho(B_2)$. Therefore $r(b) = \rho(B_1)$. By using a similar argument as above, $r(c) = \rho(C_1)$. Since $B_1$ is a proper submatrix of $C_1$, it follows that

$$r(c) = \rho(C_1) > \rho(B_1) = r(b).$$

□

We see from Theorem 7.6.2 that if we chose a cut vertex $u$ such that the value of $r(u)$ is minimized over all cut vertices, then any path $P$ that begins at $u$ that contains cut vertices other than $u$ will have the property that the values of $r(v)$ at the cut vertices in $v \in P$ will be an increasing sequence as $P$ is traversed. Observe the example from [68]:

**EXAMPLE 7.6.3** In the graph below, its vertices labeled with their values for $r(v)$:

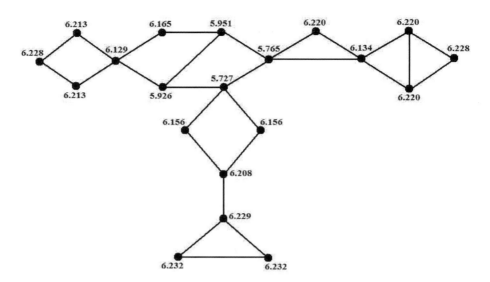

Observe that the cut vertex with the minimum value $r(v)$ is the vertex such that $r(v) = 5.727$. As we radiate out from this vertex, the values for $r(v)$ at the cut vertices strictly increase. This parallels Fiedler's results in Theorem 6.1.10.

**OBSERVATION 7.6.4** Theorem 6.1.10 implies that for each block $B \in \mathcal{G}$, the entries of the Fiedler vector corresponding to the vertices in $B$ that are not cut vertices in $\mathcal{G}$ are bounded above and below by the entries of the Fiedler vector corresponding to the vertices in $B$ that are cut vertices. However, as we can see from the example above, this idea does not hold true when considering the function $r(v)$. Observe in one of the blocks, the minimum and maximum values for $r(v)$ over the cut vertices $v$ are 5.727 and 6.129, respectively, but there is a vertex $v$ in that block such that $r(v) = 6.165$.

Now that we have an idea of how the values of $r(v)$ for the cut vertices vary, we now generalize our results from Section 6.7 to graphs that contain cut vertices. We impose the condition of $\lambda_n$ being simple because otherwise, $\lambda_n = \lambda_{n-1}$ and by Corollary 1.2.10, it would follow that $r(v) = \lambda_{n-1} = \lambda_n$ for all vertices $v$. We begin with two definitions concerning the structure of graphs that contain cut vertices:

**DEFINITION 7.6.5** Two cut vertices $u$ and $v$ are *block adjacent* in a graph $\mathcal{G}$ if there exists a path from $u$ to $v$ such that all edges on the path lie in a single block of $\mathcal{G}$.

**DEFINITION 7.6.6** A *block adjacent path of order $k$* is a sequence of cut vertices $v_1, \ldots, v_k$ such that $v_i$ is block adjacent to $v_{i+1}$ for $i = 1, \ldots, k-1$.

With these definitions, we can rephrase Theorem 7.6.2 to say that if $P$ is a block adjacent path in $\mathcal{G}$ that begins at a cut vertex $a$ in which the next cut vertex $b$ on $P$ is such that $r(b) > r(a)$, then $P$ has the property that the values of $r(v)$ at the cut vertices contained in $P$ form a strictly increasing sequence. Therefore, while there may be a set of block adjacent cut vertices where $r(v)$ is minimized, once we travel away from such vertices, the values of $r(v)$ at the cut vertices strictly increase.

Observe that for each cut vertex $v$, the matrix $L[\overline{v}]$ is (permutationally similar to) a block diagonal matrix in which each block of $L[\overline{v}]$ corresponds to a component $\mathcal{G} - v$. Therefore, for each cut vertex $v$, it follows that $r(v)$ attains its value from the block(s) of $L[\overline{v}]$ with the maximum spectral radius. As in Section 6.7, we will refer to the corresponding component(s) of $\mathcal{G} - v$ as the spectral branch(es) at $v$. We now prove a useful lemma from [68] concerning the number of spectral branches at a cut vertex $v$:

**LEMMA 7.6.7** *Let $\mathcal{G}$ be a connected weighted graph where $L$ is its Laplacian and where $\lambda_n$ is simple. Let $v$ be a cut vertex of $\mathcal{G}$. Then the multiplicity of $r(v)$ as an eigenvalue of the matrix $L[\overline{v}]$ is at least the number of spectral branches at $v$.*

**Proof**: Recall that $L[\overline{v}]$ is (permutationally similar to) a block diagonal matrix in which each block corresponds to a component of $\mathcal{G} - v$ (i.e., a branch at $v$). Since the spectral branches at $v$ correspond to the blocks of $L[\overline{v}]$ in which $r(v)$ is attained, it follows that the multiplicity of $r(v)$ as an eigenvalue of $L[\overline{v}]$ must be at least the number of spectral branches at $v$. □

We saw in Theorem 6.7.9 that if $\mathcal{G}$ is a tree, then $v$ has more than one spectral branch if and only if $r(v) = \lambda_{n-1}$. We will now prove a similar theorem from [68] for graphs in general. However, we will only prove the theorem in one direction:

**THEOREM 7.6.8** *Let $\mathcal{G}$ be a connected weighted graph where $L$ is its Laplacian and where $\lambda_n$ is simple. Let $v$ be a cut vertex of $\mathcal{G}$ that has more than one spectral branch. Then $r(v) = \lambda_{n-1}$; hence $\mathcal{G}$ is of Type A.*

**Proof**: We will assume $r(v) \neq \lambda_{n-1}$ for a cut vertex $v$ and show that $v$ must have exactly one spectral branch. First, if $r(v) = \lambda_n$, then since $\lambda_n$ is simple, it follows from Lemma 7.6.7 that $v$ has exactly one spectral branch. If $r(v) \neq \lambda_n$, then since $r(v) \neq \lambda_{n-1}$, it follows that $\lambda_{n-1} < r(v) < \lambda_n$. By Corollary 1.2.10 it follows that $r(v)$ is a simple eigenvalue of $L(v)$. Thus by Lemma 7.6.7, we conclude that $v$ has exactly one spectral branch. □

By Theorem 6.7.9, the converse of Theorem 7.6.8 is true if $\mathcal{G}$ is a tree. However, the converse is not necessarily true for graphs in general. Consider the following example from [68]:

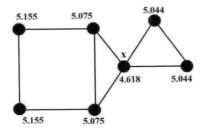

In this graph, we have $\lambda_{n-1} = 4.618$. Labeling each vertex $v$ with its value $r(v)$, we see that $r(x) = 4.618 = \lambda_{n-1}$, thus making this a graph of Type A. However, $x$ has only one spectral branch, namely the block on the left.

We should also note that unlike trees, if there exists a cut vertex $v$ such that $r(v) = \lambda_{n-1}$, then if $\mathcal{G}$ is not a tree, that cut vertex may not necessarily be the unique cut vertex such that $r(v) = \lambda_{n-1}$. Consider the following example from [68]

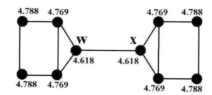

In this graph, $\lambda_{n-1} = 4.618$. Observe there are two cut vertices, $w$ and $x$, whose value $r(v) = \lambda_{n-1}$. In fact, it is possible to have a graph in which there are more than two cut vertices $v$ where $r(v) = \lambda_{n-1}$. We leave that for an exercise at the end of this section.

In Theorem 6.7.11 we showed that if $\mathcal{G}$ is a tree, there must exist two adjacent vertices $v$ and $w$ such that a spectral branch at $v$ contains $w$ and a spectral branch at $w$ contains $v$, and that this pair of vertices is unique if the tree is of Type B. We now generalize this to graphs, in general, containing cut vertices. We first need a lemma from [68]:

**LEMMA 7.6.9** *Let $\mathcal{G}$ be a weighted graph with at least one cut vertex. Then given any block $B$ of $\mathcal{G}$, there can exist at most one cut vertex $v$ of $\mathcal{G}$ lying on $B$ such that $B$ is not contained in any spectral branch at $v$.*

**Proof**: Let $B$ be a block of $\mathcal{G}$ containing cut vertices $v$ and $w$ of $\mathcal{G}$ such that no spectral branch at $v$ or $w$ contains $B$. Let $V$ and $W$ be blocks of $\mathcal{G}$ that $v$ and $w$ lie on that are spectral branches at $v$ and $w$, respectively. Let $L_V$ and $L_{B(v)}$ be the blocks of $L[\bar{v}]$ corresponding to $V$ and $B$, respectively. Similarly, let $L_W$ and $L_{B(w)}$ be the blocks of $L[\bar{w}]$ corresponding to $W$ and $B$, respectively. Note that $L_W$ is a proper submatrix of $L_{B(v)}$ and that $L_V$ is a proper submatrix of $L_{B(w)}$. Therefore:

$$r(v) = \rho(L_V) > \rho(L_{B(v)}) \geq \rho(L_W) > \rho(L_{B(w)}) \geq \rho(L_V) = r(v)$$

where the strict inequalities follow from the fact that $B$ is not on a spectral branch of $v$ or $w$. However, we have obtained the contradiction $r(v) > r(v)$. Hence for any block $B$ of $\mathcal{G}$, there can exist at most one cut vertex on $B$ in which no spectral branch at that vertex contains $B$. □

Lemma 7.6.9 leads us to the following definition:

**DEFINITION 7.6.10** *A spectral block of $\mathcal{G}$ is a block $B$ of $\mathcal{G}$ such for each cut vertex $v$ of $\mathcal{G}$ lying in $B$, the block $B$ is contained in a spectral branch at $v$.*

**EXAMPLE 7.6.11** Consider the graph from Example 7.6.3 relabeled below:

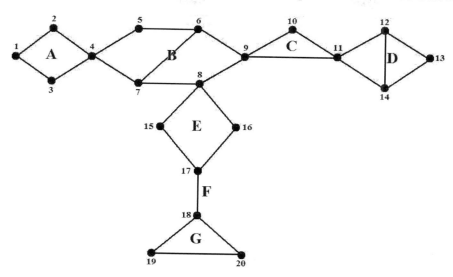

Observe that in block $A$, the cut vertex of $\mathcal{G}$ is vertex 4 and the spectral branch at this vertex does not contain $A$. In block $B$, the cut vertices of $\mathcal{G}$ are 4, 8, and 9. The spectral branch at all three of these vertices contains $B$. In block $C$, the cut vertices of $\mathcal{G}$ are 9 and 11. Observe that the spectral branch at 9 does not contain $C$ but the spectral branch at 11 does. For block $D$, the cut vertex of $\mathcal{G}$ is vertex 11 and the spectral branch at 11 does not contain $D$. For block $E$, the cut vertices of $\mathcal{G}$ are 9 and 17. Observe that the spectral branch at 9 does not contain $E$ but the spectral branch at 17 does. In block $F$, the cut vertices of $\mathcal{G}$ are 17 and 18. Observe that the spectral branch at 17 does not contain $F$ but the spectral branch at 18 does. Finally, in block $G$ the cut vertex of $\mathcal{G}$ is vertex 18 and the spectral branch at 18 does not contain $G$. Hence we see that for each block $X$, there is at most one cut vertex of $\mathcal{G}$ lying on $X$ whose spectral branch does not contain $X$, as Lemma 7.6.9 states. Moreover, every cut vertex $v$ of $\mathcal{G}$ lying in block $B$ is such that $B$ is contained in a spectral branch at $v$. Hence $B$ is a spectral block of $\mathcal{G}$.

We now prove a generalization of Theorem 6.7.11. This generalization can be found in [68]:

**THEOREM 7.6.12** *Let $\mathcal{G}$ be a weighted graph with at least one cut vertex where $L$ is the Laplacian for $\mathcal{G}$ and $\lambda_n$ is simple. Then $\mathcal{G}$ contains a spectral block. Moreover, if $\mathcal{G}$ is of Type B, then this block is unique.*

**Proof**: Suppose $\mathcal{G}$ does not contain a spectral block. Then from Lemma 7.6.9, each block in $\mathcal{G}$ contains exactly one cut vertex of $\mathcal{G}$ whose spectral branch does not contain that block. Choose a block $B_1$ of $\mathcal{G}$. Let $v_1$ be the vertex of $B_1$ whose spectral branch does not contain it. Now choose a block $B_2$, adjacent to $B_1$, such that a spectral branch at $v_1$ contains $B_2$. Next choose the cut vertex $v_2$ on $B_2$ whose spectral branch does not contain $B_2$. Choose a block $B_3$, adjacent to $B_2$, such that the spectral branch at $v_2$ contains $B_3$. Continuing in this fashion, we obtain a sequence of blocks and cut vertices: $B_1, v_1, B_2, v_2, B_3, \ldots$. Since $\mathcal{G}$ is finite, this sequence will terminate at a pendant block $B_p$ (a block containing only one cut vertex of $\mathcal{G}$). Hence $B_p$ would contain a cut vertex whose spectral branch contains it. Since $B_p$ contains only one cut vertex of $\mathcal{G}$, it follows that all cut vertices of $\mathcal{G}$ lying in $B_p$ are such that the spectral branch at the cut vertices contains $B_p$. Thus $B_p$ is a spectral block, contradicting our original assumption. Hence $\mathcal{G}$ must contain a spectral block.

To show uniqueness in the Type B case, suppose $B$ and $C$ are two spectral blocks. Letting $v \in B$ and $w \in C$ be cut vertices, consider the block adjacent path $P: v = v_1, v_2, \ldots, v_{k-1}, v_k = w$ from $v$ to $w$, and let $B_i$ be the block containing vertices $v_i$ and $v_{i+1}$ (thus $B = B_0$ and $C = B_k$). Since $B$ is a spectral block, it follows that the unique spectral branch at $v_1$ contains $B$. Since $B_1$ is not on the spectral branch at $v_1$, it follows from Lemma 7.6.9 that the unique spectral branch at $v_2$ contains $B_1$. Following this reasoning, the unique spectral branch at $v_i$ contains $B_{i-1}$ for all $i = 1, \ldots, k$. However, since $C$ is a spectral block, it follows that the unique spectral branch at $v_k$ contains $C$, and thus by similar reasoning as before, it would follow that the unique spectral branch at $v_i$ contains $B_i$ for all $i = 1, \ldots, k$. This is a contradiction. Hence we have established uniqueness. □

Theorem 6.7.11 says that if $\mathcal{T}$ is a tree of Type B, then there exits a unique pair of adjacent vertices $v$ and $w$ such that the spectral branch at $v$ contains $w$ and the spectral branch at $w$ contains $v$. Theorem 7.6.12 generalizes this result to say that if a graph is of Type B, then it necessarily has a unique set of mutually block adjacent cut vertices such that the spectral branch at each such vertices contains all of the other such vertices. In other words a graph of Type B has a unique spectral block.

**EXAMPLE 7.6.13** Consider the graph from Examples 7.6.3 and 7.6.11:

In this graph, we have $\lambda_{n-1} = 5.434$. Observe that this graph is of Type B because there do not exist any cut vertices $v$ such that $r(v) = \lambda_{n-1}$. We saw earlier that $B$ is the spectral block of this graph. Note that the cut vertices of block B are vertices 4, 8, and 9, and that the spectral branch at each of these cut vertices contains the other two as Theorem 7.6.12 states.

We see from the proof of Theorem 7.6.12 that if $m$ is a cut vertex not lying on a spectral block of $\mathcal{G}$, then the unique spectral branch must contain the spectral block(s) of $\mathcal{G}$. We state this more formally as a corollary (from [68]) which generalizes Corollary 6.7.13 to graphs.

**COROLLARY 7.6.14** *Let $\mathcal{G}$ be a connected weighted graph where $L$ is its Laplacian and $\lambda_n$ is simple. Suppose $m$ is a cut vertex that does not lie on a spectral block of $\mathcal{G}$. Then the unique spectral branch at $m$ is the branch which contains all of the spectral block(s) of $\mathcal{G}$.*

At this point we summarize our results from Theorems 7.6.8, 7.6.12, and Corollary 7.6.14 with a theorem from [68] to obtain a generalization of Theorem 6.7.14:

**THEOREM 7.6.15** *Let $\mathcal{G}$ be a connected weighted graph with at least one cut vertex where $L$ is the Laplacian for $\mathcal{G}$ and $\lambda_n$ is simple. Then*

*(a) If there exists a cut vertex $v$ having at least two spectral branches, then (i) $r(v) = \lambda_{n-1}$, and (ii) such a graph is of Type A. We call $v$ the spectral vertex of $\mathcal{G}$.*

*(b) If $\mathcal{G}$ is of Type B, then there exists a unique set of mutually block adjacent cut vertices such that the spectral branch at each such vertex contains the other such vertices. These vertices all lie on the same block of $\mathcal{G}$ known as the spectral block of $\mathcal{G}$.*

*(c) If a cut vertex $m$ is not a spectral vertex of $\mathcal{G}$, nor lies on a spectral block of $\mathcal{G}$, then the unique spectral branch at $m$ contains the all spectral vertex/block(s) of $\mathcal{G}$.*

Note that we established that the converse of (a-ii) is true when $\mathcal{G}$ is a tree, but false otherwise.

In light of Theorem 7.6.15, we can make Theorem 7.6.2 more precise in light of the results we have concerning the blocks of Type A and Type B trees. We note that as we radiate away from the cut vertices of a weighted graph $\mathcal{G}$ such that the value of $r(v)$ is minimized, the values of $r(v)$ at the subsequent cut vertices strictly increase. Thus we combine the results from Theorem 7.6.2 and Theorems 7.6.8 and 7.6.12 to obtain a generalization of Theorem 6.7.15 (see [68]):

**THEOREM 7.6.16** *Let $\mathcal{G}$ be a connected weighted graph with a cut vertex where $L$ is its Laplacian and $\lambda_n$ is simple. Then the values for $r(v)$ increases (not necessarily strictly) for cut vertices $v$ along any block adjacent path which*

*(a) begins at a spectral vertex, or*

*(b) begins at a cut vertex in a spectral block and does not contain other cut vertices of a spectral block.*

**REMARK 7.6.17** Note that in Theorem 7.6.16, that the values of $r(v)$ are not necessarily increasing strictly along said paths like they necessarily would if $\mathcal{G}$ is a tree. Consider the following graph (see [68]):

# 356 Applications of Combinatorial Matrix Theory to Laplacian Matrices of Graphs

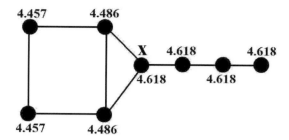

Labeling each vertex $v$ with its value $r(v)$, it can be seen that the block to the left of $x$ is the spectral block, yet the values of $r(v)$ remain constant as one travels away from $x$. It should be noted that in a graph, once the values for $r(v)$ begin to strictly increase along said block adjacent paths, the values will continue to strictly increase according to Theorem 7.6.2.

**REMARK 7.6.18** We saw in Section 6.7 that if $\mathcal{G}$ is a tree, then $r(v) < \lambda_n$ for all vertices $v \in \mathcal{G}$. However, we see from the graph in Remark 7.6.17 that this inequality does not hold for graphs in general. Observe that $\lambda_n = 4.618$ and that there exists cut vertices (and a non-cut vertex) such that $r(v) = \lambda_n$.

We now turn our attention to vertices of graphs which are not cut vertices. We begin with a theorem from [68] which concerns vertices that are neither cut vertices, nor vertices lying on a spectral block of the graph.

**THEOREM 7.6.19** *Let $\mathcal{G}$ be a connected weighted graph where $L$ is the Laplacian for $\mathcal{G}$ and where $\lambda_n$ is simple. Let $v$ be a vertex that is not a cut vertex. Suppose $v$ lies on a block $B$ of $\mathcal{G}$ that is not a spectral block of $\mathcal{G}$. Then $r(v) \geq \min r(w)$ where the minimum is taken over all vertices $w \in B$ that are cut vertices of $\mathcal{G}$.*

**Proof**: Suppose $B$ is not a spectral block of $\mathcal{G}$. Then by Theorem 7.6.2 and Lemma 7.6.9, $B$ contains a cut vertex $x$ of $\mathcal{G}$ such that $B$ is not contained in a spectral branch at $x$. Letting $L'(x)$ be the block of $L[\overline{x}]$ corresponding to a spectral branch at $x$, we see that $L'(x)$ is a positive definite submatrix of $L[\overline{v}]$. Hence:

$$r(v) = \rho(L[\overline{v}]) \geq \rho(L'(x)) = \rho(L[\overline{x}]) = r(x).$$

Thus there exists a cut vertex $x$ of $\mathcal{G}$ that lies on $B$ such that $r(v) \geq r(w)$. Hence $r(v) \geq \min r(w)$ where the minimum is taken over all vertices $w \in B$ that are cut vertices of $\mathcal{G}$. □

**REMARK 7.6.20** The condition that $B$ is not a spectral block cannot be eliminated. Observe in the graph $\mathcal{G}$ following Definition 7.6.1 that vertex $w$ lies in the spectral block of $\mathcal{G}$ yet the value for $r(w)$ is less than of that of the cut vertices of $\mathcal{G}$ that lie in the spectral block (which is the center block in this graph).

So far we have been investigating graphs in which $\lambda_n$ was simple. Therefore, it is natural to investigate graphs in which $\lambda_n$ is not simple. Clearly if $\lambda_n = \lambda_{n-1}$ then by (6.7.1) we have $r(v) = \lambda_n = \lambda_{n-1}$ for all vertices $v$. This leads us to two immediate questions:

(1) Is the converse true, i.e., if the value for $r(v)$ is constant for all vertices $v \in \mathcal{G}$, must $\lambda_n = \lambda_{n-1}$?

(2) Under what conditions will $\lambda_n = \lambda_{n-1}$?

The answer to the first question is that the converse is false. To obtain a family of counterexamples, we prove the following claim from [68]:

**CLAIM 7.6.21** *Let $\mathcal{G}$ be a connected bipartite graph in which all vertices are isomorphic. Let $L$ be its Laplacian matrix. Then the value for $r(v)$ over all vertices $v \in \mathcal{G}$ are equal, yet $\lambda_n \neq \lambda_{n-1}$.*

**Proof**: The value of $r(v)$ being constant over all vertices $v \in \mathcal{G}$ follows immediately from the fact that the vertices are all isomorphic. The inequality $\lambda_n \neq \lambda_{n-1}$ follows directly from Theorem 4.3.12. □

**EXAMPLE 7.6.22** Consider complete bipartite graph $K_{3,3}$. Observe that $r(v) = 5.449$ for all vertices $v \in K_{3,3}$, yet $\lambda_{n-1} \neq \lambda_n$ since $\lambda_n = 6$ while $\lambda_{n-1} = 3$.

**REMARK 7.6.23** Claim 7.6.21 only gives us counterexamples to the converse that is stated in Question (1); it does not claim that these are the only counterexamples.

To answer Question (2), recall from Theorem 4.1.11 that if $\lambda_k$ is an eigenvalue of $L(\mathcal{G})$ for $k = 2, \ldots, n$, then $n - \lambda_k$ is an eigenvalue of $L(\mathcal{G}^c)$, where $\mathcal{G}^c$ is the complement of $\mathcal{G}$, with the same corresponding eigenvector(s). More specifically, $\lambda_2(\mathcal{G}^c) = n - \lambda_n(\mathcal{G})$ and $\lambda_3(\mathcal{G}^c) = \lambda_{n-1}(\mathcal{G})$. Hence for the graph $\mathcal{G}$, $\lambda_n = \lambda_{n-1}$ if and only if $\lambda_2(\mathcal{G}^c)$ is not a simple eigenvalue of $L(\mathcal{G}^c)$. Therefore, in order to attempt to characterize all graphs $\mathcal{G}$ such that $\lambda_{n-1} = \lambda_n$, we will rely heavily on $\mathcal{G}^c$. When investigating $\mathcal{G}^c$, we will divide our investigation into three cases:

(a) $\mathcal{G}^c$ is not connected.

(b) $\mathcal{G}^c$ is connected and contains at least one cut vertex.

(c) $\mathcal{G}^c$ is connected but does not contain any cut vertices.

For Case (a), if $\mathcal{G}^c$ is not connected, then $\lambda_2(\mathcal{G}^c) = 0$. Hence in order to obtain $\lambda_n(\mathcal{G}) = \lambda_{n-1}(\mathcal{G})$, Theorem 4.1.11 dictates that we would need $\lambda_3(\mathcal{G}^c) = 0$ also. Hence if $\mathcal{G}^c$ is disconnected, then by Theorem 4.1.1 it follows that $\mathcal{G}^c$ must contain at least three components. This implies that if $\mathcal{G}$ is the join of at least three graphs, then $\lambda_n(\mathcal{G}) = \lambda_{n-1}(\mathcal{G})$.

**EXAMPLE 7.6.24** *[68]* To illustrate Case (a), consider the following multipartite graph $K_{1,2,3}$:

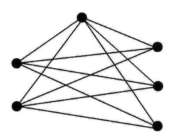

Observe that $K_{1,2,3}$ is the join of three graphs, namely the empty graphs on one, two, and three vertices. Therefore, $\lambda_n = \lambda_{n-1}$ and hence the value of $r(v)$ has the same value (namely 6) for all vertices $v$ of this graph.

For Case (b), if $\mathcal{G}^c$ is connected and contains at least one cut vertex, then in order to have $\lambda_2(\mathcal{G}^c) = \lambda_3(\mathcal{G}^c)$, we see from Theorem 7.2.4 and a generalization of Theorem 6.2.18 that $\mathcal{G}^c$ must be a graph of Type I where its characteristic vertex has at least three Perron branches.

**EXAMPLE 7.6.25** *[68]* To illustrate Case (b), consider the following graph:

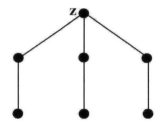

Observe that this graph is of Type I with $z$ as its characteristic vertex. Also note that $z$ has three Perron branches. Thus the complement of this graph, i.e.,

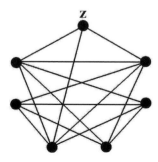

is such that $\lambda_n = \lambda_{n-1}$ and thus $r(v)$ has the same value (namely 6.618) for all vertices $v$ of this graph.

We now summarize the results of Cases (a) and (b) in the following theorem from [68]:

**THEOREM 7.6.26** *Let $\mathcal{G}$ be a graph and $L(\mathcal{G})$ be its Laplacian. Then $\lambda_n = \lambda_{n-1}$ if either of the following holds:*

*(a) $\mathcal{G}$ is the join of at least three graphs.*

*(b) $\mathcal{G}^c$ is a graph of Type I such that its characteristic vertex has at least three Perron branches.*

Moving on to Case (c), we see again that $\mathcal{G}^c$ would have to be such that $\lambda_2(\mathcal{G}^c) = \lambda_3(\mathcal{G}^c)$. But since in this case we are not allowing $\mathcal{G}^c$ to have any cut vertices, this means that in order to complete an investigation of Case (c), we would have to characterize all graphs without cut vertices such that the algebraic connectivity is not a simple eigenvalue of its Laplacian matrix. Unfortunately, no such characterization is known. However, we can again rely on the concept of characteristic vertices of Type I graphs in order to construct some examples. Often times (but not always), we can create a graph in which $\lambda_n = \lambda_{n-1}$ by taking two copies of the complement of a Type I graph, and joining the characteristic vertex of each copy of such graph with a path. For example (see [68]):

Observe that we have taken two copies of $(P_5)^c$, the complement of the path on five vertices, and joined them with a path beginning at the vertex in one copy of $(P_5)^c$ which is the characteristic vertex of $P_5$ and ending at the corresponding vertex of the other copy of $(P_5)^c$. In this graph, $\lambda_n = \lambda_{n-1} = 4.618$. Thus $r(v) = 4.618$ for all vertices in this graph.

We can construct a similar graph by taking two copies of $(P_7)^c$ and joining them with a path consisting of at least two intermediate vertices beginning at the vertex in one copy of $(P_7)^c$ which is the characteristic vertex of $P_7$ and ending at the corresponding vertex of the other copy of $(P_7)^c$. In this graph $\lambda_n = \lambda_{n-1} = 6.802$. At this point it is tempting to conjecture that this method of creating graphs in which $\lambda_n = \lambda_{n-1}$ will hold by using two copies of any Type I graph. Unfortunately, this is not the case as $P_9$ would be a counterexample. Hence it remains an open question as to which graphs of Type I we can use to create a graph in this manner in which $\lambda_n = \lambda_{n-1}$.

## Exercises:

**1.** Classify each of the following graphs $\mathcal{G}_1$ and $\mathcal{G}_2$ as Type A or Type B.

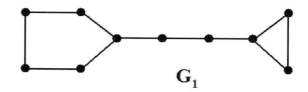

$G_1$

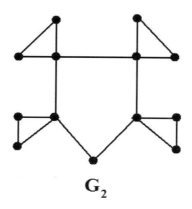

$G_2$

**2.** Construct a graph in which $\lambda_n$ is simple and there are three cut vertices $v$ such that $r(v) = \lambda_{n-1}$.

**3.** Suppose $\mathcal{G}$ is a graph with at least three cut vertices $v$ such that $r(v) = \lambda_{n-1}$. Prove that the blocks containing such vertices are all spectral blocks.

# Chapter 8

# The Group Inverse of the Laplacian Matrix

In the beginning of Chapter 7, we briefly used the group inverse of the Laplacian matrix of a graph in order to create bottleneck matrices. In this chapter, we investigate the group inverse of the Laplacian matrix much more deeply and study the combinatorial aspects behind its construction. In Section 8.1, we show how the structure of a tree can be used to construct the group inverse of the Laplacian matrix. Distances between pairs of vertices in a weighted tree will be significant in our construction of such matrices. We then use the group inverse in Section 8.2 to derive a lower bound on the algebraic connectivity of graphs. We also show when equality can occur between this lower bound and the algebraic connectivity. We then return our focus to trees in Section 8.3 and investigate which vertices of trees contribute to this lower bound. Finally, in Section 8.4 we refine our results from Section 5.2 concerning how varying the weight of the edges of a graph affects the algebraic connectivity. The group inverse turns out to be very useful in this endeavor.

## 8.1 Constructing the Group Inverse for a Laplacian Matrix of a Weighted Tree

As stated in Section 1.6 and restated in Section 7.1, the group inverse of $n \times n$ matrix $A$, when it exists, is the unique $n \times n$ matrix $X$ that satisfies all of the following:

$$
\begin{aligned}
&(i) && AXA = A \\
&(ii) && XAX = X \\
&(iii) && AX = XA
\end{aligned}
\qquad (8.1.1)
$$

Recall from Theorem 7.1.1 that if $A$ is an irreducible singular M-matrix with right null vector $x = [x_1, x_2, \ldots, x_n]^T$ and left null vector $y = [y_1, y_2, \ldots, y_n]^T$, then

$$
A^{\#} = \frac{\hat{y}^T M \hat{x}}{(y^T x)^2} xy^T + \left[ \begin{array}{c|c} M - \frac{1}{y^T x} M \hat{x} \hat{y}^T - \frac{1}{y^T x} \hat{x} \hat{y}^T M & \frac{-y_n}{y^T x} M \hat{x} \\ \hline \frac{-x_n}{y^T x} \hat{y}^T M & 0 \end{array} \right] \qquad (8.1.2)
$$

where $\hat{x} = [x_1, x_2, \ldots, x_{n-1}]^T$, $\hat{y} = [y_1, y_2, \ldots, y_{n-1}]^T$, and $M = A[\overline{n}]^{-1}$. This immediately gave rise to the formula for $L^{\#}$:

$$L^{\#} = \frac{e^T M e}{n^2} J + \begin{bmatrix} M - \frac{1}{n} MJ - \frac{1}{n} JM & -\frac{1}{n} Me \\ -\frac{1}{n} e^T M & 0 \end{bmatrix} \qquad (8.1.3)$$

The goal of this section will be to use (8.1.3) to develop a combinatorial description of the entries in $L^{\#}$ in terms of distances between vertices when $L$ is the Laplacian matrix for a weighted tree. Letting $\mathcal{T}$ be a weighted tree with vertices $1, 2, \ldots, n$, we saw from Theorem 6.2.5 that the formula for the bottleneck matrix $M$ of $\mathcal{T}$ based at vertex $v$ is

$$m_{i,j} = \sum_{e \in \tilde{P}_{i,j}} \frac{1}{w(e)}, \qquad (8.1.4)$$

where $\tilde{P}_{i,j}$ is the set of all edges $e$ which lie concurrently on the path from $i$ to $v$ and on the path from $j$ to $v$. In developing a combinatorial description of the entries in $L^{\#}$ when $L$ is the Laplacian matrix for a weighted tree, our first goal will be to determine a formula for the diagonal entries of $L^{\#}$. To this end, for a weighted tree $\mathcal{T}$ and edge $e \in \mathcal{T}$, we define $\beta_i(e)$ to be the set of vertices in the connected component of $\mathcal{T} - e$ which does not contain vertex $i$. We can now use Theorem 6.2.5 along with Corollary 7.1.2 to obtain the diagonal entries of $L^{\#}$ when $L$ is the Laplacian matrix for a weighted tree. We do so in the following lemma from [50].

**LEMMA 8.1.1** *If $L$ is the Laplacian matrix for a weighted tree $\mathcal{T}$ on $n$ vertices, then*

$$L^{\#}_{v,v} = \frac{1}{n^2} \sum_{e \in \mathcal{T}} \frac{|\beta_v(e)|^2}{w(e)} \qquad (8.1.5)$$

*for $v = 1, 2, \ldots, n$.*

**Proof**: Without loss of generality, we may assume $v = n$. By Corollary 7.1.2,

$$L^{\#}_{n,n} = \frac{e^T M e}{n^2}$$

where $M$ is the bottleneck matrix at vertex $n$. Observe that $e^T M e$ is merely the sum of the entries in $M$. Using Theorem 6.2.5, we see that

$$L^{\#}_{n,n} = \frac{1}{n^2} \sum_{e \in \mathcal{T}} \frac{f(e)}{w(e)}$$

where $f(e)$ denotes the number of times an edge $e$ is used to sum the entries of $M$. From Theorem 6.2.5, an edge $e$ is used for $m_{i,j}$ if $e$ occurs concurrently on the paths from $i$ to $n$ and from $j$ to $n$. In other words, an edge $e$ is used in the computing of $m_{i,j}$ if and only if both $i$ and $j$ lie on the same component of $\mathcal{T} - e$ and on the opposite component of $n$. Thus $f(e) = |\beta_n(e)|^2$. This proves the theorem. □

While Lemma 8.1.1 is informative, ultimately we want to be able to determine the diagonal entries of the Laplacian matrix in terms of distances between vertices in the tree. To this end, we need the following definitions.

# The Group Inverse of the Laplacian Matrix

**DEFINITION 8.1.2** The *inverse weighted distance from vertex $i$ to vertex $j$* is the sum of the reciprocals of the weights of the edges of the unique path, $P_{i,j}$ from $i$ to $j$, i.e.,

$$\tilde{d}(i,j) := \sum_{e \in P_{i,j}} \frac{1}{w(e)}. \tag{8.1.6}$$

**DEFINITION 8.1.3** Letting $\tilde{d}(i,i) = 0$ for each $i$, we define the *inverse status of a vertex $i$* as the sum

$$\tilde{d}_i := \sum_{u \in \mathcal{T}} \tilde{d}(u,i). \tag{8.1.7}$$

In our attempt to construct the group inverse of the Laplacian matrix in terms of the structural properties of a tree, we now derive a formula for the difference between diagonal entries of $L^{\#}$ in terms of the inverse status number of the corresponding vertices. We do so with a lemma from [50]:

**LEMMA 8.1.4** *Let $\mathcal{T}$ be a weighted tree on $n$ vertices. Let $v_0$ and $v_\ell$ be vertices in $\mathcal{T}$. Then*

$$\tilde{d}_{v_0} - \tilde{d}_{v_\ell} = n\left(L^{\#}_{v_0,v_0} - L^{\#}_{v_\ell,v_\ell}\right).$$

**Proof**: Let $v_1, v_2, \ldots, v_{\ell-1}$ be intermediate vertices on the path $P$ joining $v_0$ to $v_\ell$. For $1 \leq i \leq \ell$, let $e_i$ be the edge between $v_{i-1}$ and $v_i$ having weight $\theta_i$. For $0 \leq i \leq \ell$, let $t_i$ be the number of vertices, including $v_i$, whose shortest path to $P$ has terminal vertex $v_i$. For any one of the $t_i$ vertices $u$ of $\mathcal{T}$ whose shortest path to $P$ ends at $v_i$, we have that

$$\tilde{d}(u,v_0) - \tilde{d}(u,v_\ell) = \tilde{d}(v_i,v_0) - \tilde{d}(v_i,v_\ell).$$

Thus summing over all such $i$ we obtain

$$\tilde{d}_{v_0} - \tilde{d}_{v_\ell} = \sum_{i=1}^{\ell} t_i \left[\tilde{d}(v_i,v_0) - \tilde{d}(v_i,v_\ell)\right].$$

Observing that

$$\tilde{d}(v_i,v_0) - \tilde{d}(v_i,v_\ell) = \sum_{m=1}^{i} \frac{1}{\theta_m} - \sum_{m=i+1}^{\ell} \frac{1}{\theta_m},$$

we obtain

$$\tilde{d}_{v_0} - \tilde{d}_{v_\ell} = \sum_{i=0}^{\ell} t_i \left( \sum_{m=1}^{i} \frac{1}{\theta_m} - \sum_{m=i+1}^{\ell} \frac{1}{\theta_m} \right). \tag{8.1.8}$$

We see from Lemma 8.1.1 that

$$n^2 L^{\#}_{v,v} = \sum_{e \in \mathcal{T}} \frac{|\beta_v(e)|^2}{w(e)}.$$

for each vertex $v \in \mathcal{T}$. Hence

$$n^2 \left( L^{\#}_{v_0,v_0} - L^{\#}_{v_\ell,v_\ell} \right) = \sum_{e \in \mathcal{T}} \frac{1}{w(e)} (|\beta_{v_0}(e)|^2 - |\beta_{v_\ell}(e)|^2) \quad (8.1.9)$$
$$= \sum_{e \in \mathcal{T}} (|\beta_{v_0}(e)| + |\beta_{v_\ell}(e)|) \frac{|\beta_{v_0}(e)| - |\beta_{v_\ell}(e)|}{w(e)}$$

Observe that if $e \notin P$ then $|\beta_{v_0}(e)| = |\beta_{v_\ell}(e)|$, while if $e = e_m$ for some $m = 1, 2, \ldots, \ell$ then $|\beta_{v_0}(e_m)| = t_m + \ldots + t_\ell$ and $|\beta_{v_\ell}(e_m)| = t_0 + \ldots + t_{m-1}$. Therefore, we can rewrite (8.1.9) as

$$\sum_{m=1}^{\ell} (|\beta_{v_0}(e_m)| + |\beta_{v_\ell}(e_m)|) \frac{|\beta_{v_0}(e_m)| - |\beta_{v_{\ell_m}}(e_m)|}{\theta_m}$$

which equals (recalling that $t_1 + \ldots + t_\ell = n$)

$$n \sum_{m=1}^{\ell} \frac{-t_0 - \ldots - t_{m-1} + t_m + \ldots + t_\ell}{\theta_m}.$$

Simplifying the last expression, we obtain

$$n^2 \left( L^{\#}_{v_0,v_0} - L^{\#}_{v_\ell,v_\ell} \right) = n \sum_{i=0}^{\ell} t_i \left( \sum_{m=1}^{i} \frac{1}{\theta_m} - \sum_{m=i+1}^{\ell} \frac{1}{\theta_m} \right).$$

Dividing through by $n$ and then plugging into (8.1.8) gives the desired result. □

From Lemma 8.1.4, we see that $L^{\#}_{v_0,v_0} - L^{\#}_{v_\ell,v_\ell} = \frac{1}{n}(\tilde{d}_{v_0} - \tilde{d}_{v_\ell})$. We now obtain the following corollary from [50] which makes use of vector notation:

**COROLLARY 8.1.5** *Let $\mathcal{T}$ be a weighted tree on $n$ vertices with Laplacian matrix $L$. Then for some constant $c$, we have*

$$\left[ L^{\#}_{1,1} \ldots L^{\#}_{n,n} \right] = \frac{1}{n} \left[ \tilde{d}_1 \ldots \tilde{d}_n \right] + ce^T.$$

Our ultimate goal is to obtain a formula for $L^{\#}_{v,v}$ in terms of the inverse status of the vertices. This goal will be achieved when we determine the constant $c$ in Corollary 8.1.5. To do this, we need to define some notation. For each edge $e \in \mathcal{T}$ where $\mathcal{T}$ has $n$ vertices, the graph $\mathcal{T} - e$ has two connected components: $J_e$ which has $j_e$ vertices and $J_e^c$ which has $n - j_e$ vertices. Now for our theorem from [50] which determines the value of $c$:

**THEOREM 8.1.6**

$$\left[ L^{\#}_{1,1} \ldots L^{\#}_{n,n} \right] = \frac{1}{n} \left[ \tilde{d}_1 \ldots \tilde{d}_n \right] - \left( \frac{1}{2n^2} \sum_{i=1}^{n} \tilde{d}_i \right) e^T.$$

**Proof**: From Corollary 8.1.5, we obtain

$$ce^T = \begin{bmatrix} L_{1,1}^{\#} \ldots L_{n,n}^{\#} \end{bmatrix} - \frac{1}{n} \begin{bmatrix} \tilde{d}_1 \ldots \tilde{d}_n \end{bmatrix}.$$

Summing the entries in each vector, we obtain

$$nc = \sum_{i=1}^{n} L_{i,i}^{\#} - \frac{1}{n} \sum_{i=1}^{n} \tilde{d}_i$$

and hence

$$c = \frac{1}{n} \sum_{i=1}^{n} L_{i,i}^{\#} - \frac{1}{n^2} \sum_{i=1}^{n} \tilde{d}_i.$$

Thus our theorem will be proved once we show that $\sum_{i=1}^{n} L_{i,i}^{\#} = \frac{1}{2n} \sum_{i=1}^{n} \tilde{d}_i$.

From Lemma 8.1.1 we obtain

$$\sum_{i=1}^{n} L_{i,i}^{\#} = \frac{1}{n^2} \sum_{e \in \mathcal{T}} \sum_{v \in \mathcal{T}} \frac{|\beta_v(e)|^2}{w(e)}. \tag{8.1.10}$$

Using the notation that was described preceeding this theorem, we see that if $v \in J_e$ then $|\beta_v(e)|^2 = (n - j_e)^2$, while if $v \in J_e^c$ then $|\beta_v(e)|^2 = j_e^2$. Therefore

$$\sum_{e \in \mathcal{T}} \sum_{v \in \mathcal{T}} \frac{|\beta_v(e)|^2}{w(e)} = \sum_{e \in \mathcal{T}} \frac{1}{w(e)} \left( \sum_{v \in J_e} |\beta_v(e)|^2 + \sum_{v \in J_e^c} |\beta_v(e)|^2 \right)$$

$$\sum_{e \in \mathcal{T}} \frac{1}{w(e)}((n - j_e)^2 j_e + j_e^2(n - j_e)) \tag{8.1.11}$$

$$\sum_{e \in \mathcal{T}} \frac{n j_e (n - j_e)}{w(e)}.$$

From the definition of inverse status numbers, we have

$$\sum_{i=1}^{n} \tilde{d}_i = \sum_{v \in \mathcal{T}} \sum_{u \in \mathcal{T}} \sum_{e \in P_{u,v}} \frac{1}{w(e)}.$$

But for each edge $e$, there are $2j_e(n - j_e)$ unordered pairs of vertices $u$ and $v$ such that $e$ is on the path between them. Hence, each edge $e$ contributes $2j_e(n-j_e)/w(e)$ to the above sum so that

$$\sum_{i=1}^{n} \tilde{d}_i = \sum_{e \in \mathcal{T}} \frac{2j_e(n - j_e)}{w(e)}.$$

Combining this with (8.1.11) we obtain

$$\sum_{e \in \mathcal{T}} \sum_{v \in \mathcal{T}} \frac{|\beta_v(e)|^2}{w(e)} = \frac{n}{2} \sum_{i=1}^{n} \tilde{d}_i.$$

Plugging this into (8.1.10) yields the desired result. □

We now have a formula for the diagonal entries of $L^{\#}$ completely in terms of inverse status numbers. We state this more formally as a corollary from [50]:

**COROLLARY 8.1.7** *Let $\mathcal{T}$ be a weighted tree on $n$ vertices and $L$ be its Laplacian. Then for each vertex $v \in \mathcal{T}$, we have*

$$L^{\#}_{v,v} = \frac{1}{n}\tilde{d}_v - \frac{1}{2n^2}\sum_{i=1}^{n}\tilde{d}_i.$$

**OBSERVATION 8.1.8** For any two distinct vertices $i$ and $j$, it follows that $L^{\#}_{i,i} \geq L^{\#}_{j,j}$ if and only if $\tilde{d}_i \geq \tilde{d}_j$.

**EXAMPLE 8.1.9** Consider the weighted tree below:

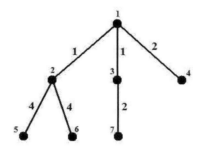

Below is a chart that gives the inverse weighted distance $\tilde{d}(i,j)$ between each pair of vertices $i$ and $j$.

|   | 1 | 2 | 3 | 4 | 5 | 6 | 7 |
|---|---|---|---|---|---|---|---|
| 1 | 0 | 1 | 1 | $\frac{1}{2}$ | $\frac{5}{4}$ | $\frac{5}{4}$ | $\frac{3}{2}$ |
| 2 | 1 | 0 | 2 | $\frac{3}{2}$ | $\frac{1}{4}$ | $\frac{1}{4}$ | $\frac{5}{2}$ |
| 3 | 1 | 2 | 0 | $\frac{3}{2}$ | $\frac{9}{4}$ | $\frac{9}{4}$ | $\frac{1}{2}$ |
| 4 | $\frac{1}{2}$ | $\frac{3}{2}$ | $\frac{3}{2}$ | 0 | $\frac{7}{4}$ | $\frac{7}{4}$ | 2 |
| 5 | $\frac{5}{4}$ | $\frac{1}{4}$ | $\frac{9}{4}$ | $\frac{7}{4}$ | 0 | $\frac{1}{2}$ | $\frac{11}{4}$ |
| 6 | $\frac{5}{4}$ | $\frac{1}{4}$ | $\frac{9}{4}$ | $\frac{7}{4}$ | $\frac{1}{2}$ | 0 | $\frac{11}{4}$ |
| 7 | $\frac{3}{2}$ | $\frac{5}{2}$ | $\frac{1}{2}$ | 2 | $\frac{11}{4}$ | $\frac{11}{4}$ | 0 |

From the above chart, we can determine the values for $\tilde{d}_i$ for each $i = 1, \ldots, 7$:

| $i$ | 1 | 2 | 3 | 4 | 5 | 6 | 7 |
|---|---|---|---|---|---|---|---|
| $\tilde{d}_i$ | $\frac{13}{2}$ | $\frac{15}{2}$ | $\frac{19}{2}$ | 9 | $\frac{35}{4}$ | $\frac{35}{4}$ | 12 |

Since $n = 7$ it follows that $\frac{1}{2n^2}\sum \tilde{d}_i = \frac{31}{49}$. Therefore, by Corollary 8.1.7 we have the following values for $L^{\#}_{i,i}$:

| $i$ | 1 | 2 | 3 | 4 | 5 | 6 | 7 |
|---|---|---|---|---|---|---|---|
| $L_{i,i}^{\#}$ | $\frac{29}{98}$ | $\frac{43}{98}$ | $\frac{71}{98}$ | $\frac{32}{49}$ | $\frac{121}{196}$ | $\frac{121}{196}$ | $\frac{53}{49}$ |

Now that we have derived a formula for the diagonal entries of $L^{\#}$ in terms of inverse statuses of the vertices, our next goal is to do the same for the off diagonal entries of $L^{\#}$. To this end, we need the following two claims from [50]:

**CLAIM 8.1.10** *Let $\mathcal{T}$ be a weighted tree and $L$ be its Laplacian. For $i \neq n$, we have*

$$L_{i,i}^{\#} = \tilde{d}(i,n) - \frac{2}{n}\sum_{e \in P_{i,n}} \frac{|\beta_n(e)|}{w(e)} + L_{n,n}^{\#}.$$

**Proof**: We will consider the quantity $L_{i,i}^{\#} - L_{n,n}^{\#}$. In light of Lemma 8.1.1, observe that if $e \notin P_{i,n}$ then its contribution to $L_{i,i}^{\#}$ is the same as that to $L_{n,n}^{\#}$. Hence in $L_{i,i}^{\#} - L_{n,n}^{\#}$ we need only consider edges in $\mathcal{T}$ that lie on the path from $i$ to $n$. For such an edge $e$, its contribution to $L_{n,n}^{\#}$ is $|\beta_n(e)|/(n*w(e))$, while its contribution to $L_{i,i}^{\#}$ is $(n-|\beta_n(e)|)/(n*w(e))$. Thus

$$L_{i,i}^{\#} - L_{n,n}^{\#} = \frac{1}{n}\sum_{e \in P_{i,j}} \frac{n-2|\beta_n(e)|}{w(e)} = \tilde{d}(i,n) - \frac{2}{n}\sum_{e \in P_{i,n}} \frac{|\beta_n(e)|}{w(e)},$$

the last equality stemming from the fact that $\tilde{d}(i,n) = \sum_{e \in P_{i,n}} 1/w(e)$. Solving for $L_{i,i}^{\#}$ proves the claim. □

**CLAIM 8.1.11** *Let $\mathcal{T}$ be a weighted tree on $n$ vertices with Laplacian matrix $L$. Then for $i,j = 1,\ldots,n$ with $i \neq j$,*

$$L_{i,j}^{\#} = \frac{1}{n^2}\sum_{e \in \mathcal{T}} \frac{|\beta_j(e)|^2}{w(e)} - \frac{1}{n}\sum_{e \in \tilde{P}_{i,j}} \frac{|\beta_j(e)|}{w(e)}.$$

**Proof**: Without loss of generality we can assume $j = n$ and $i = 1,\ldots,n-1$. From Theorem 7.1.2 we obtain

$$L_{i,n}^{\#} = \frac{e^T M e}{n^2} - \frac{1}{n}(e^T M)_i.$$

where $M$ is the bottleneck matrix at vertex $n$. From Lemma 8.1.1 we saw that $\frac{e^T M e}{n^2} = \frac{1}{n^2}\sum_{e \in \mathcal{T}} \frac{|\beta_v(e)|^2}{w(e)}$. By similar reasoning as in the proof of Lemma 8.1.1, we obtain that $(e^T M)_i = \sum_{e \in \tilde{P}_{i,n}} \frac{|\beta_n(e)|}{w(e)}$. The result follows. □

We are now able to prove a formula from [50] for the off diagonal entries of $L^{\#}$ in terms of inverse distances:

**THEOREM 8.1.12** *For $i \neq k$ where $1 \leq i, k \leq n$, we have*

$$L_{i,k}^{\#} = \frac{\tilde{d}_i + \tilde{d}_k}{2n} - \frac{1}{2}\tilde{d}(i,k) - \frac{1}{2n^2}\sum_{j=1}^{n}\tilde{d}_j.$$

**Proof**: Without loss of generality, we let $k = n$ and $1 \leq i \leq n-1$. From Claim 8.1.10, we see that

$$L_{i,i}^{\#} - L_{j,j}^{\#} = \tilde{d}(i,n) - \tilde{d}(j,n) + 2\left[\frac{1}{n}\sum_{e \in P_{j,n}}\frac{|\beta_n(e)|}{w(e)} - \frac{1}{n}\sum_{e \in P_{i,n}}\frac{|\beta_n(e)|}{w(e)}\right].$$

Observe from Claim 8.1.11 that the expression in brackets is precisely $L_{i,n}^{\#} - L_{j,n}^{\#}$. Hence

$$L_{i,i}^{\#} - L_{j,j}^{\#} = \tilde{d}(i,n) - \tilde{d}(j,n) + 2(L_{i,n}^{\#} - L_{j,n}^{\#}).$$

Therefore

$$2L_{i,n}^{\#} = L_{i,i}^{\#} - \tilde{d}(i,n) + \tilde{d}(j,n) - L_{j,j}^{\#} + 2L_{j,n}^{\#}.$$

Since $L_{i,n}^{\#}$ only depends on $i$ and $n$, it follows that $j$ is arbitrary. Thus let

$$\alpha := \tilde{d}(j,n) - L_{j,j}^{\#} + 2L_{j,n}^{\#} \tag{8.1.12}$$

and rewrite $2L_{i,n}^{\#}$ as

$$2L_{i,n}^{\#} = L_{i,i}^{\#} - \tilde{d}(i,n) + \alpha. \tag{8.1.13}$$

Using the idea that $j$ is arbitrary, we obtain from (8.1.12):

$$(n-1)\alpha = \sum_{j=1}^{n-1}\tilde{d}(j,n) - \sum_{j=1}^{n-1}L_{j,j}^{\#} + 2\sum_{j=1}^{n-1}L_{j,n}^{\#}.$$

The first summation equals $\tilde{d}_n$ by definition of $\tilde{d}_n$. Moreover, recalling that the column sums of $L^{\#}$ are zero, we know that $\sum_{j=1}^{n-1}L_{j,n}^{\#} = -L_{n,n}^{\#}$. Hence we rewrite the above as

$$(n-1)\alpha = \tilde{d}_n - \sum_{j=1}^{n-1}L_{j,j}^{\#} - 2L_{n,n}^{\#} = \tilde{d}_n - \sum_{j=1}^{n}L_{j,j}^{\#} - L_{n,n}^{\#}.$$

We can replace $L_{n,n}^{\#}$ by the formula from Corollary 8.1.7. From the proof of Corollary 8.1.7, we see that the sum of the diagonal entries in $L^{\#}$ equals $\frac{1}{2n}\sum_{j=1}^{n}\tilde{d}_j$. Therefore

$$(n-1)\alpha = -\frac{1}{n}\tilde{d}_n + \frac{1}{2n^2}\sum_{j=1}^{n}\tilde{d}_j - \frac{1}{2n}\sum_{j=1}^{n}\tilde{d}_j + \tilde{d}_n = \frac{n-1}{n}\tilde{d}_n - \frac{n-1}{2n^2}\sum_{j=1}^{n}\tilde{d}_j.$$

Therefore

$$\alpha = \frac{1}{n}\tilde{d}_n - \frac{1}{2n^2}\sum_{j=1}^{n}\tilde{d}_j.$$

Plugging this into (8.1.13) and solving for $L_{i,n}^{\#}$ yields

$$L_{i,n}^{\#} = \frac{1}{2}\left(L_{i,i}^{\#} - \tilde{d}(i,n) + \frac{1}{n}\tilde{d}_n - \frac{1}{2n^2}\sum_{j=1}^{n}\tilde{d}_j\right)$$

Using Corollary 8.1.7 to substitute for $L_{i,i}^{\#}$ and then simplifying yields the result. □

**EXAMPLE 8.1.13** Consider the tree in Example 8.1.9. In this example, we had found the diagonal entries of $L^{\#}$. Using the tables we used to find the diagonal entries along with Theorem 8.1.12, we can now find the off-diagonal entries of $L^{\#}$. Therefore, we now have the entire matrix $L^{\#}$:

$$L^{\#} = \frac{1}{196}\begin{bmatrix} 58 & -26 & 2 & 44 & -33 & -33 & -12 \\ -26 & 86 & -82 & -40 & 79 & 79 & -96 \\ 2 & -82 & 142 & -12 & -89 & -89 & 128 \\ 44 & -40 & -12 & 128 & -47 & -47 & -26 \\ -33 & 79 & -89 & -47 & 121 & 72 & -103 \\ -33 & 79 & -89 & -47 & 72 & 121 & -103 \\ -12 & -96 & 128 & -26 & -103 & -103 & 212 \end{bmatrix}$$

**OBSERVATION 8.1.14** If $\mathcal{T}$ is a tree, then for any vertex $v \in \mathcal{T}$, the entries $L_{v,i}^{\#}$ in the $v^{th}$ row of $L^{\#}$ decrease along any path beginning at $v$.

# Exercises:

**1.** Use Corollary 8.1.7 and Theorem 8.1.12 to find $L^{\#}$ for the following weighted tree:

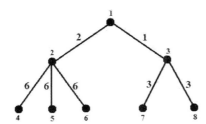

**2.** Prove Observation 8.1.14.

**3.** (See [50]) Let $L$ be the Laplacian matrix of a weighted tree on $n$ vertices. Show that $L^{\#}$ is an M-matrix if and only if for every pair of adjacent vertices $i$ and $j$ we have that

$$\tilde{d}_i + \tilde{d}_j \leq \frac{n}{w(e_{i,j})} + \frac{1}{n}\sum_{k=1}^{n}\tilde{d}_k.$$

**4.** (See [50]) Let $L$ be the Laplacian matrix for an *unweighted* tree $\mathcal{T}$ on $n$ vertices. Prove that $L^{\#}$ is an M-matrix if and only if $\mathcal{T}$ is a star.

## 8.2 The Zenger Function as a Lower Bound on the Algebraic Connectivity

In this section, we use the group inverse to derive an important lower bound for the algebraic connectivity of a graph. We then show the conditions when the algebraic connectivity of a graph is equal to this lower bound. To begin this process, we have the following definition:

**DEFINITION 8.2.1** The *Zenger* of an $m \times n$ matrix $B$ is the quantity

$$\mathcal{Z}(B) = \frac{1}{2} \max_{1 \leq i,j \leq m} \|e_i^T B - e_j^T B\|_1$$

In other words, to compute the Zenger of a matrix $B$, subtract each pair of distinct rows of B and take the one-norm of each difference. The Zenger is one half the maximum of such one-norms.

**EXAMPLE 8.2.2** Consider the matrix

$$B = \begin{bmatrix} 2 & 6 & 4 & -1 \\ 8 & 1 & -3 & 0 \\ -1 & 3 & 0 & 2 \end{bmatrix}$$

Observe that

$$\|e_1^T B - e_2^T B\|_1 = 6 + 5 + 7 + 1 = 19,$$

$$\|e_1^T B - e_3^T B\|_1 = 3 + 3 + 4 + 3 = 13,$$

$$\|e_2^T B - e_3^T B\|_1 = 9 + 2 + 3 + 2 = 16.$$

Thus $\mathcal{Z}(B) = \frac{1}{2} \bullet 19 = 9.5$.

If $B$ is an $n \times n$ matrix with constant row sums $\gamma$, then $Z(B)$ provides an upper bound for the eigenvalues of $B$ other than $\gamma$. We state this more precisely as a theorem from [74]:

**THEOREM 8.2.3** *Let $B$ be an $n \times n$ matrix with constant row sums $\gamma$, and let $\lambda \neq \gamma$ be an eigenvalue of $B$. Then $|\lambda| \leq \mathcal{Z}(B)$.*

Before we can prove Theorem 8.2.3, we need the following lemma from [74]:

**LEMMA 8.2.4** *Suppose $x \in \Re^n$, $n \geq 2$, such that $x^T e = 0$ and $x \neq 0$. Then there exist suitable values $\xi_{i,j}$ for $1 \leq i,j \leq n$ such that*

$$x = \frac{1}{2} \sum_{1 \leq i,j, \leq n} \xi_{i,j}(e_i - e_j)$$

*where $\xi_{i,j} \geq 0$, $\xi_{i,i} = 0$, and $\sum_{i,j} \xi_{i,j} = \|x\|_1$.*

**Proof**: We prove this by induction on $n$. If $n = 2$, then $x^T = [x_1, -x_1]$ for some $x_1 \neq 0$. If $x_1 > 0$ then $\xi_{1,2} = 2x_1$ and $\xi_{2,1} = 0$; if $x_1 < 0$ then $\xi_{1,2} = 0$ and $\xi_{2,1} = x_1$. Hence the lemma is true for $n = 2$. Now assume the lemma is true for $n = k$ for some $k \geq 2$. Suppose $n = k+1$. Then in $x^T = [x_1, x_2, \ldots, x_{k+1}]$, choose $p$ such that $|x_p| = \max |x_i|$, and $q$ such that $x_q \neq 0$ and $\operatorname{sgn} x_q = -\operatorname{sgn} x_p$. Let $\alpha = x_q(e_q - e_p)$, and let $y = x - \alpha$. Observe that $y_q \neq 0$, but clearly $y \neq 0$, $y^T e = 0$, and $\|y\|_1 = \|x\|_1 - 2|x_q|$. We now can apply the induction hypothesis to $\epsilon$ to obtain the required result for $x$, where $\xi_{p,q} = 2|x_q|$. □

We can now prove Theorem 8.2.3:

**Proof**: Let $z^T = [z_1, z_2, \ldots, z_n]$ be an arbitrary row vector of complex numbers. Let $x$ be an arbitrary nonzero vector in $\Re^n$ such that $x^T e = 0$. By Lemma 8.2.4 we have for appropriate $\xi_{i,j}$,

$$z^T x = \frac{1}{2} \sum_{1 \leq i,j \leq n} \xi_{i,j}(z_i - z_j)$$

Thus

$$|z^T x| \leq \frac{1}{2} \sum_{i,j} \xi_{i,j} |z_i - z_j|$$

$$\leq \frac{1}{2} \max_{i,j} |z_i - z_j| \sum_{i,j} \xi_{i,j}$$

$$= \frac{1}{2} \max_{i,j} |z_i - z_j| \|x\|_1$$

$$= \frac{1}{2} f(z) \|x\|_1 \qquad (8.2.1)$$

where $f(z) := \max_{i,j} |z_i - z_j|$. Then

$$f(Bz) = \max_{i,j} \left| \sum_{s=1}^{n} (b_{i,s} - b_{j,s}) z_s \right|. \qquad (8.2.2)$$

Observe that for any $i, j$, since $B$ has constant row sums, the vector $[(b_{i,1} - b_{j,1}), (b_{i,2} - b_{j,2}), \ldots, (b_{i,n} - b_{j,n})]$ is such that the sum of its entries is zero. Letting $x^T = [(b_{i,1} - b_{j,1}), (b_{i,2} - b_{j,2}), \ldots, (b_{i,n} - b_{j,n})]$, we see from (8.2.1) that

$$\left| \sum_{s=1}^{n} (b_{i,s} - b_{j,s}) z_s \right| \leq \frac{1}{2} f(z) \sum_{s=1}^{n} |b_{i,s} - b_{j,s}|.$$

Combining this with (8.2.2) we obtain

$$f(Bz) \leq \frac{1}{2} f(z) \max_{i,j} \sum_{s=1}^{n} |b_{i,s} - b_{j,s}|.$$

Note that if $z$ is an eigenvector of $B$ corresponding to the eigenvalue $\lambda$, then $f(Bz) = |\lambda| f(z)$. Since (8.2.2) holds for any vector $z$, it follows that for such an eigenvector $z$ that

$$|\lambda| f(z) \leq \frac{1}{2} f(z) \max_{i,j} \sum_{s=1}^{n} |b_{i,s} - b_{j,s}| \qquad (8.2.3)$$

Observe that $f(z)$ can be zero only if all entries of $z$ are identical. But if all entries of $z$ are identical, then $z$ would be the eigenvector corresponding to the eigenvalue $\gamma$. Since this contradicts the fact that $\lambda \neq \gamma$, it follows that not all entries of $z$ are identical. Thus $f(z) \neq 0$. Therefore, we divide (8.2.3) through by $f(z)$ to obtain

$$|\lambda| \leq \frac{1}{2} \max_{i,j} \sum_{s=1}^{n} |b_{i,s} - b_{j,s}| = \frac{1}{2} \max_{1 \leq i,j \leq n} \|e_i^T B - e_j^T B\|_1 = \mathcal{Z}(B)$$

which is our desired result. □

Let $L$ be the Laplacian matrix of a connected graph with $0 = \lambda_1 < \lambda_2 \leq \lambda_3 \leq \ldots \leq \lambda_n$ as its eigenvalues. Recall from Section 1.6 that the nonzero eigenvalues of $L^{\#}$ are the reciprocals of the nonzero eigenvalues of $L$ with the same corresponding eigenvectors. Additionally, zero is also an eigenvector of $L^{\#}$ with $e$ as the corresponding eigenvector (as $e$ is the eigenvector of $L$ corresponding to zero). Thus $L^{\#}$ has constant row sums of 0. Hence we can apply Theorem 8.2.3 to conclude that $1/\lambda_2 \leq \mathcal{Z}(L^{\#})$, and hence $1/\mathcal{Z}(L^{\#}) \leq \lambda_2$. Rewriting this inequality, we obtain the following corollary which gives a lower bound for the algebraic connectivity $a(\mathcal{G})$:

**COROLLARY 8.2.5** *Let $\mathcal{G}$ be a graph with $L$ as its Laplacian. Then*

$$\frac{1}{\mathcal{Z}(L^{\#})} \leq a(\mathcal{G})$$

We can now combine Corollary 8.2.5 and Theorem 5.1.10 to obtain the following string of inequalities involving the Zenger, algebraic connectivity, and vertex connectivity:

$$\frac{1}{\mathcal{Z}(L^{\#})} \leq a(\mathcal{G}) \leq v(\mathcal{G}). \qquad (8.2.4)$$

Recall Theorem 5.1.10 states that the second inequality is sharp for a noncomplete graph on $n$ vertices if and only if $\mathcal{G}$ can be written as $\mathcal{G}_1 \vee \mathcal{G}_2$, where $\mathcal{G}_1$ is a disconnected graph on $n - v(\mathcal{G})$ vertices and $\mathcal{G}_2$ is a graph on $v(\mathcal{G})$ vertices with $a(\mathcal{G}_2) \geq 2v(\mathcal{G}) - n$. The goal for the remainder of this section is to describe the conditions under which the first inequality is sharp, i.e., when $1/\mathcal{Z}(L^{\#}) = a(\mathcal{G})$. Before showing this, we will first show the conditions under which $1/\mathcal{Z}(L^{\#}) = v(\mathcal{G})$ since this necessarily requires that $a(\mathcal{G}) = v(\mathcal{G})$. We begin with the following lemma from [46].

**LEMMA 8.2.6** *Suppose that $\mathcal{G}$ is a graph on $n$ vertices with vertex connectivity $v$ that is of the form $\mathcal{G} = \mathcal{G}_1 \vee \mathcal{G}_2$ where $\mathcal{G}_1$ and $\mathcal{G}_2$ are graphs on $n - v$ vertices and $v$*

vertices, respectively, and where $\mathcal{G}_1$ is a disconnected graph. Then

$$L^{\#}(\mathcal{G}) = \left[\begin{array}{c|c} [L(\mathcal{G}_1) + vI]^{-1} - \dfrac{n+v}{vn^2}J & -\dfrac{1}{n^2}J \\ \hline -\dfrac{1}{n^2}J & [L(\mathcal{G}_2) + (n-v)I]^{-1} - \dfrac{2n-v}{(n-v)n^2}J \end{array}\right], \quad (8.2.5)$$

where the $(1,1)$ block is an $(n-v) \times (n-v)$ matrix, and the $(2,2)$ block is a $v \times v$ matrix.

**Proof**: We have

$$L(\mathcal{G}) = \left[\begin{array}{c|c} L(\mathcal{G}_1) + vI & -J \\ \hline -J & L(\mathcal{G}_2) + (n-v)I \end{array}\right],$$

where the $(1,1)$ block is an $(n-v) \times (n-v)$ matrix and the $(2,2)$ block is a $v \times v$ matrix. For the sake of notational simplicity, Let $L = L(\mathcal{G})$ and $X$ be the matrix on the right of (8.2.5). We find that $LX = XL = I - \frac{1}{n}J$, $XLX = X$, $LXL = L$, and $XL = LX$. Thus $X$ satisfies (8.1.1). □

Since computing the Zenger of a matrix involves subtracting rows, it is clear that adding a constant to all entries of a matrix will not affect the value of the Zenger. Therefore $Z(A + bJ) = Z(A)$ for any matrix $A$ and scalar $b$, and thus Lemma 8.2.6 yields immediate the corollary from [46]:

**COROLLARY 8.2.7** *Let $\mathcal{G}$ be a graph on $n$ vertices with vertex connectivity $v$ of the form $\mathcal{G} = \mathcal{G}_1 \vee \mathcal{G}_2$ where $\mathcal{G}_1$ and $\mathcal{G}_2$ are graphs on $n-v$ vertices and $v$ vertices, respectively, and where $\mathcal{G}_1$ is a disconnected graph. Then*

$$\mathcal{Z}(L^{\#}(\mathcal{G})) = \max\left\{\mathcal{Z}\left([L(\mathcal{G}_1)+vI]^{-1}\right), \mathcal{Z}\left([L(\mathcal{G}_2)+(n-v)I]^{-1}\right),\right.$$

$$\max_{\substack{1 \leq i \leq n-v \\ n-v+1 \leq j \leq n}} \left\{\frac{1}{2}\left[\left\|e_i^T((L(\mathcal{G}_1)+vI)^{-1} - \frac{1}{nv}J)\right\|_1 \right.\right. \quad (8.2.6)$$

$$\left.\left.\left. + \left\|e_j^T((L(\mathcal{G}_2)+(n-v)I)^{-1} - \frac{1}{(n-v)n}J)\right\|_1\right]\right\}\right\}.$$

In order to establish, under the conditions of Corollary 8.2.7, which of the three expressions in the right-hand side of (8.2.6) yields the maximum we need to establish more precise information about the matrices $[L(\mathcal{G}_1) + vI]^{-1}$ and $[L(\mathcal{G}_2) + (n-v)I]^{-1}$. Observe that both of these matrices are inverses of matrices in which the diagonal entries of a Laplacian matrix are perturbed. The following observation, lemma, and subsequent corollary, all from [46], give us useful information about the diagonals of the inverses of these matrices.

**OBSERVATION 8.2.8** *Let $A$ be an invertible matrix and let $x$ be a vector. Then*
$$[A + xx^T]^{-1} = A^{-1} - \frac{1}{1 + x^T A^{-1} x} A^{-1} xx^T A^{-1}$$

This formula can be verified merely by multiplying the expression on the right-hand side of the equal sign by $A + xx^T$ to obtain the identity.

**LEMMA 8.2.9** *Let $\mathcal{H}$ be a graph and $m > 0$. Form $\mathcal{H}'$ from $\mathcal{H}$ by adding an edge. Then*
$$\operatorname{diag}\left([L(\mathcal{H}') + mI]^{-1}\right) \leq \operatorname{diag}\left([L(\mathcal{H}) + mI]^{-1}\right).$$

**Proof:** $L(\mathcal{H}') = L(\mathcal{H}) + xx^T$ for some vector $x$, so
$$[L(\mathcal{H}') + mI]^{-1} = \left[L(\mathcal{H}) + mI + xx^T\right]^{-1}$$
$$= [L(\mathcal{H}) + mI]^{-1} - \frac{(L(\mathcal{H}) + mI)^{-1} xx^T (L(\mathcal{H}) + mI)^{-1}}{1 + x^T (L(\mathcal{H}) + mI)^{-1} x}$$

where the last equality follows from Observation 8.2.8, letting $A = L(\mathcal{H}) + mI$. Since $L(\mathcal{H}) + mI$ is symmetric, it follows that
$$(L(\mathcal{H}) + mI)^{-1} xx^T (L(\mathcal{H}) + mI)^{-1} = [(L(\mathcal{H}) + mI)^{-1} x][(L(\mathcal{H}) + mI)^{-1} x]^T.$$

Since multiplying any real-valued column vector by its transpose yields a matrix with nonnegative diagonal entries, it follows that $\operatorname{diag}[(L(\mathcal{H}) + mI)^{-1} xx^T (L(\mathcal{H}) + mI)^{-1}] \geq 0$. Moreover, since $[L(\mathcal{H}) + mI]^{-1}$ is positive definite, $1 + x^T (L(\mathcal{H}) + mI)^{-1} x > 0$. Thus the diagonal entries of $\frac{(L(\mathcal{H}) + mI)^{-1} xx^T (L(\mathcal{H}) + mI)^{-1}}{1 + x^T (L(\mathcal{H}) + mI)^{-1} x}$ are all nonnegative and the result follows. □

**COROLLARY 8.2.10** *Let $\mathcal{G}$ be a graph on $n$ vertices such that $\mathcal{G} = \mathcal{G}_1 \vee \mathcal{G}_2$, where $\mathcal{G}_1$ and $\mathcal{G}_2$ are graphs on $n - m$ vertices and $m$ vertices, respectively, and where $m > 0$. Then*
$$\operatorname{diag}\left([L(\mathcal{G}_1) + mI]^{-1}\right) > \frac{m+1}{nm} I.$$

**Proof:** By Lemma 8.2.9,
$$\operatorname{diag}\left([L(\mathcal{G}_1) + mI]^{-1}\right) \geq \operatorname{diag}\left([L(K_{n-m}) + mI]^{-1}\right) = \operatorname{diag}\left([nI - J]^{-1}\right)$$
$$> \operatorname{diag}\left([(n+m)I - J]^{-1}\right) = \operatorname{diag}\left(\frac{1}{n}\left[I + \frac{1}{m}J\right]\right) = \frac{m+1}{nm} I$$
□

Lemma 8.2.9 and Corollary 8.2.10 gave us necessary information about the first two quantities in the braces on the right-hand side of (8.2.6). We now consider the third quantity. We do so in the following lemma from [46]:

# The Group Inverse of the Laplacian Matrix

**LEMMA 8.2.11** *Let $\mathcal{G}$ be a graph on $n$ vertices with vertex connectivity $v$ of the form $\mathcal{G} = \mathcal{G}_1 \vee \mathcal{G}_2$ where $\mathcal{G}_1$ and $\mathcal{G}_2$ are graphs on $n - v$ vertices and $v$ vertices, respectively, and where $\mathcal{G}_1$ is a disconnected graph. Then*

$$\frac{1}{n} \leq \max_{1 \leq i \leq n-v} \left\| e_i^T \left( [L(\mathcal{G}_1) + vI]^{-1} - \frac{1}{nv}J \right) \right\|_1 \leq \frac{n-v-2}{nv} + \frac{1}{v}. \quad (8.2.7)$$

*and*

$$\frac{1}{n} \leq \max_{1 \leq j \leq v} \left\| e_j^T \left( [L(\mathcal{G}_2) + (n-v)I]^{-1} - \frac{1}{n(n-v)}J \right) \right\|_1$$

$$\leq \frac{v-2}{nv} + \frac{1}{(n-v)}. \quad (8.2.8)$$

**Proof**: For the lower bound, note that

$$[L(\mathcal{G}_1) + vI]^{-1} - \frac{1}{nv}J = (L(\mathcal{G}_1) + vI + J)^{-1}.$$

Hence

$$\max_{1 \leq i \leq n-v} \left\| e_i^T \left( [L(\mathcal{G}_1) + vI]^{-1} - \frac{1}{nv}J \right) \right\|_1$$

$$\geq \max_{1 \leq i \leq n-v} \left\{ e_i^T \left( [(L(\mathcal{G}_1) + vI]^{-1} - \frac{1}{nv}J \right) e \right\}$$

$$= \max_{1 \leq i \leq n-v} \left\{ e_i^T [L(\mathcal{G}_1) + vI + J]^{-1} e \right\} = \frac{1}{n}.$$

Next we consider the upper bound in (8.2.7). We begin by noting that according to Corollary 8.2.10, the $i$–th diagonal entry of $(L(\mathcal{G}_1) + vI)^{-1}$ is at least $(v+1)/nv > 1/nv$. Also, we have that $e_i^T(L(\mathcal{G}_1) + vI)^{-1}e = 1/v$. Now write $e_i^T(L(\mathcal{G}_1) + vI)^{-1}$ as $de_i^T + x^T$, where $x^T$ has a 0 in its $i$–th position and where $x^T \geq 0$. Then we have that

$$\left\| e_i^T \left( [(L(\mathcal{G}_1) + vI]^{-1} - \frac{1}{nv}J \right) \right\|_1 = \left\| de_i^T + x^T - \frac{1}{nv}e^T \right\|_1$$

$$= \left\| \left( d - \frac{1}{nv} \right) e_i^T + x^T - \frac{1}{nv}\left( e^T - e_i^T \right) \right\|_1$$

$$\leq \left( d - \frac{1}{nv} \right) + \left\| x^T \right\|_1 + \left\| \frac{1}{nv}\left( e^T - e_i^T \right) \right\|_1$$

$$= \left( d - \frac{1}{nv} \right) + \left( \frac{1}{v} - d \right) + \frac{n-v-1}{nv}$$

$$= \frac{n-v-2}{nv} + \frac{1}{v}.$$

proving (8.2.7). Using similar arguments, we can prove (8.2.8). Hence the proof is complete. □

We are now ready to prove the main result of this section which establishes the conditions under which $1/\mathcal{Z}(L^{\#}(\mathcal{G})) = v(\mathcal{G})$. We do this in a theorem from [46]:

**THEOREM 8.2.12** *Suppose that $\mathcal{G}$ is a noncomplete, connected graph on $n$ vertices, with algebraic connectivity $a$, vertex connectivity $v$, where $n \geq v^2$. Then $a = v$ if and only if $1/\mathcal{Z}\left(L^{\#}(\mathcal{G})\right) = v$.*

**Proof**: Since from (8.2.4) we know that $1/\mathcal{Z}(L^{\#}(\mathcal{G})) \leq a \leq v$, we see that if $1/\mathcal{Z}(L^{\#}(\mathcal{G})) = v$, then necessarily $a = v$. Now suppose that $a = v$ and we will prove that $\mathcal{Z}(L^{\#}(\mathcal{G})) = 1/v$. From Theorem 5.1.10 we know that $\mathcal{G} = \mathcal{G}_1 \vee \mathcal{G}_2$, where $\mathcal{G}_1$ is a disconnected disconnected graph on $n - v$ vertices, and $\mathcal{G}_2$ is a graph on $v$ vertices with $a(\mathcal{G}_2) \geq 2v - n$. By Corollary 8.2.7 we have have that

$$\mathcal{Z}(L^{\#}(\mathcal{G})) = \max\left\{ \mathcal{Z}\left([L(\mathcal{G}_1) + vI]^{-1}\right), \mathcal{Z}\left([L(\mathcal{G}_2) + (n-v)I]^{-1}\right), \right.$$

$$\max_{\substack{1 \leq i \leq n-v \\ n-v+1 \leq j \leq n}} \left\{ \frac{1}{2} \left\| e_i^T \left( (L(\mathcal{G}_1) + vI)^{-1} - \frac{1}{nv}J \right) \right\|_1 \right.$$

$$+ \left. \left\| e_j^T \left( (L(\mathcal{G}_2) + (n-v)I)^{-1} - \frac{1}{(n-v)n}J \right) \right\|_1 \right\} \Bigg\}.$$

Note that $(L(\mathcal{G}_2) + (n-v)I)^{-1}$ is a nonnegative matrix with row sums $1/(n-v)$ and that $n \geq v^2$ implies $n - v \geq v$. Therefore

$$\mathcal{Z}((L(\mathcal{G}_2) + (n-v)I)^{-1}) \leq \frac{1}{(n-v)} \leq \frac{1}{v}.$$

From Lemma 8.2.11 we find that for each pair $i$ and $j$,

$$\frac{1}{2}\left\| e_i^T \left( [L(\mathcal{G}_1) + vI]^{-1} - \frac{1}{nv}J \right) \right\|_1$$

$$+ \left\| e_j^T \left( [L(\mathcal{G}_2) + (n-v)I]^{-1} - \frac{1}{n(n-v)}J \right) \right\|_1$$

$$\leq \frac{1}{2}\left[ \frac{n-v-2}{nv} + \frac{1}{v} + \frac{v-2}{n(n-v)} + \frac{1}{n-v} \right]$$

$$= \frac{1}{v} + \frac{1}{2}\left[ \frac{n-v-2}{nv} - \frac{1}{v} + \frac{v-2}{n(n-v)} + \frac{1}{n-v} \right]$$

$$= \frac{1}{v} + \frac{1}{2}\left[ \frac{2(v^2 - n)}{nv(n-v)} \right] \leq \frac{1}{v},$$

the last inequality following from the hypothesis that $n \geq v^2$.

Finally, we claim that $\mathcal{Z}\left([L(\mathcal{G}_1) + vI]^{-1}\right) = 1/v$. To see this note that $(L(\mathcal{G}_1) + vI)^{-1}$ is a direct sum of positive matrices each of which corresponds to a connected component of $\mathcal{G}$ and each with constant row sums equal $1/v$. Hence $\|(e_i^T - e_j^T)(L(\mathcal{G}_1) + vI)^{-1}\|_1 \leq 2/v$ for any pair of indices $i$ and $j$. Furthermore with $i$ and $j$ corresponding to rows in different direct summands, $\|(e_i^T - e_j^T)(L(\mathcal{G}_1) + vI)^{-1}\|_1 = 2/v$. Hence $\mathcal{Z}((L(\mathcal{G}_1) + vI)^{-1}) = 1/v$, as claimed. From all that we have established it now follows that $\mathcal{Z}(L^\#(\mathcal{G})) = 1/v$, and hence $1/\mathcal{Z}(L^\#(\mathcal{G})) = v$. □

**EXAMPLE 8.2.13** Consider the bipartite graph $K_{3,7}$. Observe that $a(K_{3,7}) = v(K_{3,7}) = 3$. Since $n = 10 > v^2$, it follows from Theorem 8.2.12 that $\mathcal{Z}(L^\#) = 1/3$. To see this, label the vertices of $K_{3,7}$ so that vertices $1, 2, 3$ are in one partite set and vertices $4, \ldots, 10$ are in the other partite set. We can compute (via software) that

$$L^\# = \left[\begin{array}{c|c} \frac{1}{7}I - \frac{17}{700}J & -\frac{1}{100}J \\ \hline -\frac{1}{100}J & \frac{1}{3}I - \frac{13}{300}J \end{array}\right]$$

Thus if we take any two of the last seven rows and compute the 1-norm of the difference between these rows, we see that the 1-norm would be $2/3$ as expected. (Note: computing the 1-norm of the difference between one of the first three rows and any other row will yield a 1-norm of less that $2/3$.)

The following example shows that the condition $n \geq v^2$ cannot be eliminated.

**EXAMPLE 8.2.14** [46] Consider the graph $\mathcal{G}$ below:

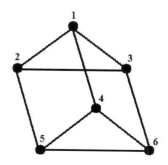

Observe that $a(\mathcal{G}) = v(\mathcal{G}) = 3$ but $n = 6 < v^2$. Using software, we can compute $L^\#$ as follows:

$$L^\# = \frac{1}{180} \left[ \begin{array}{c|c} 48I - J & 12I - 19J \\ \hline 12I - 19J & 48I - J \end{array} \right]$$

Observe that $\mathcal{Z}(L^\#) = 1/2$ where this quantity is achieved by computing the 1-norm of the difference between one row from any of the first three rows and one row from any of the last three rows. Hence $1/\mathcal{Z}(L^\#) = 2 < v(\mathcal{G})$.

## Exercise:

1. Let $\mathcal{G}$ be a maximal graph (recall Section 4.5). Prove that $1/\mathcal{Z}(L^\#(\mathcal{G})) = a(\mathcal{G}) = v(\mathcal{G})$.

## 8.3 The Case of the Zenger Equalling the Algebraic Connectivity in Trees

In this section, we turn our attention back to trees. Letting $L$ be the Laplacian matrix for a graph $\mathcal{G}$, recall from Corollary 8.2.5 that

$$\frac{1}{\mathcal{Z}(L^\#)} \leq a(\mathcal{G}). \tag{8.3.1}$$

We saw in the previous section (Theorem 8.2.12) that for a graph $\mathcal{G}$, equality occurs in (8.3.1) if and only if $1/\mathcal{Z}(L^\#) = v(\mathcal{G})$. The goal of this section is to determine the structure that a tree $\mathcal{T}$ must have in order for $1/\mathcal{Z}(L^\#) = a(\mathcal{G}) = v(\mathcal{G})$. Furthermore, we are intersted in the vertices of $\mathcal{T}$, i.e., the rows of $L^\#$ that contribute to the value of $1/\mathcal{Z}(L^\#)$. In other words, we are interested in the rows of $L^\#$ that cause the 1-norm to be maximized. Our first theorem from [51] gives us insight on which rows we should focus.

**THEOREM 8.3.1** *Let $\mathcal{T}$ be a weighted tree on $n$ vertices with Laplacian matrix $L$. Let $i$ and $j$ be vertices of $\mathcal{T}$. Then the maximum of the expression*

$$\left\| e_i^{(n)T} L^\# - e_j^{(n)T} L^\# \right\|_1 \tag{8.3.2}$$

*can only be attained at a pair of pendant vertices.*

**Proof**: Without loss of generality, we can assume that (8.3.2) is maximized for some $1 \leq i \leq n-1$ and $j = n$. Let

$$L^\# = \left[ \begin{array}{cc} (L^\#)_{1,1} & (L^\#)_{1,2} \\ (L^\#)_{2,1} & (L^\#)_{2,2} \end{array} \right]$$

where the blocks of $L^\#$ are as in (8.1.3) with $M$ being the bottleneck matrix at vertex $n$. Fix $1 \leq i \leq n-1$ and observe that

$$e_i^{(n)T}L^\# - e_n^{(n)T}L^\# = \left[e_i^{(n-1)T}M - \frac{1}{n}\left(e_i^{(n-1)T}Me^{(n-1)}\right)e^{(n-1)} \mid -\frac{1}{n}\left(e_i^{(n-1)T}Me^{(n-1)}\right)\right].$$

Note that the last entry of this $n$-vector is negative. Therefore

$$\left\|e_i^{(n)T}L^\# - e_n^{(n)T}L^\#\right\|_1 = \left\|e_i^{(n-1)T}M - \frac{1}{n}\left(e_i^{(n-1)T}Me^{(n-1)}\right)e^{(n-1)}\right\|_1 + \frac{1}{n}\left(e_i^{(n-1)T}Me^{(n-1)}\right) \quad (8.3.3)$$

Our goal is to show that $i$ and $n$ need both be pendant vertices when (8.3.2) is maximized. Suppose now that $i$ is not a pendant vertex of $\mathcal{T}$. Then there is a vertex $i_0$ which is adjacent to $i$ such that the path from $i_0$ to $n$ includes $i$. Let $\theta$ be the weight of the edge between $i$ and $i_0$ and let $S$ be the set of vertices in $\mathcal{T}$ whose path to $n$ includes the vertex $i_0$. Let $\sigma$ denote the cardinality of $S$ and let $v_S^T$ be the row vector with $1/\theta$ in position $k$ if and only if $k \in S$, and $0$ otherwise. By (8.1.4), it follows that

$$e_{i_0}^{(n-1)T}M = e_i^{(n-1)T}M + v_S^T$$

and, if $k \in S$ then the $(i,k)$ and $(i,i)$ entries of $M$ are equal, i.e.,

$$e_i^{(n-1)T}Me_k^{(n-1)} = e_i^{(n-1)T}Me_i^{(n-1)}. \quad (8.3.4)$$

Since $\left\|v_S^T\right\|_1 = \sigma/\theta$, by the triangle inequality, we have

$$\left\|e_i^{(n-1)T}M - \frac{1}{n}(e_i^{(n-1)T}Me^{(n-1)})e^{(n-1)T} + v_S^T\right\|_1 \leq \left\|e_i^{(n-1)T}M - \frac{1}{n}(e_i^{(n-1)T}Me^{(n-1)})e^{(n-1)T}\right\|_1 + \frac{\sigma}{\theta}. \quad (8.3.5)$$

Note that the $i^{th}$ entry of the vector $e_i^{(n-1)T}M$ is a largest entry in the vector. Thus the $i^{th}$ entry of $e_i^{(n-1)T}M - \frac{1}{n}(e_i^{(n-1)T}Me^{(n-1)})e^{(n-1)T}$ is positive. Since by (8.3.4), the $i^{th}$ entry of the vector $e_i^{(n-1)T}M$ equals its $k^{th}$ entry, it follows that the $k^{th}$ entry of $e_i^{(n-1)T}M - \frac{1}{n}(e_i^{(n-1)T}Me^{(n-1)})e^{(n-1)T}$ is also positive. Therefore we have equality in (8.3.5). Consequently, by calculations similar to (8.3.3) we have

$$\left\|e_{i_0}^{(n)T}L^\# - e_n^{(n)T}L^\#\right\|_1 = \frac{1}{n}\left(e_{i_0}^{(n-1)T}Me^{(n-1)}\right) + \left\|e_{i_0}^{(n-1)T}M - \frac{1}{n}\left(e_{i_0}^{(n-1)T}Me^{(n-1)}\right)e^{(n-1)T}\right\|_1$$

$$= \frac{1}{n}\left((e_i^{(n-1)T}M + v_S^T)e^{(n-1)}\right) + \left\|e_i^{(n-1)T}M + v_S^T - \frac{1}{n}\left((e_i^{(n-1)T}M + v_S^T)e^{(n-1)}\right)e^{(n-1)T}\right\|_1$$

# 380 Applications of Combinatorial Matrix Theory to Laplacian Matrices of Graphs

$$= \frac{1}{n}\left(e_i^{(n-1)T} M e^{(n-1)} + \frac{\sigma}{\theta}\right) +$$

$$\left\| e_i^{(n-1)T} M + v_S^T - \frac{1}{n}\left(e_i^{(n-1)T} M e^{(n-1)}\right) e^{(n-1)T} - \frac{\sigma}{n\theta} e^{(n-1)T} \right\|_1$$

$$\geq \frac{1}{n}\left(e_i^{(n-1)T} M e^{(n-1)} + \frac{\sigma}{\theta}\right) +$$

$$\left\| e_i^{(n-1)T} M - \frac{1}{n}\left(e_i^{(n-1)T} M e^{(n-1)}\right) e^{(n-1)T} \right\|_1 + \frac{\sigma}{\theta} - \frac{\sigma}{n\theta} \left\| e^{(n-1)T} \right\|_1$$

$$= \left\| e_i^{(n)T} L^\# - e_j^{(n)T} L^\# \right\|_1 + \frac{\sigma}{n\theta} > \left\| e_i^{(n)T} L^\# - e_n^{(n)T} L^\# \right\|_1.$$

where the last equality follows from (8.3.3). As a result we see that if $i$ is not a pendant vertex, then the expression in (8.3.2) is not maximal. Thus in order for (8.3.2) to be maximal, $i$ must be pendant, and by similar arguments, $n$ must also be pendant. □

**EXAMPLE 8.3.2** Consider the weighted tree below:

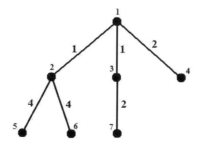

The group inverse for the Laplacian matrix is

$$L^\# = \frac{1}{196}\begin{bmatrix} 58 & -26 & 2 & 44 & -33 & -33 & -12 \\ -26 & 86 & -82 & -40 & 79 & 79 & -96 \\ 2 & -82 & 142 & -12 & -89 & -89 & 128 \\ 44 & -40 & -12 & 128 & -47 & -47 & -26 \\ -33 & 79 & -89 & -47 & 121 & 72 & -103 \\ -33 & 79 & -89 & -47 & 72 & 121 & -103 \\ -12 & -96 & 128 & -26 & -103 & -103 & 212 \end{bmatrix}$$

Observe that $\mathcal{Z}(L^\#) = 41/14$ and that this quantity is obtained from either rows 5 and 7 or rows 6 and 7. In either case, both rows correspond to pairs of pendant vertices.

Theorem 8.3.1 shows us that in computing $\mathcal{Z}(L^\#)$, we need only consider rows of $L^\#$ corresponding to pendant vertices of a tree. In order to determine when the inequality in (8.3.1) is sharp, we need some matrix-theoretic results. The first result is a lemma from [74] which we state without proof:

## The Group Inverse of the Laplacian Matrix

**LEMMA 8.3.3** *Let $z = [z_1, \ldots, z_n]^T \in C^n$. Then for any $x \in \Re^n$ with $x^T e = 0$, it follows that*

$$|z^T x| \leq \frac{1}{2} \max_{1 \leq i,j \leq n} |z_i - z_j| \, \|x\|_1 \qquad (8.3.6)$$

Dertermining when equality occurs in (8.3.1) requires knowing when equality occurs in (8.3.6). The following lemma from [51] shows when equality occurs in (8.3.6):

**LEMMA 8.3.4** *Let $x \in \Re^n$ be a vector such that $x^T e = 0$ and let $z \in C^n$. Then equality holds in (8.3.6), i.e.,*

$$|z^T x| = \frac{1}{2} \max_{1 \leq i,j \leq n} |z_i - z_j| \, \|x\|_1 \qquad (8.3.7)$$

*if and only if $z$ and $x$ can be reordered simultaneoulsy such that*

$$x = [\, x_1, \ldots, x_m, -x_{m+1}, \ldots, -x_{m+k}, 0, \ldots, 0 \,]^T$$

$$z = [\, a, \ldots, a, b, \ldots, b, c_1, \ldots, c_{n-k-m} \,]^T \qquad (8.3.8)$$

*and where*

$$\max_{1 \leq i,j \leq n} |z_i - z_j| = |a - b| \quad \text{and} \quad x_i > 0, \quad i = 1, \ldots, m+k. \qquad (8.3.9)$$

**Proof**: First suppose conditions (8.3.8) and (8.3.9) hold. Then

$$|z^T x| = \left| a \sum_{i=1}^{m} x_i - b \sum_{i=m+1}^{m+k} x_i \right|$$

$$= \frac{1}{2} \left| a \sum_{i=1}^{m+k} x_i - b \sum_{i=1}^{m+k} x_i \right|$$

$$= \frac{1}{2} |a - b| \sum_{i=1}^{m+k} x_i$$

$$= \frac{1}{2} \max_{1 \leq i,j \leq n} |z_i - z_j| \, \|x\|_1$$

where the second equality follows from the fact that $x^T e = 0$, i.e., $\sum_{i=1}^{m} x_i = \sum_{m+1}^{m+k} x_i$. This proves the sufficiency of conditions (8.3.8) and (8.3.9).

To prove the necessity of conditions (8.3.8) and (8.3.9), we will assume (8.3.7) true and prove conditions (8.3.8) and (8.3.9) by induction on the number of nonzero entries in $x$. For the base step, suppose $x$ has two nonzero entries. Then

$$x = [\, x_1, -x_1, 0, \ldots 0 \,]^T.$$

Therefore, observing that $\|x\|_1 = 2x_1$ we see that

$$\left| z^T x \right| = |a - b| \, |x_1| = |a - b| \, \|x\|_1.$$

By assumption (8.3.7), it follows that $\max_{1\leq i,j\leq n} |z_i - z_j| = |a - b|$, thus proving (8.3.9).

For the inductive hypothesis, assume that conditions (8.3.8) and (8.3.9) hold when $x$ has $m + k - 1$ nonzero entries. We now need to prove that these conditions hold when $x$ has $m + k$ nonzero entries. Assume without loss of generality that $x_1 \geq x_m$. Then
$$x = x_{m+1}\left(e_1^{(n)} - e_{m+1}^{(n)}\right) + \hat{x}$$
where
$$\hat{x} = [\, x_1 - x_{m+1},\ x_2,\ \ldots,\ x_m,\ 0,\ -x_{m+2},\ \ldots,\ -x_{m+k},\ 0,\ldots,\ 0\,]^T \tag{8.3.10}$$
has $m + k - 1$ nonzero entries. Note that $\|x\|_1 = 2x_{m+1} + \|\hat{x}\|_1$, $\hat{x}^T e^{(n)} = 0$, and $z^T x = x_{m+1}(z_1 - z_{m+1}) + z^T \hat{x}$. Since we are assuming (8.3.7), we obtain

$$\begin{aligned}
\left|z^T x\right| &= \left|x_{m+1}(z_1 - z_{m+1}) + z^T \hat{x}\right| \\
&\leq x_{m+1}|z_1 - z_{m+1}| + \left|z^T \hat{x}\right| \\
&\leq x_{m+1} \max_{1\leq i,j\leq n}|z_i - z_j| + \tfrac{1}{2}\max_{1\leq i,j\leq n}|z_i - z_j|\,\|\hat{x}\|_1 \qquad (8.3.11) \\
&= \tfrac{1}{2}(2x_{m+1} + \|\hat{x}\|_1)\max_{1\leq i,j\leq n}|z_i - z_j| \\
&= \left|z^T x\right|,
\end{aligned}$$

where both the last inequality and the last equality follow from applying (8.3.7) to $z^T \hat{x}$ and $z^T x$, respectively. (Applying (8.3.7) to $z^T \hat{x}$ is allowed by the inductive hypothesis since $\hat{x}$ has $m + k - 1$ nonzero entries.) Observe that the first and last expressions in (8.3.11) are identical, hence all expressions in (8.3.11) are equal. Thus we can conclude the following:

(i) $\left|z^T \hat{x}\right| = \tfrac{1}{2}\max_{1\leq i,j\leq n}|z_i - z_j|\,\|\hat{x}\|_1$.

(ii) $|z_1 - z_{m+1}| = \max_{1\leq i,j\leq n}|z_i - z_j|$.

(iii) The argument of $z_1 - z_{m+1}$ is the same as the argument of $z^T \hat{x}$.

From the inductive hypothesis applied to $\hat{x}$, we see that $z$ is of the form
$$z = [\, z_1,\ a,\ \ldots,\ a,\ z_{m+1},\ b,\ \ldots,\ b,\ c_1,\ \ldots,\ c_{n-k-m}\,].$$
where $\max_{1\leq i,j\leq n}|z_i - z_j| = |a - b|$. It remains to show that $z_1 = a$ and $z_{m+1} = b$. If $x_1 > x_{m+1}$, then the first entry in $\hat{x}$ is nonzero (see (8.3.10)). Since (i) holds, we

can apply the inductive hypothesis to $\hat{x}$ to obtain $z_1 = a$. Observe then that

$$\begin{aligned}
z^T \hat{x} &= a(x_1 - x_{m+1}) + a \sum_{i=2}^{m} x_i - b \sum_{i=m+2}^{m+k} x_i \\
&= a(x_1 - x_{m+1}) + a \sum_{i=2}^{m} x_i - b \left( \sum_{i=1}^{m} x_i - x_{m+1} \right) \\
&= a(x_1 - x_{m+1}) + a \sum_{i=2}^{m} x_i - b \left( x_1 + \sum_{i=2}^{m} x_i - x_{m+1} \right) \\
&= (a-b)(x_1 - x_{m+1}) + (a-b) \sum_{i=2}^{m} x_i.
\end{aligned}$$

This shows that $z^T \hat{x}$ is a positive multiple of $a-b$. Hence $arg(z_1 - z_{m+1}) = arg(z^T \hat{x})$. Using this, (ii) from above, and the fact that $\max_{1 \leq i,j \leq n} |z_i - z_j| = |a-b|$ all together yield that $z_{m+1} = b$.

Hence if $x_1 > x_{m+1}$, we conclude that $z_1 = a$ and $z_{m+1} = b$. However, if $x_1 = x_{m+1}$, then the first entry of $\hat{x}$ is zero (see (8.3.10)). In this case, we repeat the argument above using $x_1$ and $x_{m+2}$ or using $x_2$ and $x_{m+1}$ to deduce that $z_1 = a$ and $z_{m+1} = b$. □

We now use Lemma 8.3.4 together with the fact that (8.3.2) is maximized when $i$ and $j$ are pendant vertices (Theorem 8.3.1) to help us determine when equality can occur in (8.3.1). We do this in the following theorem from [51]. In this theorem, let $w(i,j)$ denote the weight of the edge between vertices $i$ and $j$.

**THEOREM 8.3.5** *Let $\mathcal{T}$ be a weighted tree with Laplacian matrix $L$. Then there is equality in (8.3.1) if and only if $\mathcal{T}$ is a Type I tree whose characteristic vertex $u$ has the property that two of its Perron branches consist of single edges adjacent to pendant vertices $v_1$ and $v_2$, and*

$$\max_{1 \leq i,j \leq n} \left\| e_i^{(n)T} L^\# - e_j^{(n)T} L^\# \right\|_1 = \frac{2}{w(u, v_1)} = \frac{2}{w(u, v_2)}. \tag{8.3.12}$$

**Proof**: First assume:

(i) $\mathcal{T}$ is a Type I tree whose characteristic vertex $u$ has the property that two of its Perron branches consist of single edges adjacent to pendant vertices, $v_1$ and $v_2$, and

(ii) The equation (8.3.12) holds.

We will show that under (i) and (ii) there is equality in (8.3.1). By (i) we have

$$a(\mathcal{T}) = w(u, v_1),$$

and by (ii) we have

$$\frac{1}{w(u, v_1)} = \frac{1}{2} \max_{1 \leq i,j \leq n} \left\| e_i^{(n)T} L^\# - e_j^{(n)T} L^\# \right\|_1.$$

Thus $1/a(\mathcal{T}) = \mathcal{Z}(L^{\#})$ which completes this direction of the proof.

Now let us assume that $1/\mathcal{Z}(L^{\#}) = a$, i.e., $\mathcal{Z}(L^{\#}) = 1/a$. For a vector $z \in C^n$, define $f(z) := \max_{1 \leq i,j \leq n} |z_i - z_j|$. Next, for any $1 \leq i, j \leq n$ let $x_{i,j}^T = e_i^T L^{\#} - e_j^T L^{\#}$. Finally, suppose $y$ is an eigenvector of $L$ corresponding to $a$. Then

$$\frac{1}{a} f(y) = \frac{1}{2} f(y) \max_{1 \leq i,j \leq n} \|x_{i,j}\|_1 . \tag{8.3.13}$$

Without loss of generality, suppose that $\max_{1 \leq i,j \leq n} \|x_{i,j}\|_1$ is attained at $i = 1$ and $j = n$. Then by Theorem 8.3.1, it follows that 1 and $n$ are pendant vertices of $\mathcal{T}$. Furthermore,

$$\frac{1}{a} f(y) = \frac{1}{a} |y_1 - y_n| = \left| x_{1,n}^T y \right| . \tag{8.3.14}$$

Putting together (8.3.13) and (8.3.14) yields

$$\left| x_{1,n}^T y \right| = \frac{1}{2} f(y) \max_{1 \leq i,j \leq n} \|x_{i,j}\|_1 .$$

so that Lemma 8.3.4 applies.

By Lemma 8.3.4, the vector $y$ is constant on the indices where $x_{1,n}^T$ is positive and constant on the indices where $x_{1,n}^T$ is negative. Moreover, if the $k_1$-th and $k_2$-th entries of $x_{1,n}^T$ are positive and negative, respectively, then $|y_{k_1} - y_{k_2}| = \max_{1 \leq i,j \leq n} |y_i - y_j|$.

By Observation 8.1.14, the entries of $e_n^T L^{\#}$ are strictly decreasing along any path which starts at $n$, while the entries of $e_1^T$ are strictly increasing along any path which ends at 1. Hence the entries of $x_{1,n}^T$ are strictly increasing as we move along the path which starts at $n$ and ends at 1. In particular, there is at most one vertex on that path for which the corresponding entry in $x_{1,n}^T$ is zero.

We saw in the proof of Theorem 8.3.1 that $L_{1,1}^{\#} - L_{n,1}^{\#}$ is positive and $L_{1,n}^{\#} - L_{n,n}^{\#}$ is negative. Also, since the sum of the entries in $y$ is zero, we can assume without loss of generality that $y_1 > 0$ and $y_n < 0$. Suppose now that $j$ is on the path from vertex 1 to vertex $n$. Then from the previous two paragraphs we have three possibilities:

(a) $L_{1,j}^{\#} - L_{n,j}^{\#} > 0$ and $y_j = y_1 > 0$.

(b) $L_{1,j}^{\#} - L_{n,j}^{\#} < 0$ and $y_j = y_n < 0$.

(c) $L_{1,j}^{\#} - L_{n,j}^{\#} = 0$ and $|y_1 - y_j|, |y_n - y_j| \leq y_1 - y_n$ (by Lemma 8.3.4).

Also note there is at most one index $j$ for which (c) holds.

Recall from Corollary 6.2.1 that if $\mathcal{T}$ is a Type I tree, then along any path starting from the characteristic vertex, the entries in $y$ are either strictly increasing and positive, or strictly decreasing and negative, or all zero. Since $y_1 > 0 > y_n$, we conclude that if $\mathcal{T}$ is a Type I tree, then the characteristic vertex must be on the path from vertex 1 to vertex $n$. Similarly, if $\mathcal{T}$ is a Type II tree, then both characteristic vertices must be on that path.

Suppose first that $\mathcal{T}$ is a Type I tree whose characteristic vertex is $k$. Then $y_k = 0$, so necessarily (c) holds for $j = k$. Note that entries of $y$ are strictly increasing along the path from $k$ to 1. Since $k$ is on the path from $n$ to 1, we saw earlier that the entries of $x_{1,n}$ are also increasing on the path from $k$ to 1. Thus for all vertices $j \neq k$ on the path from $k$ to 1, it follows that (a) holds and therefore $y_j = y_1$ for all such vertices $j$. Since the entries in $y$ are strictly increasing on the path from $k$ to 1, we see that the path from $k$ to 1 can only be of length one. Hence we conclude that $k$ is adjacent to 1, a pendant vertex. By similar arguments, we see that $k$ is also adjacent to $n$, a pendant vertex. Thus we see that the characteristic vertex $k$ has two Perron branches which consist of single edges adjacent to the pendant vertices 1 and $n$. It follows from (8.1.3) and (8.1.4) that

$$x_{1,n}^T = \left[ \frac{1}{w(1,k)}, 0, \ldots, 0, -\frac{1}{w(1,k)} \right]$$

and hence

$$\max_{1 \leq i,j \leq n} \left\| e_i^{(n)T} L^\# - e_j^{(n)T} L^\# \right\|_1 = \frac{2}{w(1,k)}.$$

To complete the proof, we need to show that $\mathcal{T}$ cannot be a Type II tree. Suppose that $\mathcal{T}$ is a Type II tree. Let $i$ and $j$ be the characteristic vertices of $\mathcal{T}$ where the entries of $y$ are increasing along the path from $i$ to 1 and decreasing along the path from $j$ to $n$. But then there is at most one vertex on the path from 1 to $n$ for which (c) can hold. So either $y_i = y_1$ or $y_j = y_n$, both of which yield contradictions since characteristic vertices cannot be pendant. Consequently, if $\mathcal{T}$ is a Type II tree, then $1/\mathcal{Z}(L^\#) \neq a$. □

In the unweighted case, the only unweighted tree of Type I whose characteristic vertex has the property that two of its Perron branches consist of edges adjacent to pendant vertices is the star. Hence if $1/\mathcal{Z}(L^\#) = a(\mathcal{T})$ then $\mathcal{T}$ is a star. Conversely, if $\mathcal{T}$ is a star then $a(\mathcal{T}) = 1$ and by simple calculations we obtain $\max_{1 \leq i,j \leq n} \left\| e_i^{(n)T} L^\# - e_j^{(n)T} L^\# \right\|_1 = 2$, and hence $\mathcal{Z}(L^\#) = 1/a(\mathcal{T})$. Thus we just proved the following corollary from [51]:

**COROLLARY 8.3.6** *Let $\mathcal{T}$ be an unweighted tree with Laplacian matrix $L$ and algebraic connectivity $a$. Then $1/\mathcal{Z}(L^\#) = a(\mathcal{T})$ if and only if $\mathcal{T}$ is a star.*

In the weighted case, the conditions for equality in (8.3.1) are more relaxed. We will see from the corollary below from [51] that any tree having two pendant vertices adjacent to the same vertex can admit a weighting where equality in (8.3.1) is achieved.

**COROLLARY 8.3.7** *Let $\mathcal{T}$ be a tree on $n$ vertices with Laplacian matrix $L$. Suppose $\mathcal{T}$ has two pendant vertices which are adjacent to the same vertex. Then there is a weighting of $\mathcal{T}$ such that $1/\mathcal{Z}(L^\#) = a(\mathcal{T})$.*

**Proof**: Without loss of generality we can relabel the vertices of $\mathcal{T}$ so that vertices $n$ and $n-1$ are pendant and adjacent to vertex $n-2$. Now weight the edges from

$n$ and $n-1$ to $n-2$ with $\epsilon > 0$ and weight every other edge of of $\mathcal{T}$ with 1. We will use the results of Theorem 8.3.5 to show that for a choice of $\epsilon$ sufficiently small, we have $1/\mathcal{Z}(L^{\#}) = a(\mathcal{T})$.

Consider the branches of $\mathcal{T}$ at vertex $n-2$. The branches at $n-2$ containing vertices $n$ and $n-1$ have a Perron value of $1/\epsilon$, while every other branch at $n-2$ yields a Perron value which is independent of $\epsilon$. As a result, when $\epsilon$ is sufficiently small, we find that $\mathcal{T}$ is a Type I tree with characteristic vertex $n-2$ and that the branches at $n-1$ containing $n$ and $n-1$ are Perron branches.

Now let $L$ be the Laplacian matrix of this newly weighted tree. It remains to show that

$$\max_{1 \leq i,j \leq n} \left\| e_i^{(n)T} L^{\#} - e_j^{(n)T} L^{\#} \right\|_1 = \frac{2}{\epsilon}. \tag{8.3.15}$$

As in the proof of Theorem 8.3.5, we find that

$$e_{n-1}^{(n)T} L^{\#} - e_n^{(n)T} L^{\#} = \left[ 0, \ldots, 0, \frac{1}{\epsilon}, -\frac{1}{\epsilon} \right]$$

and so

$$\left\| e_{n-1}^{(n)T} L^{\#} - e_n^{(n)T} L^{\#} \right\|_1 = \frac{2}{\epsilon}.$$

To complete the proof, we need to show that the maximum in (8.3.15) is attained at vertices $n-1$ and $n$. We do this by considering the vectors $e_i^{(n)T} L^{\#} - e_n^{(n)T} L^{\#}$, $e_{n-1}^{(n)T} L^{\#} - e_i^{(n)T} L^{\#}$, and $e_i^{(n)T} L^{\#} - e_j^{(n)T} L^{\#}$, for $1 \leq i, j \leq n-2$. Considering the first of these vectors, let $M$ be the bottleneck matrix of $\mathcal{T}$ at vertex $n$. Looking at the $i^{th}$ row of $M$, we see from Theorem 6.2.5 that there exists a positive vector $x \in \Re^{n-1}$ that is independent of $\epsilon > 0$ such that

$$e_i^{(n-1)T} M = x^T + \frac{1}{\epsilon} e^{(n-1)T}.$$

Hence

$$e_i^{(n)T} L^{\#} - e_n^{(n)T} L^{\#} = \left[ \frac{1}{n\epsilon} e^{(n-1)T} + x^T - \frac{x^T e^{(n-1)}}{n} e^{(n-1)T} \mid -\frac{n-1}{n\epsilon} - \frac{x^T e^{(n-1)}}{n} \right].$$

It follows that for $\epsilon > 0$ sufficiently small,

$$\left\| e_i^{(n)T} L^{\#} - e_n^{(n)T} L^{\#} \right\|_1 = \left\| \frac{1}{n\epsilon} e^{(n-1)T} + x^T - \frac{x^T e^{(n-1)}}{n} e^{(n-1)T} \right\|_1 + \frac{n-1}{n\epsilon} + \frac{x^T e^{(n-1)}}{n}$$

$$\leq \left\| \frac{1}{n\epsilon} e^{(n-1)T} \right\|_1 + \left\| x^T \right\|_1 + \left\| \frac{x^T e^{(n-1)}}{n} e^{(n-1)T} \right\|_1 + \frac{n-1}{n\epsilon} + \frac{x^T e^{(n-1)}}{n}$$

$$= \frac{n-1}{n\epsilon} + x^T e^{(n-1)} + \frac{(n-1)x^T e^{(n-1)}}{n} + \frac{n-1}{n\epsilon} + \frac{x^T e^{(n-1)}}{n}$$

$$= \frac{2(n-1)}{n\epsilon} + 2x^T e^{(n-1)} < \frac{2}{\epsilon}.$$

An analogous argument also shows that

$$\left\| e_i^{(n)T} L^{\#} - e_{n-1}^{(n)T} L^{\#} \right\|_1 < \frac{2}{\epsilon}$$

when $\epsilon > 0$ is sufficiently small. Finally, suppose that $1 \leq i, j \leq n-2$. Then again letting $M$ be the bottleneck matrix at vertex $n$, we have

$$e_i^{(n)T} L^\# - e_j^{(n)T} L^\# = \left[ \left( e_i^{(n-1)T} - e_j^{(n-1)T} \right) M \left( I - \frac{1}{n} J \right) \mid \right.$$
$$\left. - \frac{1}{n} \left( e_i^{(n-1)T} - e_j^{(n-1)T} \right) M e^{(n-1)} \right].$$

Since by Theorem 6.2.5 there are positive vectors $y, z \in \Re^{n-1}$ independent of $\epsilon > 0$, such that $e_i^{(n-1)T} M = y^T + (1/\epsilon) e^{(n-1)}$ and $e_j^{(n-1)T} M = z^T + (1/\epsilon) e^{(n-1)}$, we find that $\left\| e_i^{(n)T} L^\# - e_j^{(n)T} L^\# \right\|_1$ is independent of $\epsilon$. By similar arguments as earlier, we deduce that

$$\left\| e_i^{(n)T} L^\# - e_j^{(n)T} L^\# \right\|_1 < \frac{2}{\epsilon}$$

when $\epsilon > 0$ is sufficiently small. Thus for such $\epsilon$, the expression

$$\max_{1 \leq i,j \leq n} \left\| e_i^{(n)T} L^\# - e_j^{(n)T} L^\# \right\|_1$$

is maximized at $i = n-1$ and $j = n$. This completes the proof. □

## Exercises:

**1.** Find the pair of vertices $i$ and $j$ in the tree $\mathcal{T}$ below so that $\left\| e_i^{(n)T} L^\# - e_j^{(n)T} L^\# \right\|_1$ is maximized.

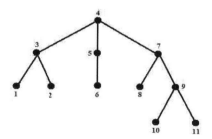

**2.** Find a weighting of the tree $\mathcal{T}$ below so that $1/\mathcal{Z}(L^\#) = a(\mathcal{T})$.

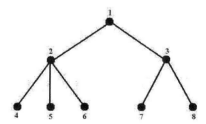

## 8.4 Application: The Second Derivative of the Algebraic Connectivity as a Function of Edge Weight

Recall from Theorem 5.2.10 that if $A$ and $E$ are positive semi-definite matrices with a common null vector, then setting $A_t := A + tE$, we have that $\lambda_2(A_t)$ is an increasing function of $t$. The immediate corollary to this was that if $a(\mathcal{G})$ is a simple eigenvalue of $L(\mathcal{G})$, then increasing the weight of an edge causes the algebraic connectivity to increase in a concave down fashion. In this section, we use the group inverse of a variation of the Laplacian matrix to refine these results. We begin with a matrix-theoretic lemma from [15] which gives us the derivative of the eigenvector corresponding to $\lambda_2$.

**LEMMA 8.4.1** *Let $A$ and $E$ be $n \times n$ positive semidefinite matrices with a common null vector and suppose that $\lambda_2(A) > 0$ is a simple eigenvalue of $A$. Let $A_t = A + tE$ and let $v_t$ be an eigenvector of $A_t$ corresponding to $\lambda_2(A_t)$. If $Q_t = \lambda_2(A_t)I - A_t$, then $(v_t)' = Q_t^\# E v_t$.*

**Proof**: Considering the eigenvalue-eigenvector relationship $A_t v_t = \lambda_2(A_t) v_t$ and differentiating both sides with respect to $t$, we obtain

$$(A_t)' v_t + A_t (v_t)' = [\lambda_2(A_t)]' v_t + [\lambda_2(A_t)](v_t)'$$

hence

$$A_t (v_t)' = -(A_t)' v_t + [\lambda_2(A_t)]' v_t + [\lambda_2(A_t)](v_t)'. \qquad (8.4.1)$$

Noting that $Q_t(v_t)' = [\lambda_2(A_t)](v_t)' - A_t(v_t)'$, we substitute in (8.4.1) to obtain

$$Q_t(v_t)' = (A_t)' v_t - [\lambda_2(A_t)]' v_t.$$

Therefore, recalling that $Q_t^\# v_t = 0$ we obtain

$$\begin{aligned}(v_t)' &= Q_t^\#(A_t)' v_t - [\lambda_2(A_t)]' Q_t^\# v_t \\ &= Q_t^\#(A_t)' v_t \\ &= Q_t^\#(A + tE)' v_t \\ &= Q_t^\# E v_t\end{aligned}$$

□

**OBSERVATION 8.4.2** *[15]* We should note that since we are taking the group inverse of $Q_t$ (since $Q_t^{-1}$ does not exist), in general $(v_t)' = Q_t^\#(A_t)' v_t + \alpha v_t$ where $\alpha$ is the constant of normalization. However, since $e$ is the common null vector of $A$ and $E$, it follows that $\alpha = 0$.

We now use Lemma 8.4.1 to refine Theorem 5.2.10 by showing the conditions under which the second derivative is equal to zero. We do this in a theorem from [48].

**THEOREM 8.4.3** *Let $A$, $E$, $A_t$, and $v_t$ be as in Lemma 8.4.1. Then there exists an interval $[0, t_0)$ such that*

$$\frac{d^2 \lambda_2(A_t)}{dt^2} \leq 0 \tag{8.4.2}$$

*for all $t \in [0, t_0)$. Moreover, equality holds if and only if $v_t$ is also an eigenvector for $E$*

**Proof**: Since $\lambda_2(A)$ is a simple eigenvalue of $A$, it follows from Lemma 5.2.1 that there is a $t_0 > 0$ such that for all $t \in [0, t_0)$, both $\lambda_2(A_t)$ and $v_t$ are analytic functions of $t$. Therefore, differentiating the eigenvalue-eigenvector relation $A_t v_t = \lambda_2(A_t) v_t$ twice with respect to $t$, we find that

$$A_t(v_t)'' + 2E(v_t)' = [\lambda_2(A_t)]'' v_t + 2[\lambda_2(A_t)]'(v_t)' + \lambda_2(A_t)(v_t)''. \tag{8.4.3}$$

Letting $Q_t = \lambda_2(A_t) I - A_t$, we know from Lemma 8.4.1 that $(v_t)' = Q_t^\# E v_t$. Multiplying both sides of (8.4.3) on the left by $v_t^T$, substituting $Q_t^\# E v_t$ in for $(v_t)'$, and solving for $[\lambda_2(A_t)]''$, we obtain

$$[\lambda_2(A_t)]'' = 2 (E v_t)^T Q_t^\# (E v_t). \tag{8.4.4}$$

At this point we should note that we already proved in Theorem 5.2.10 that $[\lambda_2(A_t)]'' \leq 0$ when $t \in [0, t_0)$. However, the goal of this theorem is to refine Theorem 5.2.10 by distinguishing the cases of equality and inquality there. To this end, observe that there are two nonnegative eigenvalues of $Q_t$: $\lambda_n(Q_t) = \lambda_2(A_t) > 0$ and $\lambda_{n-1}(Q_t) = 0$, and that the remaining eigenvalues $\lambda_{n-i}(Q_t) = \lambda_2(A_t) - \lambda_i(A_t)$, for $i = 3, \ldots, n$, are negative. Observe that if $w$ is a null vector common to both $A$ and $E$, then $w$ is an eigenvector of $Q_t^\#$ corresponding to $1/(\lambda_2(A_t))$. Since the remaining eigenvalues of $Q_t^\#$ are nonpositive, it follows that $Q_t^\#$ is negative semidefinite when restricted to the orthogonal complement of $w$. Since $v_t$, and hence $E v_t$, is orthogonal to $w$, it follows that $(E v_t)^T Q_t^\# (E v_t) \leq 0$, hence proving (8.4.2).

Recalling that $v_t$ is an eigenvector of $Q_t$ (and hence $Q_t^\#$) corresponding to the eigenvalue 0, it follows that if $v_t$ is an eigenvector for $E$, then $E v_t = c v_t$ for some real number $c$. Hence

$$(E v_t)^T Q_t^\# (E v_t) = c^2 (v_t)^T Q_t^\# (v_t) = 0.$$

However, if $v_t$ is not an eigenvector of $E$, then $E v_t$ is a linear combination of $v_t$ and the eigenvectors of $Q_t$ corresponding to the negative eigenvalues. This immediately gives us $(E v_t)^T Q_t^\# (E v_t) \leq 0$, but since $E v_t$ is not a scalar multiple of $v_t$, it follows that the inequality in (8.4.2) in this case is strict. □

We now apply Theorem 8.4.3 to graphs in which we increase the weight of the edge joining vertices $i$ and $j$. We obtain an explicit formula for $(L + t E^{(i,j)})''$ where

$E^{(i,j)}$ is the matrix with 1 in the $(i,i)$ and $(j,j)$ entries, $-1$ in the $(i,j)$ and $(j,i)$ entries, and 0 elsewhere. Moreover, we can determine the conditions this second derivative equals zero. We do this in the following corollary from [48].

**COROLLARY 8.4.4** *Let $\mathcal{G}$ be a connected weighted graph on $n$ vertices with Laplacian matrix $L$. Suppose that $a(\mathcal{G})$ is a simple eigenvalue of $L$. For a fixed pair of distinct vertices $i$ and $j$, let $L_t = L + tE^{(i,j)}$ and let $Q_t = a(\mathcal{G}_t)I - L_t$. Then there exists $t_0 > 0$ such that for all $t \in [0, t_0)$ we have*

$$(a(\mathcal{G}_t))'' = 2((v_t)_i - (v_t)_j)^2((Q_t^\#)_{i,i} + (Q_t^\#)_{j,j} - 2(Q_t^\#)_{i,j}) \leq 0 \quad (8.4.5)$$

*Moreover, equality holds if and only if either $(v_t)_i = (v_t)_j$ or $e_i - e_j$ is a Fiedler vector for $L_t$.*

**Proof**: We know from Theorem 5.2.2 that there exists a number $t_0 > 0$ such that in the interval $[0, t_0)$, both $a(\mathcal{G}_t)$ and $v_t$ are analytic functions of $t$. Substituting $E^{(i,j)}$ for $E$ and $a(\mathcal{G}_t)$ for $\lambda_2(A_t)$ in (8.4.4) yields the equality in (8.4.5). The inequality in (8.4.5) is a direct consequence of Theorem 8.4.3. Moreover, observe from Theorem 8.4.3 that $(a(\mathcal{G}_t))'' = 0$ if and only if $v_t$ is an eigenvector of $E^{(i,j)}$. This occurs if and only if either $(v_t)_i = (v_t)_j$ or if $v_t$ is a scalar multiple of $e_i - e_j$. □

**EXAMPLE 8.4.5** Observe in Example 5.2.4 we begin with the graph $K_5 - e$ and note that $a(K_5 - e) = 3$. It can be easily seen that if $i$ and $j$ are the unique pair of nonadjacent vertices, then $e_i - e_j$ is an eigenvector corresponding to the algebraic connectivity. Observe that if we join vertices $i$ and $j$ by and edge of weight $t$, we see that increasing $t$ by increments of one causes the algebraic to increase by two. Thus the algebraic connectivity is a linear function of $t$ (i.e., $(a(\mathcal{G}_t))' = 2$), hence $(a(\mathcal{G}_t))'' = 0$ as Corollary 8.4.4 states.

**EXAMPLE 8.4.6** Consider the graph $\mathcal{G}_t$ given below

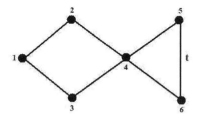

Observe that $(v_t)_5 = (v_t)_6$ for all $t$. Thus by Corollary 8.4.4 we expect $(a(\mathcal{G}_t))'' = 0$. Observe that for all $t$ we have $a(\mathcal{G}_t) = 0.764$. Since $a(\mathcal{G}_t)$ is constant, it follows that $(a(\mathcal{G}_t))' = 0$ and hence $(a(\mathcal{G}_t))'' = 0$ as expected from Corollary 8.4.4.

At this point, we know that 0 is an upper bound on $(a(\mathcal{G}_t))''$ and we know the conditions when this upper bound is achieved. We now turn our attention to finding a lower bound on $(a(\mathcal{G}_t))''$. To do this, we recall two matrix theoretic results which we state as a lemma:

**LEMMA 8.4.7** *Let $A$ be a positive semidefinite $n \times n$ matrix with eigenvalues $\lambda_1, \ldots, \lambda_n$ and corresponding orthonormal eignenvectors $x_1, \ldots, x_n$. Then*

*(i) $\sum_{i=1}^{n} \lambda_i x_i x_i^T = A$, and*

*(ii) $\sum_{i=1}^{n} x_i x_i^T = I$.*

We now prove our result from [48] concerning the lower bound on $(a(\mathcal{G}_t))''$:

**THEOREM 8.4.8** *Let $\mathcal{G}$ be a connected weighted graph on $n$ vertices with Laplacian matrix $L$. Suppose that $a(\mathcal{G})$ is a simple eigenvalue of $L$. Fix a distinct pair of vertices $i$ and $j$, put $L_t := L + tE^{(i,j)}$, and suppose that $[0, t_0)$ is an interval in which $a(\mathcal{G}_t)$ is analytic. Then*

$$(a(\mathcal{G}_t))'' \geq \frac{-2}{\lambda_3(L_t) - a(\mathcal{G}_t)} \qquad (8.4.6)$$

*for all $t \in [0, t_0)$. Equality holds if and only if $L_t$ can be written as*

$$L_t = \lambda_3(L_t)\left(I - \frac{1}{n}J\right) - (\lambda_3(L_t) - a(\mathcal{G}_t))v_t v_t^T,$$

*where $v_t$ has the property that $(v_t)_i - (v_t)_j = 1$.*

**Proof**: Let $v_t$ be a Fiedler vector of $\mathcal{G}_t$ normalized so that $\|v_t\|_2 = 1$ and $(v_t)_i - (v_t)_j \geq 0$. Letting $Q_t = a(\mathcal{G}_t)I - L_t$, recall from Corollary 8.4.4 that

$$(a(\mathcal{G}_t))'' = 2((v_t)_i - (v_t)_j)^2((Q_t^{\#})_{i,i} + (Q_t^{\#})_{j,j} - 2(Q_t^{\#})_{i,j}). \qquad (8.4.7)$$

For each $p = 3, \ldots, n$, let $w_{p,t}$ be (orthogonal) eigenvectors of $L_t$ corresponding to $\lambda_p(L_t)$ normalized so that $w_{p,t}^T w_{p,t} = 1$. Thus by Lemma 8.4.7(i) (also [5] and [9])

$$Q_t^{\#} = aw_{1,t} w_{1,t}^T + \sum_{p=3}^{n} \frac{-1}{\lambda_p(L_t) - a(\mathcal{G}_t)} w_{p,t} w_{p,t}^T.$$

Multiplying both sides on the left and right by $(e_i - e_j)^T$ and $e_i - e_j$, respectively, and noting that all entries in $aw_{1,t}w_{1,t}^T$ are the same, we obtain

$$\begin{aligned}(Q_t^{\#})_{i,i} + (Q_t^{\#})_{j,j} - 2(Q_t^{\#})_{i,j} &= \sum_{p=3}^{n} \frac{-1}{\lambda_p(L_t) - a(\mathcal{G}_t)}((w_{p,t})_i - (w_{p,t})_j)^2 \\ &\geq \frac{-1}{\lambda_3(L_t) - a(\mathcal{G}_t)} \sum_{p=3}^{n}((w_{p,t})_i - (w_{p,t})_j)^2.\end{aligned} \qquad (8.4.8)$$

By Lemma 8.4.7(ii), $(1/n)J + v_t v_t^T + \sum_{p=3}^{n} w_{p,t} w_{p,t}^T = I$. Thus considering the $i^{th}$ diagonal entry of each matrix we have

$$\frac{1}{n} + (v_t)_i^2 + \sum_{p=3}^{n}(w_{p,t})_i^2 = 1. \qquad (8.4.9)$$

Similarly for the corresponding $j^{th}$ diagonal entry we have

$$\frac{1}{n} + (v_t)_j^2 + \sum_{p=3}^{n} (w_{p,t})_j^2 = 1. \quad (8.4.10)$$

Summing the corresponding $(i,j)$ and $(j,i)$ entries of each matrix we have

$$\frac{2}{n} + 2(v_t)_i(v_t)_j + 2\sum_{p=3}^{n} (w_{p,t})_i(w_{p,t})_j = 0. \quad (8.4.11)$$

Adding (8.4.9) and (8.4.10) and then subtracting (8.4.11) we obtain

$$((v_t)_i - (v_t)_j)^2 + \sum_{p=3}^{n} ((w_{p,t})_i - (w_{p,t})_j)^2 = 2$$

or equivalently,

$$\sum_{p=3}^{n} ((w_{p,t})_i - (w_{p,t})_j)^2 = 2 - ((v_t)_i - (v_t)_j)^2.$$

Plugging this into (8.4.8) it follows that

$$(Q_t^\#)_{i,i} + (Q_t^\#)_{j,j} - 2(Q_t^\#)_{i,j} \geq \frac{-1}{\lambda_3(L_t) - a(\mathcal{G}_t)}[2 - ((v_t)_i - (v_t)_j)^2].$$

Thus by (8.4.7) we have

$$(a(\mathcal{G}_t))'' \geq \left(\frac{-1}{\lambda_3(L_t) - a(\mathcal{G}_t)}\right) 2((v_t)_i - (v_t)_j)^2 [2 - ((v_t)_i - (v_t)_j)^2]$$

$$\geq \frac{-2}{\lambda_3(L_t) - a(\mathcal{G}_t)}.$$

If equality holds in (8.4.6), then from the argument above, we see that it necessarily follows that $\lambda_3(L_t) = \ldots = \lambda_n(L_t)$ and $((v_t)_i - (v_t)_j)^2[2 - ((v_t)_i - (v_t)_j)^2] = 1$, from which it follows that $(v_t)_i - (v_t)_j = \pm 1$. But since we took $(v_t)_i - (v_t)_j$ to be nonnegative, we see that $(v_t)_i - (v_t)_j = 1$. Further, by Lemma 8.4.7(i) we have

$$L_t = a(\mathcal{G}_t) v_t v_t^T + \lambda_3(L_t) \sum_{p=3}^{n} w_{p,t} w_{p,t}^T.$$

Thus we obtain the desired form for $L_t$ by noting that $(1/n)J + v_t v_t^T + \sum_{p=3}^{n} w_{p,t} w_{p,t}^T = I$. Finally, if $L_t = \lambda_3(L_t)(I - (1/n)J) - (\lambda_3(L_t) - a(\mathcal{G}_t))v_t v_t^T$, where $v_t^T v_t = 1$ and $v_t^T e = 0$, and $(v_t)_i - (v_t)_j = 1$, then equality holds in (8.4.6). □

**OBSERVATION 8.4.9** *[48]* Given a suitable natural number $n$, and suitable real numbers $\lambda_3$ and $a(\mathcal{G})$ such that $\lambda_3 > a(\mathcal{G}) > 0$, observe that constructing a weighted graph whose Laplacian matrix $L = \lambda_3(L)(I - (1/n)J) - (\lambda_3(L) - a(\mathcal{G}))vv^T$ is equivalent to constructing a vector $v$ such $\|v\|_2 = 1$, $v^T e = 0$, and $v_i - v_j = 1$, and with the property that for each pair of distinct indices $p$ and $q$ we have $-\lambda_3/n - (\lambda_3 - a(\mathcal{G}))v_p v_q \leq 0$ (since $L$ is an M-matrix, thus the off-diagonal entries must be nonpositive).

**EXAMPLE 8.4.10** *[48]* Suppose that for $n \geq 3$, we choose $\lambda_3$ and $a(\mathcal{G})$ where $\lambda_3 > a(\mathcal{G}) > 0$ and

$$\frac{a(\mathcal{G})}{\lambda_3} \geq \frac{\sqrt{n/(n-2)} - 1}{\sqrt{n/(n-2)} + 1}. \tag{8.4.12}$$

Let $x = (1 - \sqrt{(n-2)/n})/2$ and let $v$ be a vector such that $v_i = x$, $v_j = x - 1$, and the remaining entries of $v$ are all $(1 - 2x)/(n - 2)$. Note that $\|v\|_2 = 1$, $v^T e = 0$ and $v_i - v_j = 1$. By using (8.4.12) we can also observe that for each pair of distinct indices $p$ and $q$ that $\lambda_3/n - (\lambda_3 - a(\mathcal{G}))v_p v_q \leq 0$. Hence the weighted graph whose Laplacian matrix $L = \lambda_3(L)(I - (1/n)J) - (\lambda_3(L) - a(\mathcal{G}))vv^T$ is a graph where equality holds in (8.4.6).

## Exercises:

1. (See [48]) Let $\mathcal{G}$ be a connected weighted graph on $n$ vertices with Laplacian matrix $L$. Suppose that $a(\mathcal{G})$ is a simple eigenvalue of $L$. Fix a pair of distinct vertices $i$ and $j$. Let $L_t = L + tE^{(i,j)}$ and $Q_t = a(\mathcal{G}_t)I - L_t$. Suppose that $[0, t_0)$ is an interval in which $a(\mathcal{G}_t)$ remains a simple eigenvalue. Show that

$$\lambda_3(L_t) - a(\mathcal{G}_t) \geq \frac{1}{\mathcal{Z}(Q_t^\#)}.$$

for all $t \in [0, t_0)$.

2. (See [48]) Let $L$, $L_t$, and $Q_t$ be as in the previous exercise, fixing a pair of distinct vertices $i$ and $j$. Suppose $[0, t_0)$ is an interval in which $a(\mathcal{G}_t)$ is analytic. Show

$$\frac{d(v_t)_i}{dt} - \frac{d(v_t)_j}{dt} = ((v_t)_i - (v_t)_j)((Q_t^\#)_{i,i} + (Q_t^\#)_{j,j} - 2(Q_t^\#)_{i,j})$$

for all $t \in [0, t_0)$. What does this tell us about $|(v_t)_i - (v_t)_j|$ as a function of $t$ in $[0, t_0)$?

3. (See [48]) Let $L$ be the Laplacian matrix for a connected graph $\mathcal{G}$ on $n$ vertices. Fix a pair of distinct vertices $i$ and $j$ and let $L_t = L + tE^{(i,j)}$. Prove the following:

(i) If $n = 2$, then $a(\mathcal{G}_t) \to \infty$ and $t \to \infty$.

(ii) If $n \geq 3$, then $a(\mathcal{G}_t)$ converges at $t \to \infty$ and

$$\lim_{t \to \infty} a(\mathcal{G}_t) = \min\{z^T L z \mid z^T z = 1, \, z^T e = 0, \, z_i = z_j\}.$$

# Bibliography

[1] N. Alon and V.D. Milman $\lambda_1$, Isoperimetic inequalities for graphs and superconcentrators. *Journal of Combinatorial Theory*, B38: 73–88, 1985.

[2] W.N. Anderson and T.D. Morley. Eigenvalues of the Laplacian of a graph. *Linear and Multilinear Algebra*, 18:141–145, 1985.

[3] Hua Bai. The Grone-Merris Conjecture. *Transactions of the American Mathematical Society*, 363:4463–4474, 2011.

[4] R.B. Bapat and S. Pati. Algebraic connectivity and the characteristic set of a graph. *Linear and Multilinear Algebra*, 45:247–273, 1998.

[5] A. Ben-Israel and T.N. Greville. Generalized Inverses: Theory and Applications. Academic Press, New York, 1973.

[6] A. Berman and R.J. Plemmons. Nonnegative Matrices in the Mathematical Sciences. SIAM, Philadelphia, 1994.

[7] N. Biggs. Algebraic Graph Theory, Second Edition. Cambridge University Press, Cambridge, U.K., 1993.

[8] B. Bollobas. Random Graphs, Second Edition. Cambridge University Press, Cambridge, U.K., 2001.

[9] S.L. Campbell and C.D. Meyer, Jr. Generalized Inverses of Linear Transformations. Dover Publications, New York, 1991.

[10] S. Chaiken. A combinatorial proof of the all minors matrix tree theorem. *SIAM Journal of Algebra and Discrete Mathematics*, 3(3):319–329, 1982.

[11] G. Chartrand and O. Oellermann. Applied and Algorithmic Graph Theory. McGraw-Hill, New York, 1993.

[12] G. Chartrand and P. Zhang. Introduction to Graph Theory. McGraw-Hill, New York, 2005.

[13] D.J. de Solla Price. Networks of scientific papers. *Science* 149:510–515, 1965.

[14] D.J. de Solla Price. A general theory of bibliometric and other cumulative advantage processes. *Journal of American Society Information Sciences* 27:292–306, 1976.

[15] E. Deutsch and M. Neumann. On the first and second order derivatives of the Perron vector. *Linear Algebra and its Applications* 71:57–76, 1985.

[16] C. H. Edwards and D. E. Penney. Differential Equations and Boundary Value Problems - Computing and Modeling, Fourth Edition. Prentice Hall, Upper Saddle River, NJ, 2008.

[17] L. C. Evans. Partial Differential Equations. American Mathematical Society, Providence, RI, 1998.

[18] S. Fallat and S.J. Kirkland. Extremizing algebraic connectivity subject to graph theoretic constraints. *Electronic Journal of Linear Algebra* 3:48–74, 1998.

[19] S. Fallat, S.J. Kirkland, J.J. Molitierno, and M. Neumann. On graphs whose Laplacian matrices have distinct integer eigenvalues. *Journal of Graph Theory*, 50:162–174, 2005.

[20] S. Fallat, S.J. Kirkland, and S. Pati. Maximizing algebraic connectivity over unicyclic graphs. *Linear and Multililnear Algebra* 51:221–241, 2003.

[21] S. Fallat, S.J. Kirkland, and S. Pati. Minimizing algebraic connectivity over connected graphs with fixed girth. *Discrete Mathematics* 254:115–142, 2002.

[22] S. Fallat, S.J. Kirkland, and S. Pati. On graphs with algebraic connectivity equal to minimum edge density. *Linear Algebra and its Applications* 373:31–50, 2003.

[23] Fan Yizheng. On spectral integral variations of graphs. *Linear and Multilinear Algebra*, 50:133–142, 2002.

[24] Fan Yizheng. Spectral integral variations of degree maximal graphs. *Linear and Multilinear Algebra*, 51:147–154, 2003.

[25] Yi-Zheng Fan, Shi-Cai Gong, Yi Wang, and Yu-Bin Gao. On trees with exactly one characteristic element. *Linear Algebra and its Applications* 412:233–242, 2007.

[26] I. Faria. Permanental roots and the star degree of a graph. *Linear Algebra and its Applications* 64:255–265, 1985.

[27] M. Fiedler. Bounds for eigenvalues of doubly stochastic matrices. *Linear Algebra and its Applications*, 5:299–310, 1972.

[28] M. Fiedler. Algebraic connectivity of graphs. *Czechoslovak Mathematical Journal*, 23:298–305, 1973.

[29] M. Fiedler. Eigenvectors of Acyclic Matrices *Czechoslovak Mathematical Journal*, 25:607–618, 1975.

[30] M. Fiedler. A property of eigenvectors of nonnegative symmetric matrices and its applications to graph theory. *Czechoslovak Mathematical Journal*, 25:619–633, 1975.

[31] M. Fiedler. Laplacian of graphs and algebraic connectivity. *Combinatorics and Graph Theory*, 25:57–70, 1989.

[32] M. Fiedler and V. Ptak. On matrices with non-positive off-diagonal elements and positive principal minors. *Czechoslovak Mathematical Journal*, 22(87):382–400, 1962.

[33] C. Godsil and G. Royle. Algebraic Graph Theory. Springer-Verlag, New York, 2001.

[34] G.H. Golub and C.F. Van Loan. Matrix Computations, Third Edition. Johns Hopkins University Press, Baltimore, 1996.

[35] R. Grone and R. Merris. Algebraic connectivity of trees. *Czechoslovak Mathematical Journal*, 37(112):660–670, 1987.

[36] R. Grone and R. Merris. Ordering trees by algebraic connectivity. *Graphs and Combinatorics*, 6:229–237, 1990.

[37] R. Grone, R. Merris, and V.S. Sunder. The Laplacian spectrum of a graph. *SIAM Journal of Matrix Analysis and Applications*, 11:218–238, 1990.

[38] S.L. Hakimi. On the realizability of a set of integers as the degrees of the vertices of a graph. *SIAM Journal of Applied Math*, 10:496–506, 1962.

[39] V. Havel. A remark on the existence of finite graphs. *Casopis Pest. Mat.*, 80:477–480, 1955.

[40] O.J. Heilmann and E.H. Lieb. Theory of monomer-dimer systems. *Communications of Mathematics and Physics*, 25:190–232, 1972.

[41] R. A. Horn and C. R. Johnson. Matrix Analysis. Cambridge University Press, Cambridge, U.K., 1985.

[42] S. J. Kirkland. Constructions for type I trees with nonisomorphic Perron branches. *Czechoslovak Mathematical Journal*, 49(124): 617–632, 1999.

[43] S. J. Kirkland. A bound on the algebraic connectivity of a graph in terms of the number of cutpoints. *Linear and Multilinear Algebra*, 47: 93–103, 2000.

[44] S. J. Kirkland. An upper bound on the algebraic connectivity of graphs with many cutpoints. *Electronic Linear Algebra*, 8: 94–109, 2001.

[45] S. J. Kirkland and S. Fallat. Perron Components and Algebraic Connectivity for Weighted Graphs. *Linear and Multilinear Algebra* 44:131–148, 1998.

[46] S. J. Kirkland, J. J. Molitierno, M. Neumann, and B. L. Shader. On graphs with equal algebraic and vertex connectivity. *Linear Algebra and its Applications*, 341: 45–56, 2002.

[47] S. J. Kirkland and M. Neumann. Algebraic connectivity of weighted trees under perturbation. *Linear and Multilinear Algebra*, 42: 187–203, 1997.

[48] S. J. Kirkland and M. Neumann. On algebraic connectivity as a function of an edge weight. *Linear and Multilinear Algebra*, 52: 17–33, 2004.

[49] S. J. Kirkland, M. Neumann, and B. L. Shader. Characteristic vertices of weighted trees via Perron values. *Linear and Multilinear Algebra*, 40:311–325, 1996.

[50] S. J. Kirkland, M. Neumann, and B. L. Shader. Distances in weighted trees and group inverse of Laplacian matrices. *SIAM Journal of Matrix Analysis and Applications*, 18(4): 827–841, 1997.

[51] S. J. Kirkland, M. Neumann, and B. L. Shader. On a bound on algebraic connectivity, the case of equality. *Czechoslovak Mathematical Journal*, 48: 65–76, 1998.

[52] S. J. Kirkland, M. Neumann, and B. L. Shader. Bounds on the subdominant eigenvalue involving group inverses with applications to graphs. *Czechoslovak Mathematical Journal*, 48(123):1–20, 1998.

[53] K. Kuratowski. Sur le problème des courbes gauches en topologie. *Fund. Math.*, 15: 271–283, 1930.

[54] P. Lancaster. Theory of Matrices. Academic Press, New York, 1969.

[55] J.S. Li and Y.L. Pan. de Caen's inequality and bounds on the largest Laplacian eigenvalue of a graph. *Linear Algebra and its Applications*, 328:153–160, 2001.

[56] J.S. Li and X.D. Zhang. A new upper bound for the eigenvalues of the Laplacian matrix of a graph. *Linear Algebra and its Applications*, 265:93–100, 1997.

[57] J.S. Li and X.D. Zhang. On the Laplacian eigenvalues of a graph. *Linear Algebra and its Applications*, 285:305–307, 1998.

[58] R. Merris. Characteristic vertices of trees. *Linear and Multilinear Algebra*, 22:115–131, 1987.

[59] R. Merris. Degree maximal graphs are Laplacian integral. *Linear Algebra and its Applications*, 199:381–389, 1994.

[60] R. Merris. Laplacian matrices of graphs: a survey. *Linear Algebra and its Applications*, 197,198:143–176, 1994.

[61] R. Merris. A survey of graph Laplacians. *Linear and Multilinear Algebra*, 39:19–31, 1995.

# Bibliography

[62] R. Merris. A note on Laplacian graph eigenvalues. *Linear and Multilinear Algebra*, 285:33–35, 1998.

[63] B. Mohar. Isoperimetric Numbers of Graphs. *Journal of Combinatorial Theory* B47: 274–291, 1989.

[64] B. Mohar. Eigenvalues, Diameter, and Mean Distance in Graphs. *Graphs and Combinatorics* 7: 53–64, 1991.

[65] B. Mohar. Laplace eigenvalues of graphs – a survey. *Discrete Mathematics* 109: 171–183, 1992.

[66] J. J. Molitierno. The spectral radius of submatrices of Laplacian matrices for trees and its comparison to the Fiedler vector. *Linear Algebra and its Applications* 406: 253–271, 2005.

[67] J. J. Molitierno. On the algebraic connectivity of graphs as a function of genus. *Linear Algebra and its Applications* 419: 519–531, 2006.

[68] J. J. Molitierno. The spectral radius of submatrices of Laplacian matrices for graphs with cut vertices. *Linear Algebra and its Applications* 428: 1987–1999, 2008.

[69] J. J. Molitierno and M. Neumann. The algebraic connectivity of two trees connected by an edge of infinite weight. *Electronic Journal of Linear Algebra* 8:1–13, 2001.

[70] S. Pati. The third smallest eigenvalue of the Laplacian matrix. *Electronic Journal of Linear Algebra* 8:128–139, 2001.

[71] G. Ringel. Map Color Theorem. Springer-Verlag, New York, 1974.

[72] G. Ringel and J.W.T. Youngs. Solution to the Headwood map-coloring problem. *Proceedings of the National Acadedemy of Science, USA* 60:438–445, 1968.

[73] E. Ruch and I. Gutman. The branching extent of graphs. *Journal of Combinatorics and Information System Sciences*, 4:285–295, 1979.

[74] E. Seneta. Non-negative Matrices and Markov Chains, Second Edition. Springer-Verlag, New York, 1981.

[75] Jin-Long Shu, Yuan Hong and Kai Wen-Ren. A sharp upper bound on the largest eigenvalue of the Laplacian matrix of a graph. *Linear Algebra and its Applications* 347: 123–129, 2002.

[76] W. So. Rank One Perturbation and its Application to the Laplacian Spectrum of a Graph. *Linear and Multilinear Algebra*, 46:193–198, 1999.

[77] Dragan Stevanovic. Bounding the largest eigenvalue of trees in terms of the largest vertex degree. *Linear Algebra and its Applications* 360: 35–42, 2003.

[78] Douglas B. West. Introduction to Graph Theory, Second Edition. Prentice Hall, Upper Saddle River, NJ, 2001.

[79] J.H. Wilkinson. The Algebraic Eigenvalue Problem. Oxford University Press, New York, 1965.

[80] Chai Wah Wu. Synchronization in Complex Networks of Nonlinear Dynamical Systems. World Scientific Publishing Co., Hackensack, NJ, 2007.

[81] Choujun Zhan, Guanrong Chen, and Lam F. Yueng. On the distribution of Laplacian eigenvalues versus node degrees in complex networks. *Physica A* 389: 1779–1788, 2010.

[82] $http://3.bp.blogspot.com/_swn7VcF-Vqc/TCpcMmi8qII/AAAAAAAAHw/3QtMkZsikpY/s1600/part1(6).png$

# Index

**A**

Adjacency matrices, 91, 92–93, 96, 128, 130
Algebraic connectivity of graphs of fixed girth, 308, 309–310, 311, 312, 313–315, 316
Algebraic connectivity of Laplacian matrices. *See also* Laplacian matrices
    diameter, related to, 187
    edge density, 192, 193–194
    edge weight, as function of, 180–184
    fixed edges, 187
    genus $k$ graphs, 205, 206–207, 208
    genus zero graphs, of, 205
    isoperimetric number, 195–196
    lower bounds of, 189–191
    mean distance, related to, 187, 188
    monotonical increases of, 179–180
    origins of term, 174
    overview, 173, 174
    perturbing branches, effect of, 239–242, 243
    planar graphs and; *see under* Planar graphs
    smallest, 175–176
    upper bounds of, 176, 185

**B**

Bethe trees, 132
Bipartite graphs, 51, 52, 53, 143, 377. *See also* Graphs
Block adjacent vertices, 350–351, 357
Bottleneck matrices, 211
    Fiedler vector, 228
    graphs, for; *see* Bottleneck matrices for graphs
    joining two trees by edge of infinite weight, 256–259, 260–262
    negative Perron vector, 341–342
    non-Perron branches, 248, 249, 250, 251, 340
    Perron value of, 225, 227, 230, 232–233, 235, 238–239, 248, 251, 252, 311
    Type I trees, 219–221, 235–237, 247–248, 249, 251, 252, 262
    Type II trees, 219–221, 233, 252, 253
    unweighted trees, 233–234, 243
    weighted trees, 221–222, 223, 225, 228–230, 237–238
Bottleneck matrices for graphs
    connected weighted graphs, 284–285, 286–289
    overview, 283–284
    Type I weighted graph, 302–303
    Type II weighted graph, 298–299, 300–301, 304–305

**C**

Cauchy-Binet Theorem, 97–98
Cauchy-Shwarz inequality, 108–109, 130, 135
Characteristic valuation of vertices, 211, 212, 213–215, 216–218, 302
Cliques, 49
Connected weighted graph, 390
Courant-Fischer Minimax Principle, 11, 12–13

**D**

Degree sequences, 65–66, 67, 69–70
    equality in, 72–73
    nonincreasing, of nonnegative integers, 71, 72

Depth-first search algorithm, 59

Directed graphs, 20

Dirichlet's Principle, 108–109

Doubly stochastic matrices
    cosine function, role of, 30
    definition, 28
    eigenvalues of, 28, 32–33
    examples, 28, 29
    irreducibility of, 33
    lemmas pertaining to, 30–32, 33

**E**

Eigenvalues
    Laplacian matrices, of, 119, 120, 121, 123, 161, 163
    Laplacian matrices, of, upper bounds, 126–128
    location of, on a matrix, 8–10
    symmetric matrix, of a, 10

Eigenvectors of Laplacian matrices, 124, 141–142

Embedded graphs, 81
    2-cell, 87–88

Equivalent norms, 4

Euclidean norms, 2

Euler's Formula, 82, 87

**F**

Ferres-Sylvester diagram, 70, 71, 72, 73, 74, 78

Fiedler vector, 228, 319, 343

Forests, 60, 284–285. *See also* Trees

**G**

Generalized inverses, 34–35. *See also* Moore-Penrose inverse

Genus $k$ graphs, 205, 206–207, 208

Genus zero graphs, 205

Gersgorin Disc Theorem, 9, 10, 26, 119

Gram-Schmidt Process, 112

Graphs
    bipartite; *see* Bipartite graphs
    block of, 61
    bottleneck matrices for; *see* Bottleneck matrices for graphs
    bridges of, 63
    cliques; *see* Cliques
    complements of, 46–47
    complete graphs, 49
    connectivity of, 42, 43, 56, 60–61, 64–65, 80, 83, 88, 160, 390
    cube, 50, 51
    degree sequences of; *see* Degree sequences
    edges, 40, 42, 47, 64
    embedded; *see* Embedded graphs
    empty, 49
    forest, 60; *see also* Trees
    genus k, 82
    join of, 48
    k-bipartite; *see* K-bipartite graphs
    Laplacian realizable; *see* Laplacian realizable graphs
    maximal, 71, 75–77, 79, 80, 151–152, 158, 160, 161
    noncomplete maximal, 79–80
    nonplanar; *see* Nonplanar graphs
    nonseparable, 61
    overview, 39–40
    paths, 42, 50
    planar; *see* Planar graphs
    r-regular, 41
    random; *see* Random graphs
    regular, 41
    split, 50
    trees; *see* Trees
    unicyclic; *see* Unicyclic graphs
    unweighted, 43, 121–122
    vertexes; *see* Vertices
    walks, 42
    weighted; *see* Weighted graphs
    wheel, 50

# Index

Grone-Merris Conjecture, 145–146, 147–150, 151
Group inverse of Laplacian matrices
    constructing, 361–362, 363
    distances between vertices, 363
    intermediate vertices, 363
    overview, 361
    vector notation, 364
    weighted tree, 364, 365, 366, 367

## I

Incidence matrices, 93, 94, 96, 100
Induction hypothesis, 88
Inverses, generalized. *See* Generalized inverses
Irreducible matrices, 20, 21
Isoperimetric number, 195–196

## K

$K$ vectors, 263, 265–266, 339–340
$K$-bipartite graphs, 53
$K$-genus graphs, 82
Kuratowski's Theorem, 85, 86

## L

Laplacian integral graphs, 155, 156, 159
    integer eigenvalues, distinct, 163–164, 166–168
Laplacian matrices, 27, 34, 91
    algebraic connectivity of; *see* Algebraic connectivity of Laplacian matrices
    continuous version of, 104–105, 108
    eigenvalue distribution, 136, 137, 138–139
    eigenvalues, 119, 120, 121, 123, 125, 161, 163
    eigenvalues of, upper bounds of, 126–128
    eigenvectors, 124, 141–142
    graph representations of, 108–109; *see also* Laplacian integral graphs; Laplacian realizable graphs
    group inverse of; *see* Group inverse of Laplacian matrices
    inverse variation of, 388–390
    maximal graphs and, 151–152
    networks, 114, 116–118
    perturbation of, 373–374
    relation to other matrices, 95
    spectral radius of submatrices, 252, 253–354, 273, 274–277, 280, 346–349
    symmetric nature of, 119
    temperature function, used to determine, 105–106
    trees, for, 132–133, 134–135, 140, 141, 142, 268–269, 278, 279, 280
    weighted graphs, generalized to, 96
Laplacian realizable graphs, 164, 165, 168

## M

M-ary Bethe trees, 58
M-matrices, 24, 25–26, 330, 392
    nonsingular, 27
    singular, 27
    symmetric irreducible, 265
Matrices, bottleneck. *See* Bottleneck matrices
Matrix norms
    definition, 4, 5
    usage, 1, 5
Matrix Tree Theorem, 91, 142, 285
    overview, 97–98
    proof of, 100–101
    spanning trees, 100–101, 103
    theorem related to, 98–99
    weighted graphs, extending to, 102
Matrix, spectral radius of. *See* Spectral radius of a matrix
Max norms, 3
Minimal vertex cut, 63
Moore-Penrose inverse, 35–36

Multipartite graphs, 358

**N**
Newman-Watts model (NW), 115, 116
Nonnegative irreducible matrices, 21–23
Nonnegative matrices, 15, 16, 19, 20, 21, 376
   definition, 1
   doubly stochastic; *see* Doubly stochastic matrices
Nonplanar graphs, 83, 84, 85, 86. *See also* Planar graphs

**P**
Pendant vertices, 55, 56, 57
Perron vector, 18, 225, 227, 230, 232–233, 235, 238–239, 248
   bottleneck vector, of, 311
   graphs, in, 290, 291–294, 297–298, 305–305, 306
   positive matrix, of, 331–332
Perron-Frobenius theory, 1, 15, 17–18
Planar graphs. *See also* Nonplanar graphs
   algebraic connectivity of, 197–199, 200, 201, 202–204
   definition, 82
   edges, number of, 88
   maximal planar, 84, 199, 203, 204
   noncomplete planar, 197–198
   outerplanar, 198
   subdivisions, 85
   upper bounds of, 83

**R**
Random graphs, 114–115
Rational Zero Theorem, 142
Rayleigh-Ritz equations, 10–11, 12
Reducible matrices, 19, 20

**S**
Scale-free networks, 116–117
Schur triangularization theorem, 6

Spanning trees, 100, 101
Spectral radius of a matrix, 1
   corrollaries pertaining to, 7
   lemma pertaining to, 5, 6
   sizes of, 6
Split graphs, 50
Square matrices, 36–37
Sum norms, 2
Symmetric matrices, 10, 13–14, 139, 153, 154–155, 180, 181

**T**
Trees
   2-simple, 263, 264
   bottleneck matrices for; *see under* Bottleneck matrices
   definition, 55
   $k$ simple, 263, 271, 272
   $k$ vectors, 263, 265–266
   Laplacian matrices for; *see under* Laplacian matrices
   m-ary Bethe, 58
   pendant vertices of, 55, 56, 57
   rooted, 57, 58–59
   spanning, 57, 60
   structure of, 55
   Type I trees, 219–221, 235–237, 247–248, 249, 251, 252, 262
   Type II trees, 219–221, 233, 252, 253
   unweighted trees, 233–234, 243
   weighted trees, 221–222, 223, 225, 228–230, 237–238

**U**
Unicyclic graphs, 318, 319–320, 321–322
   algebraic connectivity, maximizing (with fixed girth), 322, 323–324, 325–326, 327–328
Unitary matrix, 11

**V**
Vector norms

continuous functions of, 3  
definition, 1  
Euclidean norms; *see* Euclidean norm  
max norms; *see* Max norms  
sum norms; *see* Sum norms  
usage, 1–2  
Vertex connectivity, 63  
Vertices  
    block adjacent vertices; *see* Block adjacent vertices  
    characteristic valuation of; *see* Characteristic valuation of vertices  
    cube, of a, 51  
    cut, 60, 61  
    degree of, 40  
    degree sequences of; *see* Degree sequences  
    distances between, 43  
    incident to, 40  
    neighborhood of, 40  
    pendant; *see* Pendant vertices  
    r degree, 41  
    sum of degrees of, 40  

## W

Watts-Strogatz model (WS), 115, 116  
Weierstrass theorem, 3  
Weighted graphs, 96, 102  

## Z

Z-matrices, 24, 25, 26  
Zenger function, 370, 371–372, 373  
    algebraic connectivity in trees, equalling, 378–380, 381–385